AF739472

IMAGING OF NON-TRAUMATIC ISCHEMIC AND HEMORRHAGIC DISORDERS OF THE CENTRAL NERVOUS SYSTEM

IMAGING OF NON-TRAUMATIC ISCHEMIC AND HEMORRHAGIC DISORDERS OF THE CENTRAL NERVOUS SYSTEM

Edited by

Mohammad Sarwar, M.D.

PROFESSOR OF RADIOLOGY
UNIVERSITY OF ILLINOIS COLLEGE OF MEDICINE AT CHICAGO
CHIEF OF NEURORADIOLOGY
UNIVERSITY OF ILLINOIS COLLEGE OF MEDICINE AT CHICAGO
AND COOK COUNTY HOSPITAL
CHICAGO, IL 60612

Co-edited by

Solomon Batnitzky, M.D.

PROFESSOR OF DIAGNOSTIC RADIOLOGY AND
CHIEF OF NEURORADIOLOGY
DEPARTMENT OF DIAGNOSTIC RADIOLOGY
THE UNIVERSITY OF KANSAS MEDICAL CENTER
KANSAS CITY, KANSAS 66103

Kluwer Academic Publishers

BOSTON DORDRECHT LONDON

Distributors for North America:
Kluwer Academic Publishers
101 Philip Drive
Assinippi Park
Norwell, Massachusetts 02061 USA

Distributors for all other countries:
Kluwer Academic Publishers Group
Distribution Center
Post Office Box 322
3300 AH Dordrecht, THE NETHERLANDS

Library of Congress Cataloging-in-Publication Data
Imaging of non-traumatic ischemic and hemorrhagic disorders of the central nervous system/editor, Mohammad Sarwar; co-editor, Solomon Batnitzky.
p. cm.
Includes index.
ISBN 0-89838-417-6
1. Cerebrovascular disease—Diagnosis. 2. Brain—Hemorrhage —Diagnosis. 3. Brain—Blood-vessels—Imaging. I. Sarwar, Mohammad, 1945- . II. Batnitzky, Solomon.
[DNLM: 1. Central Nervous System Diseases—radiography. WL 141 I31]
RC388.5.I42 1989
616.8'10757—dc19
DNLM/DLC
for Library of Congress 88-13598
CIP

Copyright © 1989 by Kluwer Academic Publishers, Boston

All rights reserved. No part of this publication may be reproduced, stored in a retrieval system or transmitted in any form or by any means, mechanical, photocopying, recording, or otherwise, without the prior written permission of the publisher, Kluwer Academic Publishers, 101 Philip Drive, Assinippi Park, Norwell, Massachusetts 02061.

Printed in the United States of America

CONTENTS

Dedication

This book is dedicated to the loving memory of Mannie M. Schechter, a pioneer and great neuroradiologist.

CONTRIBUTING AUTHORS

Solomon Batnitzky, M.D.
Professor of Diagnostic Radiology and Chief of Neuroradiology
Department of Diagnostic Radiology
The University of Kansas Medical Center
Kansas City, Kansas 66103

Glendon G. Cox, M.D.
Associate Professor of Diagnostic Radiology
Department of Diagnostic Radiology
The University of Kansas Medical Center
Kansas City, Kansas 66103

Roger A. Hyman, M.D.
Chief, Division of Neuroradiology
North Shore University Hospital
Manhasset, New York 11030

Joy A. Johnson, M.D.
Staff Radiologist, Baptist Medical Center
Formerly Assistant Professor of Diagnostic Radiology
Chief, Section of Pediatric Radiology
Department of Diagnostic Radiology
The University of Kansas Medical Center
Kansas City, Kansas 66103

Jan Malat, M.D.
Fellow, Department of Neuroradiology
Johns Hopkins University School of Medicine
720 Rutland Avenue
Baltimore, Maryland 21205

John H. McMillan, M.D.
Assistant Professor of Diagnostic Radiology
Department of Diagnostic Radiology
The University of Kansas Medical Center
Kansas City, Kansas 66103

Michael A. Mikhael, M.D., Ph.D.
Chief, Section of Neuroradiology
Director, Magnetic Resonance Imaging
Evanston Hospital
Evanston, Illinois 60201

Stanton J. Rosenthal, M.D.
Associate Professor of Diagnostic Radiology
Department of Diagnostic Radiology
The University of Kansas Medical Center
Kansas City, Kansas 66103

Steven F. Sands, M.D.
Fellow in Neuroradiology
North Shore University Hospital
Manhasset, New York 11030

Mohammad Sarwar, M.D.
Professor of Radiology
University of Illinois College of Medicine of Chicago
Chief of Neuroradiology
University of Illinois College of Medicine at Chicago and Cook County Hospital
Chicago, Illinois 60612

Chat Virapongse, M.D.
Chief, Radiology
V.A. Medical Center
1601 Archer Road
Gainesville, Florida 32602

Robert D. Zimmerman, M.D.
Associate Professor of Radiology
Section of Neuroradiology
The New York Hospital
Cornell Medical Center
New York, New York 10021

PREFACE

The advances in neuroimaging are occurring at a dizzying pace. It is difficult for trainees in radiology and others in neurosciences-related disciplines to keep abreast of the new developments. It is especially important to design neuroimaging protocols to evaluate various neurological diseases. It therefore seems highly desirable that review articles be readily available that comb through the plethora of literature and provide state-of-the-art information on neuroimaging of neurological diseases. It is this goal that we wish to accomplish by bringing out a series of volumes, each dealing with a single theme. The first one is in your hands.

We wish to express our deepest gratitude to the distinguished contributors, who have done an outstanding job. We equally thank our publisher. Comments are welcome.

MS
SB

IMAGING OF NON-TRAUMATIC ISCHEMIC AND HEMORRHAGIC DISORDERS OF THE CENTRAL NERVOUS SYSTEM

1. MAGNETIC RESONANCE IMAGING OF INTRACRANIAL HEMORRHAGE

Robert D. Zimmerman

Historical Background

The advent of magnetic resonance imaging led to attempts to define the appearance of hemorrhage using this new technique. Early reports focused on hematomas studied with T1-weighted (T1 W) inversion recovery (IR) Scans performed on resistive MR imagers. These early studies demonstrated that hematomas were hyperintense to the brain due to T1 shortening and were easily detected when they were nearly isodense and thus difficult to visualize on computerized tomography (CT) [1–5]. With experience, however, it became clear that hyperintensity required several days to develop—and when studied acutely (under three days), hemorrhage was often difficult to detect because it was nearly isointense to the brain on T1 W IR and spin-echo (SE) scans [5, 6]. Initially, changes in intensity on T2 weighted (T2 W) scans received much less attention [6] because these sequences often produced images of poor quality, and were time-consuming and difficult to obtain in uncooperative patients. Technical improvements in intermediate-field-strength (.35–.6T) scanners and the introduction of high-field-strength (1.5T) units made high-quality T2 W imaging possible. Interest rose markedly when it was noted that, at high field strength, acute hematomas (one to seven days) were hypointense on T2 W scans [7]. It was initially claimed that this effect would make possible the diagnosis of acute intracranial hemorrhage (AIH) on MR, but only for scans obtained at high field strength. At approximately the same time, other authors demonstrated that T1 shortening effects were more obvious at ultralow field strengths (.02T), leading to speculaton that acute hemorrhage would be more easily detected on these scanners as well [8]. These confusing and seemingly contradictory claims are not merely of academic interest. The economic implications are enormous. If MR (or at least a particular type of MR) can detect AIH with the same sensitivity and specificity as CT, it might become possible to replace all hospital-based CT scanners with MR units. If, however, MR is inferior to CT in the detection of hemorrhage, hospitals would still be required to maintain CT scanners, since the demonstration of hemorrhage is of paramount diagnostic and therapeutic importance in a patient with acute neurologic ictus.

The confusion and controversy surrounding the appearance of AIH on MR arises from the complex nature of this phenomenon. Most attempts to explain intensity of AIH have been extrapolations from limited clinical and/or experimental data. This approach worked well with CT since temporal density changes are dependent upon essentially one variable, the concentration of the globin fraction of the hemoglobin molecule [9] (figure 1–1). This does not mean that the evaluation of hemorrhage on CT is always simple. Factors such as hematocrit, clot formation and retraction, and hematoma location all affect density, but the basic temporal changes are quickly and rapidly understood. Furthermore, extrinsic technical factors (e.g., type of CT scanner or specific scanning protocol) have no effect on density. With MR, the intensity of blood is much more variable, and it changes rapidly in a bidirectional manner in the first postictal week [10] (figure 1–1). In a sense, the complex findings of AIH on MR reflect more accurately than CT the complex pathophysiologic and biochemical changes that occur after extravasation of blood. Included among these changes are the paramagnetic effects of various blood break-down products [11]. Since an understanding of these phenomena and their effects on Mr intensity is fundamental to the appreciation of the MR features of hemorrhage, they will be reviewed first.

Paramagnetism (figure 1–2)

Paramagnetic substances become magnetic when placed within a strong magnetic field (all of MRI is therefore dependent upon the paramagnetism of protons). Substances are paramagnetic when they

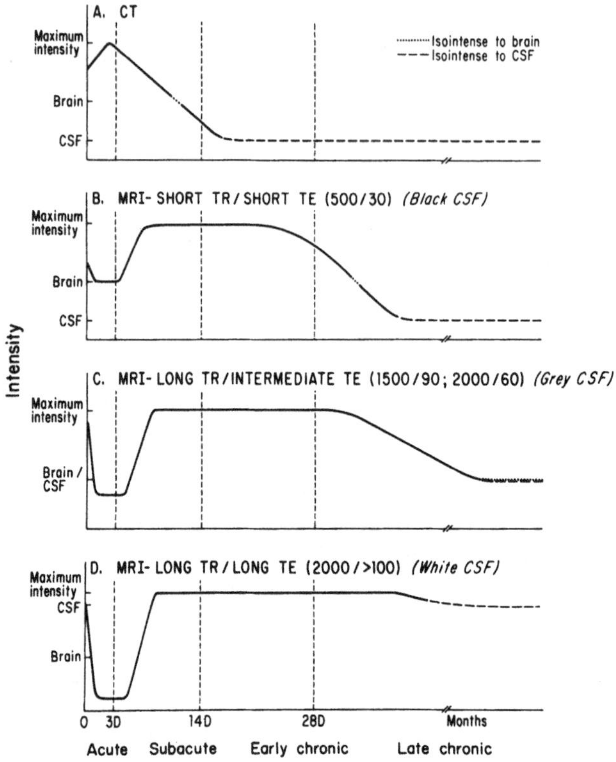

FIGURE 1–1. Time–intensity curves for CT and MR of intracranial hemorrhage. CT density and MR intensity relative to the brain are plotted versus time. On CT, hematomas are midly hyperdense to the brain for several hours. After clot formation, they become denser, reaching maximal density within the first 24 hours. Thereafter, density gradually decreases, with hematomas passing through an isodense phase between the second and third week to become hypodense by the end of the fourth week. On STR/STE MR, hematomas are initially mildly hyperintense to the brain, but intensity diminishes to iso- to mild hypointensity over the first 24 hours. On the third to fourth day, intensity rapidly increases, becoming markedly hyperintense to the brain. Hematomas remain hyperintense for approximately one to two months, then over the next several weeks, intensity gradually diminishes, passing through a second isointense phase to become hypointense to the brain and isointense to CSF. On LTR/ITE-LTE scans, the clots are initially hyperintense to the brain, but within the following 24 hours, there is a marked decrease in intensity as the hematomas rapidly pass through an isointense phase to become hypointense to the brain (darker on LTR/LTE scans). At about four to five days, intensity once again increases, and the lesions pass through the second isointense phase to become hyperintense to the brain. After one to three months, there is a gradual decrease in intensity, with hematomas once again becoming isointense to CSF. Review of the graphs indicate several important features, including

1. No single intensity is characteristic of hemorrhage, since a full range of intensity is encountered on all pluse sequences;
2. Hematomas becomes isointense at several points in time, so contrast between the hematoma and the adjacent brain is variable;
3. Therefore, no single pulse sequence provides contrast at all times or allows for characterization of the hemorrhagic nature of the lesion;
4. However, a combination of multiple pulse sequences always produces some contrast between the lesion and the adjacent brain; and
5. The pattern of internsity on multiple pulse sequences may allow for characterization of the lesion as a hemorrhage.

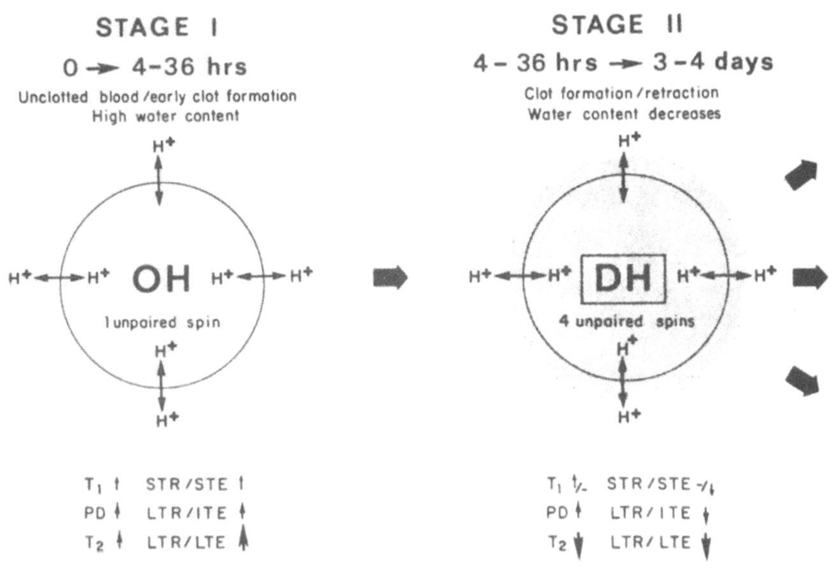

FIGURE 1-2A. Effects of water content and hemoglobin state on intensity.
Stage I (initial 0–6/24 hours)—Hematoma prior to and during early clot formation has high water and protein content. These are major determinants of intensity, since hemoglobin is predominantly in the form of diamagnetic (no unpaired electron) oxyhemoglobin, which produces no PRE. T1 is intermediate between the brain and CSF, proton density is high, and T2 is prolonged. On STR/STE scans, AIH is mildly hyperintense to the brain because the high PD and long T2 are combined with only minimal T1 prolongation. Hyperintensity is more marked on LTR/ITE-LTE scans. Stage II (6/24 hours–3/4 days)—Clot formation and retraction cause a decrease in the amount of water, shortening T1 and T2. The major determinant of intensity at this stage is paramagnetic deoxyhemoglobin located within intact RBCs. It produces heterogeneous magnetic fields, with stronger fields within the RBCs where the deoxyhemoglobin is located than in the extracellular space. The protons which freely diffuse across the cell membrane dephase rapidly, producing T2 shortening.

contain unpaired spins (electrons, protons, and even neutrons) [12–14].

The greatest degree of paramagnetism is produced by unpaired electrons. There are two paramagnetic relaxation enhancement (PRE) effects that are important in the evaluation of hemorrhagic lesions [15]. The first, Proton–Electron Dipole–Dipole (PEDD) interaction, produces T1 (and to a lesser extent T2) shortening. PEDD is dependent upon the molecular structure of the paramagnetic substance and is unrelated to its biologic distribution. If the molecular structure is such that the unpaired electrons can come within three angstroms of the precessing proton, and if they have appropriate resonant frequencies, they may improve coupling between protons, which in turn facilitates the exchange of energy between the excited protons and their molecular environment (the lattice). This causes rapid longitudinal relaxation or T1 shortening. The second paramagnetic effect is preferential T2 shortening (PT2) [15–17]. This phenomenon may occur with any paramagnetic substance since PT2 is dependent upon the biological distribution of the paramagnetic substance and not on its molecular structure. Preferential T2 shortening occurs when the paramagnetic substance is heterogeneously distributed within the sample. For instance, in acute hemorrhage, paramagnetic deoxyhemoglobin is located purely within intact red blood cells (RBCs). This alters the magnetic susceptibility (the ratio between the applied external magnetic field and the internal field generated within the sample). Within the intact RBCs, the presence of paramagnetic deoxyhemoglobin produces stronger magnetic fields than in the adjacent extracellular space, a situation that in turn generates magnetic gradients between the intracellular and extracellular spaces (figure 1–2). The protons freely diffuse across the cell membrane, and therefore they experience these gradients as external field inhomogeneities [7, 17]. Thus, a group of protons whose theoretical T2 is the same (isochromates) will dephase rapidly, relative to each other, because they

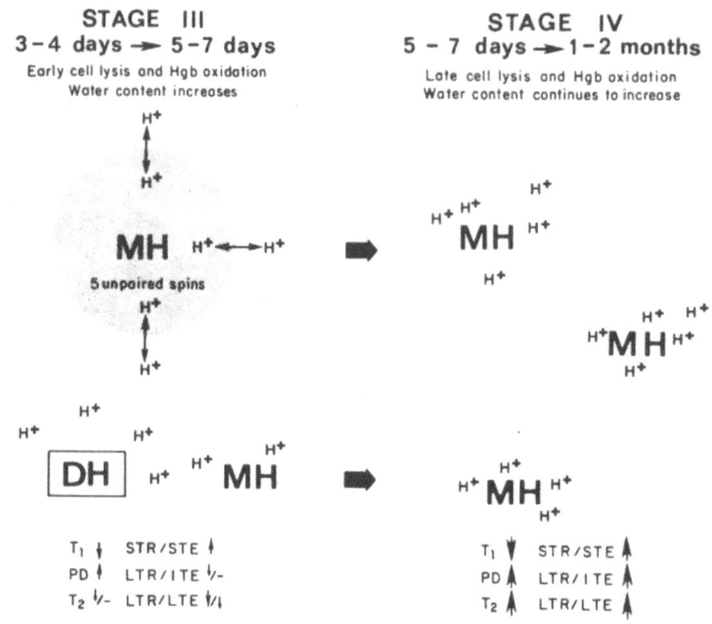

FIGURE 1–2B. Stage III (3/4–6/7 days)—Multiple effects occur. In the presence of oxygen, deoxyhemoglobin is converted to methemoglobin, which produces T1 shortening due to PEDD interaction. Hematomas, therefore, become hyperintense on STR/STE scans. If the methemoglobin is intracellular, T2 shortening will also occur because it will remain heterogeneously distributed. With RBC lysis, methemoglobin and deoxyhemoglobin become extracellular and homogeneously distributed, eliminating the T2 shortening effect. Intensity increases on LTR scans because of the high PD and long T2 of this proteinaceous fluid. With cell and clot lysis, water content also increases, further prolonging T2. Stage IV (5/6 days)—the RBCs are lysed. A solution of methemoglobin is present, producing T1 shortening and hyperintensity on STR scans. T2 is prolonged because of the proteinaceous nature of the fluid collection.

experience different and changing magnetic environments and therefore have different precessional frequencies. This phenomenon will be striking on T2 W images, but it will be even more evident on gradient-echo (GE) scans [15, 18], which utilize a reversal of the polarity in the slice-selecting gradient (*z*-axis) to rephase protons rather than the 180-degree radiofrequency pulse used in SE scanning. GE sequences are less capable of removing the effects of external magnetic field inhomogeneities. Therefore, the presence of heterogeneously distributed paramagnetic substances creates inhomogeneities that the GE scanning technique cannot eliminate. This makes GE highly sensitive to the presence of even small amounts of heterogeneously distributed paramagnetic substances [18].

These two types of proton relaxation enhancement may be present singularly or in combination during various stages of hematoma evolution, depending on the presence and concentration of various hemoglobin breakdown products and the degree of RBC lysis. Since these changes are time-related, an understanding of the MRI of hemorrhage is facilitated by dividing hemorrhage into acute (under one week), subacute (one to four weeks), and chronic (greater than four weeks) phases (figure 1–1). Furthermore, since there are rapid changes during the acute phase, this first week is further divided into four stages (figure 1–2). The appearance of blood will be described for commonly used SE sequences, including

1. Short TR/short TE (STR/STE)—400–700/20–30: *T1 W*
2. Long TR/short TE (LTR/STE)—2000–3000/20–30: *proton density (PD)*
3. Long TR/intermediate TE (LTR/ITE)—2000–3000/40–60: *intermediate T2 W*
4. Long TR/long TE (LTR/LTE)—2000–3000/70–120: *heavily T2 W*

In addition, features of hemorrhage on GE scans will also be reviewed for each stage. Other factors that affect intensity will be assessed, including a) field strength, b) origin (location) and cause of hemorrhage, and c) rehemorrhage.

Acute Intracranial Hemorrhage (0–7 days)

STAGE I: 0–6/72 HOURS (FIGURES 1–2–1–4)
The intensity of freshly extravasated blood has not been well established, since very few patients have been scanned within the first few hours of ictus. Prior to clot formation, it seems likely that the intensity of AIH is similar to that of normal intravascular blood, but evaluation of the signal within large vascular channels is of no help since intraluminal intensity is dependent upon flow effects rather than upon tissue parameters (PD, T1, and T2) [19]. In vitro measurements indicate that T1 and T2 are prolonged and that PD is high relative to brain tissue [20, 21], since blood is a protenaceous viscous fluid with more water than the normal brain. On the basis of these tissue parameters, freshly extravasated blood is expected to be hypointense to the brain on T1 W images (STR/STE and IR) and hyperintense to the brain on PD (LTR/STE) and T2 W (LTR/ITE-LTE) scans. However, based on the review of published clinical images [22], our own experience at intermediate field strength [10], and findings from in vitro animal experiments [23], it appears that AIH is mildly hyperintense to the brain on STR/STE scans (figures 1–3, 1–4). This subtle hyperintensity is a

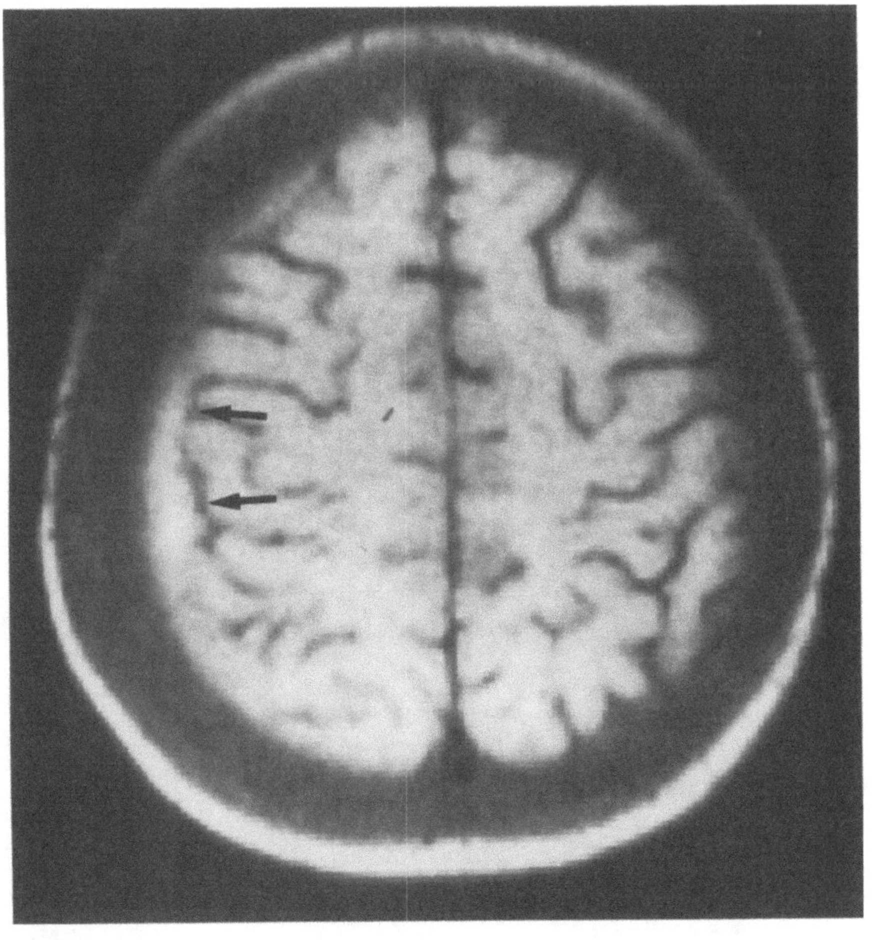

FIGURE 1–3A

FIGURE 1–3. Evolution of acute hemorrhage: comparison of different field strengths. MR scan at .6 Tesla obtained 16 hours after trauma reveals an acute subdural hematoma that is minimally hyperintense to the brain on STR/STE (arrows) (A) and hyperintense on LTR/LTE (B) (Stage I). Follow-up study at .6T, four days after ictus reveals that the clot is more hyperintense on STR-STE (C), while hypointensity has developed on LTR/ITE scan (D) (Stage III). LTR/ITE scan performed at 1.5T (E) immediately after obtaining 0.6 T scans (shown in C and D) again shows hypointensity.

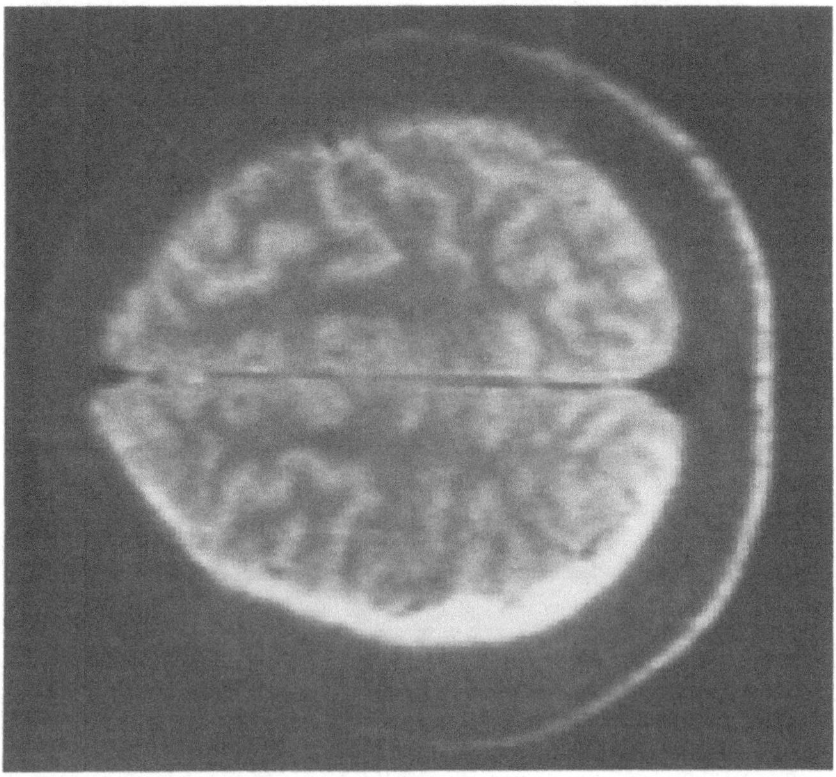

FIGURE 1–3B

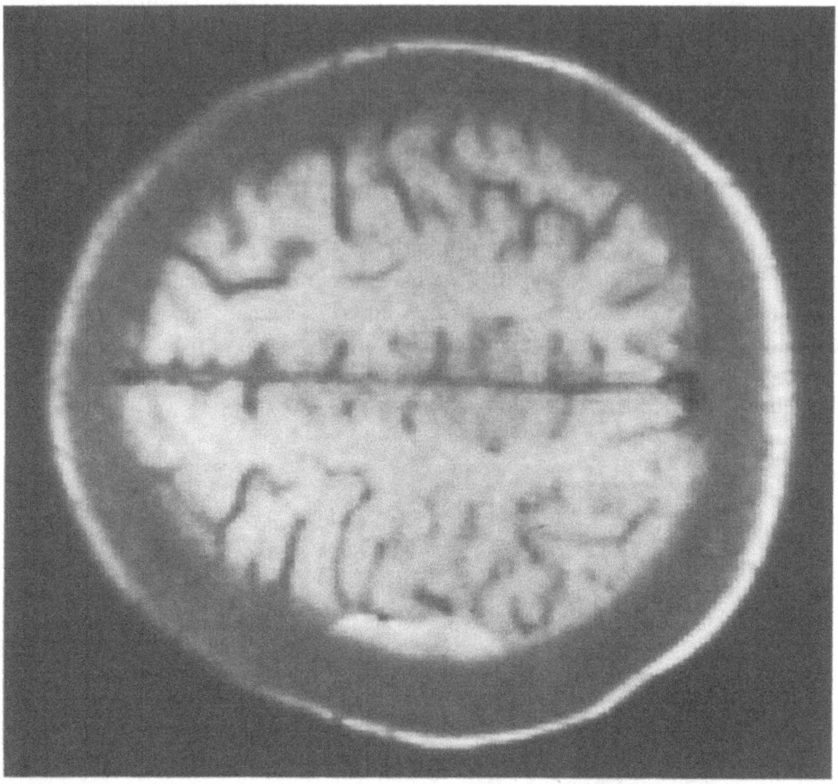

FIGURE 1–3C

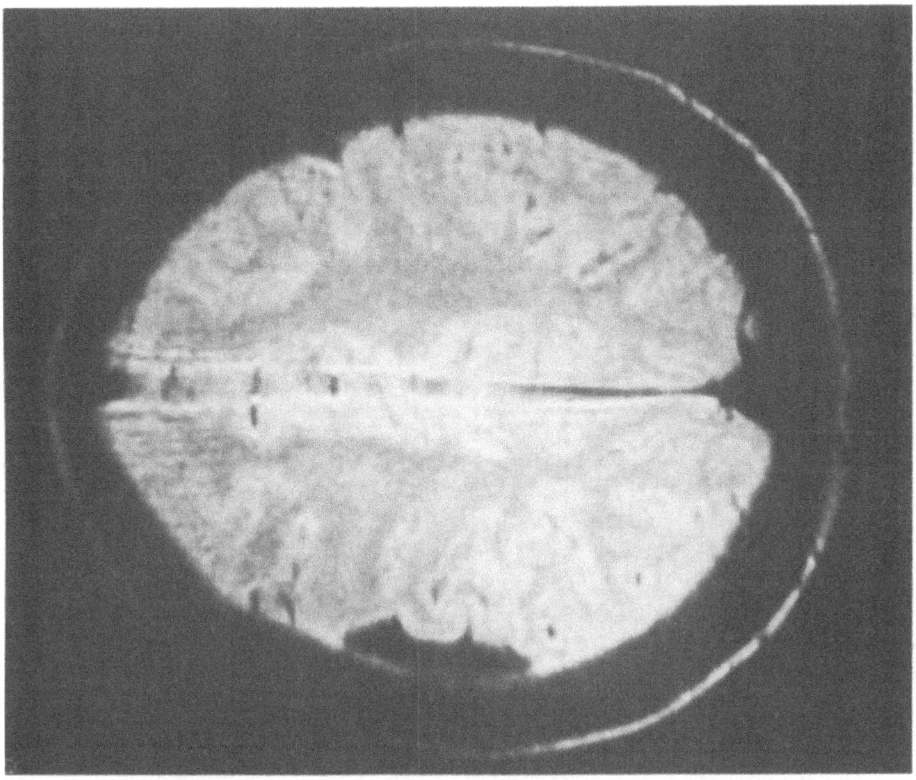

FIGURE 1–3E

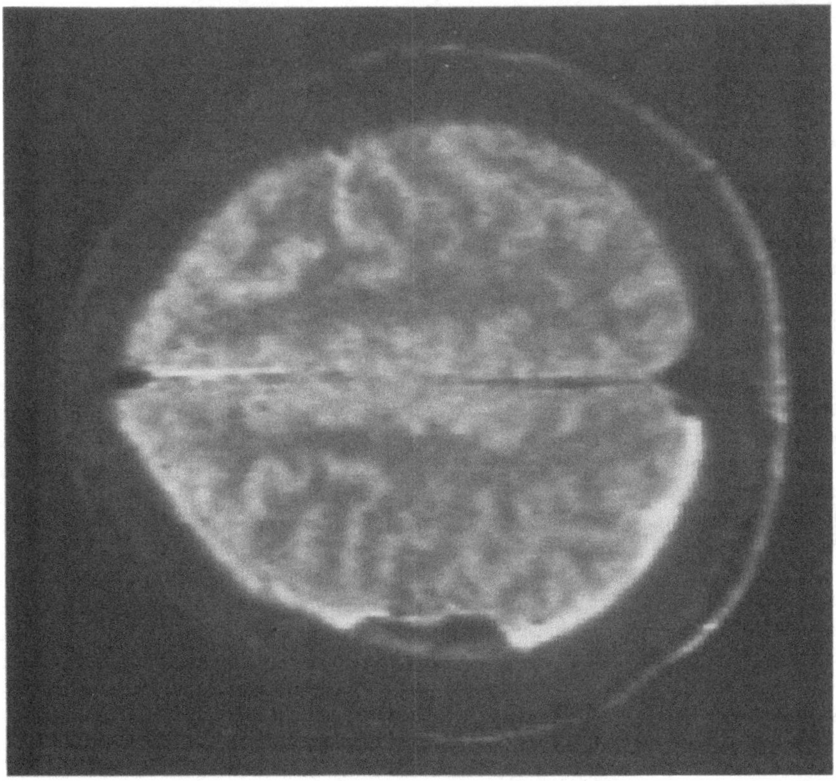

FIGURE 1–3D

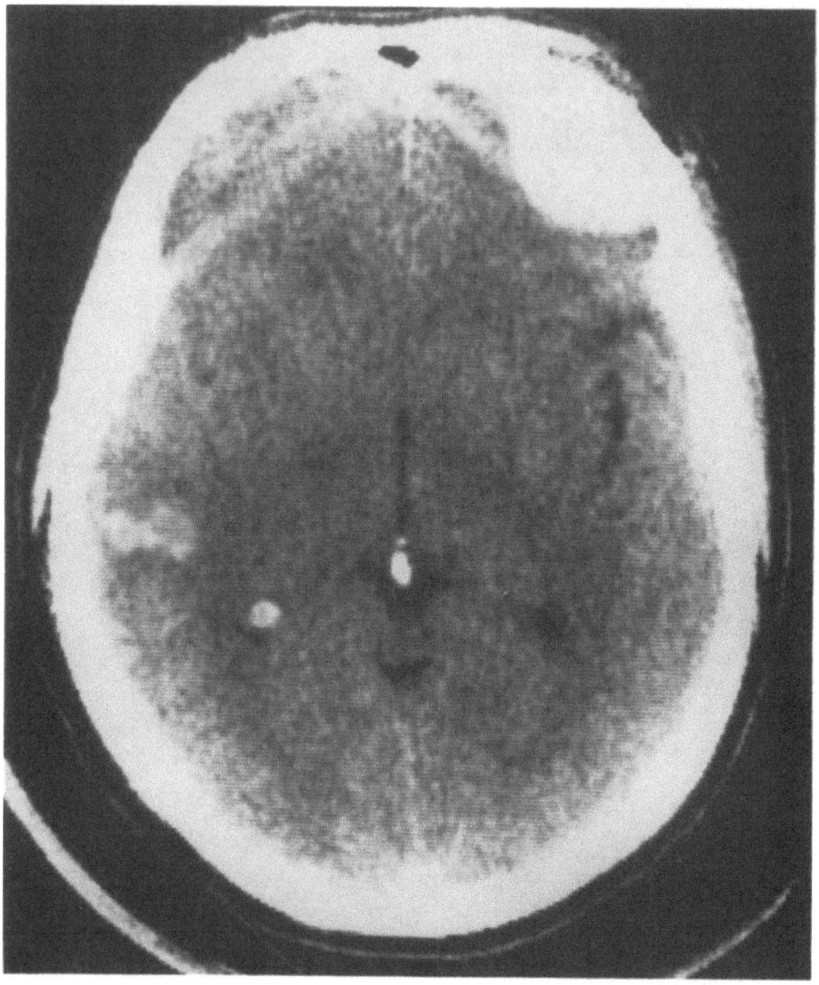

FIGURE 1–4A

FIGURE 1–4. Acute hematomas in multiple compartments: serial changes at .6T. CT (A) performed seven hours after head trauma reveals a dense right temporal hemorrhagic contusion and a dense left frontal epidural hematoma. MRI on the same day (B–E) reveals that the parenchymal and epidural hematomas are mildly hyperintense to the brain on STR/STE scans (B & C). Note the presence of a small right mildly hyperintense subdural hematoma that cannot be detected on CT (arrows). On LTR/LTE (D & E), all three hematomas are hyperintense to the brain. Note the hypointense rim at the medial margin of the epidural hematoma representing displaced dura. Follow-up CT (F) at 24 hours reveals no change. On MR, at the same time (G–J), parenchymal contusion (G) and subdural hematoma show a decrease in intensity, becoming mildly hypointense to the brain while the epidural hematoma (H) remains hyperintense to the brain on STR/STE. On LTR/LTE scans, both the parenchymal contusion and the subdural hematoma (I) show areas of marked hypointensity (Stage II), but the epidural hematoma (J) remains predominantly hyperintense, with small foci of hypointensity noted centrally (Stage I). CT at six days (K) shows no significant change in hematoma density, although edema surrounding the parenchymal contusion has increased. MR performed on the same day reveals that the parenchymal and subdural hematomas are hyperintense on STR/STE scans (L), while the epidural hematoma remains mildly hypointense (M). On LTR/LTE, (N) the hemorrhagic contusion has mixed intensity (Stage III), and subdural hematoma is now hyperintense to the brain (Stage IV). Epidural hematoma (O) is markedly hypointense to the brain (Stage II). Follow-up MRI at seven weeks reveals that the subdural hematoma has resolved. On STR/STE, parenchymal contusion is seen as an area of residual cortical hyperintensity with focal atrophy (P) while the epidural hematoma (Q) has decreased in size to become a subtle area of hyperintensity adjacent to the minimally displaced hypointense dura (arrow). On LTR/LTE, the hemorrhagic contusion is mildly hypointense (arrow) (R), presumably secondary to hemosiderin within macrophages. The minimally displaced dura is still noted at the site of previous epidural hematoma (arrow) (S).

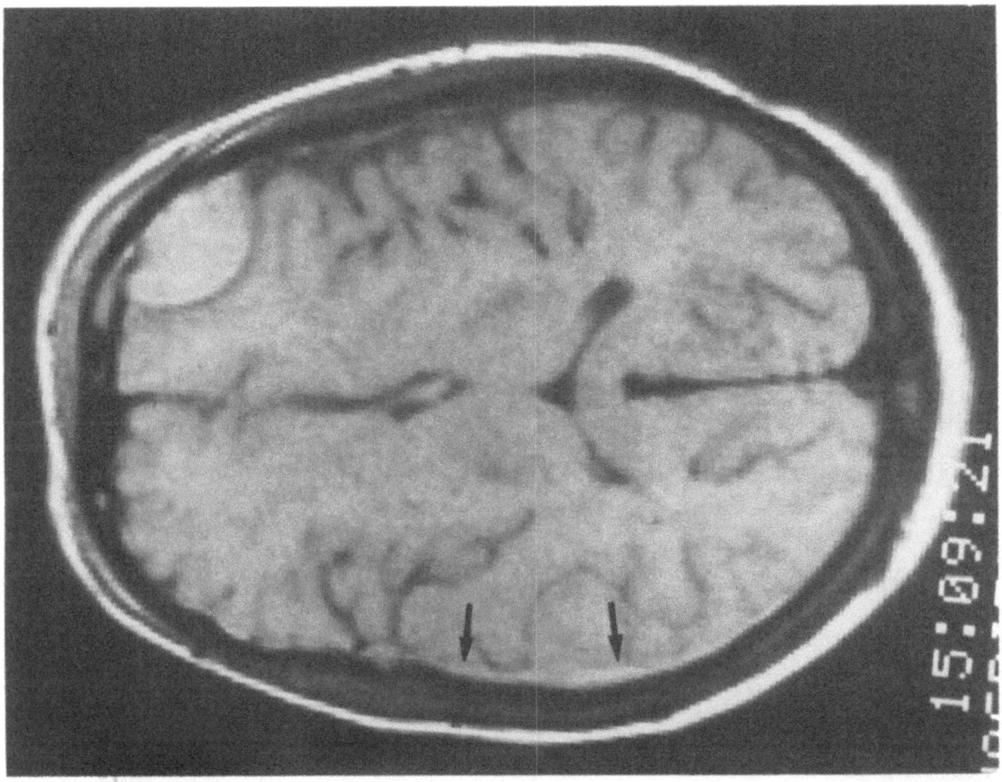

FIGURE 1–4C

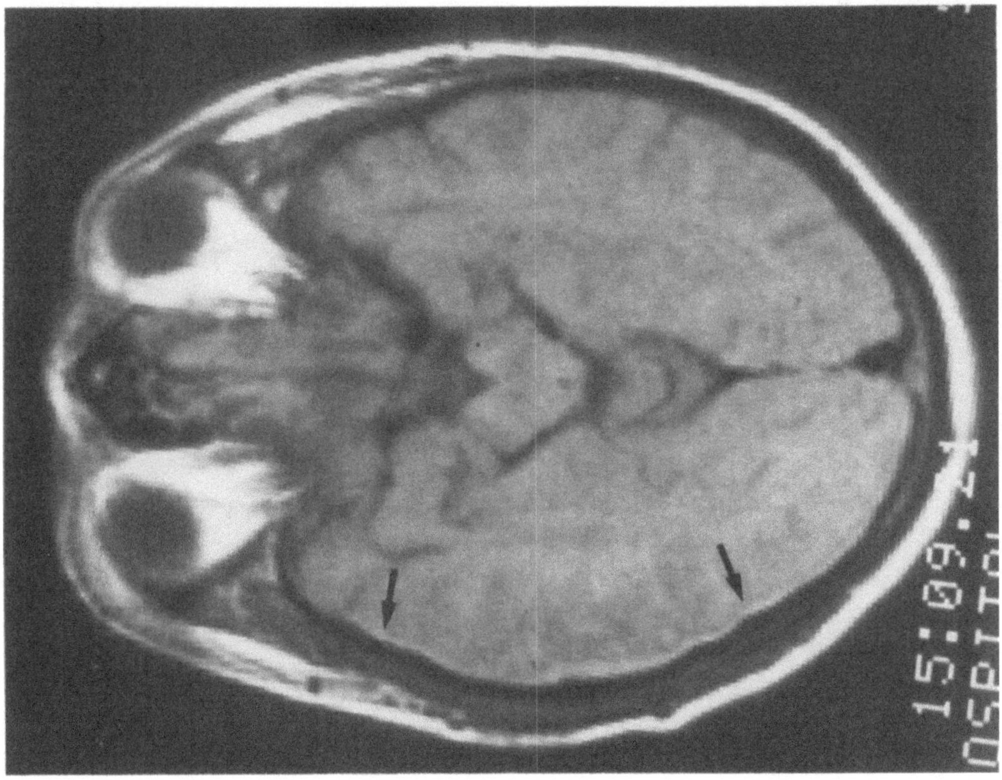

FIGURE 1–4B

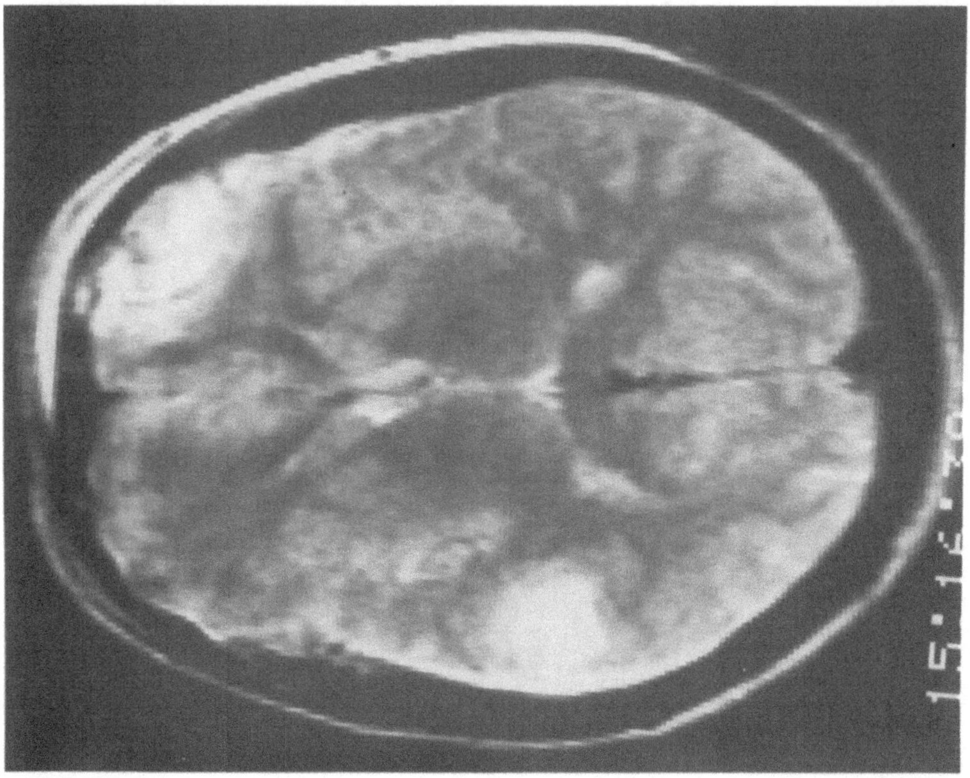

FIGURE 1–4E

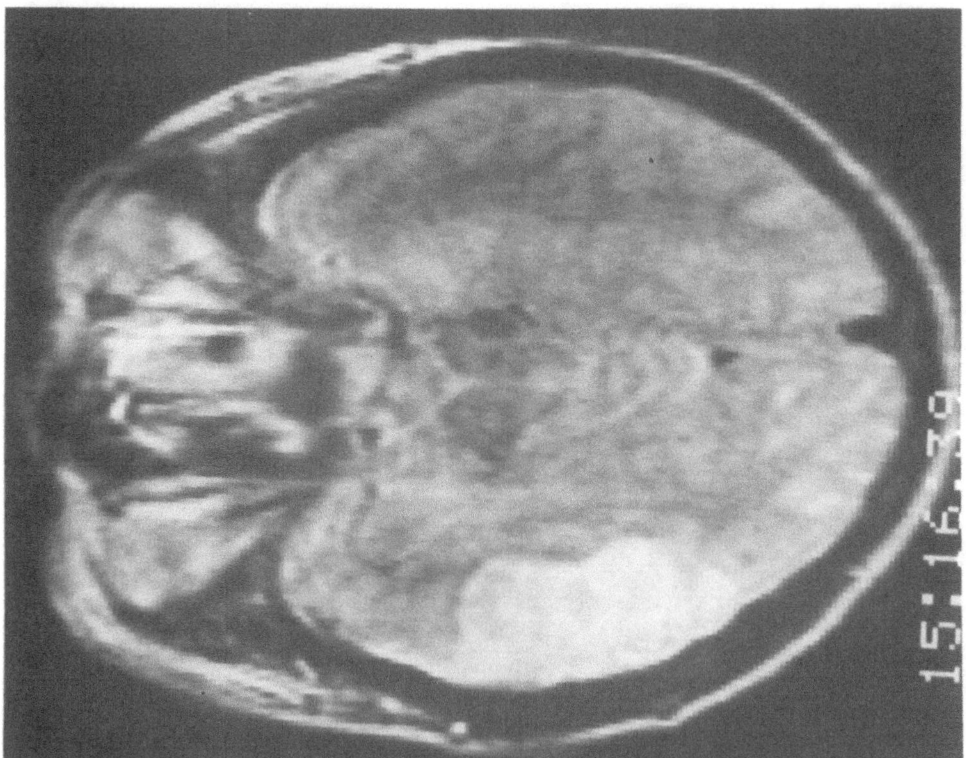

FIGURE 1–4D

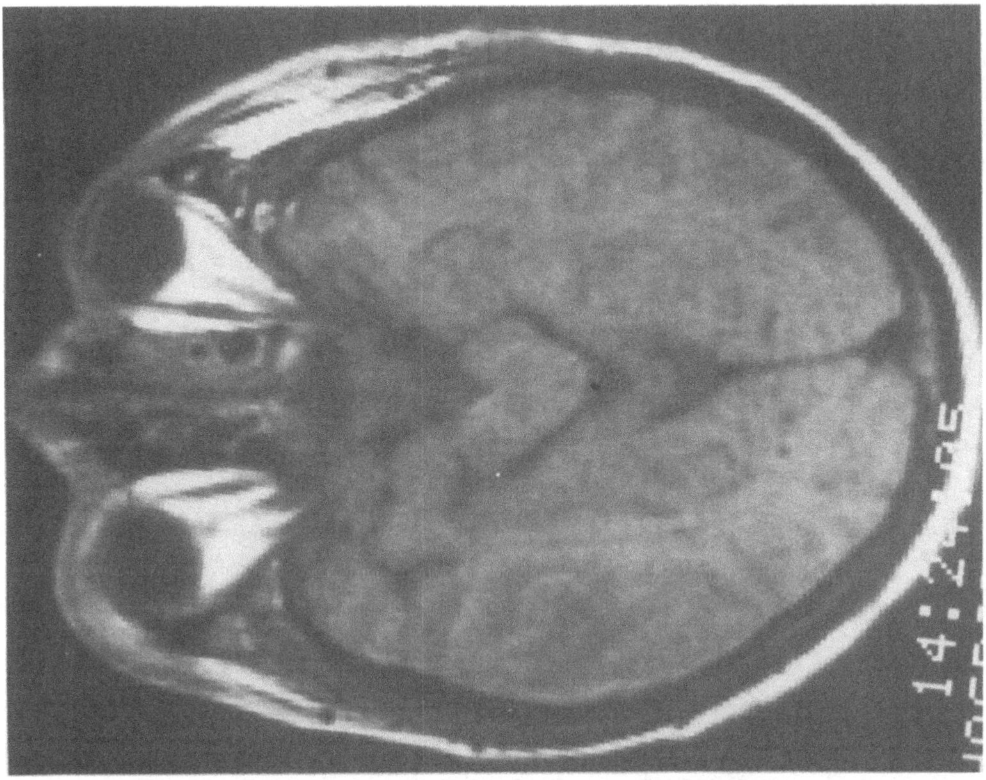

FIGURE 1–4G

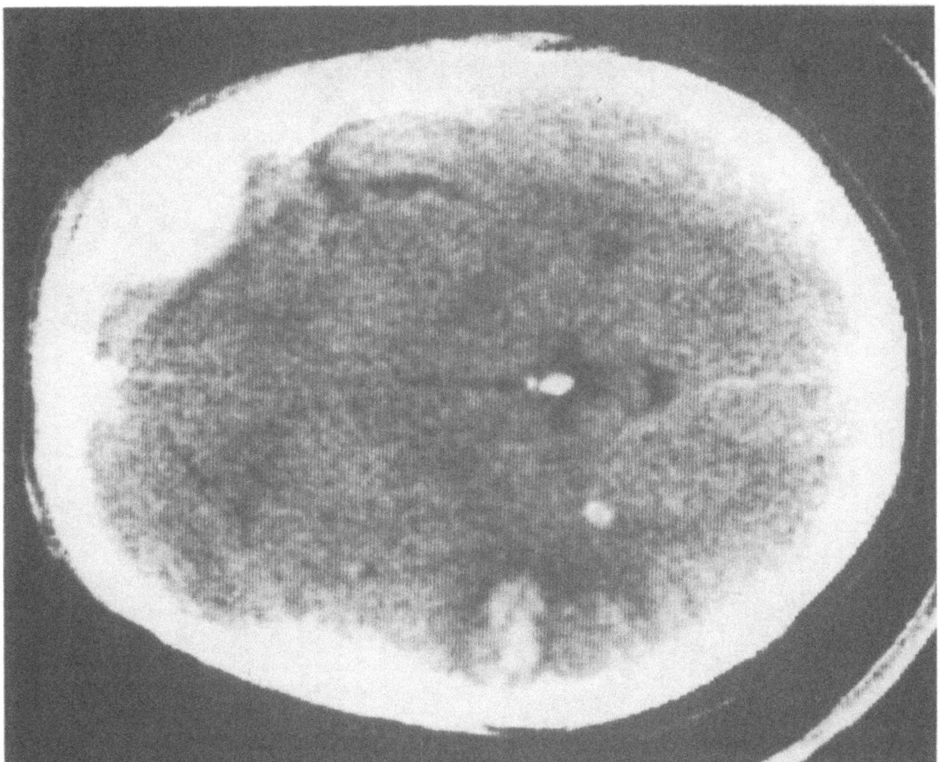

FIGURE 1–4F

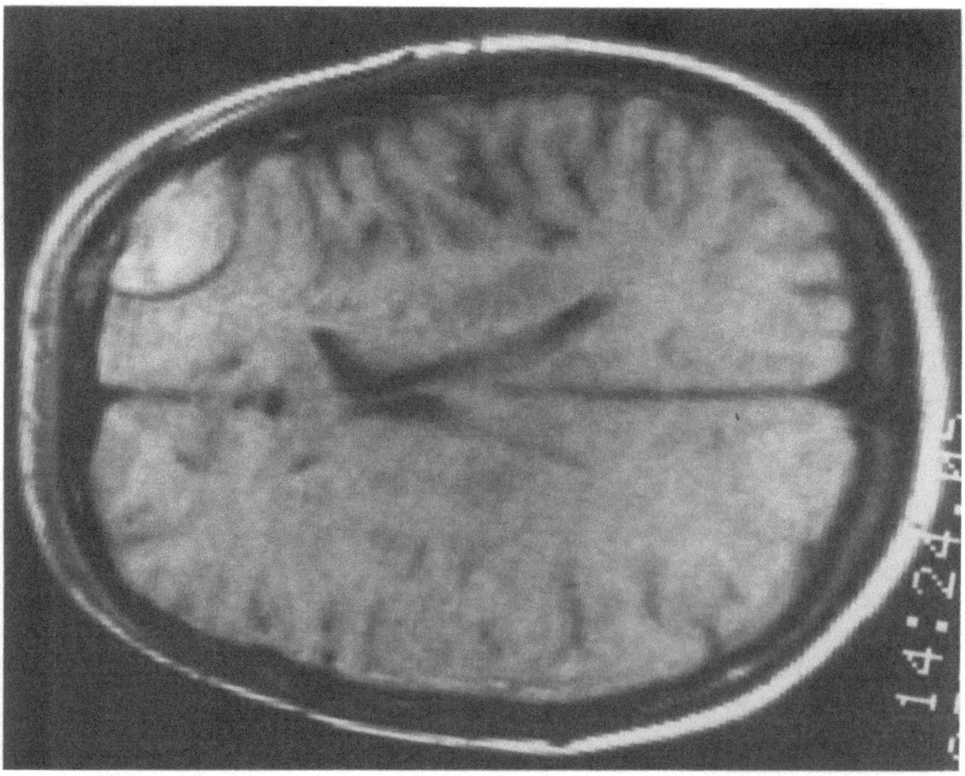

FIGURE 1–4H

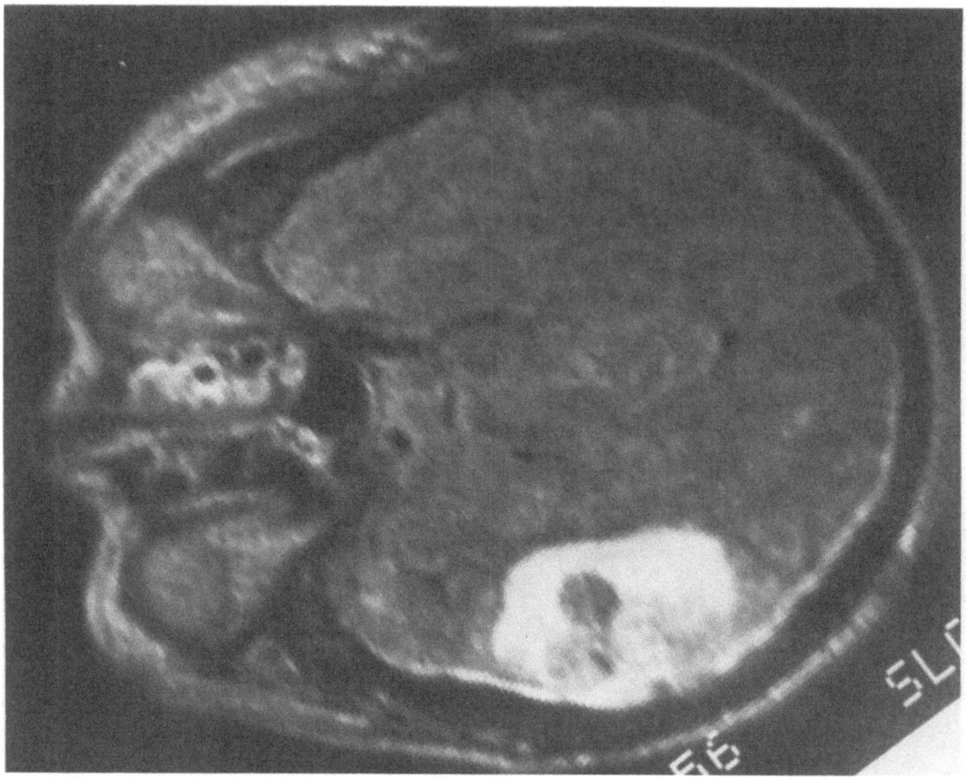

FIGURE 1–4I

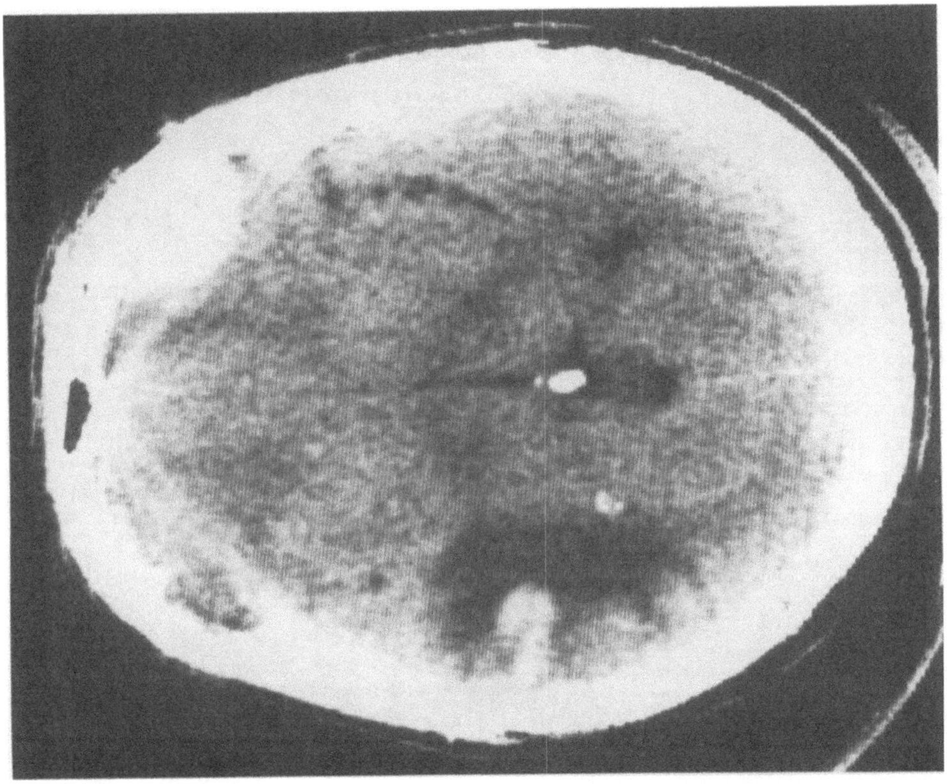

FIGURE 1–4K

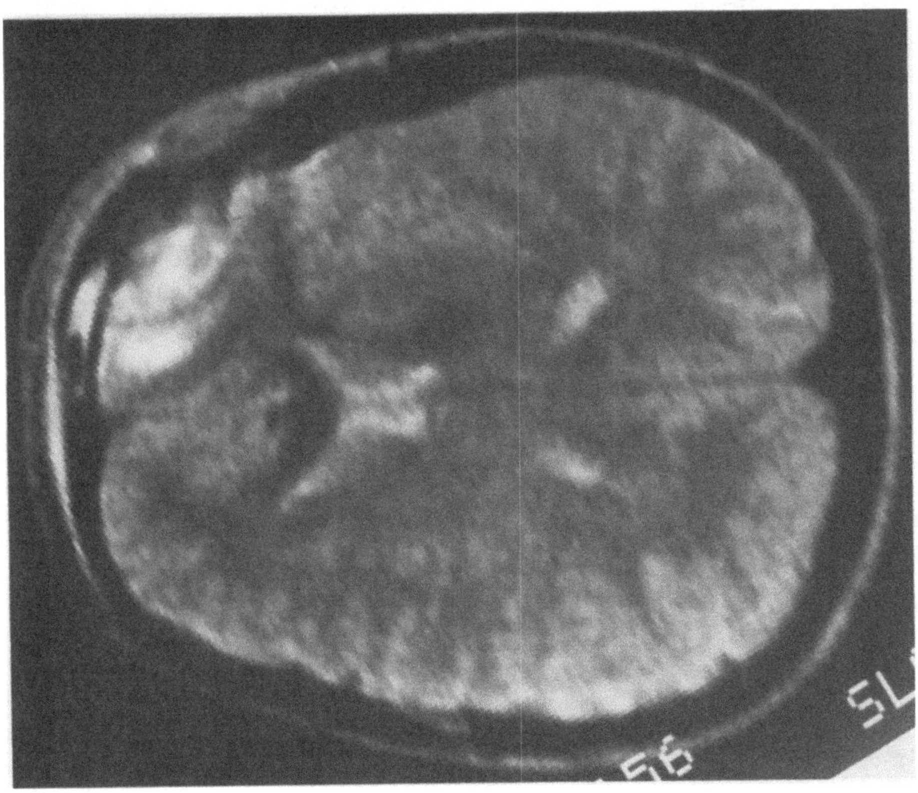

FIGURE 1–4J

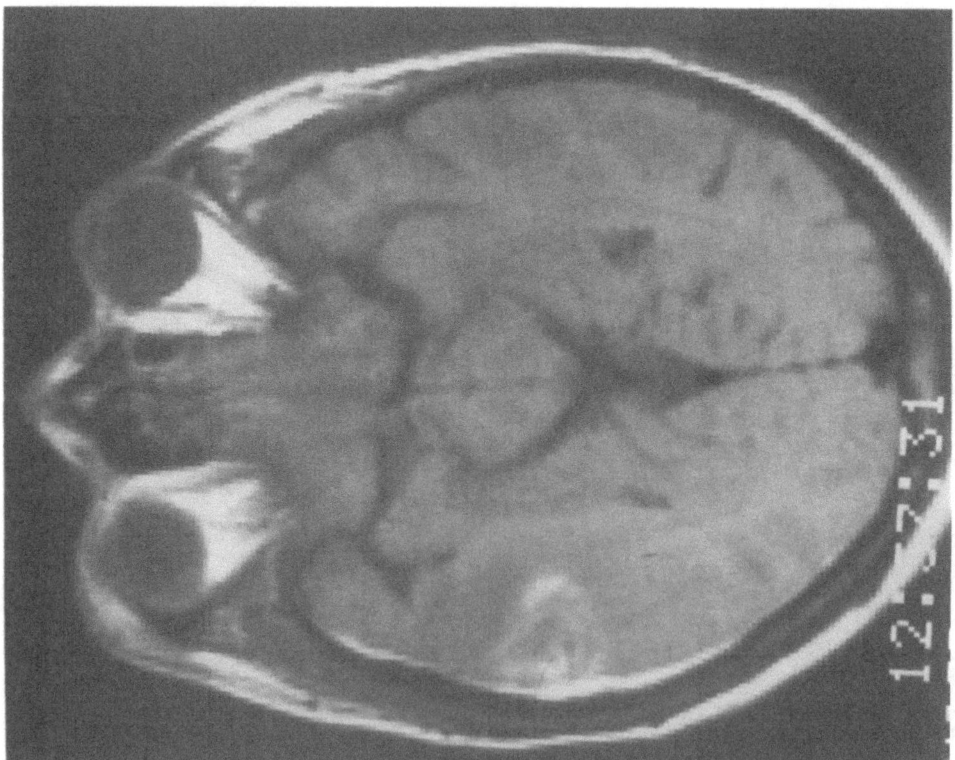

FIGURE 1–4L

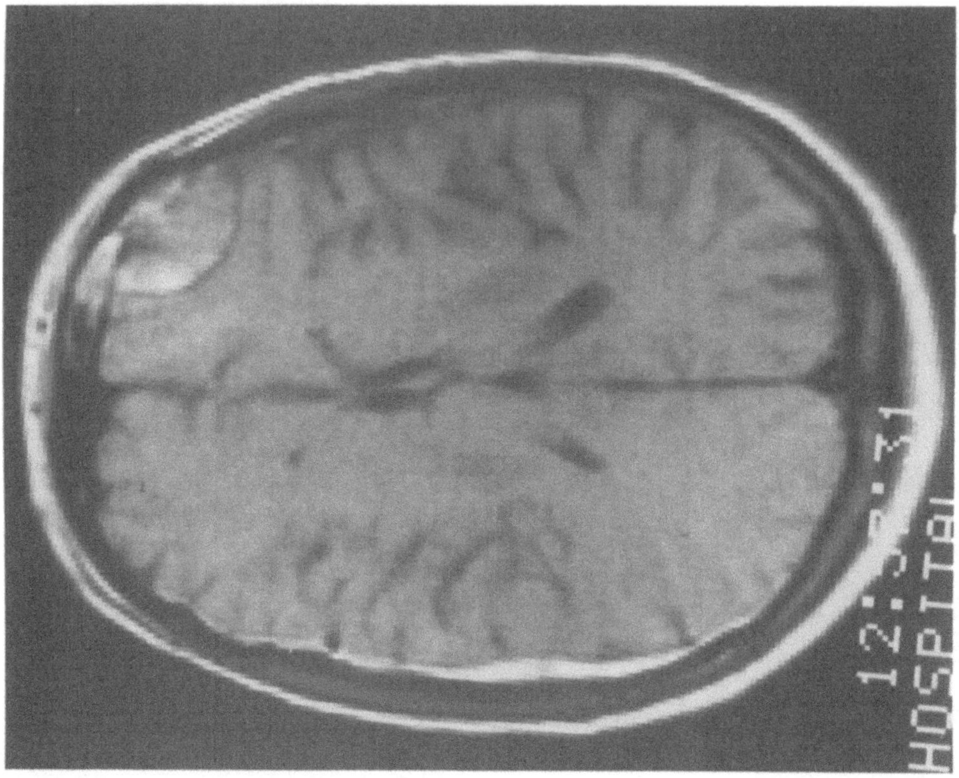

FIGURE 1–4M

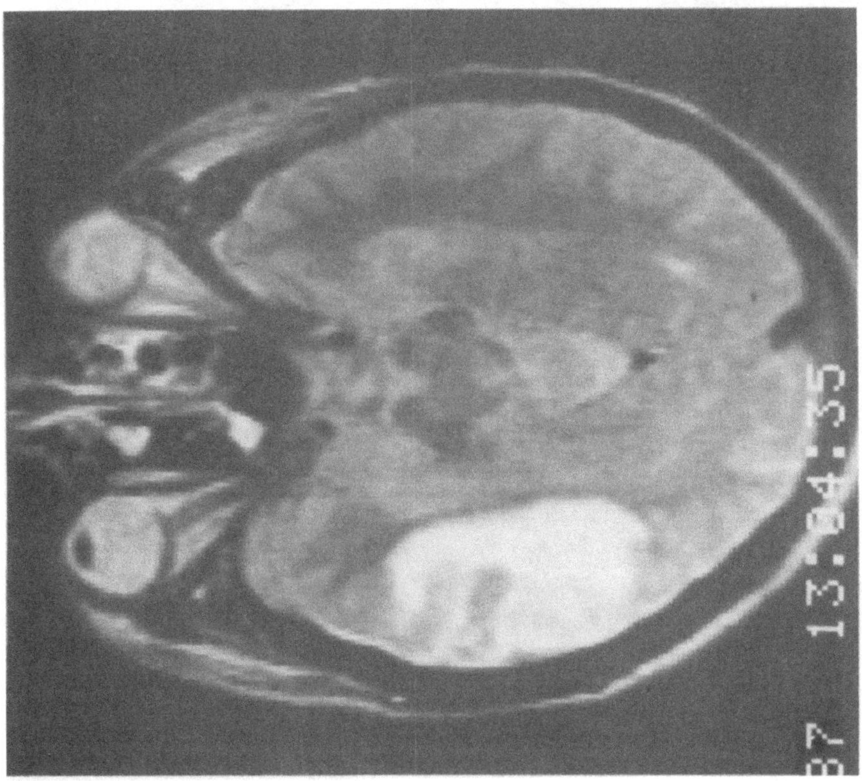

FIGURE 1–4N

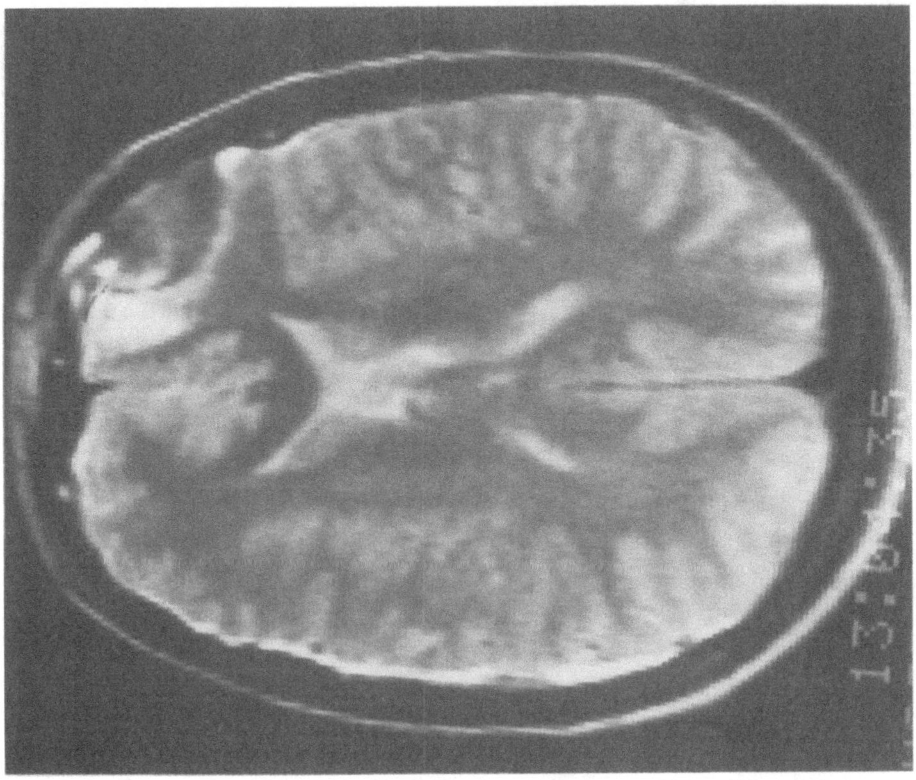

FIGURE 1–4O

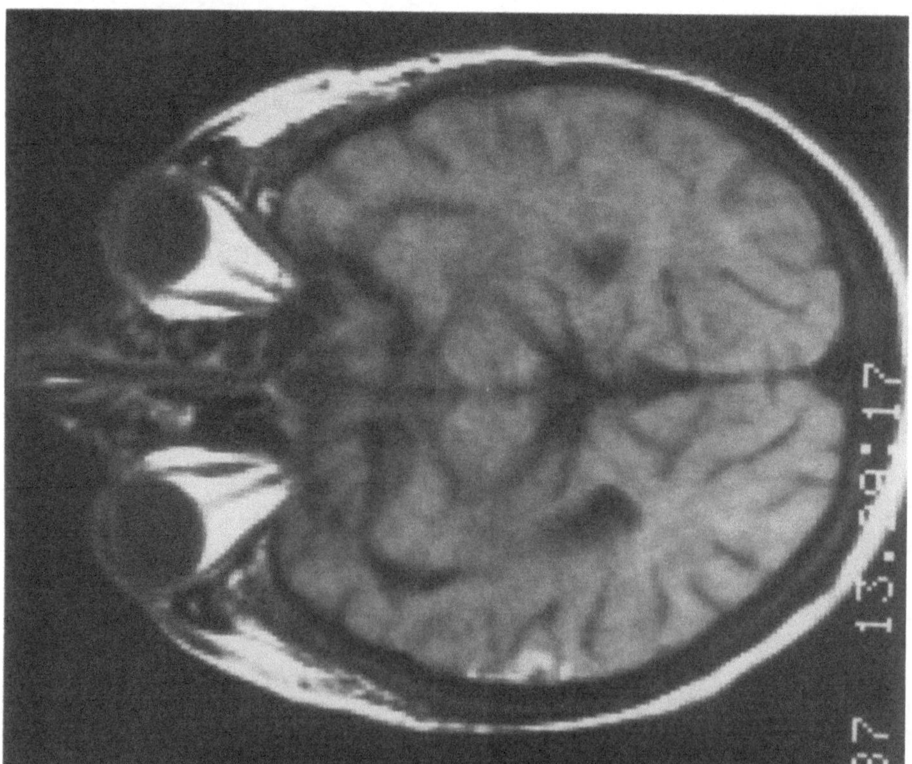

FIGURE 1–4P

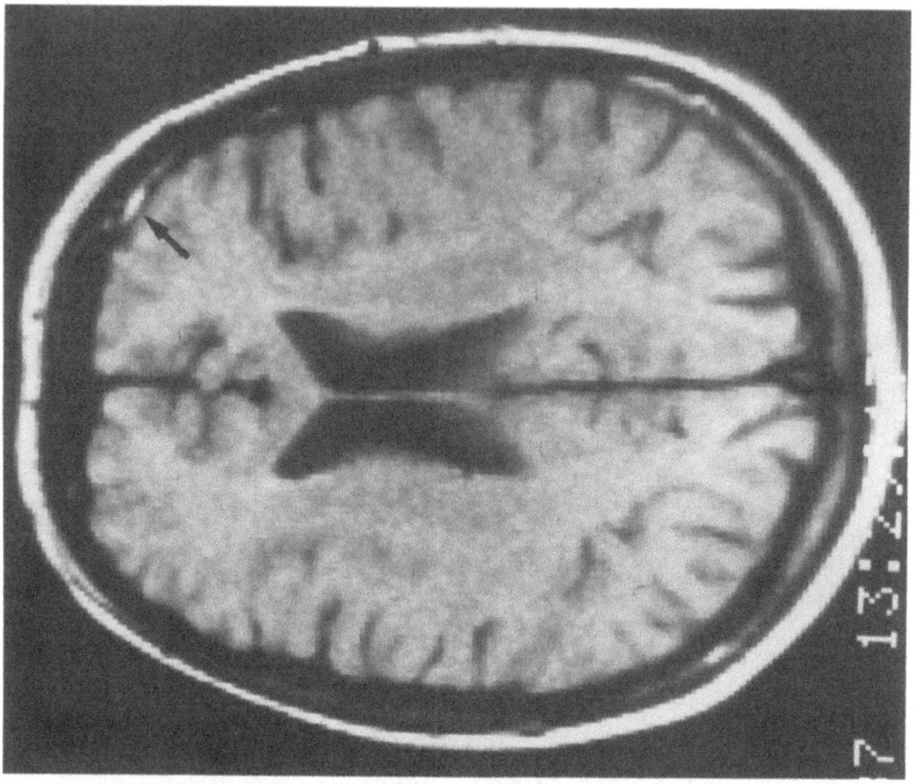

FIGURE 1–4Q

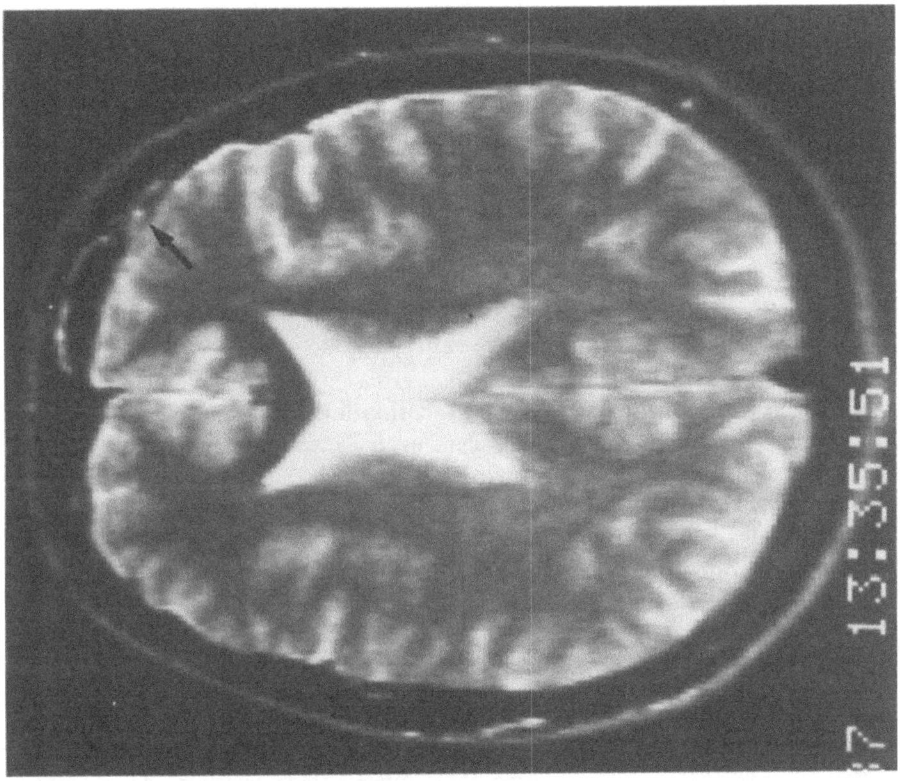

FIGURE 1–4S

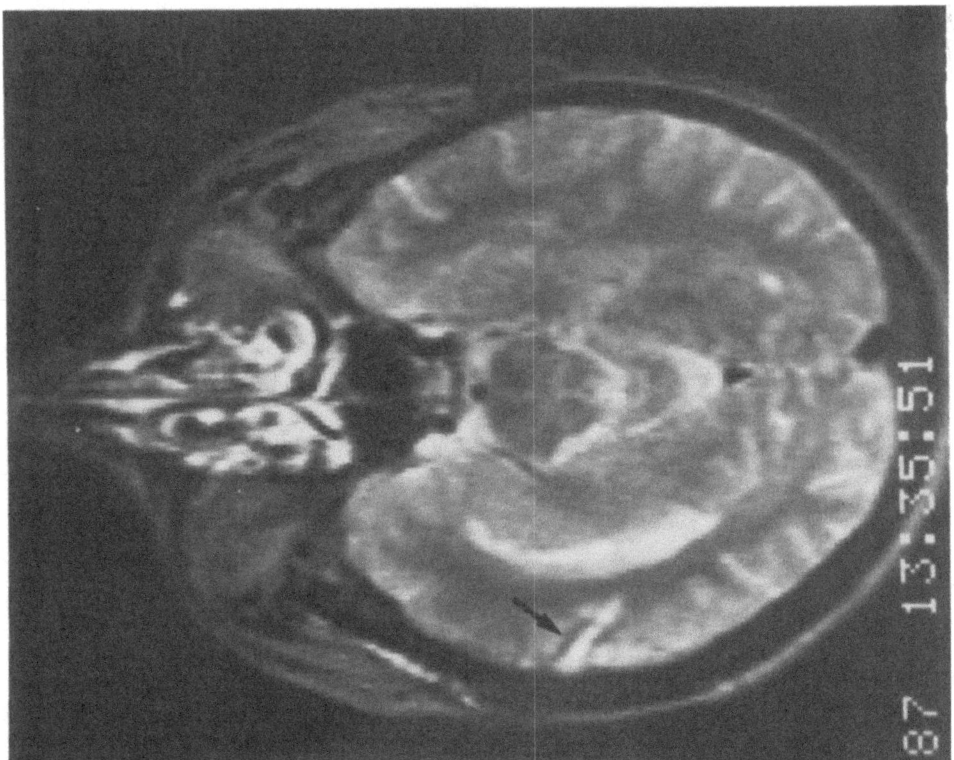

FIGURE 1–4R

manifestation of the relatively weak T1 W of STR/STE SE scans. The high protein concentration of viscous blood leads to mild T1 prolongation, because protein binding shortens the T1 of water. The same phenomenon has been observed in highly proteinaceous neoplastic cysts (e.g., colloid cyst and craniopharyngiomas). When T1 is only mildly prolonged, relative to the normal brain, the intensity of fluid is dominated by the high PD [24] and long T2 of blood. On LTR scans, AIH is hyperintense to the brain as predicted, and intensity increases as TE is prolonged (figures 1–3, 1–4).

A subtle rim of hypointensity may also be encountered on LTR scans at the margin of a hyperintense AIH [10, 18] (figure 1–4E) The cause of this rim is unclear. One possibility is that there are differences in magnetic susceptibility between the hematoma and the adjacent brain. Under these circumstances, intrapixel variation in the magnetic field at the margin of the hematoma could produce T2 shortening at the border of the hemorrhage [18]. When GE scans are performed during Stage I, AIH displays more prominent peripheral hypointensity and central hypointensity as well [18]. Although experience with this technique is limited, current data suggest that most nonhemorrhagic lesions do not become hypointense [18].

Clinically, the pattern of mild homogeneous hyperintensity on STR scans and hyperintensity with a subtle rim of hypointensity on LTR/ITE-LTE scans is characteristic and relatively specific, because most lesions that are hyperintense on LTR scans are slightly hypointense on STR scans [10]. In addition, the presence of hypointensity on GE scans may further increase specificity.

STAGE II: 6/72 HOURS–3/4 DAYS
(FIGURES 1–2, 1–4–1–6)

AIH intensity at this stage is well established because many patients have been studied [5–8, 10, 18], and in vivo animal data [20, 23] and in vitro data [20] have been obtained that confirm these findings. Hematoma intensity decreases on all pulse sequences during this stage. On STR/STE scans, a subtle decrease causes the AIH to become iso-to mildly hypointense to white matter. Hypointensity is more dramatic on LTR/ITE-LTE scans and becomes progressively more apparent as TE is lengthened [7, 10].

Development of hypointensity is the result of T2 shortening (even the subtle hypointensity noted on STR/STE T1 W scans is the result of T2 shortening, not T1 prolongation). In vitro studies have demonstrated that the T2-shortening effect is induced by intracellular paramagnetic deoxyhemoglobin with its four unpaired electrons [17]. Oxyhemoglobin is diamagnetic because it contains no unpaired electrons. The molecular structure of deoxyhemoglobin is such that PEDD T1 shortening does not occur, since the water protons cannot get close enough to the unpaired electrons that reside within the deoxyhemoglobin molecule [7, 15]. T2 shortening, however, can occur when deoxyhemoglobin is heterogeneously distributed. When purely intracellular, deoxyhemoglobin molecules act as magnets, generating stronger fields within the RBCs than those present in the surrounding extracellular space, and thereby setting up magnetic gradients between the intra- and extracellular spaces (figure 1–2A and 1–2B). In vitro studies have shown that this effect is dependent upon a number of factors, including the square of the deoxyhemoglobin concentration, the interecho time, and the square of the field strength [17]. The square relationship to deoxyhemoglobin concentration probably accounts for the rapid development of hypointensity and the marked sensitivity of MR to even small amounts of hemorrhage. The dependence on field strength, however, does not correlate completely with clinical experience. Based on in vitro data, it was initially predicted that hypointensity would be encountered only at high field strengths (figure 1–5) [7]; however, hypointensity is routinely seen in AIH at intermediate field strengths (.35–.6T) (figures 1–3, 1–4, 1–6) (10, 15, 18], and relative hypointensity can even be identified at low (.15T) [5] and ultralow (.02T) [8] field strengths. This suggests that other as yet unknown factors may contribute to the development of hypointensity. For instance, Haymen et al. have recently demonstrated that T2 shortening occurs in vitro within RBC-free fibrin clots [25]. While the precise effect of field strength on intensity remains the subject of controversy, indirect comparison of clinical images (e.g., compare figures 1–5 and 1–6) indicates that there is greater hypointensity at high field strength, albeit to a lesser extent than that predicted on the basis of in vitro data. The greater sensitivity of high-field-strength MR imagers to susceptibility effects may be manifested by more rapid development of hypointensity (clinical cases with hypointensity have been reported as early as six hours after hemorrhage [26] at high field strength) and by prolonged persistence of hypointensity (7–14 days). Of course, to determine the precise effect of field strength, direct comparisons would be necessary, but these would have to be done within a very short period of time since the intensity of AIH may change dramatically in as little as 24 hours. Such comparisons are obviously difficult to obtain, since few institutions have multiple MR systems of different field strengths, and even if two scanners were available, it would be difficult to justify performing consecutive MR scans in patients who are acutely ill and in need of care for monitoring. We have evaluated one clinically stable patient with a subdural hematoma four days after injury at .6 and 1.5T, and in this one instance, the degree and extent of hypointensity were similar (figure 1–3). To further study this phenomenon, we have also evaluated experimental AIH in a dog model [23]. Even in this relatively controlled setting, hematoma intensity varies greatly, but it does appear that hypointensity is more pronounced and does develop somewhat earlier at high field strengths. Other factors, however—in particular, the size of the hematoma and the source of the hemorrhage (arterial

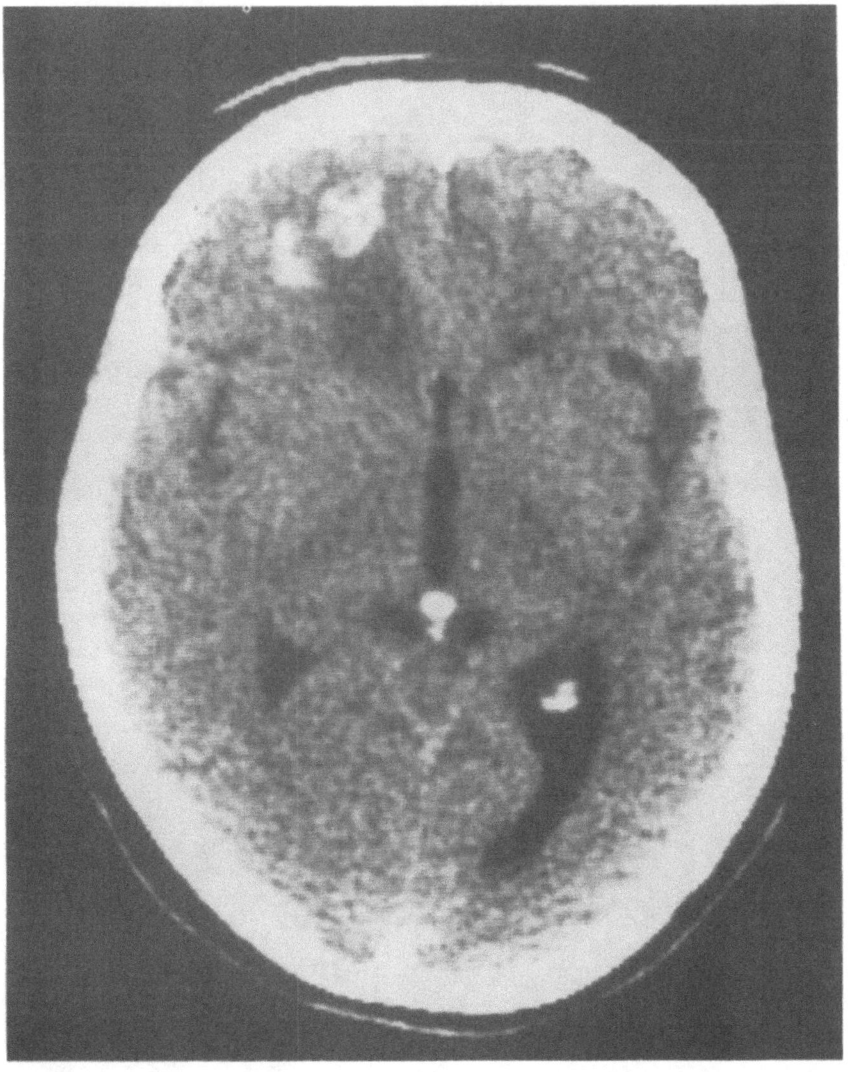

FIGURE 1–5A

FIGURE 1–5. Acute hemorrhage studied at 1.5T with follow-up examinations at 1.5T and .6T. CT scan (A) reveals a bilobed, right frontal hematoma. On sagittal STR/STE at 1.5T, the AIH is mildly hypointense to the brain (arrow) (B). On axial LTR/LTE (C), the hemorrhage produces bilobed frontal hypointensity with hyperintense surrounding edema (Stage II). A follow-up scan at five weeks at 1.5T (D & E) reveals the hematoma to be mildly hypointense on STR/STE scans with focal hypointensity on LTR/ITE. Repeat examination two weeks later at .6T (F & G) reveals only subtle hypointensity relative to adjacent white matter on LTR/ITE at the site of prior hemorrhage (arrow). Hypointensity is more marked, however, on GE scan (500/30/30 degrees) at .6T.

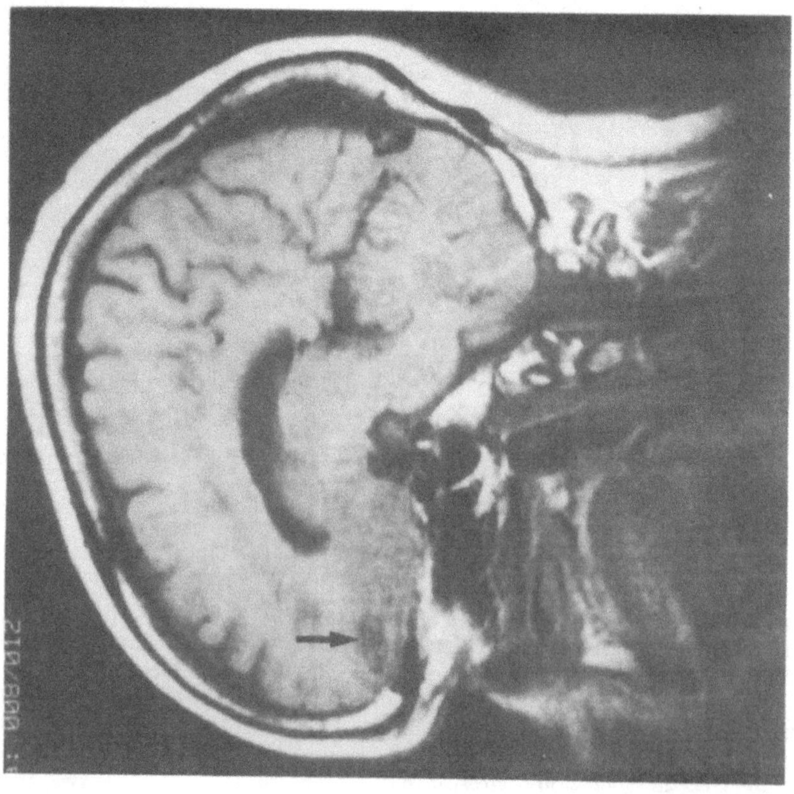

FIGURE 1–5B

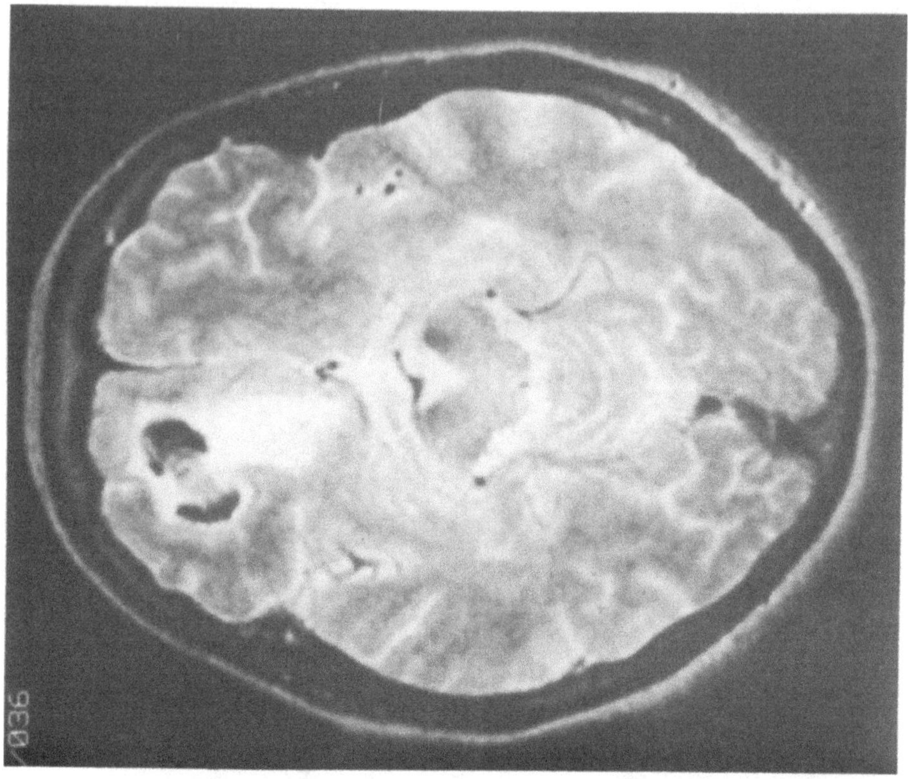

FIGURE 1–5C

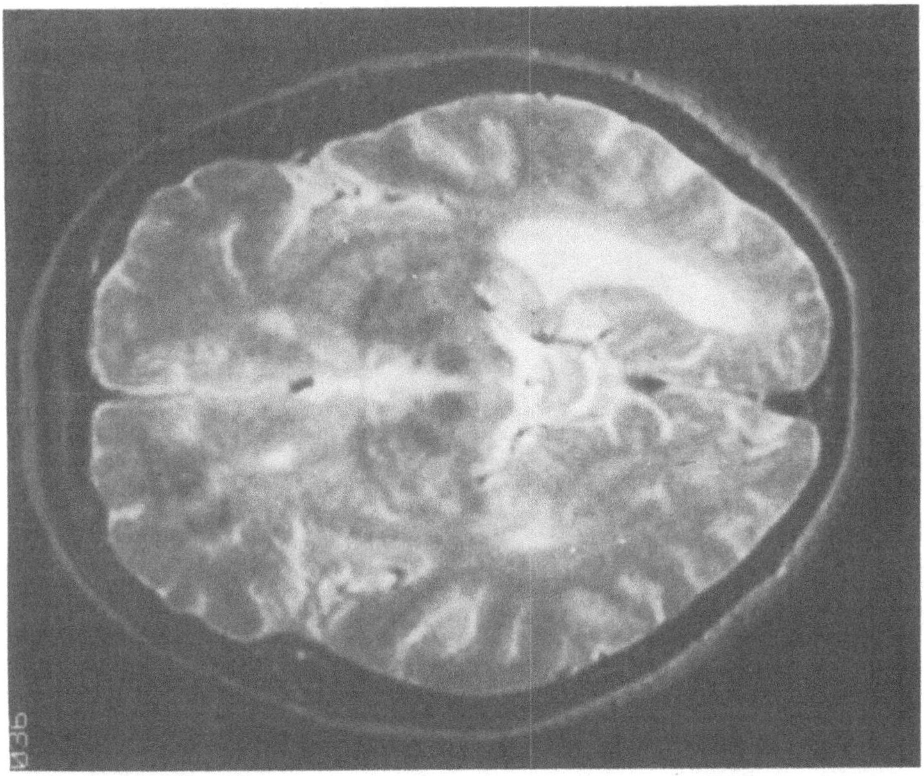

FIGURE 1–5E

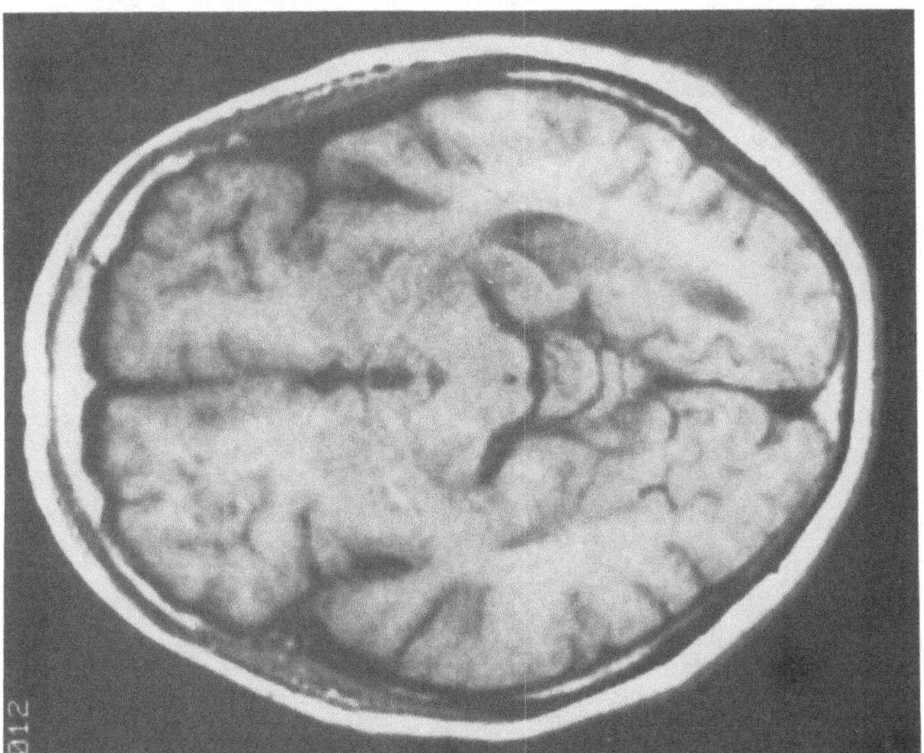

FIGURE 1–5D

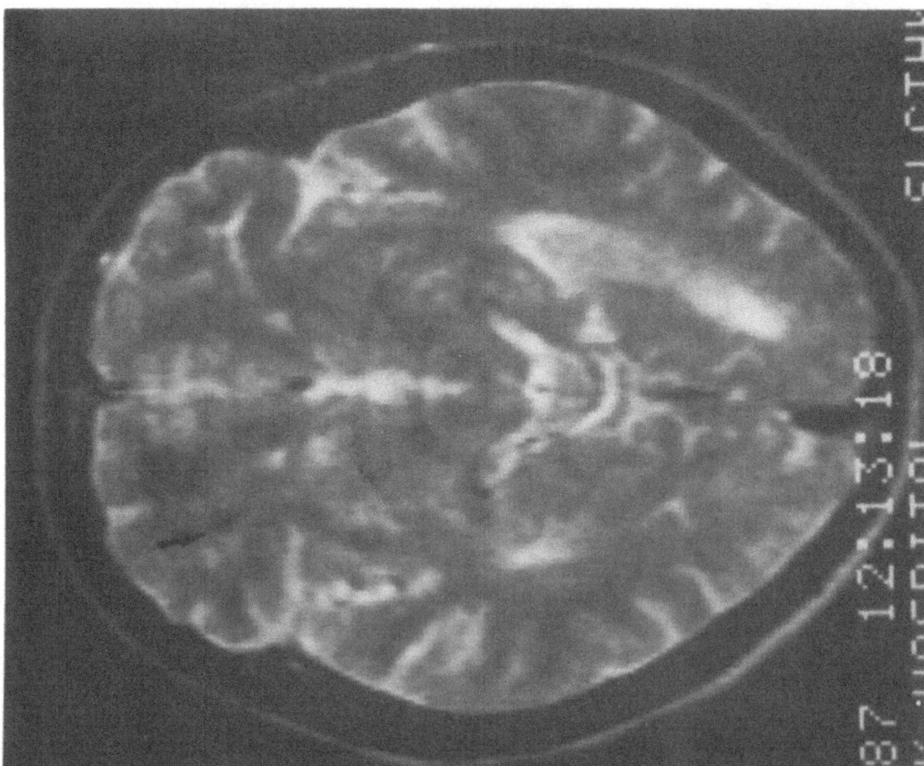

FIGURE 1–5F

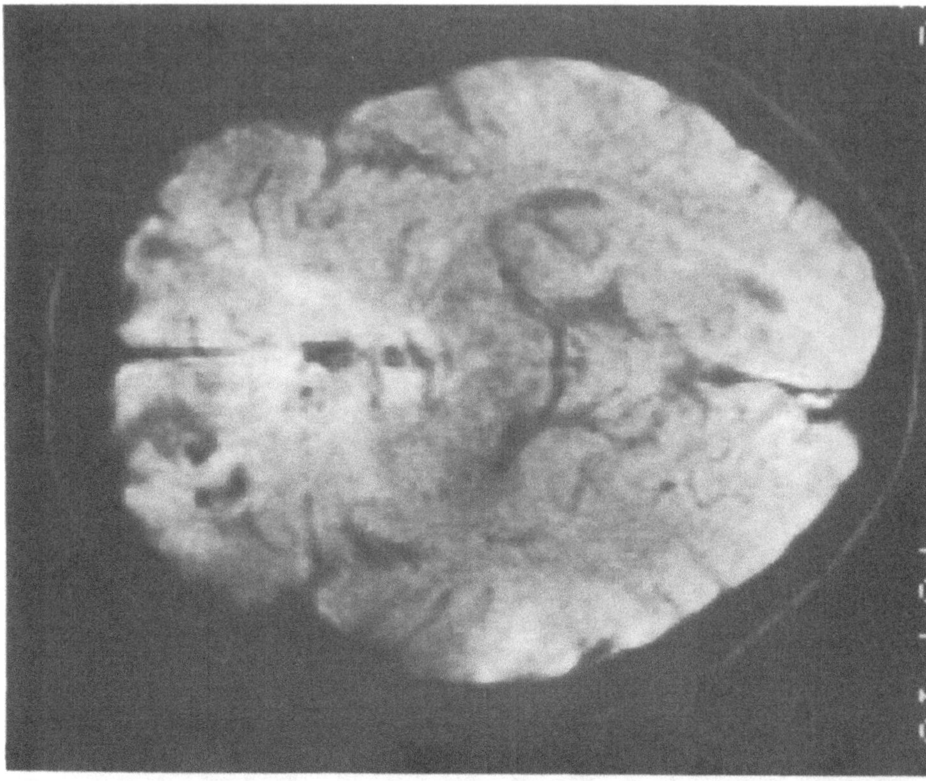

FIGURE 1–5G

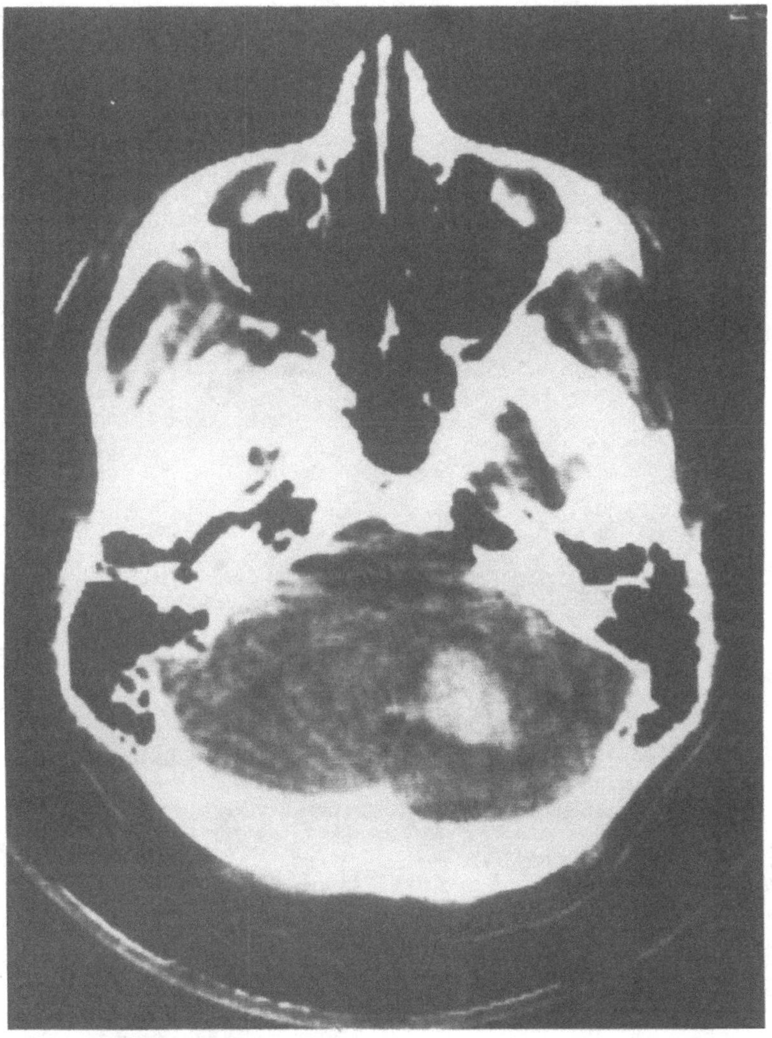

FIGURE 1–6A

FIGURE 1–6. Acute parenchymal and intraventricular hemorrhage on CT in a 25-year-old woman with a cerebellar AVM demonstrates a dense triagular hematoma in the cerebellar hemisphere (A) and intraventricular blood at the level of foramen of Monro (B). STR/STE (C), obtained 60 hours after hemorrhage and six hours after CT, reveals that the cerebellar hematoma is predominantly isointense with a few scattered foci of hyperintensity. The Fourth ventricular hemorrhage is hyperintense (arrow). On LTR/LTE scan (D), the center of the hematoma is markedly hypointense to gray and white matter (Stage II), while the periphery of the hematoma and surrounding edema are mildly hyperintense and merge with each other. Intraventricular clot is not visible. GE (500/30/30 degree) at the same time (E) reveals the entire hematoma to be hypointense. Note focus of hypointensity within the fourth ventricle (arrow) representing intraventricular hemorrhage. Follow-up MR 24 hours later reveals that the cerebellar hematoma has increased in intensity. It is now hyperintense on STR/STE (F) and less hypointense on LTR/LTE (G) (Stage III). On GE (H), hypointensity is unchanged and intraventricular hemorrhage is once again identified. LTR/ITE (I) at the level of the foramen of Monro demonstrates the intraventricular clot to be nearly isointense to CSF and difficult to detect. On GE at the same level, at the same time (J), the hypointense intraventricular clot is more easily seen.

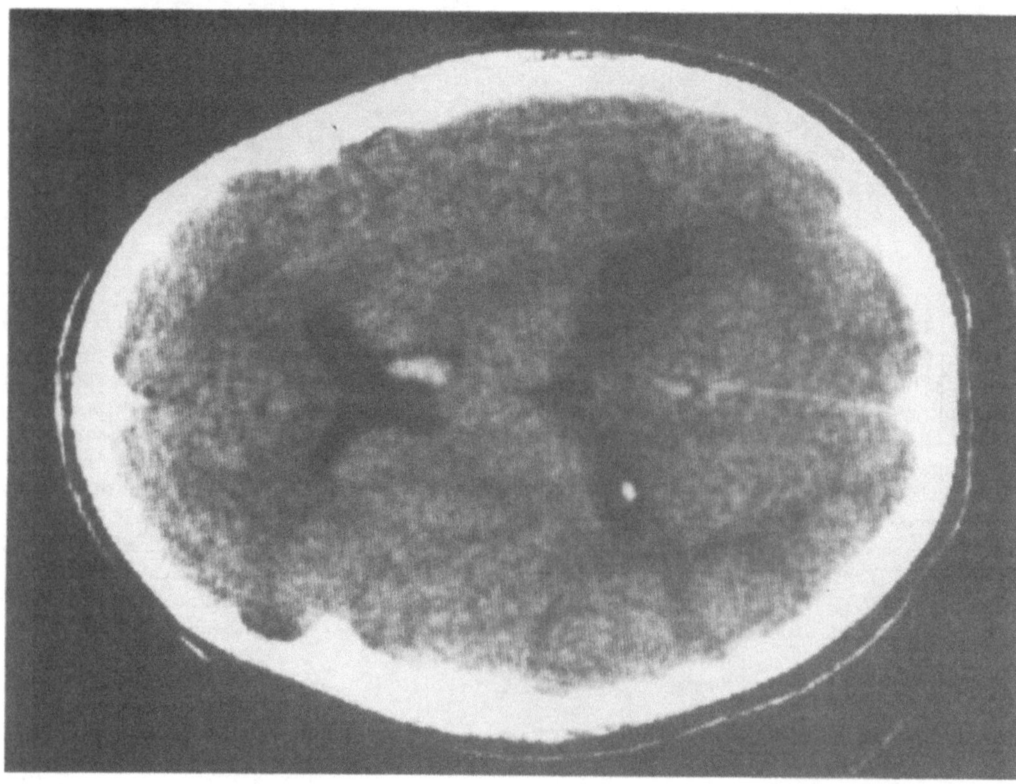

FIGURE 1–6B

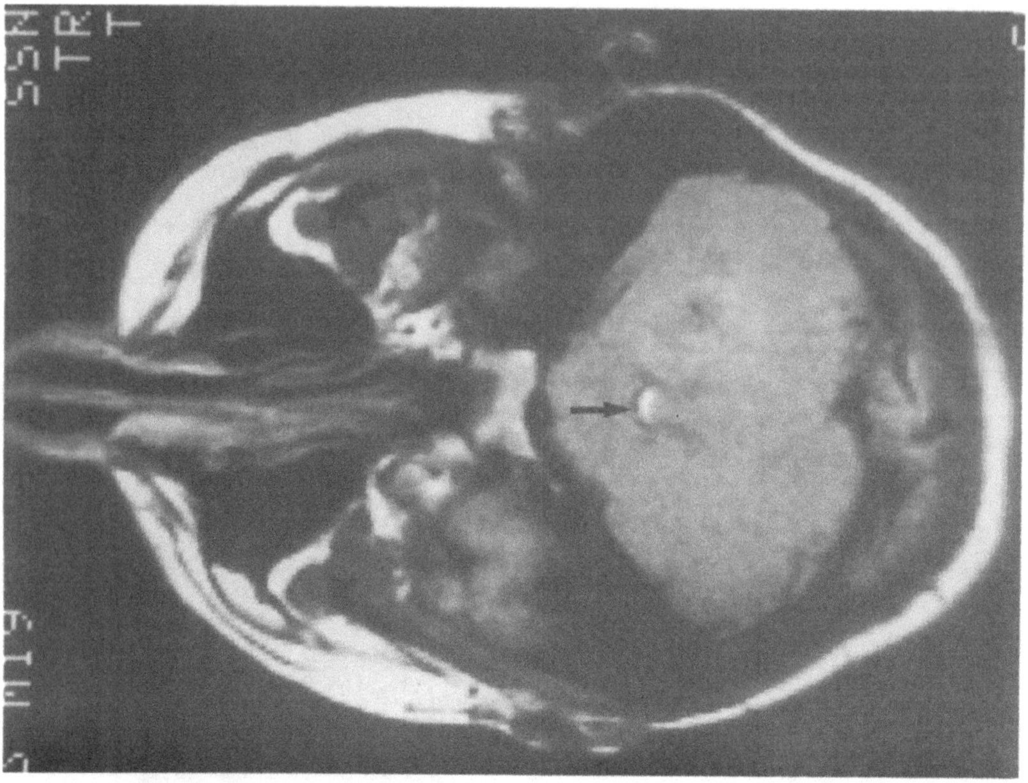

FIGURE 1–6C

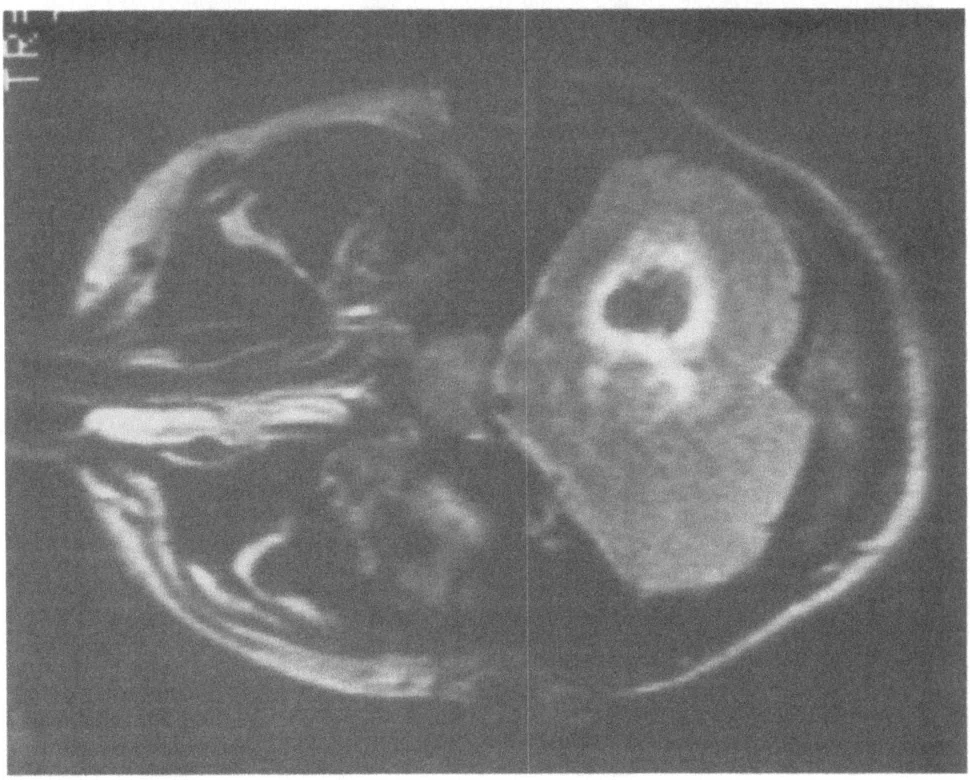

FIGURE 1–6D

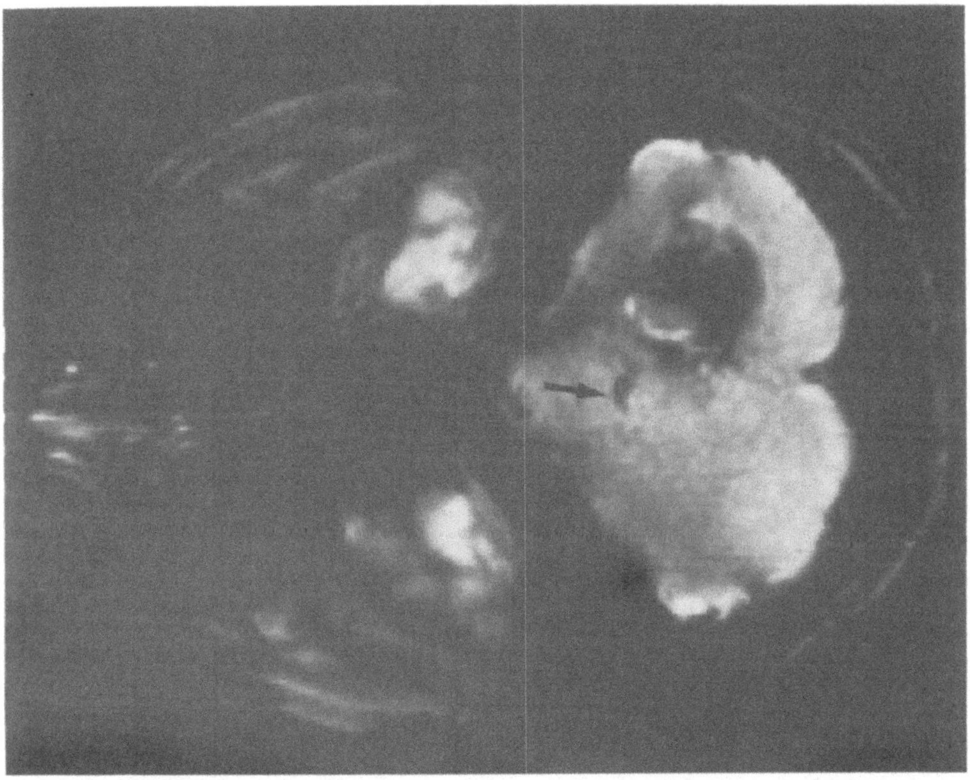

FIGURE 1–6E

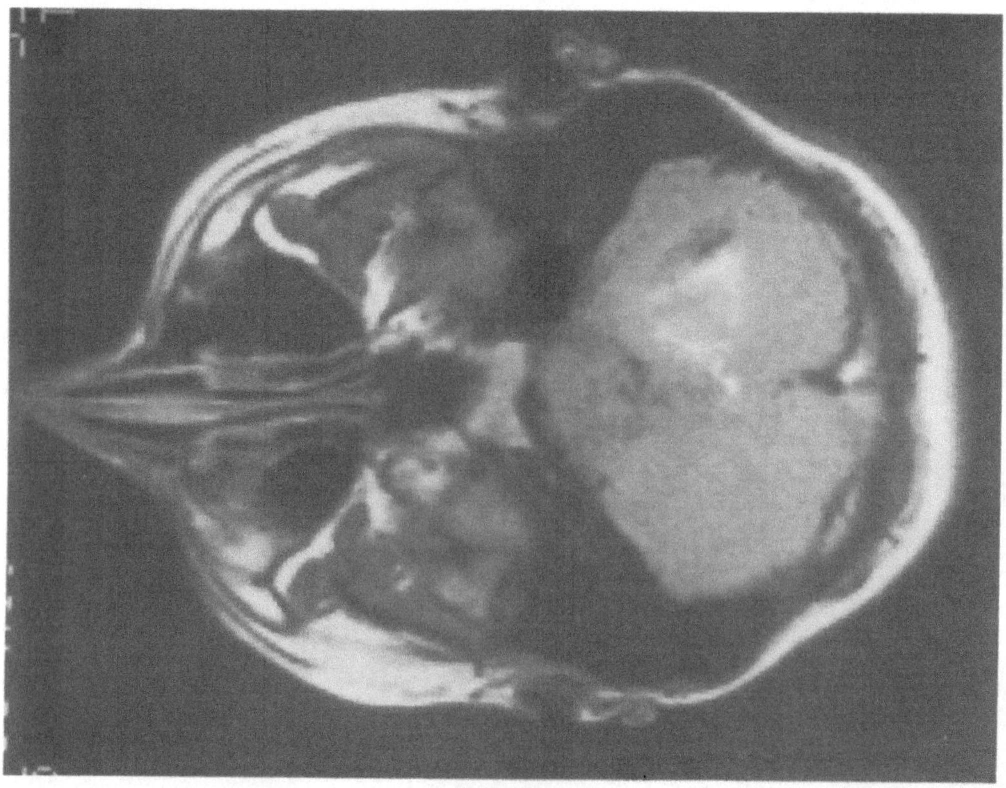

FIGURE 1-6F

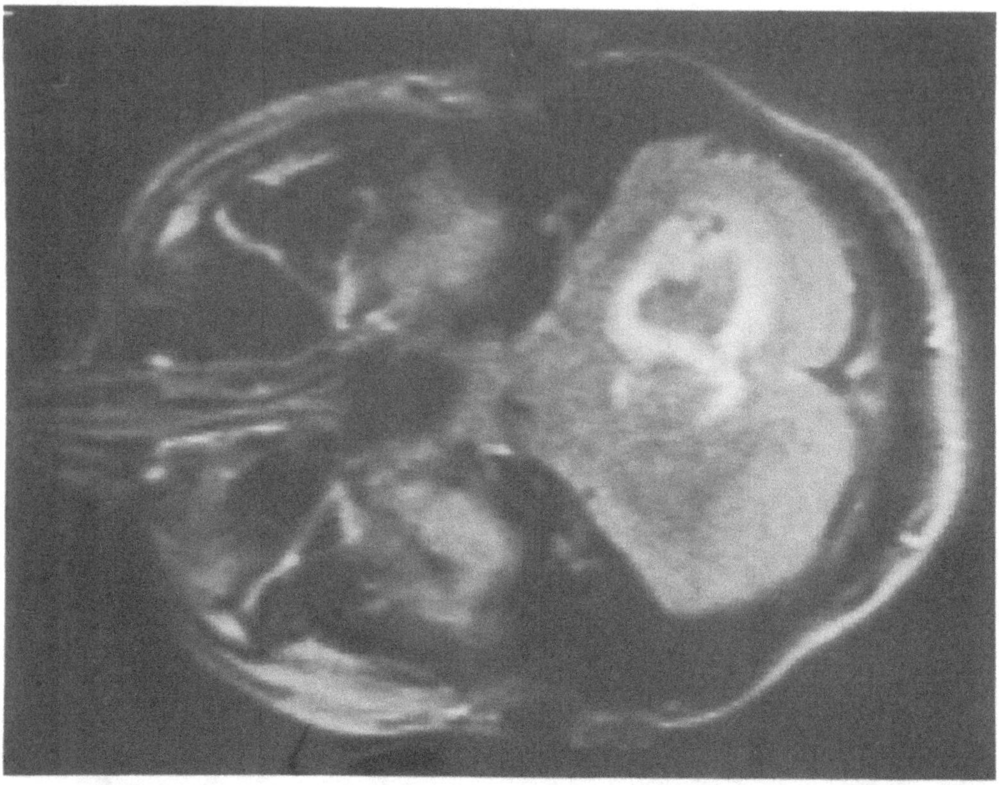

FIGURE 1-6G

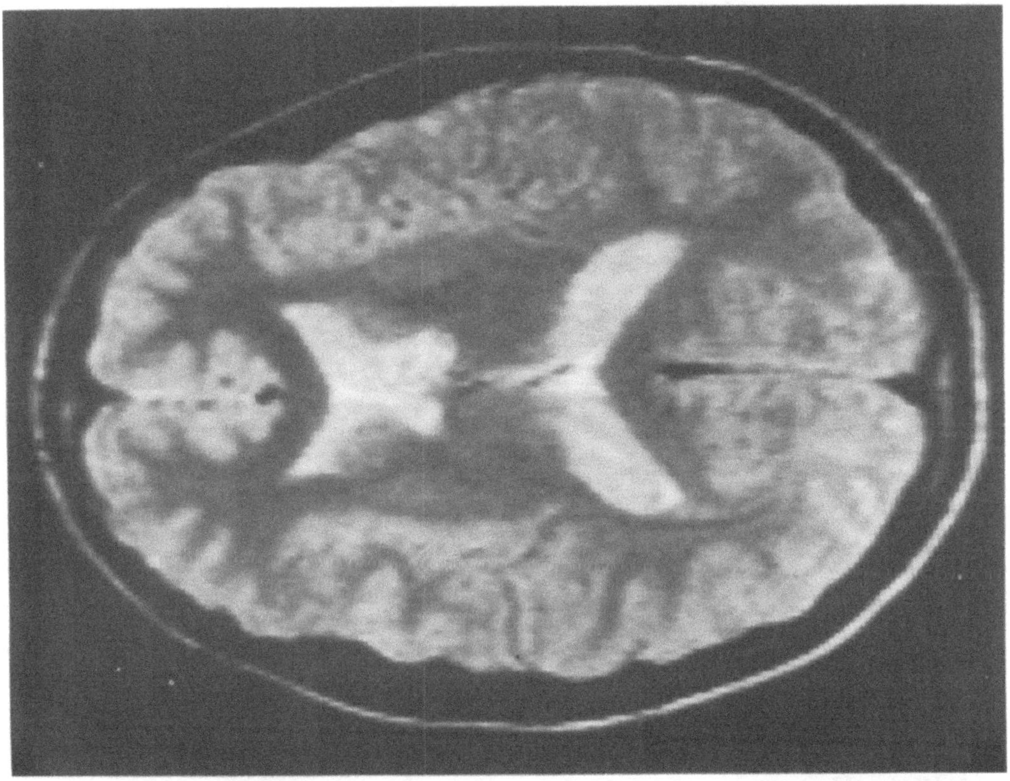

FIGURE 1–6I

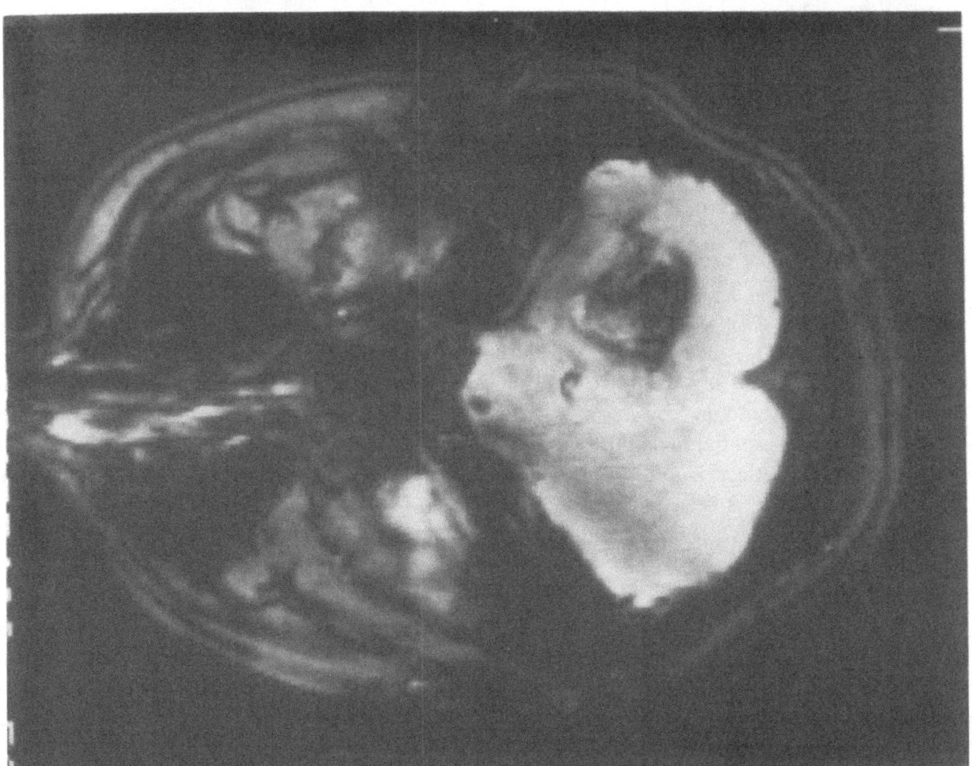

FIGURE 1–6H

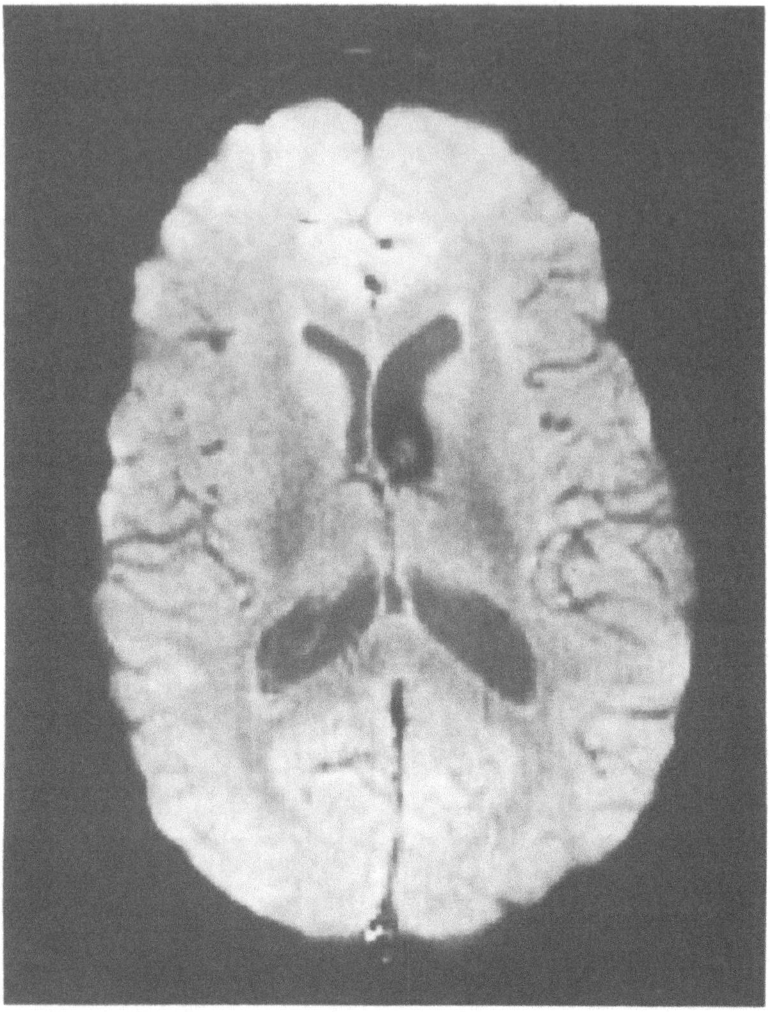

FIGURE 1–6J

or venous)—may dramatically affect the timing and extent of intensity changes.

GE scans add little to the diagnosis of AIH once the hematoma has become hypointense on SE scans. The hypointensity will be more prominent (figure 1–6) on SE sequences, but the effect is often so strong that artifactual hypointensity is seen beyond the extent of the actual hematoma (as identified on CT).

The specificity of MR during this stage is unknown. Heavily calcified lesions may also be hypointense on LTR/ITE-LTE scans [27]. In addition, we have encountered noncalcified, nonhemorrhagic intra- and extraaxial granulomatous lesions that are similarly hypointense [28]. Another problem at this stage, at least at intermediate field strengths, is that the extent of hypointensity on MR may not correlate with the extent of the hemorrhage as identified on CT, and thus the true size of the hematoma cannot be accurately evaluated (the hyperintense component of the hematoma is often at its periphery, where it merges with the adjacent hyperintense edema; see figure 1–6) [10].

STAGE III: 3/4 DAYS–5/6 DAYS
(FIGURES 1–3, 1–4, 1–6, 1–7)

During this stage, intensity rapidly increases on STR/STE, and there is a more variable increase in intensity on LTR scans. Hyperintensity on STR scans is the result of paramagnetic T1 shortening produced by methemoglobin, an oxidative breakdown product of deoxyhemoglobin that appears in significant con-

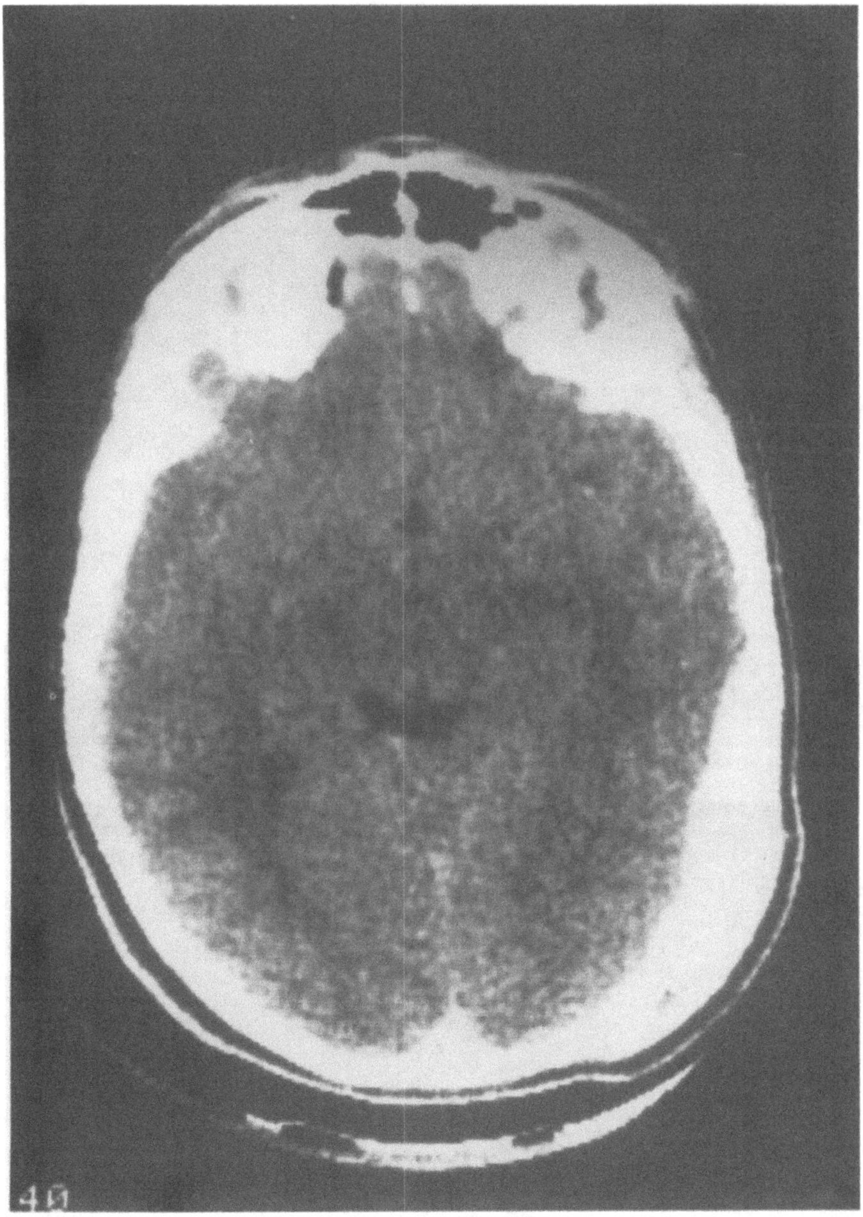

FIGURE 1–7A

FIGURE 1–7. Acute epidural hematoma studied at .5T. CT (A) three days after trauma reveals a local hyperdense extraaxial (subdural or epidural) mass adjacent to the left temporal lobe. On STR/STE (B), the lesion is hyperintense, while on LTR/LTE (C), intensity is heterogeneous with areas of hyper- and hypointensity seen within the clot (Stage III). The hypointense medial rim (B & C) represents displaced dura, indicating the epidural location of the hematoma.

centrations on approximately the third postictal day [11, 15]. This paramagnetic T1 shortening occurs regardless of whether the methemoglobin is heterogeneously distributed (intracellular) or homogeneously distributed (intra- and extracellular), and therefore it is not dependent on RBC integrity (figure 1–2B). It results from the PEDD relaxation effect that occurs because the molecular structure of methemoglobin is such that its unpaired electrons can come

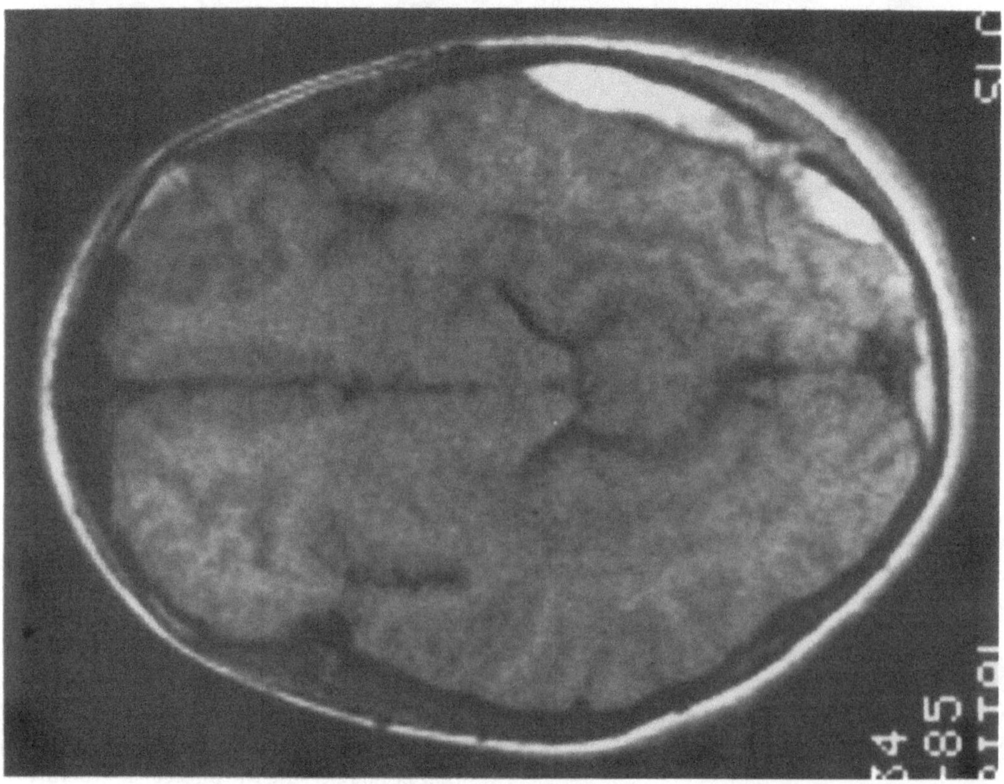

FIGURE 1–7B

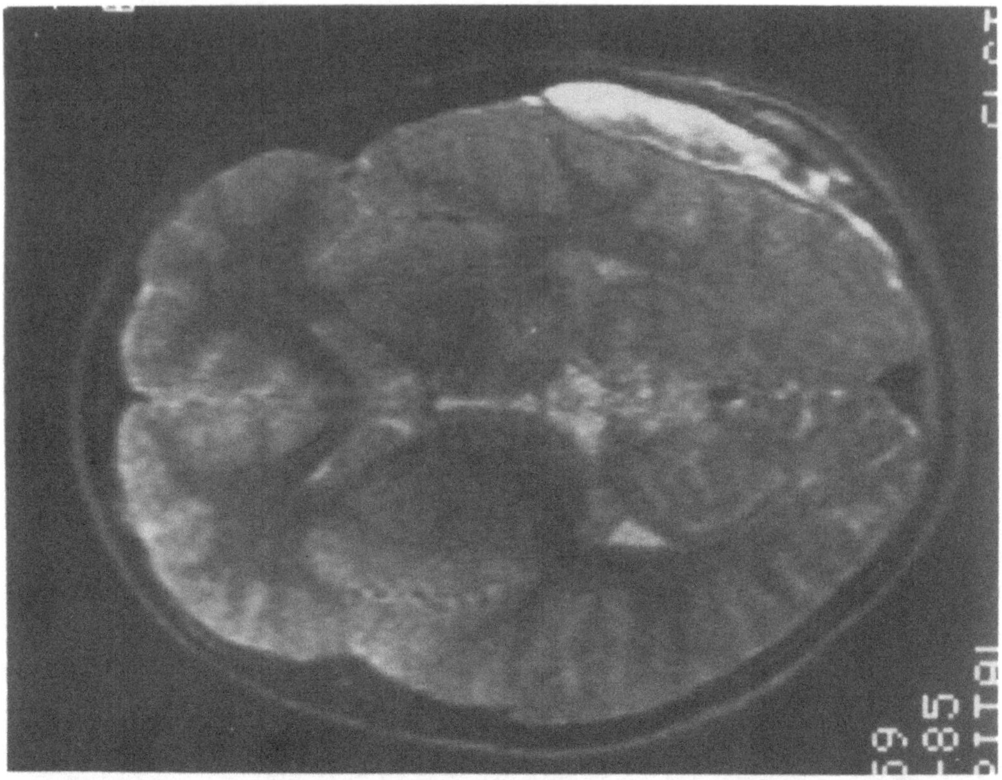

FIGURE 1–7C

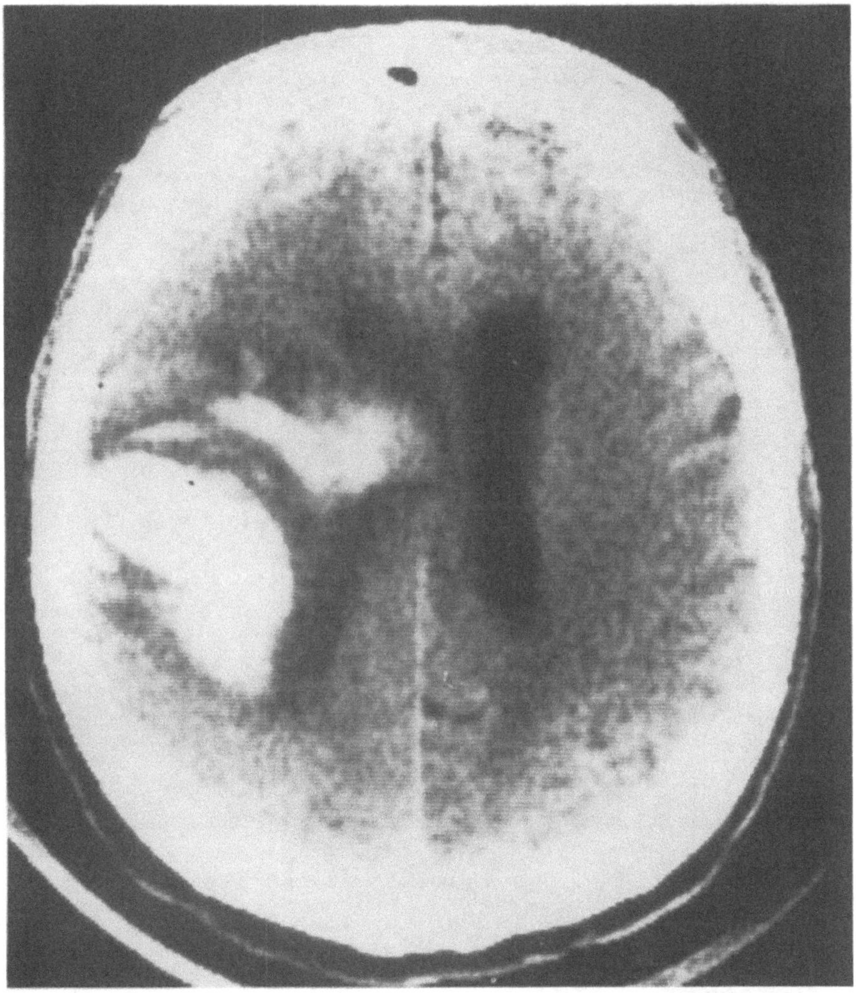

FIGURE 1–8. Parenchymal hematoma with serial changes in the late acute, subacute, and chronic stage studied at .6T. CT (A) one week after spontaneous right frontal hematoma secondary to congophylic angiopathy demonstrates a large bilobed acute hyperdense hematoma. On STR/STE at nine days (B), the hematoma is hyperintense at its periphery and isointense at its center. LTR/LTE at the same time (C) shows heterogeneous intensity. The hematoma is hyperintense near its periphery and slightly less intense at the center. A subtle hypointense rim is seen at the margins of the hematoma (arrows). Follow-up CT (D) at two weeks reveals a slight decrease in density. MR one day later reveals, that the hematoma is relatively unchanged on STR/STE (E). On LTR/LTE (F), the center of the hematoma has become more hyperintense despite lack of change on STR/STE. The peripheral hypointense rim has become more prominent. Follow-up CT at six weeks (G) shows decrease in density of the hematoma. STR/STE (H) at the same time shows very little change in intensity compared to the prior study. Note that the center of the hematoma has not become hyperintense. On LTR/LTE (I), central portions of the two lobes of the hematoma are diffusely hyperintense. The peripheral hypointense rim has become more prominent.

in close proximity (approximately three angstroms) to precessing protons. Because its production requires oxygen, methemoglobin is in highest concentration near the periphery of the hematoma, and it is absent or in low concentration at the hypoxic center of the hematoma. This accounts for the peripheral hyperintensity seen in most hemorrhages (figures 1–4, 1–6–1–8) [5, 10, 22]. The center of the hematoma may eventually become hyperintense, but in many large clots this component remains relatively isointense to the brain (figure 1–8).

On LTR scans, intensity is more variable than in the previous stages. Findings range from hypo- to hyperintensity, and many individual clots show a markedly heterogeneous appearance [10]. There is also poor correlation between findings seen on STR/

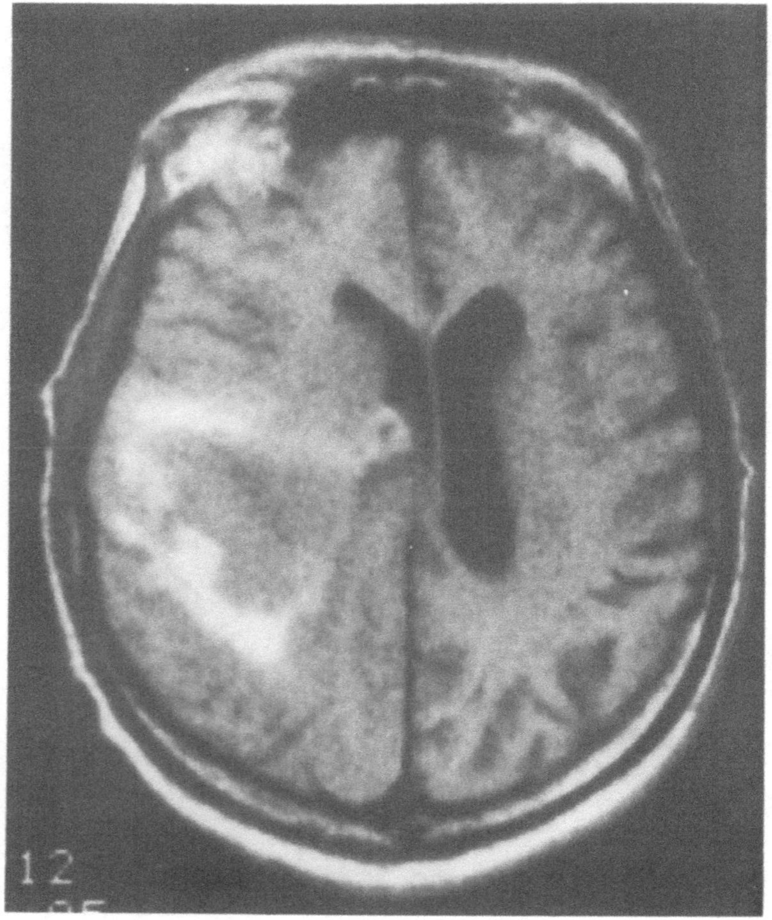

FIGURE 1–8B

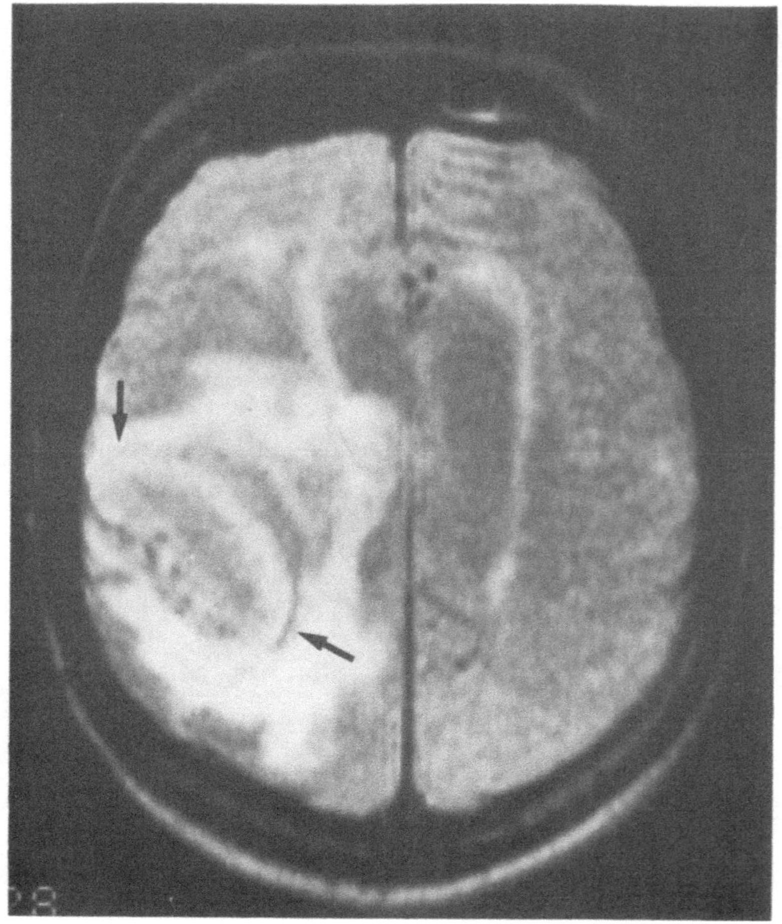

FIGURE 1–8C

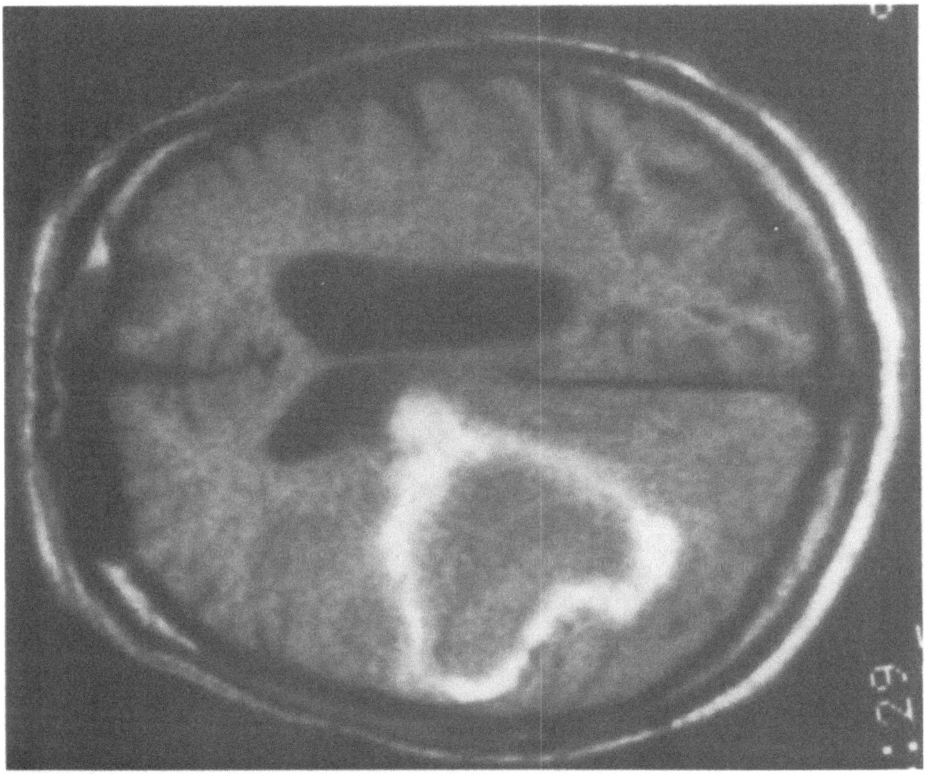

FIGURE 1–8E

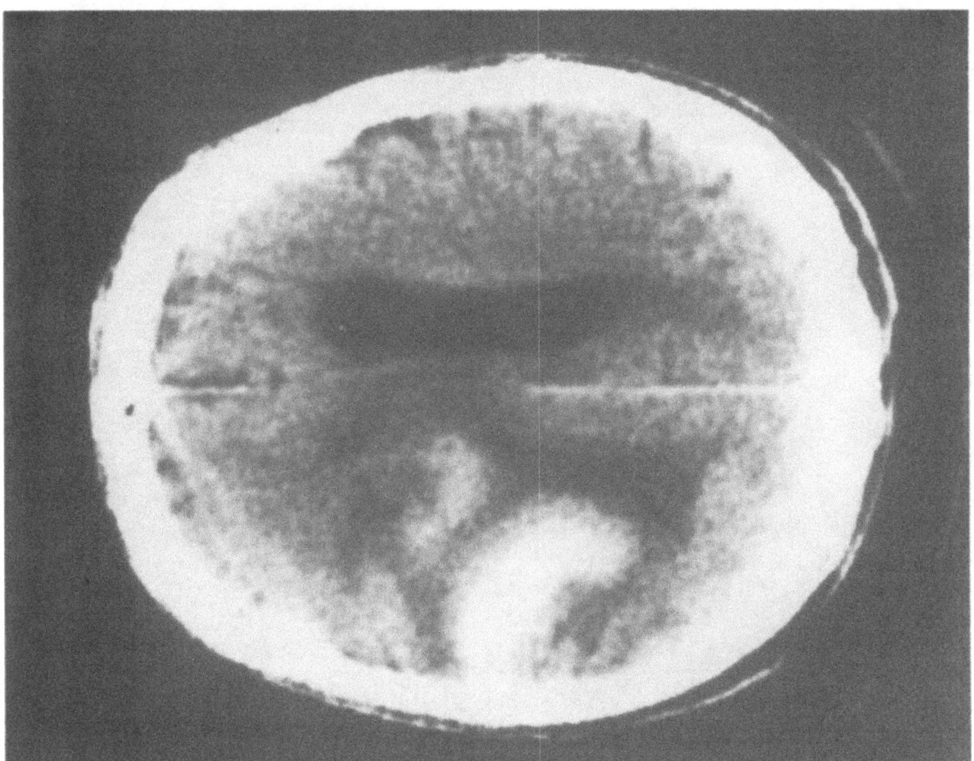

FIGURE 1–8D

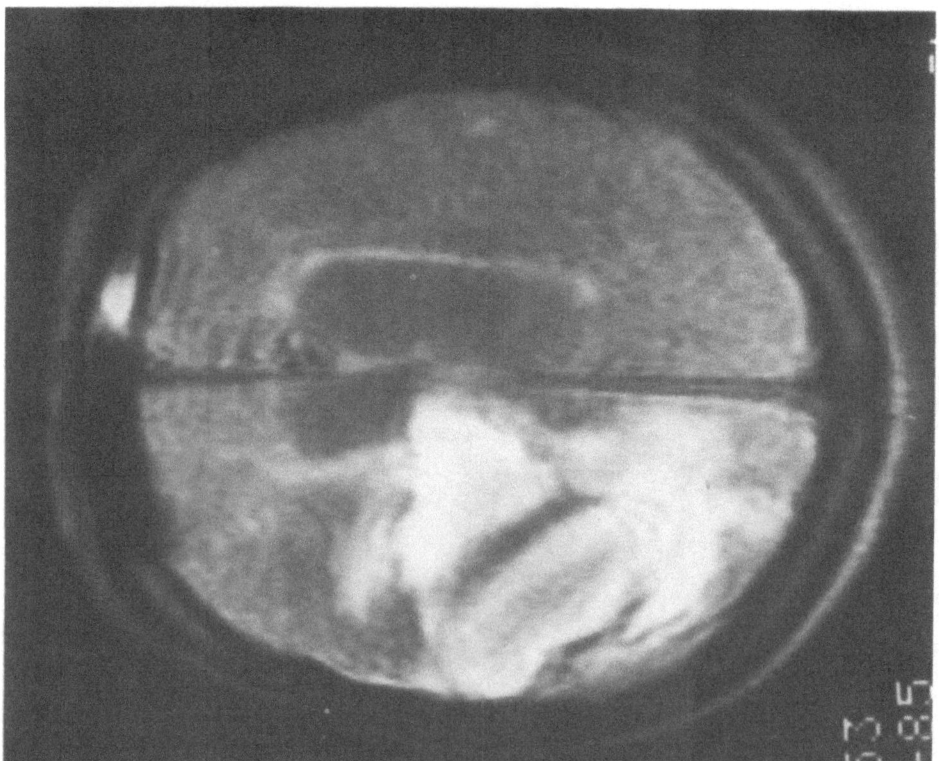

FIGURE 1–8F

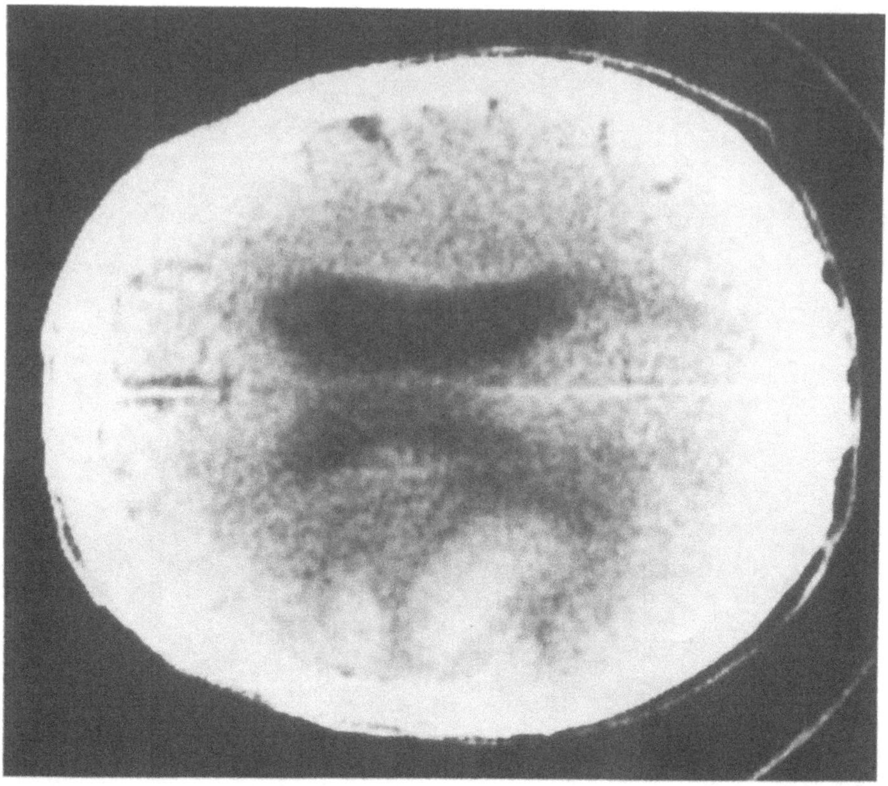

FIGURE 1–8G

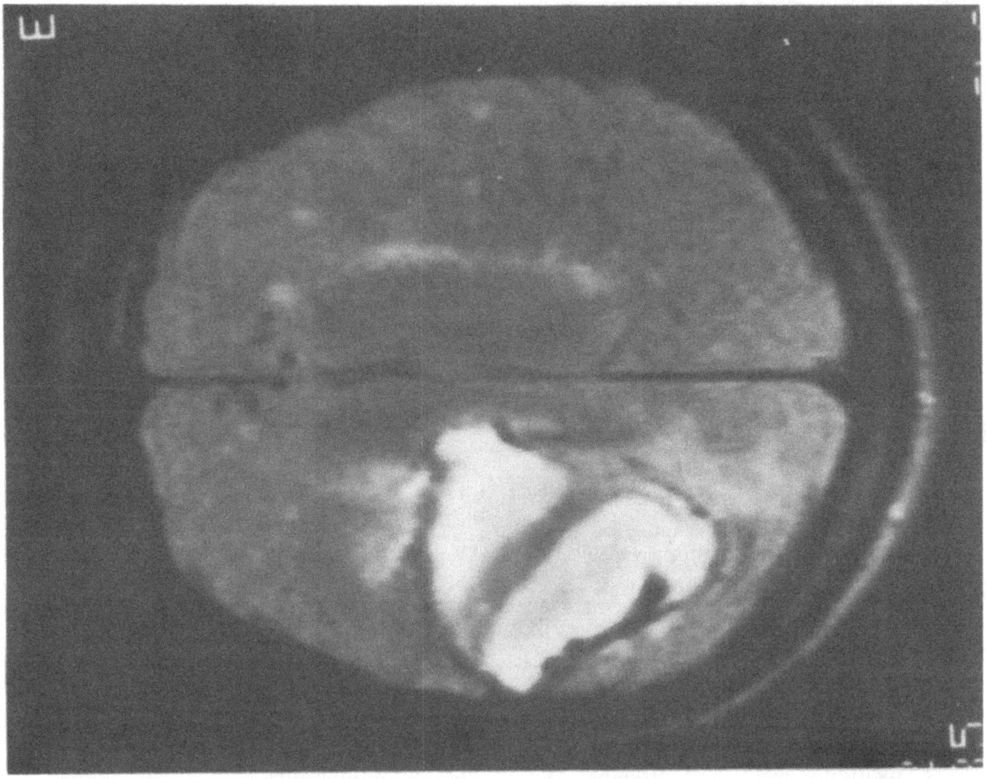

FIGURE 1–8I

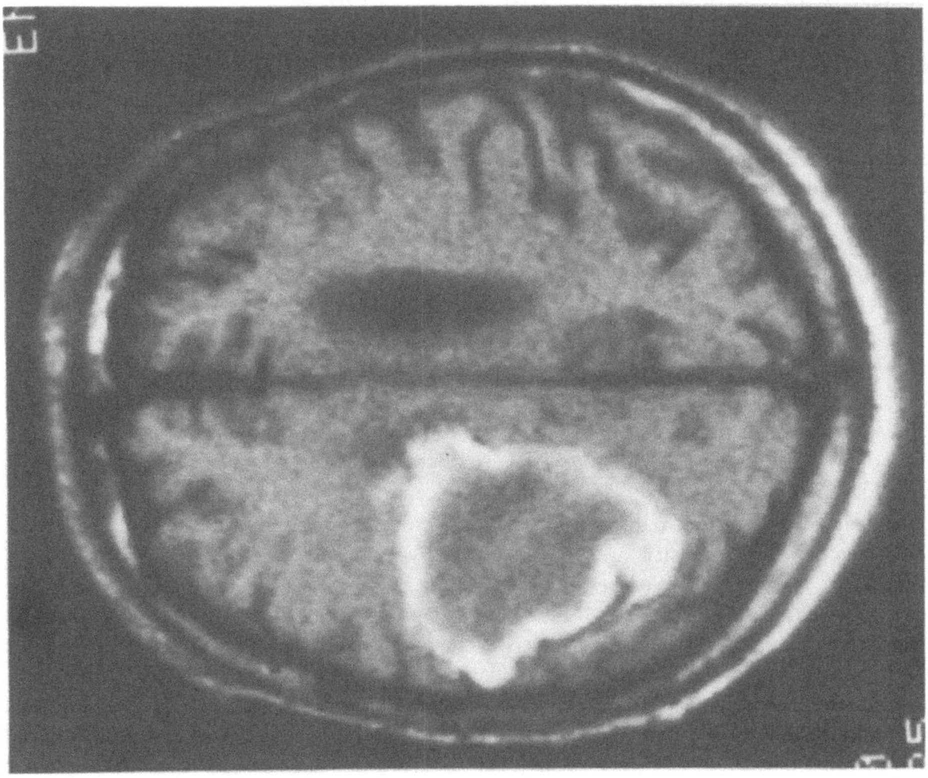

FIGURE 1–8H

STE and those seen on LTR/ITE-LTE scans. Zones of hypointensity on LTR scans overlap more homogeneous areas of hyperintensity on STR scans. During this stage, there is a trend toward increasing intensity on LTR scans, but this increase is delayed relative to that seen on STR/STE scans by approximately one to two days (figure 1–7). The increase on LTR scans is not the result of the methemoglobin effect [10, 24]. T1 shortening produced by methemoglobin does not contribute to the intensity of AIH relative to white matter on LTR scans, since both the hemorrhage and the brain have undergone essentially complete T1 recovery. (T1 effects on MR are dependent on choosing a TR that allows for more complete recovery of longitudinal magnetization of one tissue relative to another. Therefore, TR 2000 scans are essentially free of T1 effects for intracranial soft tissues, including gray matter, white matter, and AIH; see figure 1–9 (29, 30].) The cause of increasing intensity encountered on LTR/ITE-LTE scans during this stage is most directly related to RBC lysis. Once this occurs, both deoxy- and methemoglobin become homogeneously distributed, and the T2-shortening effect of both of these paramagnetic substances is eliminated [10] (figure 1–2D). The T2 of the AIH becomes dependent upon more prosaic factors, including water and protein content. The hematoma acts likes other proteinaceous fluids with a longer T2 than the brain and a shorter T1 than CSF [29], and this combination produces hyperintensity relative to the brain and CSF on LTR scans. Cell and clot lysis during this stage lead to increasing water within the hematoma, contributing to the increase in intensity on LTR scans. GE studies show persistent hypointensity because they are more sensitive to the small amounts of residual intracellular met- or deoxyhemoglobin [18]. Thus, these studies may be helpful in differentiating late acute hematomas from subacute and chronic hemorrhages, which do not show persistent hypointensity on GE scans.

Specificity during Stage III AIH is variable. Hyperintensity on STR scans is typical of hemorrhage but may also be seen in fat-containing lesions [31, 32]. Intensity on LTR scans is therefore important in differentiating hemorrhage from fat. When marked hypointensity or very heterogeneous intensity are encountered, the diagnosis of hemorrhage may be made with confidence. However, if the patient is studied just as the hematoma begains to increase in intensity, moderate hypoinentsity on LTR scans may be seen, which mimics the intensity patterns of normal fat-containing structures (e.g., subcutaneous intraorbital fat) and simple lipomatous masses [10, 31, 32]. Differentiation is generally possible, since the anatomic configuration and location of lipomas and hematomas are usually different, and hematomas virtually always have edema and mass effect, findings not often encountered in lipomas. GE scans may also be helpful, since marked hypointensity would suggest that the lesion is hemorrhagic.

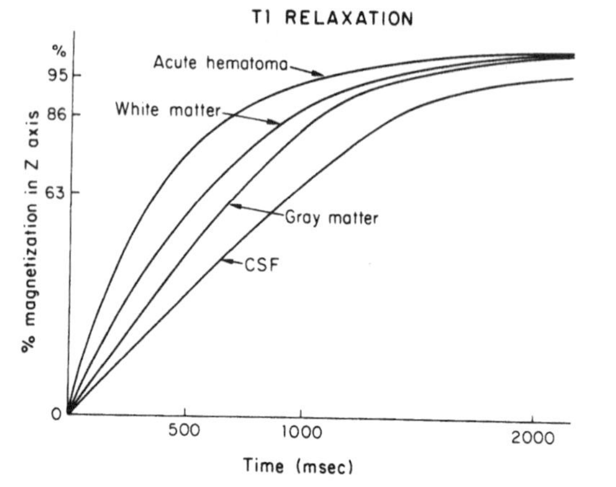

FIGURE 1–9. T1 relaxation curves for acute hemorrhage and normal intracranial structures. Exponental T1 recovery curves for CSF, gray matter, white matter, and acute hemorrhage are plotted. With short TR (e.g., 500–750 msec), the more rapid T1 relaxation of acute hemorrhage leads to relative hyperintensity when compared to white matter and gray matter. However, with a TR of 2000, there is nearly complete recovery of longitudinal magnetization, and therefore intensity differences cannot be attributed to differences in T1 between hemorrhage and normal gray and white matter.

STAGE IV: 5–7 DAYS
(FIGURES 1–4, 1–8)

Hematomas are hyperintense on both STR and LTR scans [10]. T1 is short because of the presence of extracellular methemoglobin, and T2 is prolonged because of the high water content. Thus, the hematomas appear to be hyperintense on STR scans, especially near their periphery, and are more homogeneously hyperintense on LTR scans. GE scans during Stage IV tend to show residual hypointensity.

The intensity patterns during Stage IV are characteristic and relatively specific. Similar intensities have, however, been encountered in intracranial dermoids [31]. Unlike simple lipomas, the dermoids remain hyperintense on LTR scans, presumably because they contain constituents other than fat that prolong T2. Dermoids may usually be differentiated from hematomas based on their anatomic configuration and location (e.g., suprasellar), and because the lesions tend to lack edema and have only minimal mass effect. Clinical signs of recent hemorrhage are absent.

OTHER DETERMINANTS OF INTENSITY

The role of field strength has been previously discussed. To review, in vitro data indicates that T2 shortening should be proportional to the square of the field strength [17], leading some investigators to suggest that hypointensity will only be encountered at high field strenghts [7]. This prediction has not been borne out in clinical practice, since hypointensity is routinely encountered at intermediate field strengths (figures 1–3, 1–4, 1–6) and is even seen at low field strengths [5, 8]. Based on indirect comparison of images from multiple cases and limited experience with direct comparisons (figure 1–3) of clinical and experimental images [23], it appears that hypointensity develops more quickly and persists for a longer time at high field strengths. Thus, Stage II may begin within 8–12 hours and Stage III may persist for 7–9 days.

The source of hemorrhage has an effect on intensity, especially in the first one to two days. Arterial hemorrhages (e.g., epidural hematoma and ruptured aneurysm) have a lower initial deoxyhemoglobin concentration [33], and thus hypointensity may take longer to develop than in venous (e.g., subdural) or mixed (e.g., parechymal) hemorrhages. (The effect of hematoma location on intensity is nicely demonstrated in figure 1–4).

Hematoma location has two important implications for diagnosis. First, location determines the anatomic configuration of AIH, which profoundly affects our ability to detect and characterize hemorrhage. Epidural and subdural hematomas are always more easily visualized and more accurately characterized on MR than on CT, regardless of their intensity [34, 35] (figures 1–4, 1–7). The surface of the brain and its relationship to the adjacent calvarium are visualized on MR in a manner that is impossible with CT, allowing for direct demonstration of a hematoma with typical morphology interposed between the calvarium and the brain. Parenchymal hematomas virtually always produce anatomic distortion, but the findings are nonspecific; thus, although a lesion is clearly identified, the diagnosis of hemorrhage cannot be made on morphologic features alone. Subarachnoid and intraventricular hemorrhage usually produce no anatomic distortion, and thus diagnosis is totally dependent upon intensity changes relative to normal CSF [10].

Hematoma location also directly affects intensity, especially when it extends into spinal fluid spaces (subarachnoid and intraventricular). Spinal fluid impedes clot formation, and the production and flow of CSF dilute and rapidly wash away red blood cells, preventing many of the time-dependent changes in hematoma intensity from being visualized. In addition, it has recently been established that pulsations of CSF have a direct effect on intensity, a situation which may mask the changes produced by hemorrhage [36]. Even when clots do form, serial changes in intensity are different (figure 1–6) due the higher oxygen concentration of CSF, which leads to a lower deoxyhemoglobin concentration with less resultant paramagnetic T2 shortening (33). Hypointensity, therefore, may be absent or only minimally present on LTR/ITE-LTE (figure 1–6). In these cases, hypointensity may be detected on GE scans, in particular with intraventricular hemorrhage; however, these scans are unfortunately of little value in detecting basal subarachnoid hemorrhage because phase-shift effects produce artifactual basal hypointensity, which mimics the appearance of subarachnoid hemorrhage.

In massive subarachnoid hemorrhage, frank clots may form in the subarachnoid space, and these may be detected as areas of increased intensity on STR/STE scans because of the presence of methemoglobin. The oxygen concentration of CSF may, in fact, cause methemoglobin to form more rapidly than in other intracranial hemorrhages, and therefore hyperintensity may appear more quickly in subarachnoid and intraventricular hematomas. In addition, since hyperintensity on STR scans may persist for weeks, a subacute subarachnoid clot, where the blood becomes isodense to the brain after the first week, will be appreciated more easily on MR than CT.

The final factor that may affect AIH intensity is the presence of recurrent hemorrhage, which often produces marked intensity heterogeneity on both STR and LTR scans. Because the admixture of old and new hemorrhage produces almost random intensity combinations, it is often difficult to determine the time of the most recent hemorrhage or to detect new hemorrhage. Recurrent hemorrhage occurs in various clinical circumstances including trauma, congophylic angiopathy, neoplasms (primary and metastatic), and vascular malformations, in particular cryptic arteriovenous malformations (cavernous angiomas and capillary telangiectasias). These lesions may therefore be difficult to distinguish from each other on the basis of MR intensity features alone [37–40] (figure 1–11).

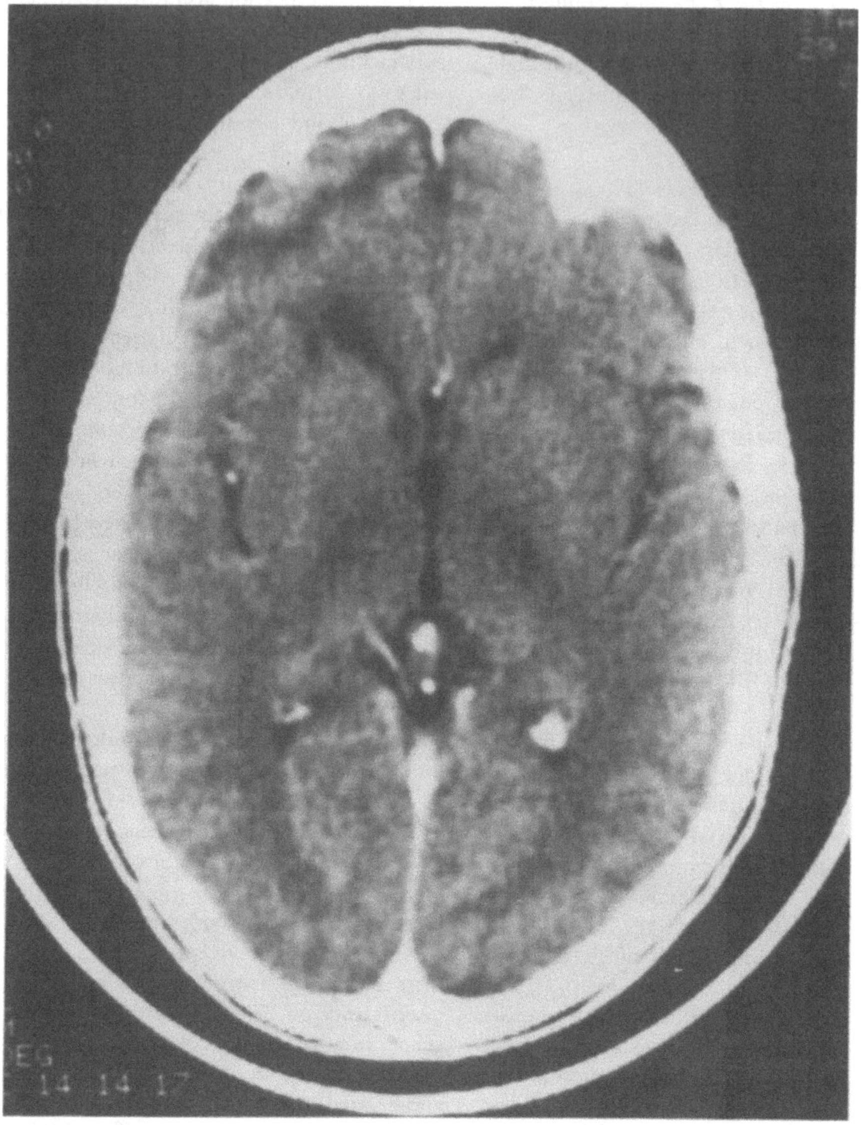

FIGURE 1–10A

FIGURE 1–10. Chronic parenchymal hemorrhage studied at 1.5T. CT (A) in a patient who sustained a traumatic right frontal hematoma several years ago reveals an irregular nonspecific hypodensity in the right frontal lobe with some atrophic dilitation of the right frontal horn. On STR/STE (B), the lesion appears hypointense. Note that focal atrophy is better visualized on MR than CT. On LTR/ITE (C), marked hypointensity is present at the margin of the lesion, indicative of old hemorrhage.

Subacute and Chronic Hemorrhage

Changes in the subacute and chronic stages of hemorrhage are less complex and more easily understood, but once again, multiple factors affect intensity. The most important factor is location. In parenchymal hemorrhage (figure 1–8), the majority of the clot remains hyperintense on all pulse sequences for a minimum of four weeks. Over several months, the hematoma diminishes in size and intensity as it is absorbed. Eventually, the central component of the hematoma may completely disappear or remain as a small focus that is hypointense on STR scans and iso- to hyperintense on long TR scans (figure 1–10). The most important feature of subacute and chronic parenchymal hematoma is the presence of a hypointense peripheral rim which is more marked on LTR/ITE-LTE scans and on GE scans (figures 1–4, 1–5, 1–8, 1–10, 1–11). This hypointense rim begins to

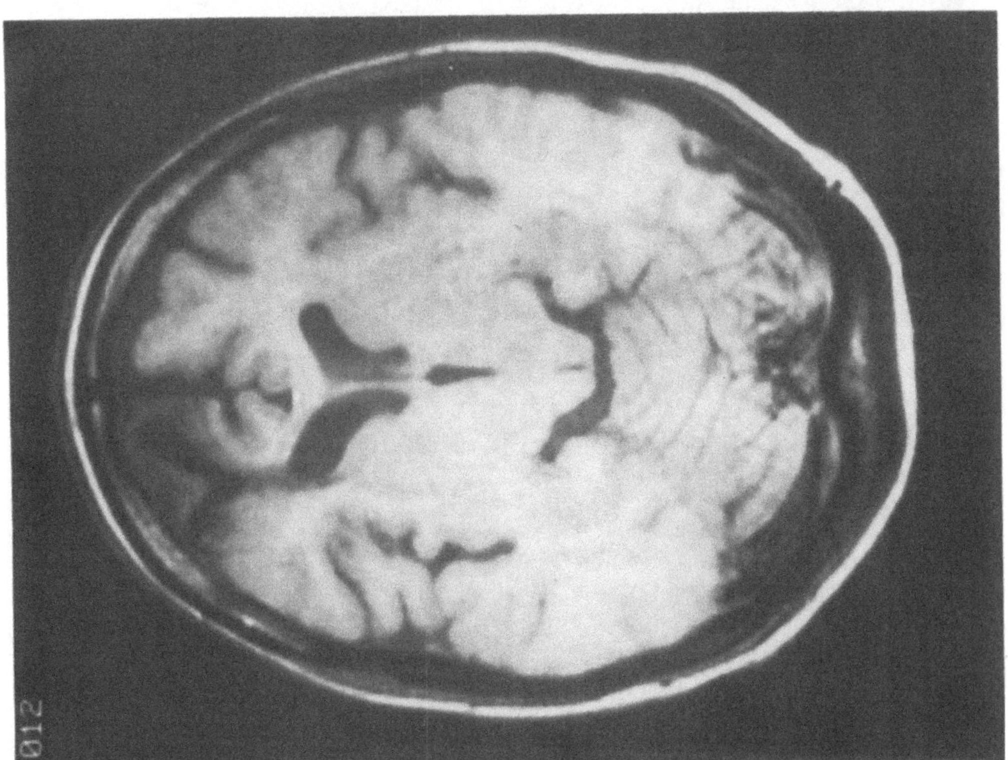

FIGURE 1–10B

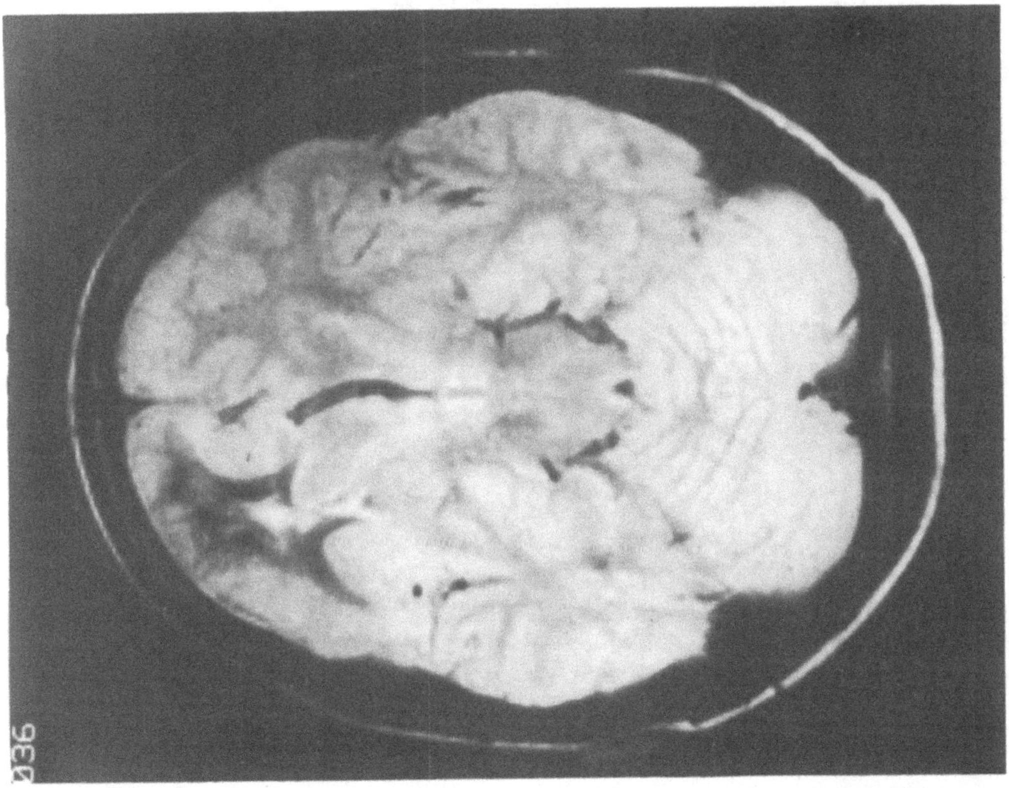

FIGURE 1–10C

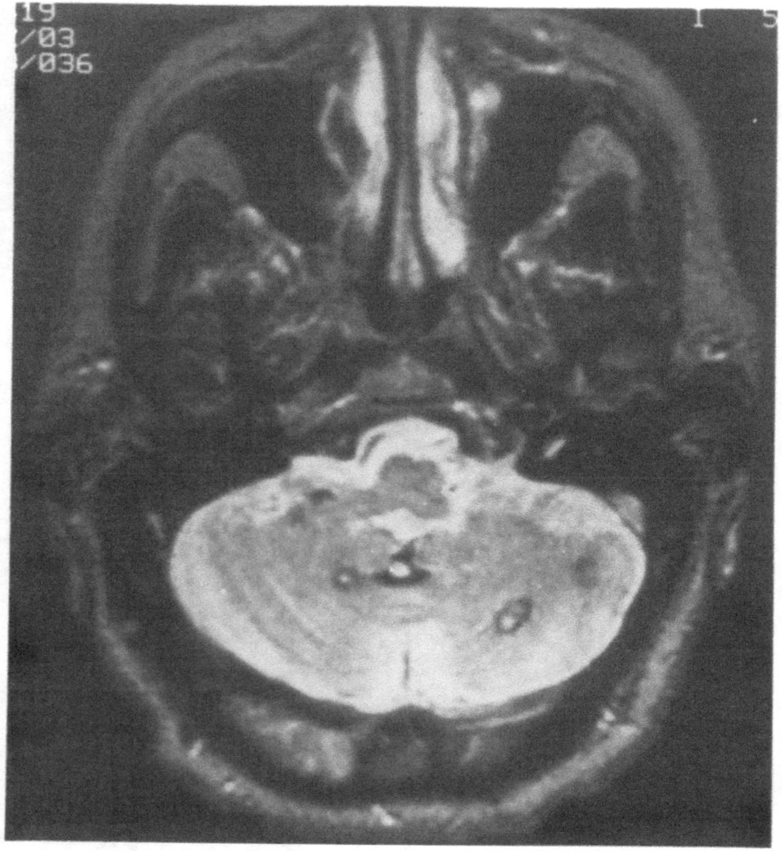

FIGURE 1–11A

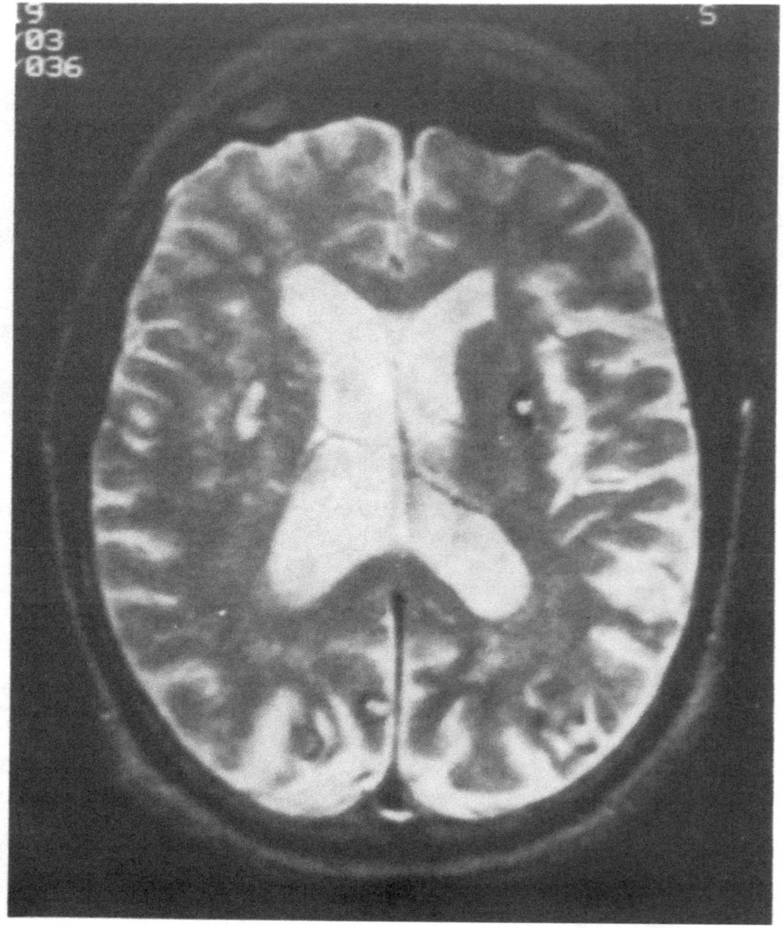

FIGURE 1–11B

FIGURE 1–11. Multiple cavernous hemangiomas studied at 1.5T. LTR/LTE at the level of the cerebellum (A) and lateral ventricles (B) in a 68-year-old man demonstrates multiple hypointense lesions. Most of the lesions have central areas of hyperintensity. Numerous other similar-appearing lesions were identified on additional scans. MR features are typical of those seen with multiple cavernous angiomas of the brain.

appear by the end of the first week and becomes progressively more prominent over the next several weeks (figure 1–8). It has been ascribed to the presence of paramagnetic hemosiderin [7, 15] and other paramagnetic iron-containing hemoglobin breakdown products within macrophages in the capsule of the hematoma. Since these macrophages are trapped within the brain by the blood–brain barrier, the hypointensity persists indefinitely as a sign of old hemorrhage [7, 15]. As the central portion of the hematoma decreases in size, the hypointense rim becomes more prominent until it may eventually be seen as an irregular stellate or serpiginous hypointense focus (figures 1–4, 1–5, 1–10, 1–11) that may, on occasion, be difficult to distinguish from a vascular malformation.

The extent to which this hypointense rim develops is dependent upon a number of variables. They are reported to be more apparent at high field strength [7], and we have been able to confirm this finding in several cases in which direct comparisons have been made (figure 1–5). Another factor which may affect

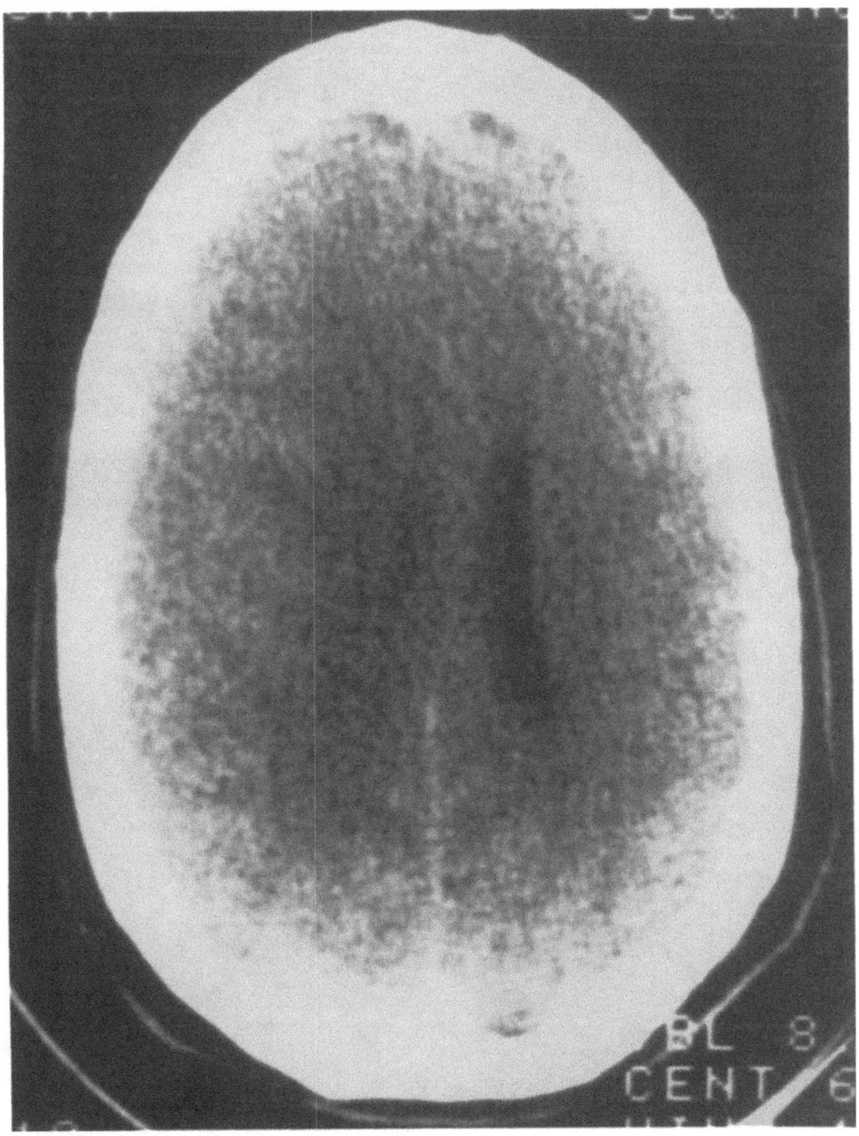

FIGURE 1–12A

FIGURE 1–12. Subacute subdural hematoma studied at .5T. CT (A) reveals an isodense, right subdural hematoma. There is buckling and displacement in the white at the corticomedullary junction. On MR, the hematoma is markedly hyperintense of STR/STE (B) and LTR/LTE (C).

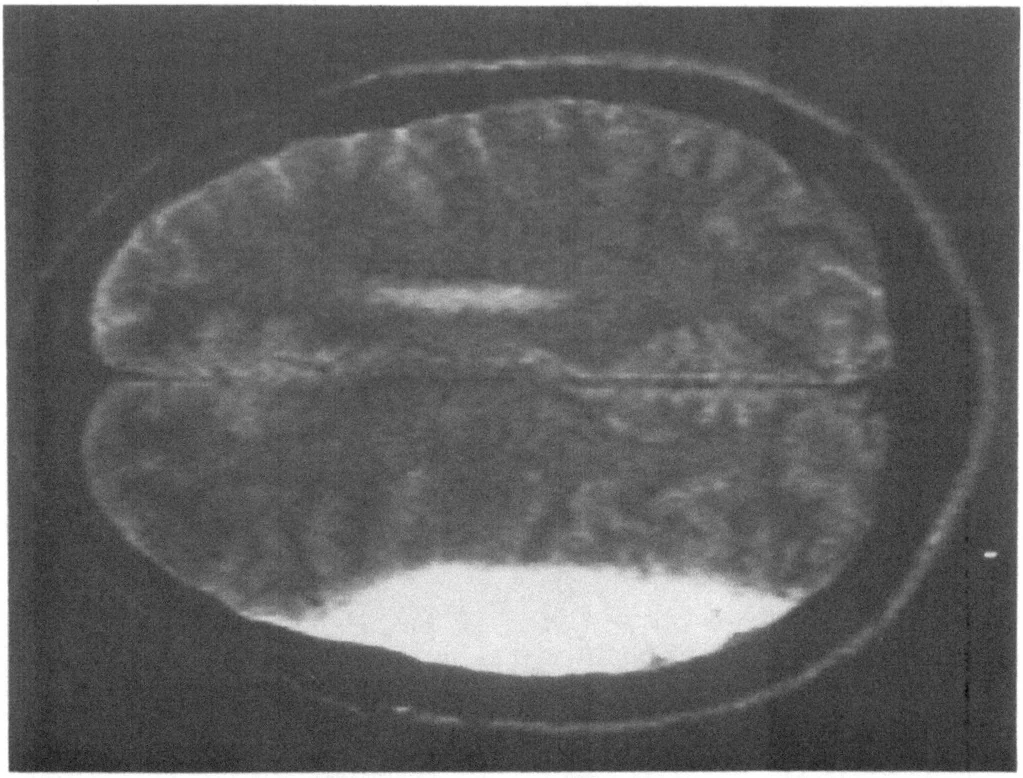

FIGURE 1–12C

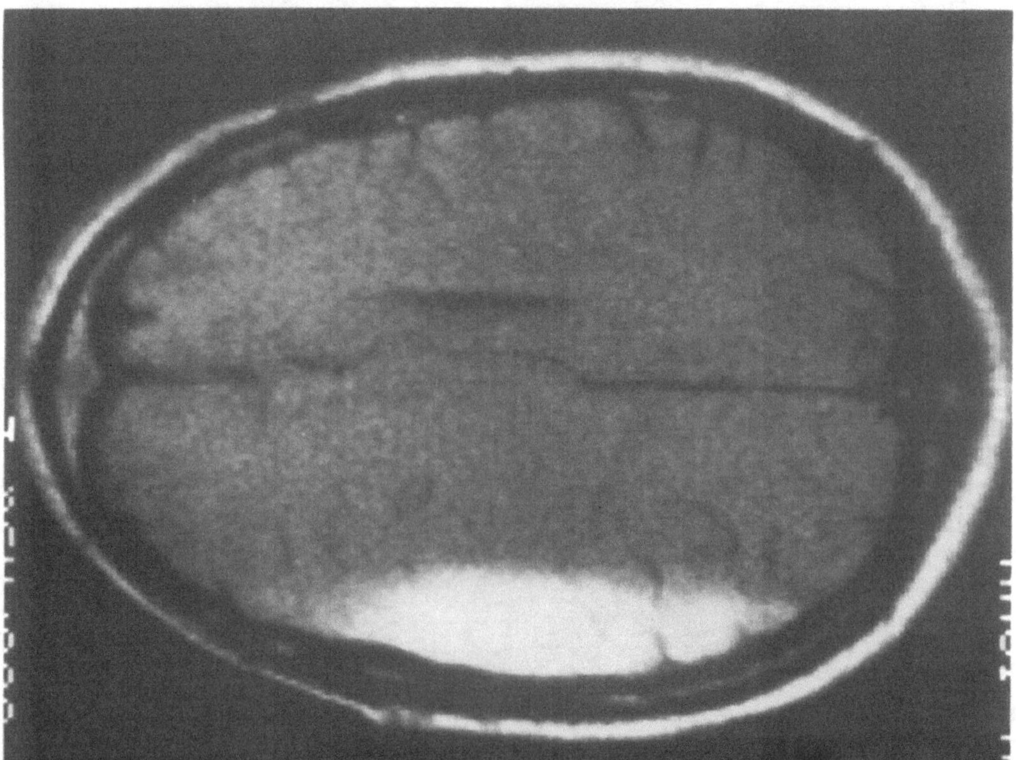

FIGURE 1–12B

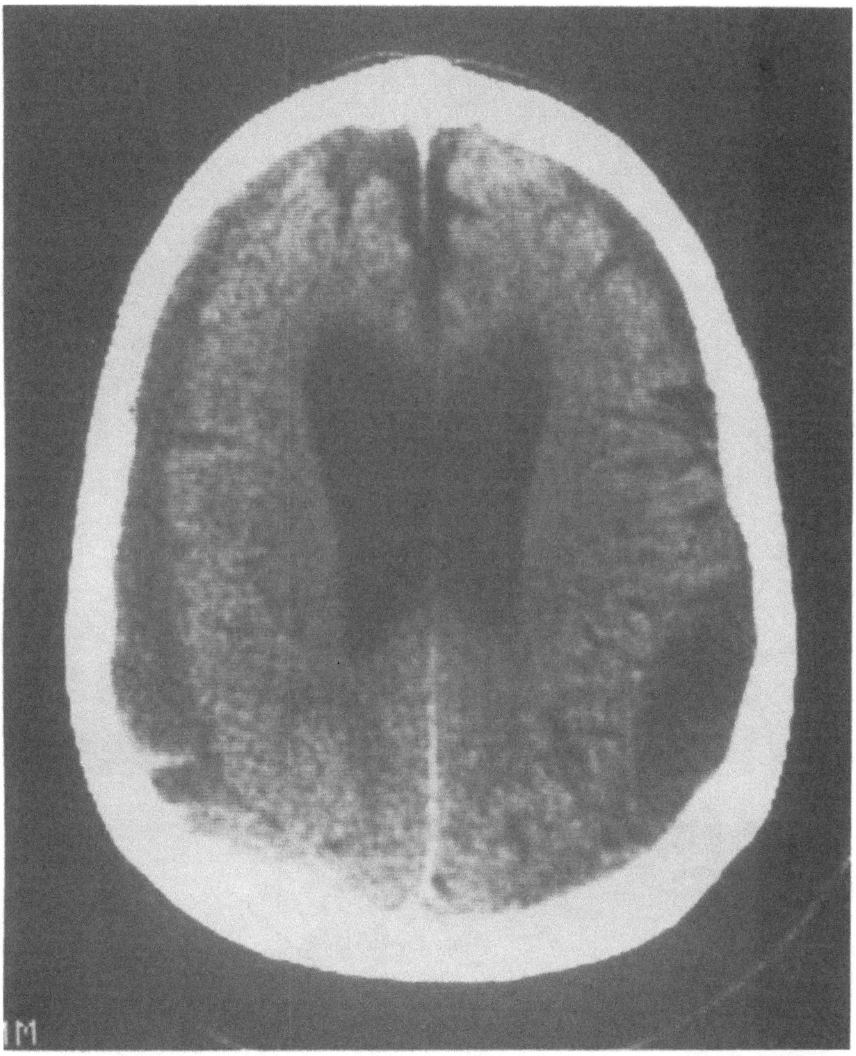

FIGURE 1–13A

FIGURE 1–13. Chronic subdural hematomas studied at .5T. On CT (A), bilateral hypodense subdural hematomas are seen. On STR/STE (B) on the same day, the right hematoma is hyperintense to the brain while the left hematoma is isointense to the brain. On LTR/ITE (C), both hematomas are hyperintense. These findings suggest that the right hematoma has undergone the most recent episode of hemorrhage.

visualization of the hypointense rim is the initial size of the hematoma: larger clots produce more prominent dark rims. There are two exceptions to this. First, a large hematoma may develop little or no hypointense rim if it is surgically removed or spontaneously evacuates itself into the subarachnoid space or ventricle. Second, small lesions may develop dramatic hypointensity (at any field strength) if there are recurrent episodes of hemorrhage. This is particularly common in cavernous angiomas (figure 1–11) and capillary telangiectasias (cryptic arteriovenous malformations) [37, 38]. These lesions show variable amounts of hyperintensity on STR scans, presumably representing areas of recent hemorrhage.

The intensity of epidural hematoma changes in a manner similar to that seen in the parenchymal clots; however, a medial hypointense rim is always seen adjacent to these extraaxial hematomas, regardless of the time from ictus (figures 1–4, 1–7). This rim is not due to the presence of hemosiderin but represents the displaced fibrous dura, which is hypointense on all pulse sequences due to the small number of mobile protons and the short T2 [41]. Since it is not

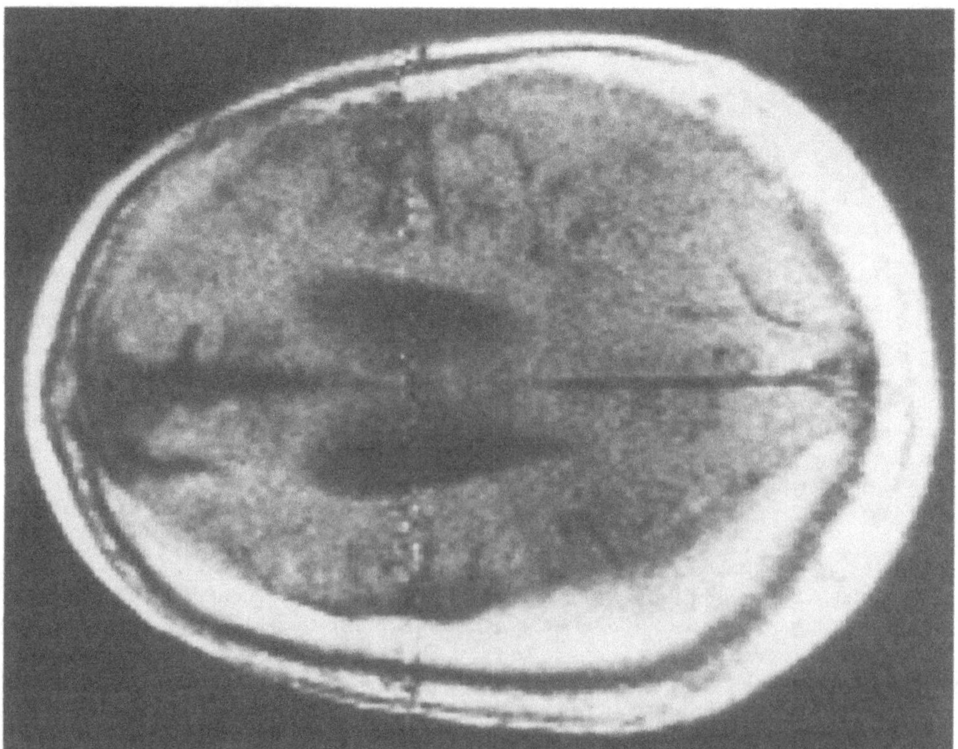

FIGURE 1 – 13B

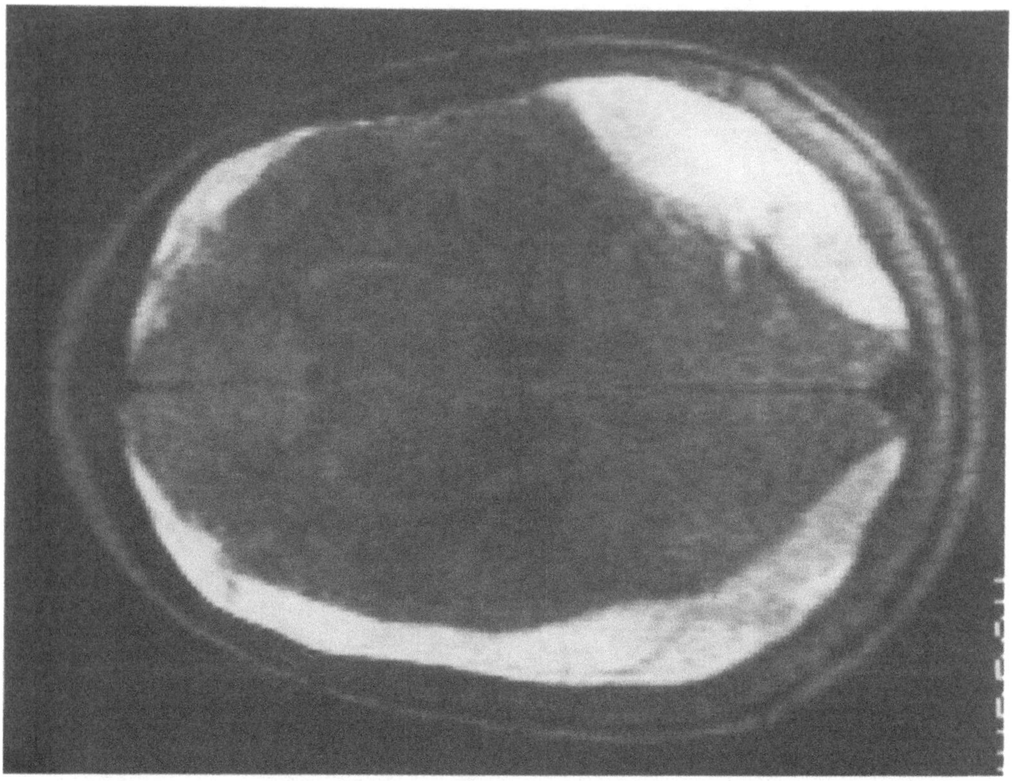

FIGURE 1 – 13C

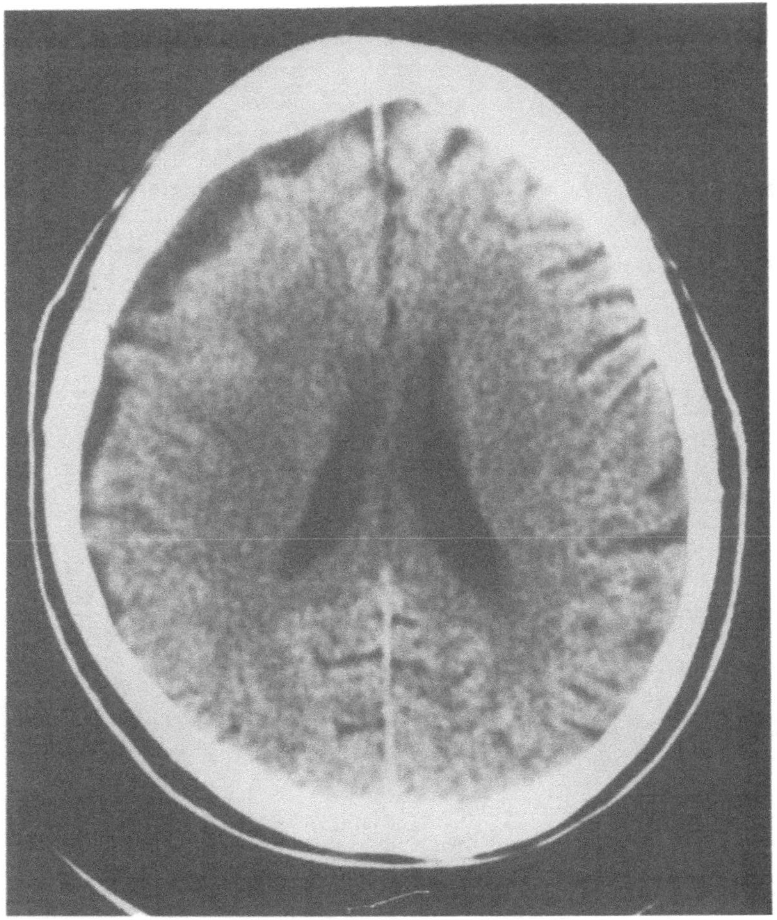

FIGURE 1–14A

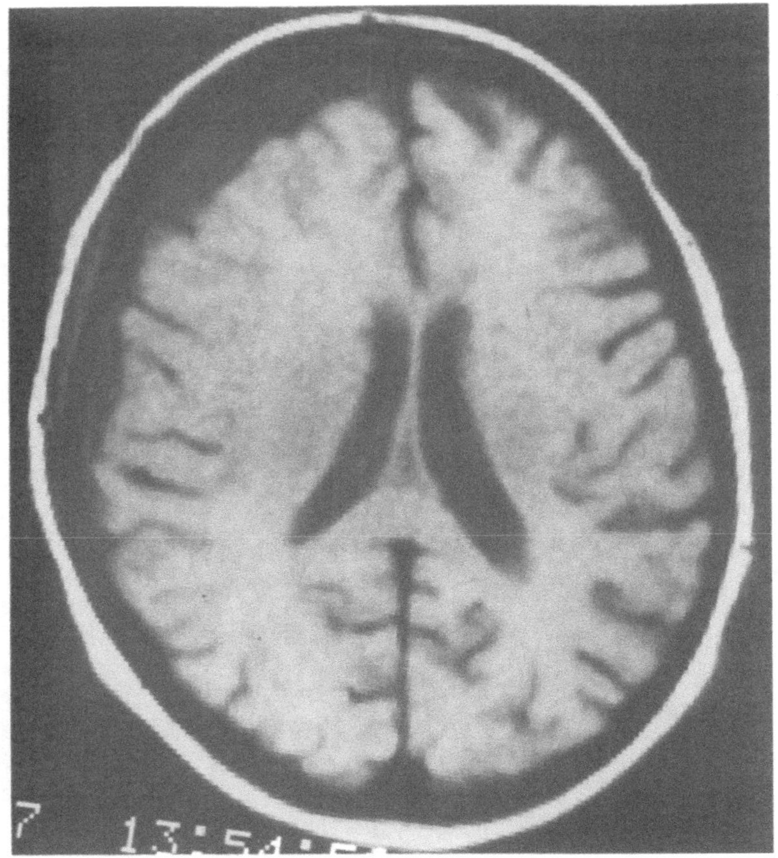

FIGURE 1–14B

FIGURE 1–14. Subdural hygroma studied at .6T. CT (A) reveals a hypodense right subdural collection. MR performed the same day reveals the collection to be isointense to spinal fluid on STR/STE (B), LTR/ITE (C), and LTR/LTE (D). Note that gyral compression is more apparent on MR than CT, allowing for differentiation of subdural hygroma from dilated subarachnoid space. No veins are noted within the collection as would occur with dilated subarachnoid spaces (see figure 1–16).

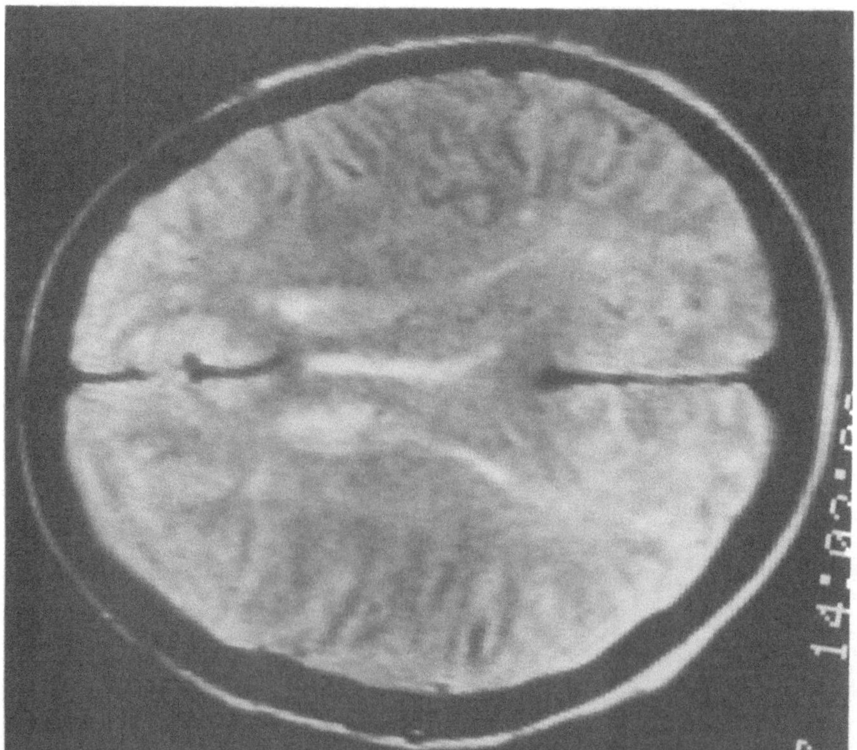

FIGURE 1–14C

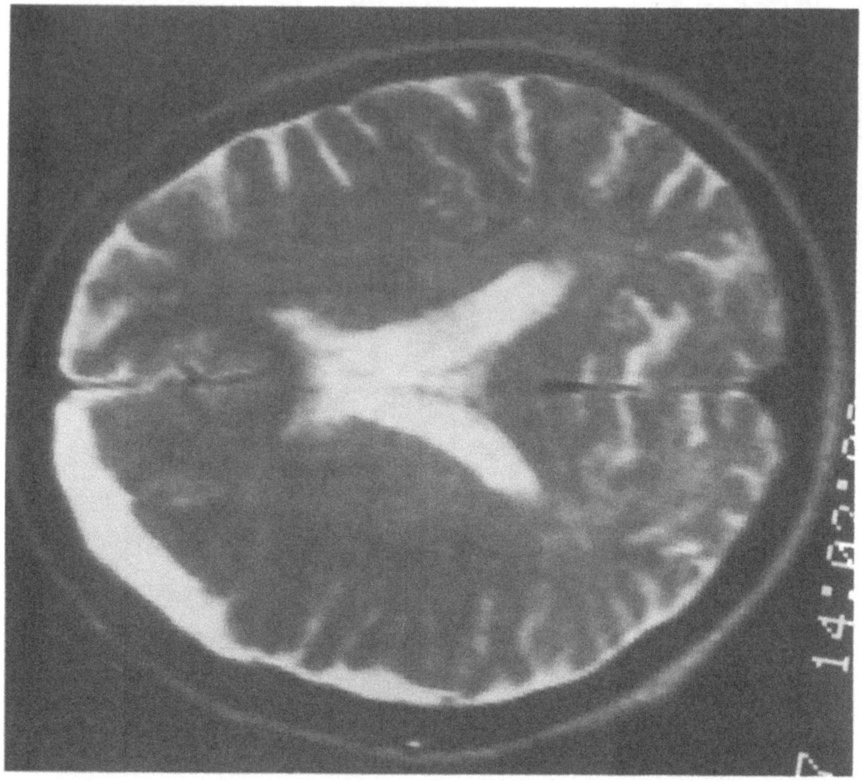

FIGURE 1–14D

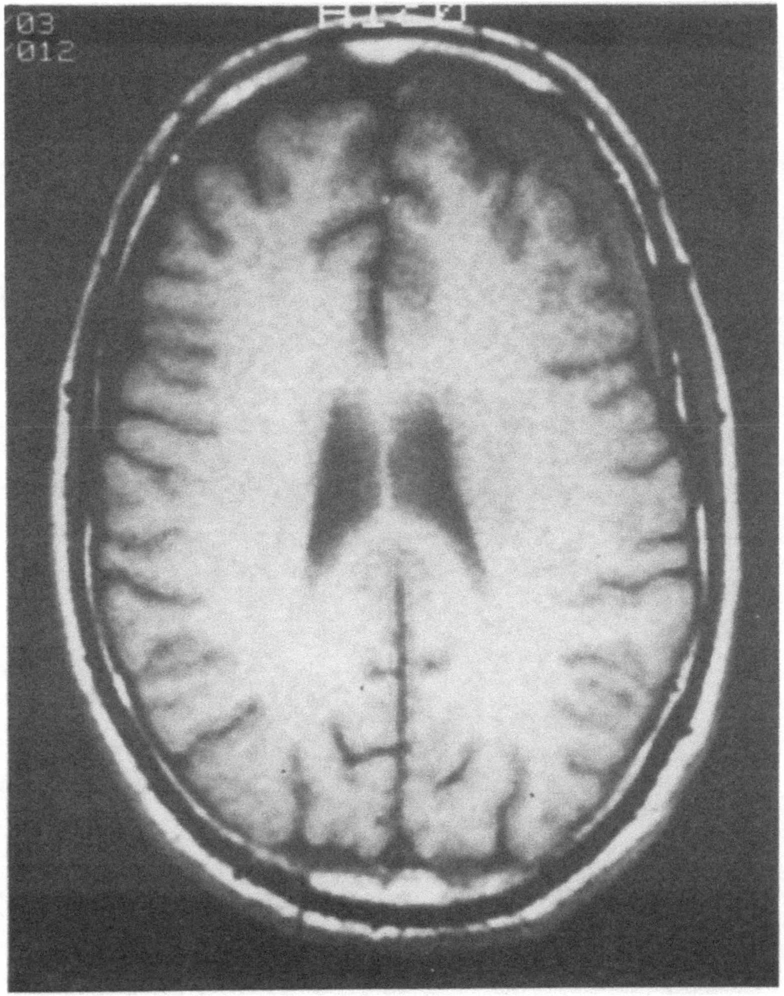

FIGURE 1–15A

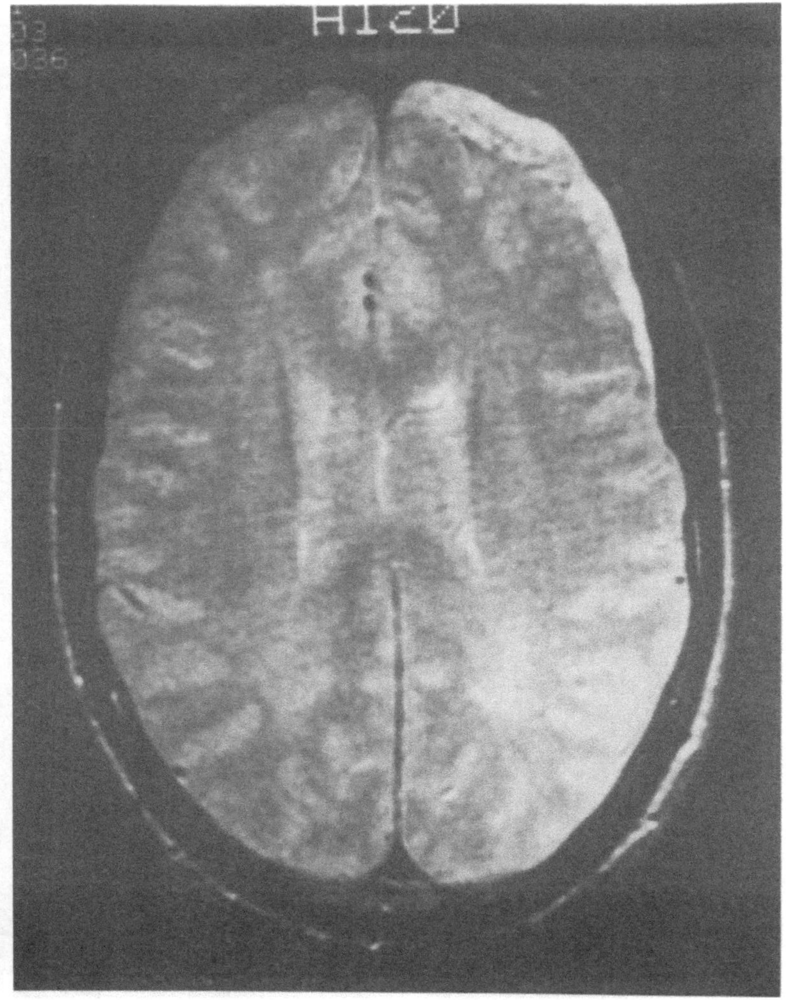

FIGURE 1–15B

FIGURE 1–15. Recurrent hemorrhage in chronic subdural hematoma studied at 1.5T. STR/STE (A) reveals a left frontal subdural collection that is hypointense to the brain and minimally hyperintense to CSF. On LTR/ITE (B), the hematoma is mildly hyperintense to the brain and CSF. On a follow-up examination (C & D) three weeks later, the hematoma has increased in intensity on both STR/STE (C) and LTR/ITE (D), indicating recurrent hemorrhage.

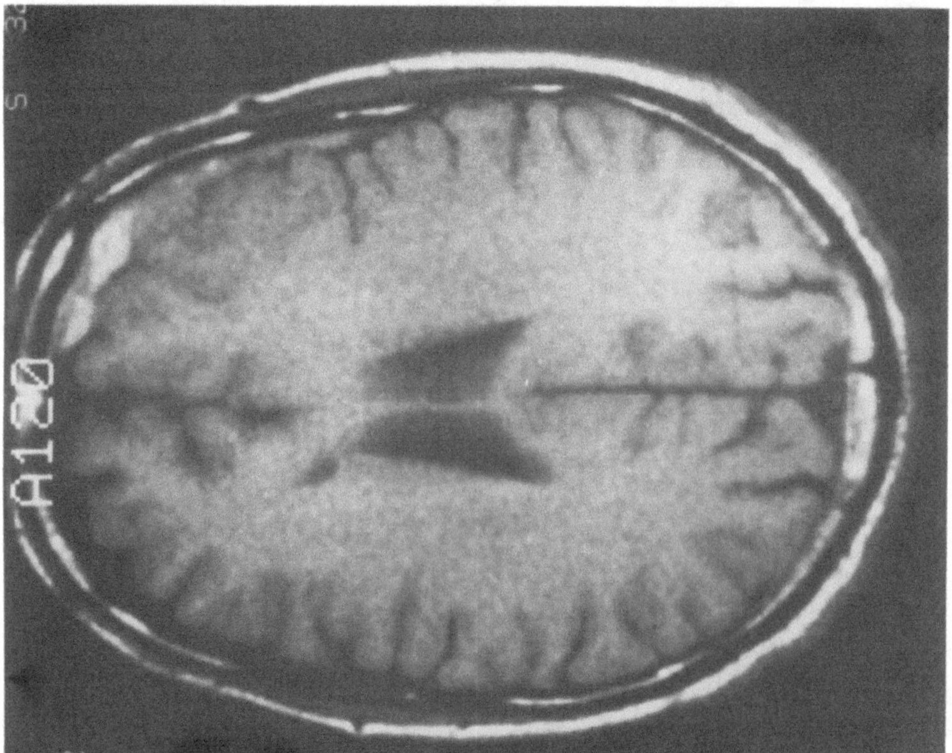

FIGURE 1–15C

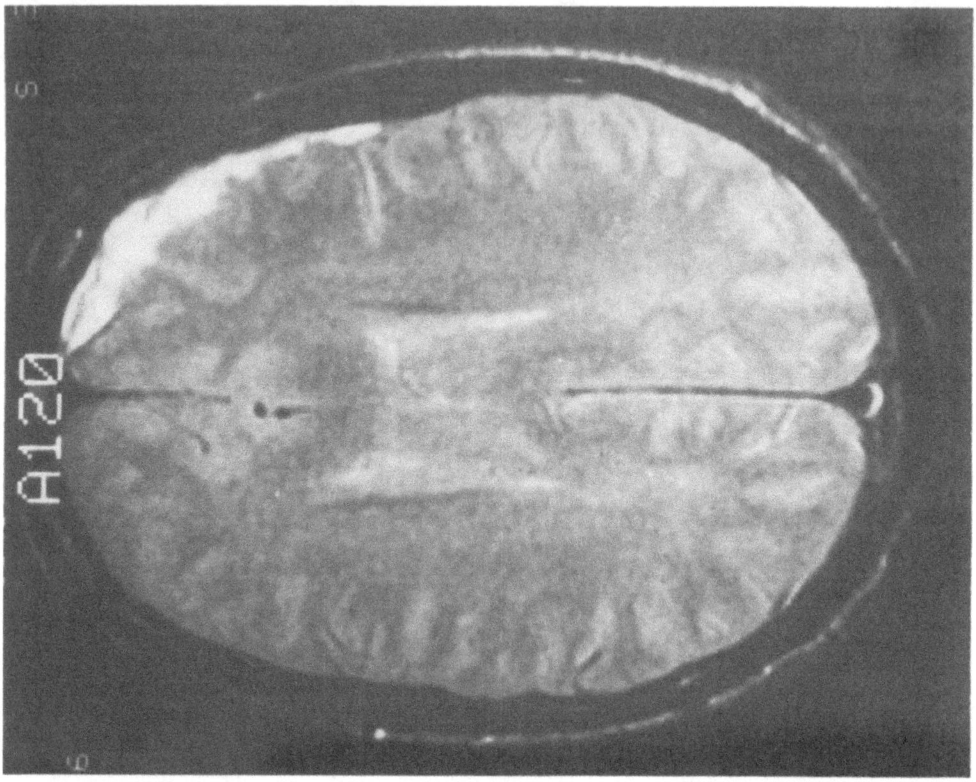

FIGURE 1–15D

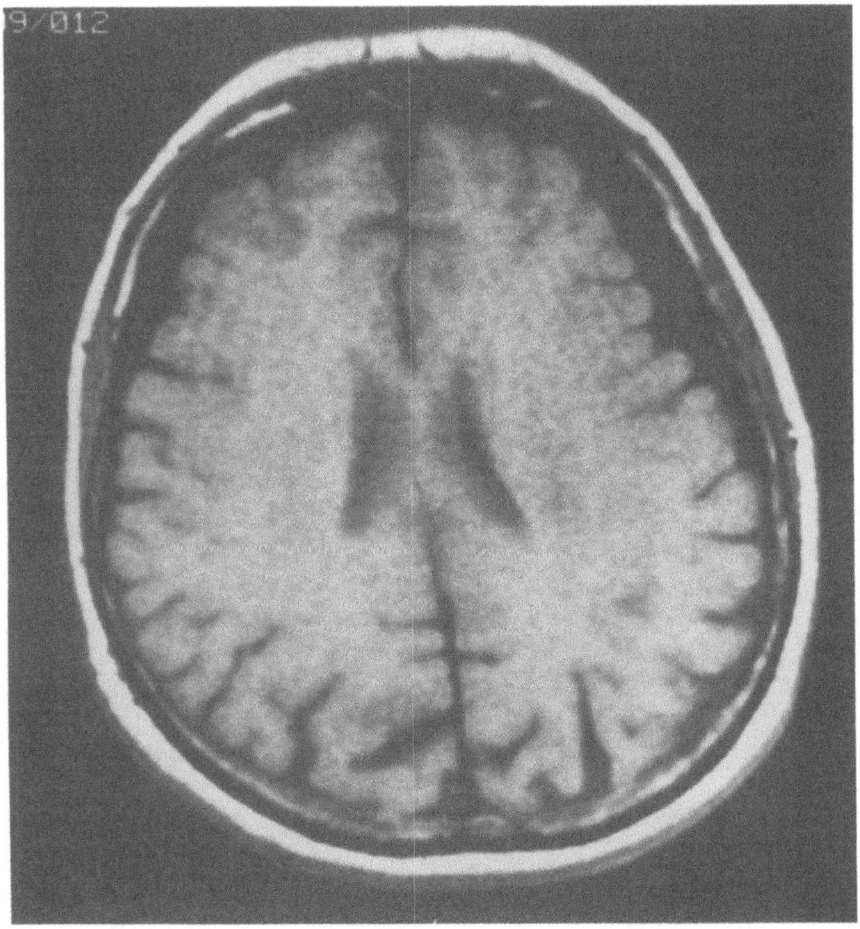

FIGURE 1–16A

FIGURE 1–16. Demonstration of dilated subarachnoid space (atrophy). On STR/STE (A) there are convexity extraaxial CSF intensity collections that remain isointense to spinal fluid on LTR/ITE (B) and LTR/LTE (C). Dilated subarachnoid spaces can be distinguished from chronic subdural hematomas or effusions because the cortical veins are seen to wander throughout the subarachnoid spaces (arrows) and because there is no evidence of cortical compression.

dependent on the presence of hemosiderin, the rim does not undergo serial time-dependent changes. Also, it is not field strength dependent and does not become more apparent on GE scans. As the hematomas resorb, dural displacement decreases and eventually the hypointense rim becomes indistinct as the dura merges with the hypointense bony cortex (figure 1–4).

Subdural hematomas undergo somewhat different changes than parenchymal and epidural hematoma [34, 35]. In some cases, hematomas decrease in size and resolve within a few weeks of hemorrhage while they are hyperintense on both STR and LTR scans. In other cases, the lesions persist, probably secondary to multiple episodes of rehemorrhage; in these instances, MR intensity is variable, and its relation to CT density is unpredictable. Many CT isodense subdural hematomas are hyperintense to the brain on STR and LTR scans (figure 1–12). They decrease in intensity initally on STR and then on LTR scans (figures 1–1, 1–13) Eventually, the hematoma may become isointense to spinal fluid on all pulse sequences, mimicking the appearance of subdural hygroma (figure 1–14). In some cases, serial studies of subdural hematomas may show an increase in intensity on STR and/or LTR scans, most likely indicative of recent recurrent hemorrhage into a chronic subdural hematoma (figure 1–15). Subdural hematomas usually lack the hypointense rims seen at the margin of the

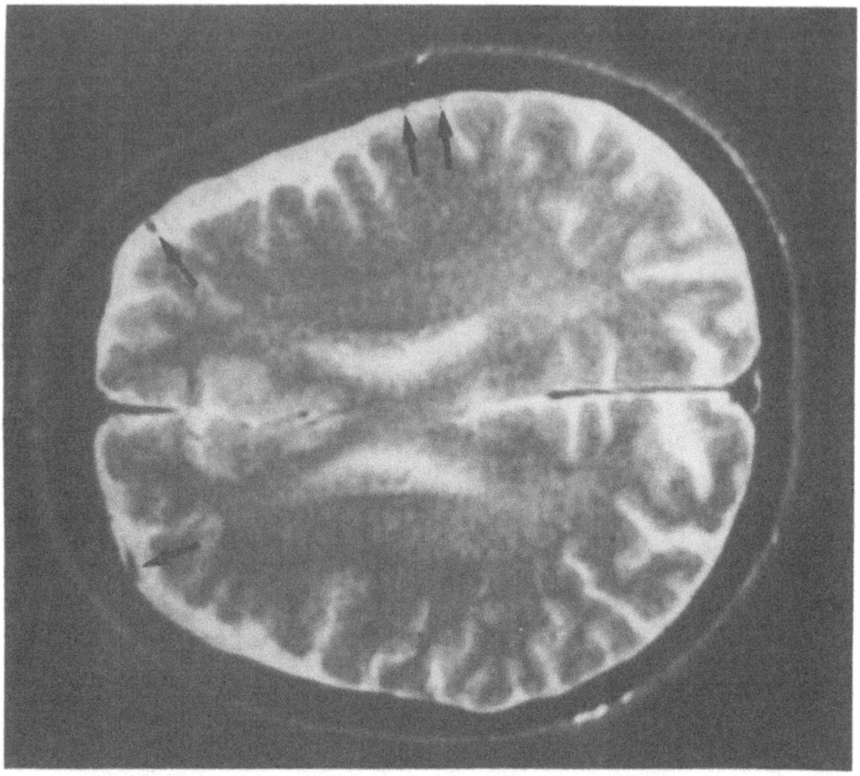

FIGURE 1–16C

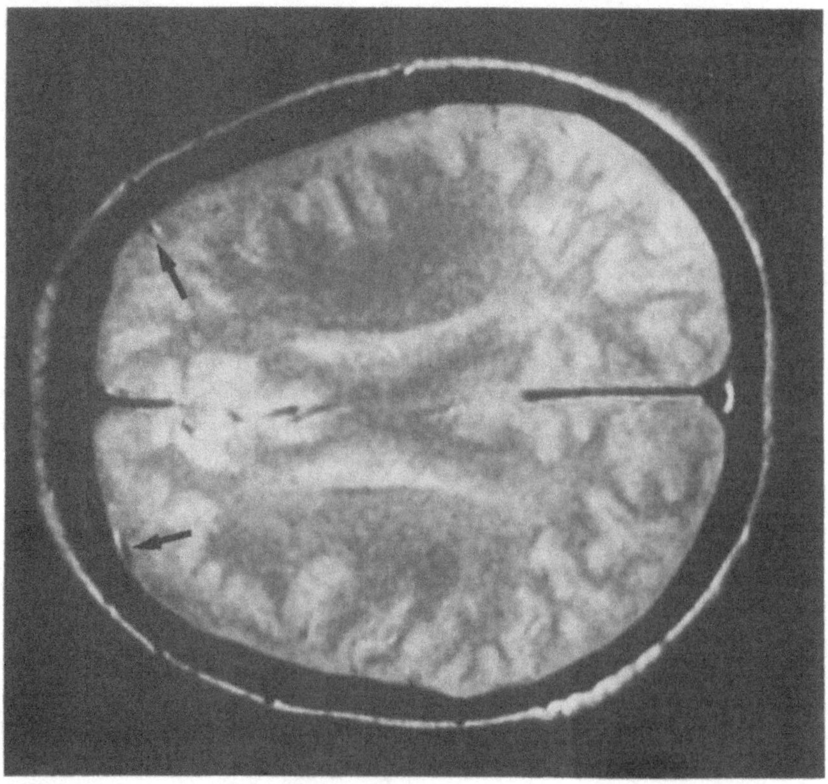

FIGURE 1–16B

parenchymal hematomas, presumably because in the absence of the blood–brain barrier, hemosiderin-laden macrophages are not permanently trapped in the hematoma membrane [7].

Throughout the subacute and chronic phase, MR is far superior to CT in detecting and delineating the extent of hemorrhage. On CT, hematomas become isodense to the brain between 10 and 28 days of icuts, and thereafter are nonspecifically hypodense. On MR, hemorrhages in all locations remain hyperintense throughout this period and are easily identified and more easily characterized as hemorrhagic. Recurrent hemorrhage into chronic subdural hematoma is more easily identified on MR than CT (figure 1–15). Finally, chronic subdural hematomas are isodense to CSF and, in the absence of mass effect, may be difficult to differentiate from atrophic dilitation of the subarachnoid spaces. On MR, this distinction is more readily made. In most cases, the fluid is hyperintense to CSF and therefore directly identifiable as a subdural collection. Even when the fluid is isointense to CSF, the subdural location can be ascertained on MR on an anatomic basis. Subtle gyral compression is more easily detected on MR, and displaced cortical veins are seen because of their flow-related signal void. Medial displacement of these veins is a classic angiographic sign of subdural hemorrhage that is readily identified on MR, especially when MR is performed in the coronal projection. Dilated subarachnoid spaces, by contrast, show no evidence of gyral compression, and the cortical veins wander throughout the subarachnoid space and are not closely applied to the brain surface (figure 1–16).

References

1. Ambrose J. Computerized transverse axial scanning (tomography). (2) Clinical application. *Br J Radiol* 1973;46:1023–1047.
2. New PFJ, Scott WR, Schnur JA, et al. Computerized axial tomography with EMI scanner. *Radiology* 1974; 110:109–123.
3. Bailes DR, Young IR, Thomas DJ, et al. NMR imaging of the brain using spin echo sequences. *Clin Radiol* 1982;33:395–414.
4. Bydder GM, Steiner RE, Young IR, et al. Clinical MR imaging of the brain: 140 cases. *Am J Radiol* 1982;139:215–236.
5. Sipponen JT, Sepponen RE, Sivula A. Nuclear magnetic resonance (NMR) imaging of intracerebral hemorrhage in the acute and resolving phases. *J Comput Axial Tomogr* 1983;7:954–959.
6. DeLaPaz RL, New PFJ, Buonanno FS, et al. NMR imaging of intracranial hemorrhage. *J Comput Axial Tomogr* 1983;8:599–607.
7. Gomori J, Grossman RI, Goldberg HI, et al. Intracranial hematomas: imaging by high field MR. *Radiology* 1985;157:87–92.
8. Sipponen JD, Sepponen RE, Tanttu JI, Sibula A. Intracranial hematomas studied by MR imaging at 0.17 and 0.02T. *J Comput Axial Tomogr* 1985;9: 698–704.
9. New PF, Aronow S. Attenuation measurements of whole blood and blood fraction in computed tomography. *Radiology* 1976;21:635–640.
10. Zimmerman RD, Heier LA, Snow RB, Liu DPC, Kelly AB, Deck MDF. Acute intracranial hemorrhage: intensity changes on sequential MR scans at 0.5T. *AJNR* 1988;9:47–57. AJR 1988;150: 651–662.
11. Bradley WG, Schmidt PC. Effect of methemoglobin formation on the MR appearance of subarachnoid hemorrhage. *Radiology* 1985;156:99–103.
12. Bloembergen. Proton relaxation times in paramagnetic solutions. *J Chem Phys* 1975;27:575–582.
13. Slichter CP. *Principles of Magnetic Resonance.* New York: Harper and Rowe, 1963: 137–183.
14. McNamara MT. Paramagnetic contrast media for magnetic resonance imaging of the central nervous system. In: Brant-Zawadski M, Norman D (eds). *Magnetic Resonance Imaging of the Central Nervous System.* New York: Raven Press, 1987: 97–105.
15. Bradley WG. MRI of hemorrhage and iron in the brain. In: Stark DD, Bradley WG (eds). *Magnetic Resonance Imaging.* St. Louis: CV Mosby, 1988: 359–375.
16. Brittenham GM, Farrell EE, Harris JW, et al. Magnetic susceptibility measurements of human iron stores. *N Engl J Med* 1982;307:1671–1675.
17. Thulborn KR, Waterton JC, Matthews PM, Morrada GK. Oxygenation dependence of the transverse relaxation time of water protons in whole blood at high fields. *Biochim Biophys Acta* 1982;174:265–270.
18. Edelman RR, Johnson K, Buxton R, et al. MR of hemorrhage: a new approach. *AJNR* 1986;7: 751–756.
19. Bradley WG, Waluch V. Blood flow: magnetic resonance imaging. *Radiology* 1985;154:443–450.
20. Dichiro G, Brooks RA, Girton ME, et al. Sequential MR studies of intracerebral hematomas in monkeys. *AJNR* 1986;7:193–199.
21. Chakers DW, Bryan RN. Acute subarachnoid hemorrhage: in vitro comparison of magnetic resonance and computed tomography. *AJNR* 1986;7:223–228.
22. Dooms GC, Uske A, Brant-Zawadski M, et al. Spin echo MR imaging of intracranial hemorrhage. *Neuroradiology* 1986;28:132–138.
23. Zimmerman RD, Markisz J, Heier LA, et al. Effect of magnetic field strength on the intensity of experimentally created acute intracerebral hemorrhage (AIH). (Abstract.) *AJNR* 1987;8:936.
24. Hackney DB, Atlas SW, Grossman RI, Gomori JM, Goldberg HJ, Zimmerman RA, Bilaniuk LT. Subacute intracranial hemorrhage: contribution of spin density to

the appearance on spin-echo MR images. *Radiology* 1987;165:199–202.

25. Hayman LA, McArdle C, Taber K, et al. Role of clot formation in MR imaging of blood. (Abstract.) *Radiology* 1987;165(P):402.
26. Goldberg HG. Personal communication.
27. Oot RF, New PF, Pile-Spellman, et al. Detection of intracranial calcification by MR. *AJNR* 1986; 7:801–810.
28. Becker RD, Zimmerman RD, Sze G, Haimes AB, Deck MDF. MR of subdural empyema and other extra-axial inflammatory lesions. (Abstract.) *AJNR* 1987;8:940.
29. Bradley WG, Crooks LE, Newton TH. Physical principles of NMR. In: Newton TH, Potts PG, (eds). *Modern Neuroradiology: Advanced Imaging Techniques*, Volume II. San Francisco: Clavadel Press, 1983: 15–62.
30. Wehrli FW, Mac Fall J, Newton TH. Parameters determining the appearance of NMR images. In: Newton TH, Potts P6 (eds). *Modern Neuroradiology: Advanced Imaging Techniques*, Volume II. San Francisco: Clavadel Press, 1983: 81–118.
31. Newton DR, Larson TC, Dillon WP, Newton PH. Magnetic resonance characteristics of cranial epidermoid, dermoid and teratomatous tumors (Abstract.) *AJNR* 1987;8:945.
32. Brant-Zawadski M, Davis PO, Crooks LE, et al. NMR demonstration of cerebral abnormalities. Comparison with CT. *AJNR* 1983;4:117–124.
33. Grossman RI, Kemp SS, Ip CY, et al. Importance of oxygenation in the appearance of acute subarachnoid hemorrhage on high field magnetic resonance imaging. *Acta Radiol* 1986;369 (Suppl):56–58.
34. Snow RB, Zimmerman RD, Gandy SE, Deck MDF. Comparison of magnetic resonance imaging in computed tomography in the evaluation of head injury. *Neurosurgery* 1986;18:45–52.
35. Kelly A, Zimmerman RD, Snow RB, Gandy SE, Deck MDF. Comparison of MR and CT in the imaging of head trauma. Experience in 100 patients. *AJNR* 1988;9:699–708.
36. Rubin JB, Enzmann DR. Imaging of spinal fluid CSF pulsation by 2DFT magnetic resonance: significance during clinical imaging. *AJNR* 8:297–306.
37. Heier LA, Amster JL, Zimmerman RD, Deck MDF. Focal recurrent hemorrhage of magnetic resonance at 0.5 Tesla: an aid to the diagnosis of cryptic cerebral vascular malformations. *Acta Radiol* 1986;369 (Suppl):113–115.
38. Gomori JM, Grossman RI, Goldberg HJ, Hackney DB, Zimmerman RA, Bilaniuk LT. Occult cerebral vascular malformations: high field MR imaging. *Radiology* 1986;158:707–713.
39. Sze G, Krol G, Olsen W, et al. Hemorrhagic neoplasms: MR mimics of occult vascular malformations. *AJNR* 1987;8:795–802.
40. Destian S, Sze G, Krol G, Zimmerman RD, Deck MDF. The MR evolution of hemorrhagic neoplasms. *AJNR* (in press).
41. Han JS, Kaufman B, Alfidi RJ, et al. Head trauma evaluated by magnetic resonance in computed tomography: a comparison. *Radiology* 1984;150:71–77.

2. INTRACRANIAL ANEURYSMS

Mohammad Sarwar

An intracranial aneurysm is a localized sacculation or dilatation of a blood vessel. Intracranial congenital (berry) aneurysms, which constitute over 90% of the intracranial aneurysms, are the commonest non-traumatic cause of subarachnoid hemorrhage (SAH) with or without associated parechymal clot. Although histologically benign, they cause catastrophic complications. Of the 28,000 cases of SAH in the USA, approximately 80% of which are caused by aneurysms, 10,000 die outright even before they reach the hospital. Of the remaining 18,000 that receive treatment, 8,000 suffer a variable severity of morbidity or die. Only 10,000 remain without any neurological deficit. The concurrence of parenchymal clot adversely affects the rate of survival or quality of neurological recovery.

Classification of Intracranial Aneurysms

Intracranial aneurysms are classified into five types: a) congenital, b) traumatic, c) arteriosclerotic or atherosclerotic, d) infective (*mycotic*), and e) neoplastic (for example, associated with atrial myxoma and chorioepithelioma).

ETIOLOGY AND PATHOGENESIS

The definite etiology of congenital intracranial aneurysms has not been established. The fact that they are so rarely seen at birth and in early age suggests that there are additional acquired factors that play a dominant role in further weakening the already defective wall of these aneurysms. The occasional concurrence of coarctation of aorta and polycystic kidney disease and their familial occurrence lend support to their developmental origin. The basic congenital defect is characterized by a weakness in the tunica media at arterial bifurcation where the aneurysm classically occurs [1–4]. Defective collagen and elastica [5] and disintegration of the internal elastic membrane that might contain liposome like granules are important factors in the pathogenesis of the aneurysms [6, 7]. It also has been suggested that these aneurysms are abnormal dilation of embryonic vessels that were destined for extinction. Atherosclerosis and hypertension, which are present in 50% and 60%, respectively, of these patients, explain why the aneurysms commonly manifest themselves clinically later in life.

AGE AND SEX

Intracranial aneurysms are rarely recognized clinically in early life. About 60% are encountered in the fifth to sixth decade. The female-to-male ratio is 3:2.

INCIDENCE AND MULTIPLICITY

The incidence of aneurysms in the general population is 4–6% [8], but only 20% rupture. They are frequently multiple. McKissock et al. in 1964 [9] studied 17 reports and described the incidence of multiple aneurysms at 5–17% in angiographic and 21–33% in autopsy series. Depending upon the quality of the cerebral angiograms and the assiduity with which they are sought, the recently reported incidence of multiple aneurysms recognized at angiography has ranged from 30% [10] to 33.5% [11]. When multiple, they are usually two or three in number and rarely more than six or seven. In a recent report, Nehls et al. [11] noted that over half their patients had three or more aneurysms. Multiple aneurysms are far more common in females than in males, with the reported female-to-male ratio ranging from 2:1 to 5:1 [11].

LOCATION

Whereas, in the autopsy series, middle cerebral artery (MCA) bifurcation aneurysms are the commonest, the anterior communicating artery (A.CO.A) aneurysms are more frequently encountered clinically. This discrepancy is explained by the higher tendency of the A.CO.A aneurysms to rupture.

The aneurysms may be divided into two groups. *Group A* includes aneurysms of the anterior Circle of Willis and constitutes 85% of all congenital intracranial aneurysms. *Group B* comprises aneurysms

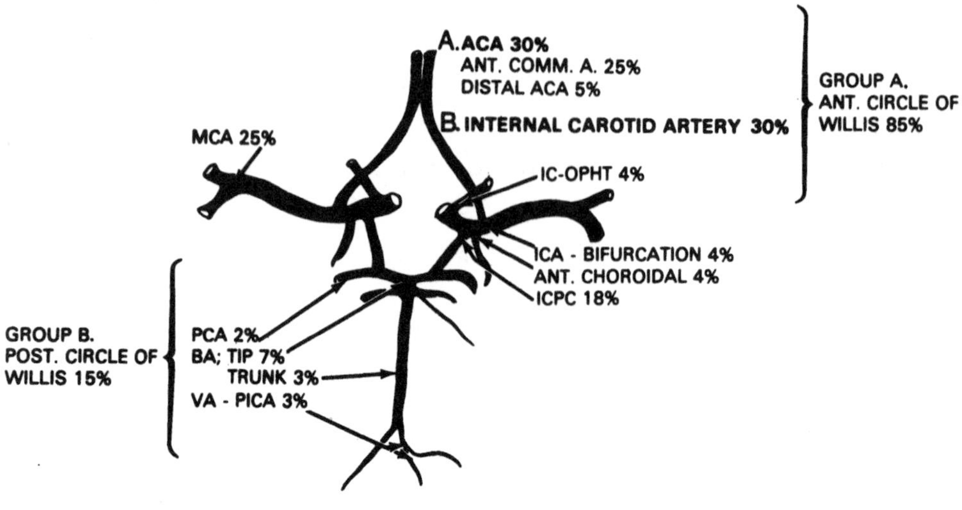

FIGURE 2-1. Diagram to show the distribution of congenital aneurysms at the Circle of Willis (after [12]).

of the vertebrobasilar system, with an incidence of 15%. The incidence of aneurysms [12] that belong to each group of Circle of Willis is shown in figure 2-1.

SIZE AND RUPTURE

The aneurysm size varies. The ruptured aneurysm usually measures 5–15 mm. An aneurysm rarely bleeds when its size is 5 mm or less [13]; in this situation it could be clinically followed. The critical size for rupture is 10 mm or more. In 80% of cases, the bleeding occurs at the apex of the aneurysm.

Natural History of Aneurysms

Until recently, it was believed that once an aneurysm has bled, its rebleed rate is 3% per year, mortality rate is 2% per year, and the chance of rebleed peaks at day 7. However, Jane et al. [14] have recently collected data from the Atkinson Morley's Hospital, Wimbledon, England, the Cooperative Aneurysm Study, the International Cooperative Study on the Timing of Aneurysm Surgery, and analysis of incidental aneurysms in Charlottesville, Virginia. According to them, the highest rebleed rate occurs at day 1, and at least 50% of aneurysms rebleed during the six months after the initial hemorrhage. Thereafter the rate diminishes at 3% per year. They also reported that patients with unruptured incidential aneurysms bleed at a rate of 1% per year.

Forty-five percent of patients with SAH die after first bleed and 60% after rebleed. Factors associated with rebleeding comprise poor neurological status at initial presentation, hypertension, older age, aneurysm pointing upward, short broad aneurysm, and aneurysm that exhibits enlargement on serial angiograms [15].

Complications of a Ruptured Aneurysm

Complications are a) SAH and/or parenchymal and/or intraventricular hemorrhage, b) spasm, cerebral infarction, and arterial wall myonecrosis, c) hydrocephalus, d) increased intracranial pressure, e) hypothalamic dysfunction, f) cardiac abnormalities (arrhythmia, subendocardial hemorrhage, and myocardial infarction), g) hypertension, and h) vitreous hemorrhage.

SUBARACHNOID HEMORRHAGE AND/OR PARENCHYMAL AND/OR INTRAVENTRICULAR HEMORRHAGE

Subarachnoid hemorrhage is located predominantly at the base of the brain where the Circle of Willis is located. The subarachnoid hemorrhage alone is less likely to be fatal unless it is associated with parenchymal and/or intraventricular hemorrhage and/or cerebral infarction. About 85% of the patients who die from aneurysm rupture have sustained parenchymal and/or intraventricular clot [12].

Subdural hematoma following aneurysm rupture is unusual. Its incidence is 1–3%. When encountered, it is most likely a consequence of a ruptured MCA aneurysm. The previous bleed and consequent arachnoidal tethering of the aneurysm is believed to make it more vulnerable to bleed into a potentially capacious subdural compartment.

SPASM

Spasm is a dreaded complication of SAH [16] that adversely affects the neurological status. The spasm represents a state of altered activity of the blood vessels as a result of contact of blood with the arterial adventitia. The mere physical contact of a blood clot and its breakdown products with arterial adventitia is not the only cause of arterial spasm; there are other contributing factors. The mechanical factors include periarterial nerve stimulation and arterial compression with smooth muscle contraction. The chemical factors include serotonin, prostaglandins, and catecholamines.

The angiographic chronology of arterial spasm varies. It has been demonstrated at 15 minutes in experimental SAH in animals [17, 18]. It is infrequently recognized angiographically in humans within three days of SAH and is most often seen between days 10 and 17 [19] (figure 2–2).

The spasm commonly involves the major arterial trunks at the skull base. Whereas it may be generalized, it is more pronounced in arteries that are located close to the ruptured aneurysm. The mere presence of angiographic spasm, which is seen in about two thirds of all cases of SAH [20], should not arouse unnecessary clinical concern until it has caused a significant reduction in arterial lumen. It is generally agreed that cerebral vasospasm of the basal vessels that reduces the arterial lumen to more than 50% is associated with a reduction in cerebral blood flow [21, 22], thereby enhancing the likelihood of the development of cerebral infarction.

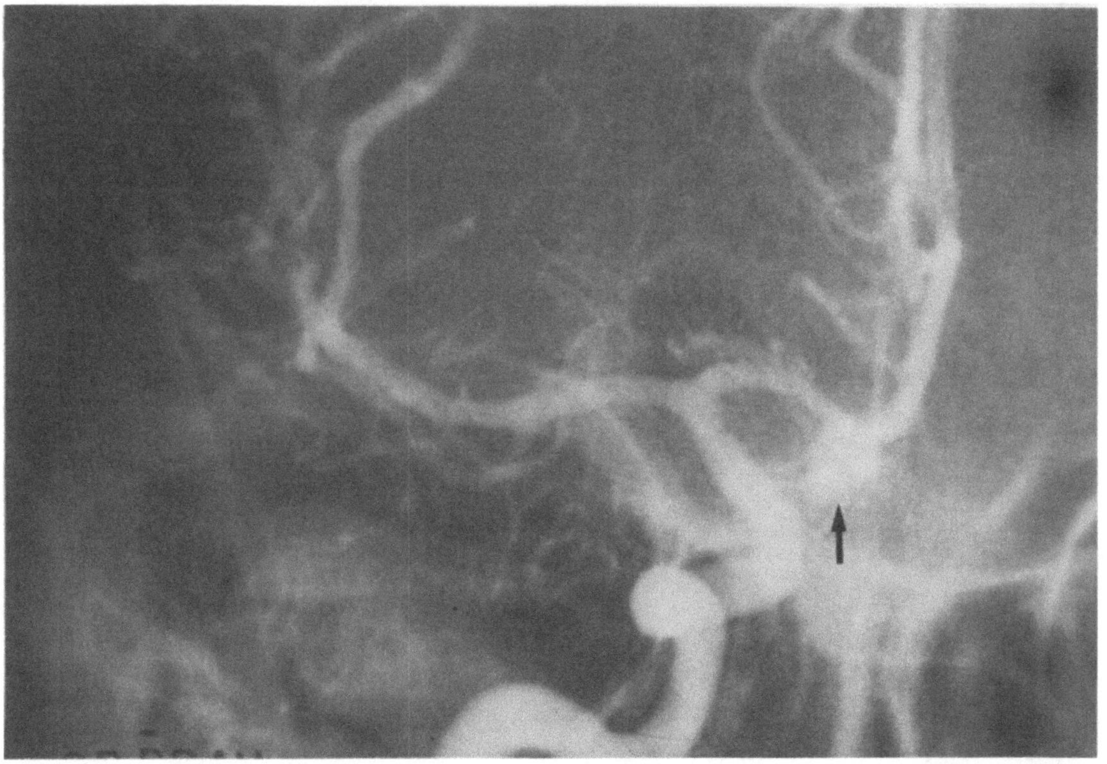

FIGURE 2–2A

FIGURE 2–2. Spasm chronology. Right carotid angiogram, anteroposterior views, (A) at day 3, (B) at day 6, and (C) at day 22 (in this illustration the aneurysm has been clipped) after sustaining SAH. Note that the spasm has manifested at day 6 and disappeared by day 22. Arrows in (A) and (B) are at aneurysm. Arrowheads in (B) are at spasm. The filling defect within the aneurysm in (B) could suggest luminal thrombus and/or poor filling of the lumen due to decreased flow caused by spasm. This degree of severe spasm can lead to nonopacification of the aneurysm. This situation therefore calls for repeat angiography in one to two weeks if no aneurysm has been demonstrated on the initial bilatered carotid and bilateral vertebral angiograms.

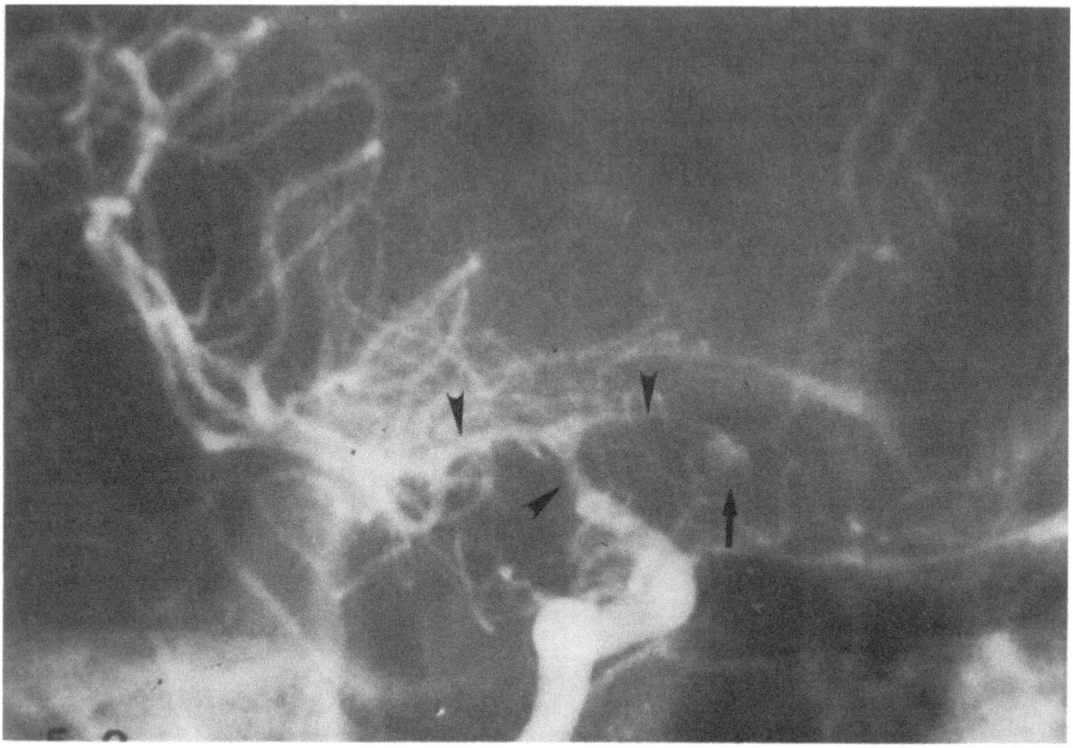

FIGURE 2–2B

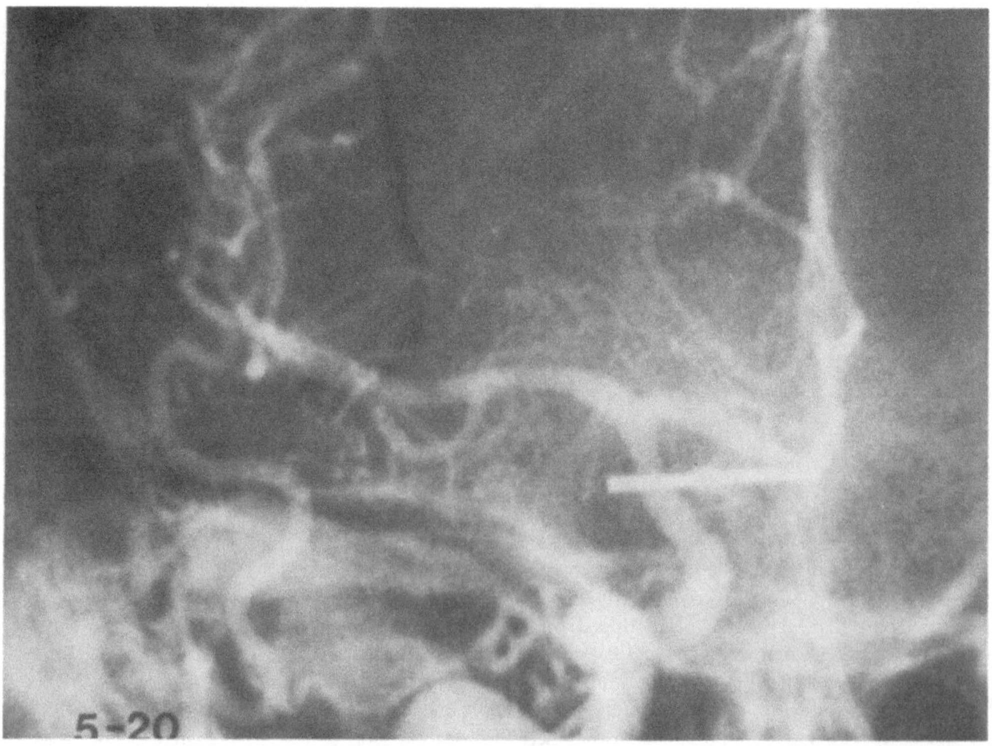

FIGURE 2–2C

Cerebral blood flow (CBF) of or less than 20 ml per minute per 100 gm brain (normal 50) is likely to cause ischemic cerebral infarction [23, 24]. According to Knucky et al. [24], patients with reduction of CBF and diffuse blood on computerized tomography (CT) scan have a greatly increased probability of developing cerebral ischemia. Similarly, the greater the amount of SAH visualized on CT, the greater the chance of developing arterial spasm. Cerebral angiography is not believed to cause spasm per se.

Cerebral Infarction. Cerebral infarction follows spasm at a rate of about 20–36% [25]. However, at least two thirds of patients who die from SAH complications have cerebral infarction. Dislodged thrombi from aneurysms, that are usually large, only rarely cause embolic infarction [26, 27, 28].

Myonecrosis. Myonecrosis of arterial walls has been documented experimentally in dogs after SAH. The muscle necrosis and the consequent repairing process cause intimal edema, thrombosis, and subendothelial granulation tissue, leading to arterial narrowing that may be permanent. Such a phenomenon is rarely recognized in the human on follow-up angiograms.

HYDROCEPHALUS

The acute hydrocephalus develops due to mechanical blockage of the subarachnoid spaces [29]. It has a deleterious effect on the neurological status of the patient. Van Gijn et al. [30] reported a 20% incidence of acute hydrocephalus studied in 174 patients. Acute hydrocephalus in patients who survived 24 hours was defined as a bicaudate index above the 95th percentile for age, measured on CT scans that were obtained within 72 hours of the initial SAH. These authors observed that concurrent intraventricular clot was more likely to cause acute hydrocephalus than cisternal clot alone.

The hydrocephalus that develops after days to weeks is due to arachnoidal adhesions. Hydrocephalus following SAH that requires shunting because of delayed neurologic deterioration or enlarging ventricle has an incidence of 14% [13].

The larger the SAH, the more likely it is for hydrocephalus to develop. Also, hydrocephalus is observed more commonly following rupture of midline aneurysms [31].

Black et al. [31] studied the relationship of hydrocephalus and vasospasm after SAH in 87 patients. They noted that vasospasm and mild hydrocephalus have a concurrence rate of 62%. It is possible that hydrocephalus impairs clearance of blood from CSF pathways, thereby allowing longer contact of blood and its breakdown products with arterial walls, thereby accounting for higher concurrence of spasm and hydrocephalus. There is poor correlation between mild ventricular enlargement and neurological deterioration. However, progressive increase in ventricular size usually is of clinical importance and warrants ventricular shunting.

Infective (Mycotic) Aneurysms

Infective aneurysms are the consequence of aneurysmal dilatation of an artery secondary to weakening of its wall from an inflammatory cause, whether embolic in origin or by local spread of infection (e.g., meningocerebritis, infective cavernous sinus thrombosis) [32]. They are commonly located on peripheral branches of MCA, since MCA is almost a direct continuation of the internal carotid artery. This situation allows infected emboli from the heart to have a relatively direct route to MCA branches. Infective aneurysms should be especially suspected in drug addicts and in patients with subacute bacterial endocarditis (SBE), septicemia, and systemic infection.

The clinical presentation of infective aneurysms is commonly by hemorrhage, infarction, or both. Since the parent artery, which harbors the aneurysm, and the surrounding arteries show weakening of their walls, these aneurysms carry great risk of rupture at the time of applying surgical clips. They are therefore best managed by antibiotic therapy, since the majority of them thrombose following conservative treatment.

Giant Intracranial Aneurysms

An aneurysm is defined as giant when its angiographically opacified lumen is larger than 2.5–3 cm. Giant aneurysms account for approximately 2.5–5% of all intracranial aneurysms [33, 34]. These aneurysms rarely present clinically with SAH, because repeated small hemorrhages in the past seem to invite laminated thrombosis, which subsequently protects hemorrhage into the subarachnoid space or the parenchyma. Giant aneurysms cause their symptoms primarily by pressure on adjacent cranial nerves and cerebral tissue, and may cause seizure disorder. Cranial nerves are involved in at least 66% of the cases [35]. The second through the sixth cranial nerves are most commonly involved. Occasionally these aneurysms may cause transient ischemic attacks (TIAs) by cerebral embolization of dislodged thrombotic debris that is present within their lumen.

Enlargement of Intracranial Aneurysms

What causes aneurysmal enlargement? It is quite reasonable to suspect that age, arterioslerosis, and hypertension further weaken the already defective wall of the aneurysm [36] and thereby cause it to enlarge. A situation is probably created wherein turbulence occurs in the aneurysm. The turbulence, stasis, and eddy currents in the lumen of the aneurysm lead to poor nutrition of the aneurysmal wall. Turbulence causes more destruction of the arterial wall than the laminar flow. This leads to a vicious circle of arterial wall weakening and enlargement of the aneurysmal lumen.

Dolichoectasia

Dolichoectasia is an anomaly characterized by elongation and dilatation of the involved arteries [37]. The disease commonly involves the vertebrobasilar system, the basilar artery being involved far more commonly. Less frequently, the carotid arteries at the skull base may be involved either in common with the vertebrobasilar system or alone. The underlying pathogenesis of this condition is not fully understood. However, histologically it is characterized by a defect in the tunica media with disruption of the internal and external elastic membrane. The disease is not related to arteriosclerosis. It is not certain whether dolichoectasia is a developmental anomaly.

This condition manifests itself clinically by compression of the brain stem, lower cranial nerves, or both. Such compression depends on the size of the involved artery. Commonly there is a variable degree of associated thrombus formation that may calcify. The lesion might slowly enlarge in size due to repeated small hemorrhages that are contained within the surrounding thrombus and collagen capsule [38]. Depending on compression of the fourth ventricle, hydrocephalus may ensue.

Infundibular Widening

Infundibular widening is a small dilatation at the origin of the posterior communicating artery. Whether this dilatation is a normal variant or an aneurysm is debated. Hassler and Saltzman [39] believe that this dilatation represents an aneurysm because histologically it has features of an aneurysm. It has a typical triangular lumen, the apex being away from the internal carotid artery (figure 2–3). Usually the posterior communicating artery can be traced beyond its apex; if not, distinguishing it from an aneurysm may be difficult. It is generally agreed, however, that if this dilatation measures more than 3 mm it should be considered an aneurysm. The dilatation can enlarge to assume the typical morphology of an aneurysm.

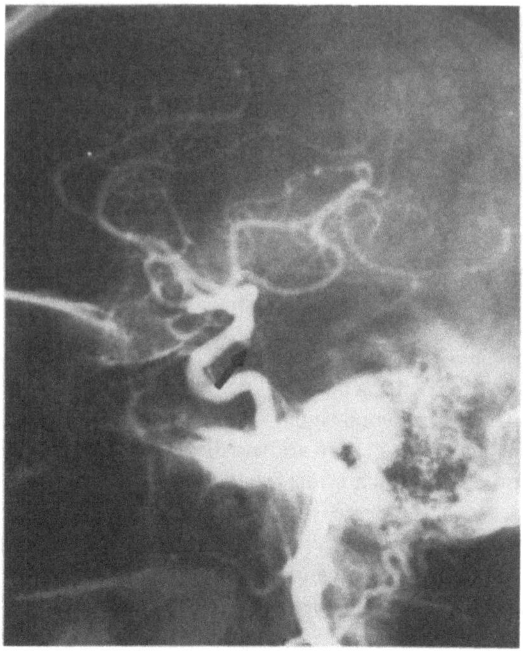

FIGURE 2–3. Infundibular widening. Note that it has a broad base at the origin from the internal carotid artery. The posterior communicating artery emanates from its apex (arrow).

Neuroradiology of Intracranial Aneurysms

Whereas clinical diagnosis of SAH from a ruptured intracranial aneurysm is easy, the exact clinical localization of the culprit aneurysm is difficult. The ability by clinical signs alone to localize the ruptured aneurysm varies from as low as 7% [11] to about one third of the cases [40, 41]. Neuroimaging and cerebral angiography therefore remain the preeminent modalities that provide the requisite information for clinical care of the patients afflicted with aneurysmal SAH.

Computerized Tomography of Intracranial Aneurysms

Plain computerized tomography (CT) depicts blood in acute SAH as a high-density lesion, be that in subarachnoid space, parenchyma, or ventricular system. The most definite clue to the ruptured aneurysm is the presence of a localized parenchymal clot in the neighborhood of the Circle of Willis where a particular aneurysm is known to occur (figures 2–4, 2–5, 2–6). An excessive amount of SAH at a certain part of the Circle of Willis carries the same connotation but with less certainty. To-and-fro pulsatile movement of CSF may cause excessive SAH to occur away from the ruptured aneurysm; this circumstance is more likely to occur in the vertebrobasilar system aneurysms (figure 2–7). Although one may detect acute SAH on CT in 86% of cases, the ability of CT to correctly localize the aneursym that has ruptured is only 45% [11].

If CT scan is done a few days after ictus or if SAH is small in amount, CT may appear normal. In such cases, careful evaluation should be made of the basal cisterns. If they are not visualized at all or visualized less optimally, one should suspect isodense SAH. In such cases, diagnosis of SAH should be documented with lumbar puncture.

Contrast CT (CCT) usually does not disclose the culprit aneurysm if it is small and is buried in excessive SAH and/or parenchymal clot. However, if the aneurysm is larger than 5–8 mm, it may be recognized when the SAH surrounding it is small and the parenchymal clot minimal in size or absent (figure 2–8). Detection of the aneurysm on CCT depends upon the assiduity with which the aneurysm is sought; its rate of detection varies from 36–76% [42]. Whereas CT detection of the aneurysm does not provide information that the neurosurgeon can use to plan surgical treatment, it may provide him valuable information for timing his surgery. For example, if the ruptured aneurysm happens to be at the anterior communicating artery or on the vertebrobasilar system, the neurosurgeon may decide to delay surgery until the patient's clinical status is optimal, since surgery on these aneurysms carries relatively high risk. Also, the recognition of the ruptured aneurysm on CCT may determine the timing of cerebral angiography. For example, for anterior-communicating-artery aneurysms the neurosurgeon may want to wait but may wish to do surgery as soon as possible if the aneurysm is such that it is easily accessible and carries little morbidity, as in the case of a posterior-communicating-artery aneurysm. In the case of a giant aneurysm, dynamic contrast CT may aid in the assessment of aneurysmal neck and of whether or not aneurymal neck is buried within the thrombosed portion of the aneurysm (figure 2–9).

Cerebral Angiography of Intracranial Aneurysms

Cerebral angiography at present remains the definitive study that provides the requisite information for appropriate surgical management to the neurosurgeon. It delineates the morphology of the aneurysm in terms of site, size, multiplicity, and associated spasm. However, CT, MRI, and cerebral angiography are complementary in the full radiological assessment of a ruptured aneurysm.

When digital angiography appeared on the scene several years ago, there were great expectations that it would replace conventional film-screen cerebral angiography. Whereas intravenous digital cerebral angiography has no role in the surgical management of the patient with SAH, arterial digitial cerebral angiography (ADCA) has a definite and valuable role in workup of SAH. The main reason ADCA has not been widely accepted by neurosurgeons is that it is somewhat inferior to conventional angiography in the matter of spatial resolution. The neck and the entire morphology of the aneurysm are not as clearly and reliably shown on ADCA as on conventional angiography. Also, neurosurgeons are very reluctant to let old habits wither fast.

The recommendation of this author is as follows. If by clinical signs and by CT there are sufficient clues to the ruptured aneurysm, then do conventional angiography and demonstrate the presence of that aneurysm using routine anterior–posterior and lateral views and supplementing them with oblique and other views to be tailored by the angiographer's ingenuity and mental alertness. Perform ADCA to exclude other incidental aneurysms. If available, one must perform magnification and subtraction conventional cerebral angiography. Regarding the vertebrobasilar system, one should catheterize the larger vertebral artery (usually the left). One should inject a slightly larger amount (6 cc/sec and for a total of 9 cc) of contrast medium into the vertebral artery to enhance the chance of opacifying retrograde the contralateral vertebral artery proximal to the origin of the posterior inferior cerebellar artery. If this cannot be accomplished, one must then selectively opacify the contralateral vertebral artery. In older and hypertensive patients, in whom major arterial trunks originating from the aortic arch may be toruous and difficult to catheterize selectively, one should resort to ADCA, which usually provides satisfactory opacification of intracranial vasculature.

If by clinical signs and by CT there is no clue to the ruptured aneurysm, one should then resort to

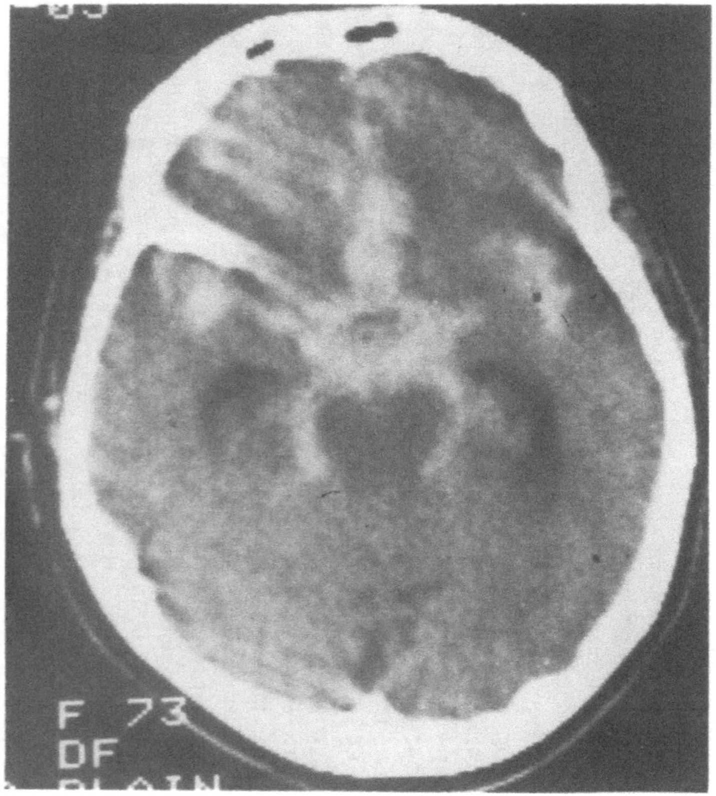

FIGURE 2–4A

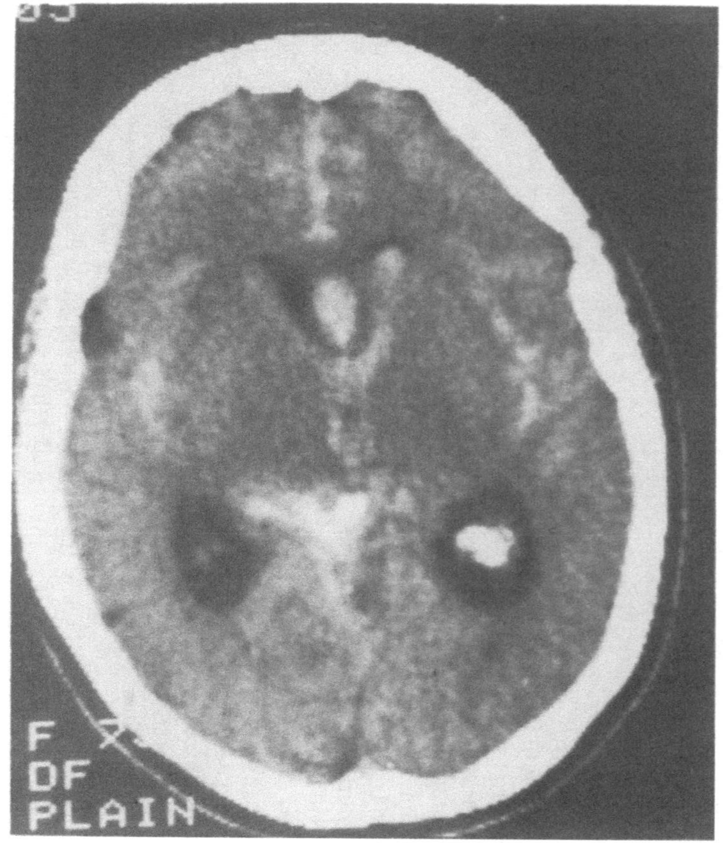

FIGURE 2–4B

FIGURE 2–4. Anterior communicating artery aneurysm rupture. (A)–(C) Axial plain CT images caudad to cephalad. Note the typical appearance of blood in the distended interhemispheric fissure and in the adjoining rectus gyri, septum pellucidum, third ventricle, and frontal horns. There is SAH in the basal CSF spaces as well. There is mild hydorcephalus due to physical plugging of the subarachnoid spaces by blood. (D) Left carotid angiogram, oblique view. The anterior communicating artery aneurysm (arrowhead) and the spasm of the supraclinoid portion of the internal carotid artery (arrow) are well shown.

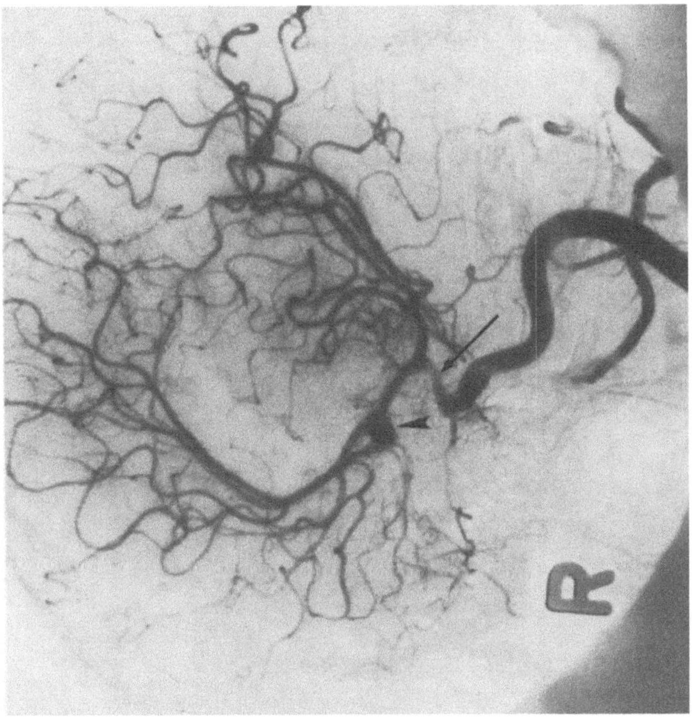

FIGURE 2–4D

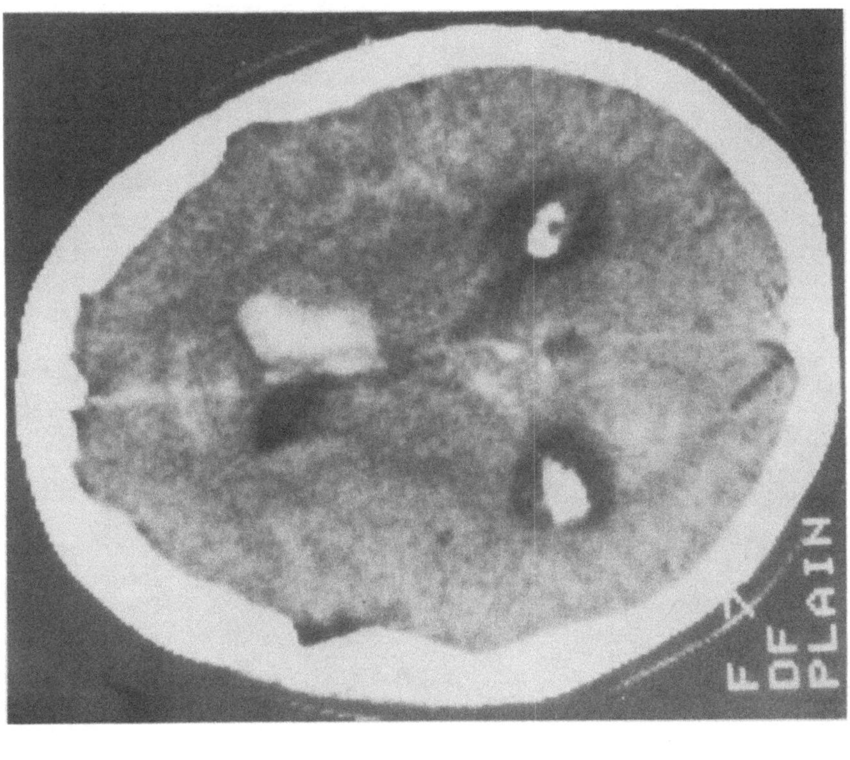

FIGURE 2–4C

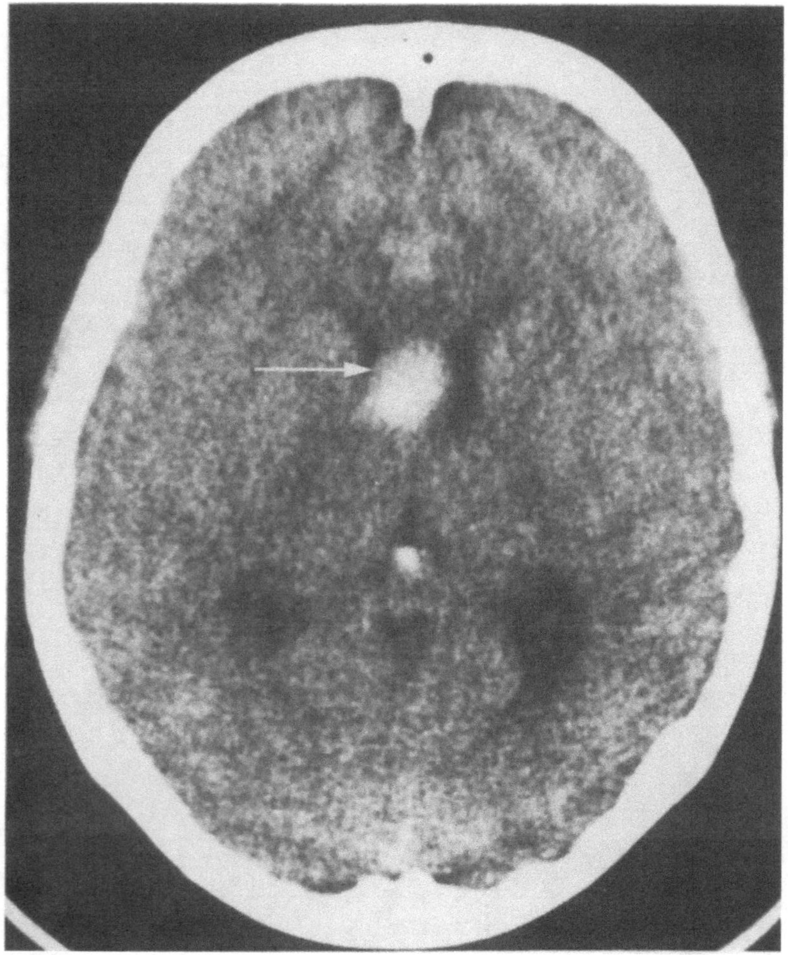

FIGURE 2–5A

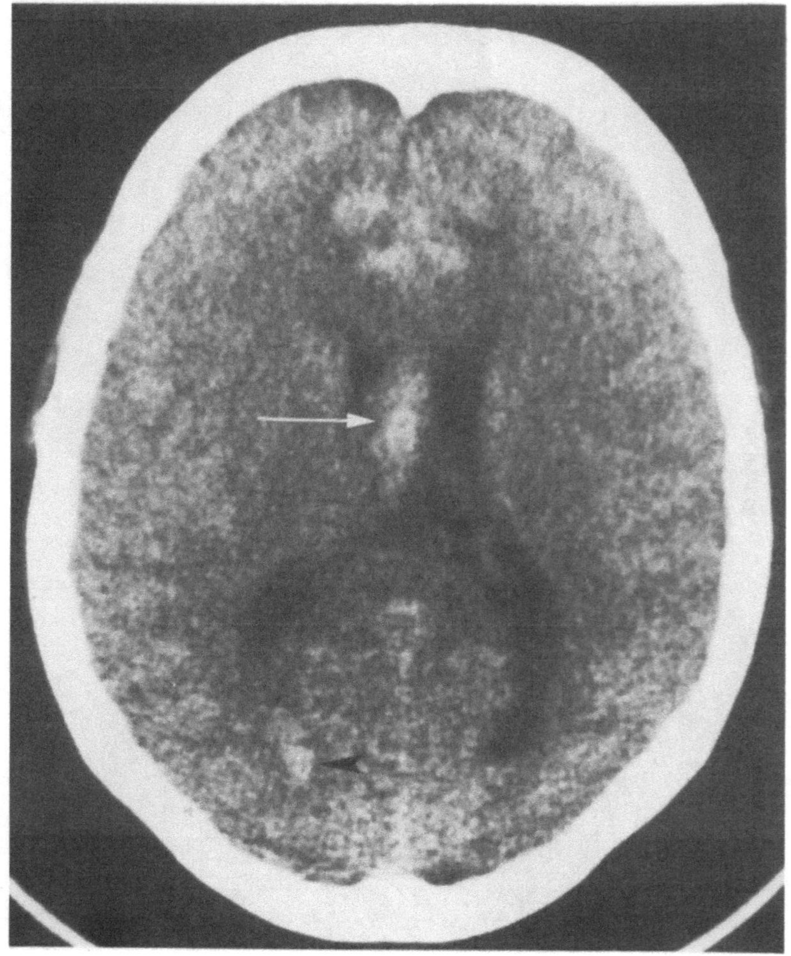

FIGURE 2–5B

FIGURE 2–5. Ruptured pericallosal artery aneursym. (A, B) Contiguous images 10 mm apart at the level of genu and body of corpus callosum. Note clot in the region of the septum pellucidum, genu, and body of corpus callosum (arrows). There is excessive blood in the adjacent cisterns as well. Part of the blood has ruptured into the right lateral ventricle and has traveled to the dependent part of the right occipital horn (arrowhead). Had this been due to a ruptured anterior communicating artery aneurysm, one would have seen blood emanating into the lower frontal interhemispheric fissure in (C), which is not the case. (D) Left carotid angiogram, lateral view. The 7-mm-size aneurysm is shown. Note the nipple (arrow) indicating the site of rupture at the dome of the aneurysm.

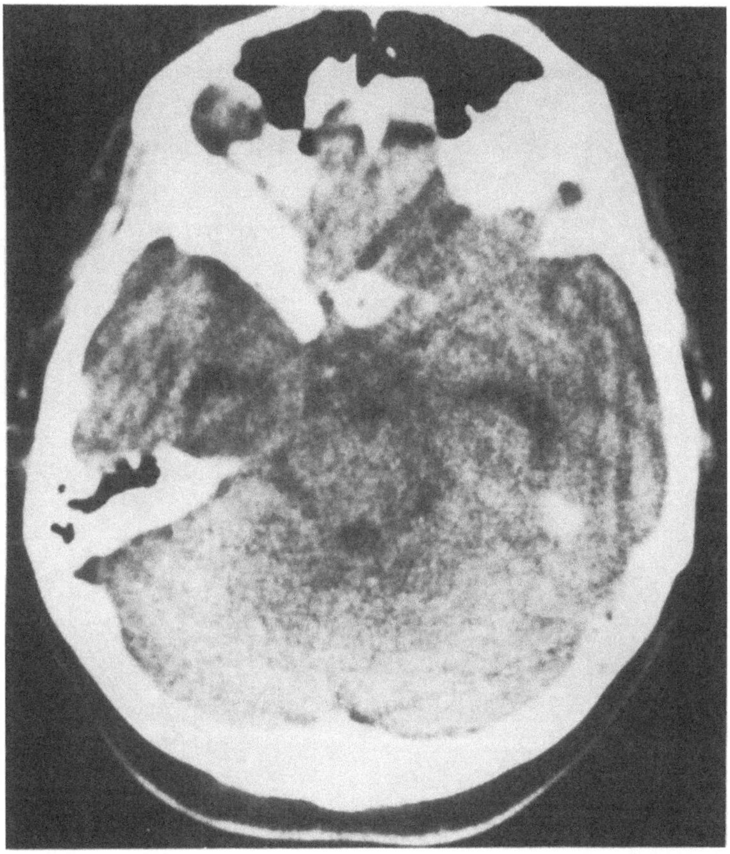

FIGURE 2–5C

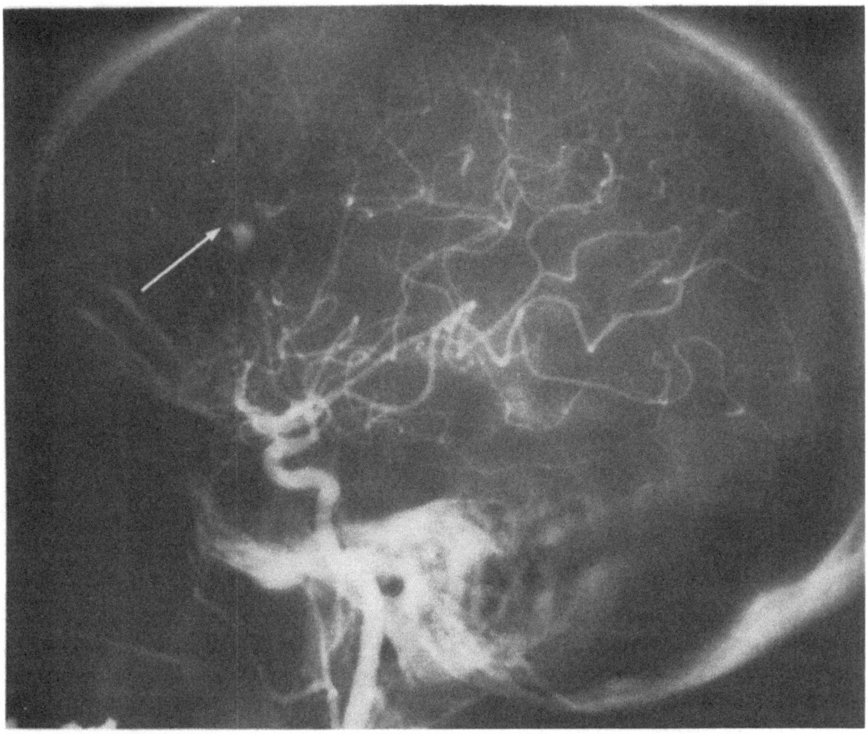

FIGURE 2–5D

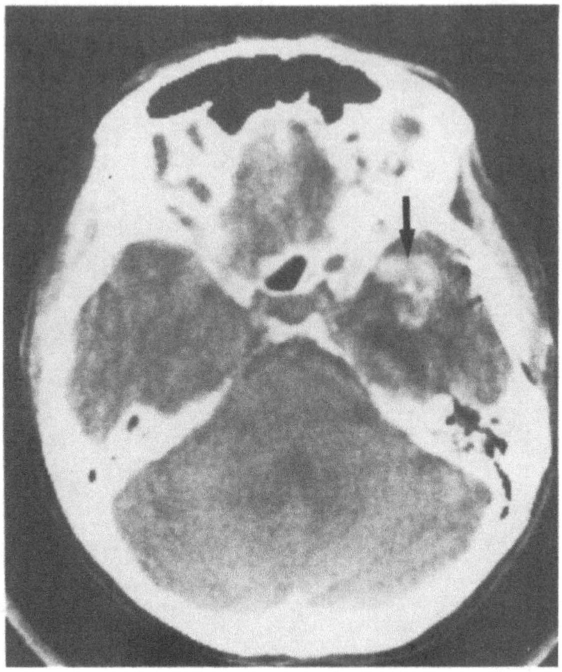

FIGURE 2–6A

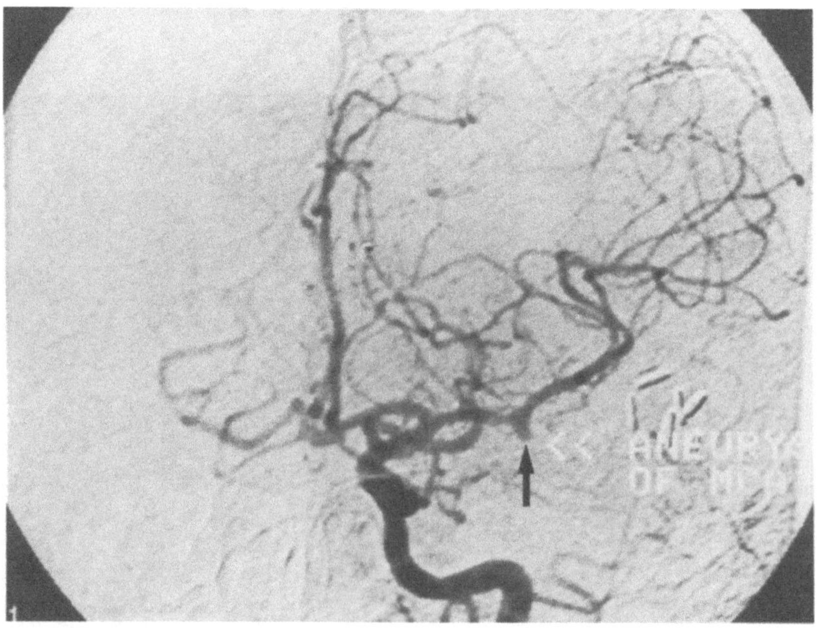

FIGURE 2–6B

FIGURE 2–6. Middle cerebral artery aneurysm rupture. A 61-year-old female presented with severe headache. (A) Plain CT shows localized clot (arrow) in the anterior temporal lobe. There is minimal subarachnoid hemorrhage because the aneurysm had been coated ten years earlier; this coating perhaps resulted in containment of the clot within the scarred brain and caused less dispersion of blood in the subarachnoid spaces. (B) Anteroposterior view of intraarterial digital angiogram. The aneurysm (arrow) is well shown.

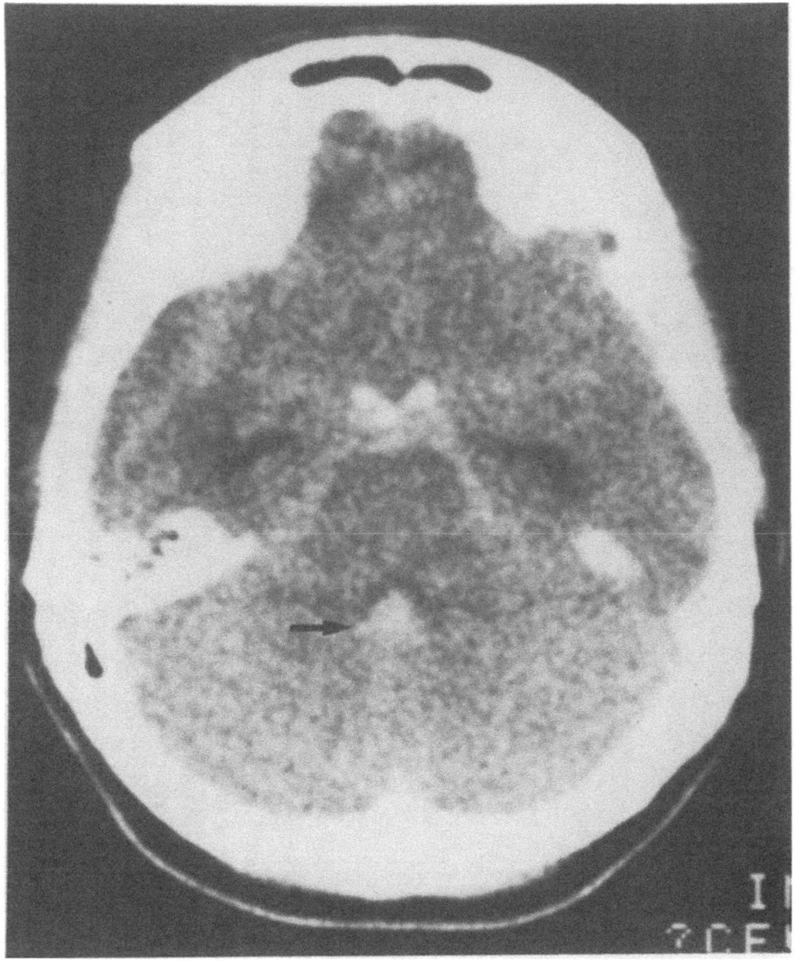

FIGURE 2–7A

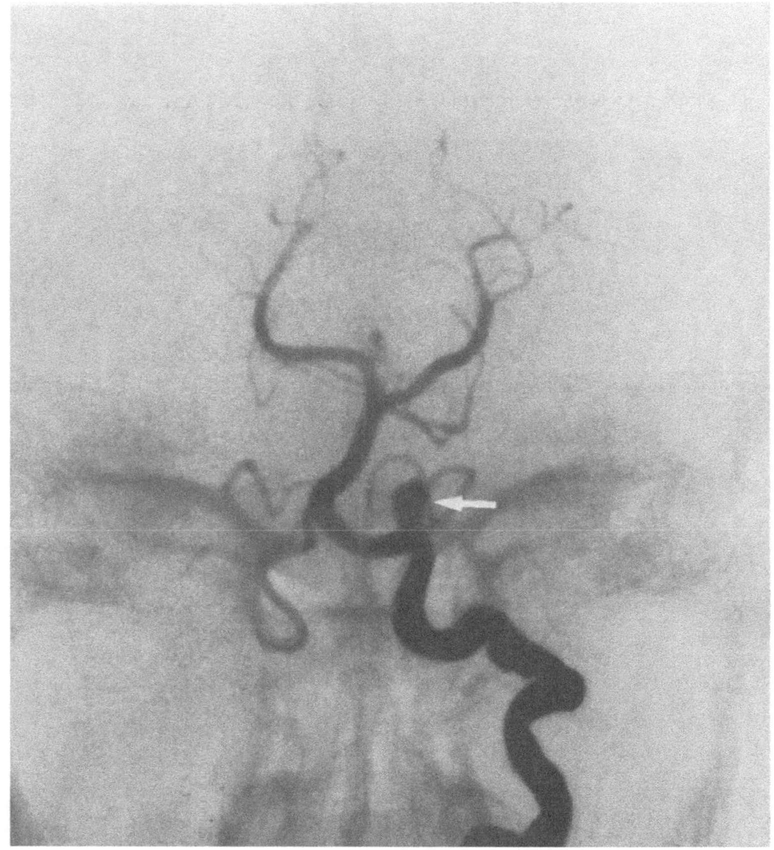

FIGURE 2–7B

FIGURE 2–7. Ruptured posterior inferior cerebellar artery aneurysm. This 50-year-old female presented with headache, nausea, and vomiting. (A) Plain CT. There is subarachnoid hemorrhage around the pons and in the fourth ventricle (arrow). There was no blood in the rest of the ventricular system. Such an isolated collection of blood in the fourth ventricle is highly suggestive of a ruptured vertebrobasilar system aneurysm. The to-and-fro pulsatile CSF flow allows retrograde entry of the blood through outlet foramina of the fourth ventricle into it. (B) Left vertebral angiogram, anterior–posterior view. The culpri: aneurysm (arrow) is well shown.

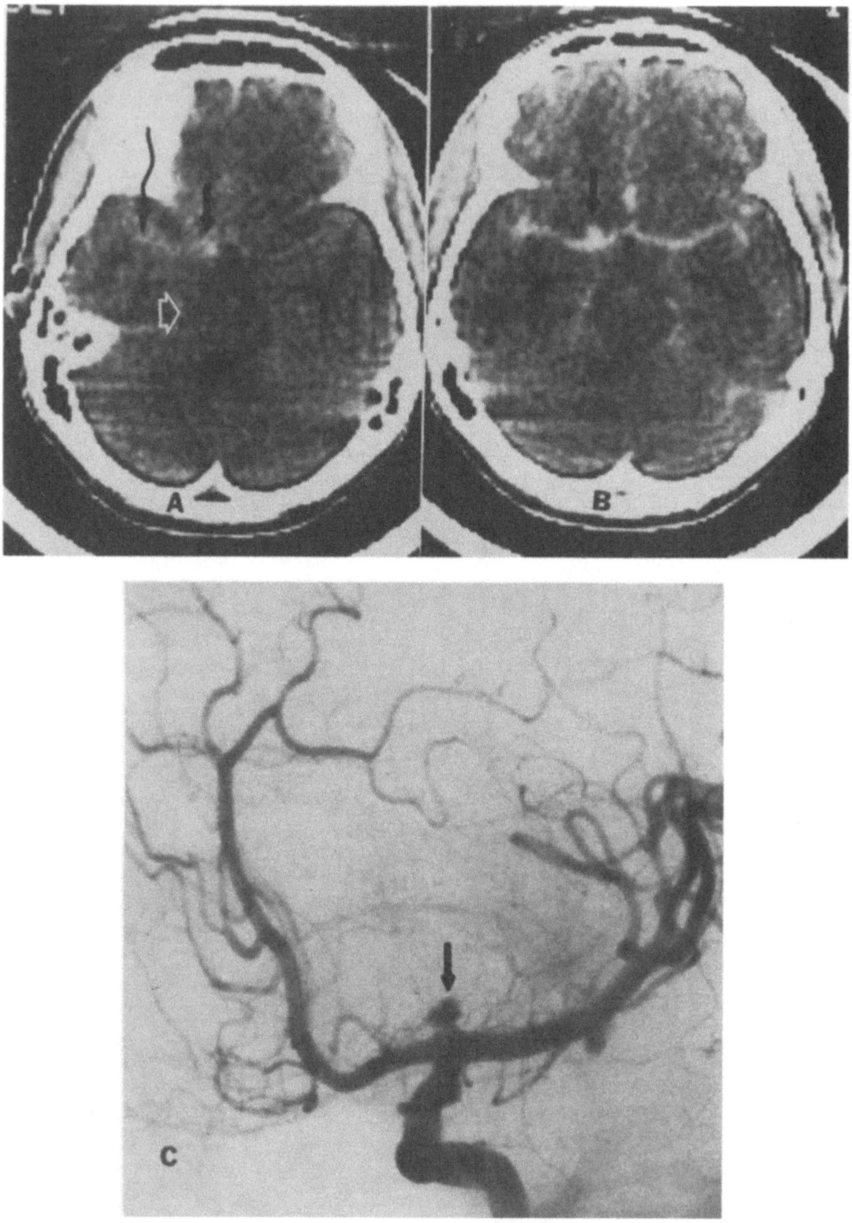

FIGURE 2–8. A mixed and hyperdense clot from a ruptured internal carotid artery bifurcation aneurysm. (A) A small clot is noted at the expected site of the internal carotid artery bifurcation (straight arrow). The adjoining sylvian fissure also contains slightly hyperdense clot (curved arrow). The isodense component of the subarachnoid hemorrhage in the cisterns around the brainstem is evidenced by their nonvisualization (open arrow). In (B), which is a contrast CT scan, the lumen of the aneurysm is now visualized (arrow). Somewhat shaggy contrast enhancement in the subarachnoid spaces is related to granulation tissue that is partaking in the phagocytosis of RBCs. In (C) the ruptured aneurysm (arrow) is shown; note its slight irregularity and multilocularity.

conventional cerebral angiography to opacify all four vessels so that, in the case of the patient harboring multiple aneurysms, there is a sufficiently satisfactory demonstration of the morphology of all aneurysms to enable one to decide which aneurysm has ruptured. The angiographic criteria [43] for detecting the culprit aneurysm in the presence of multiple aneurysms, in order of decreasing confidence level, are as follows: a) The largest and the most irregular (figure 2–10) or multilocular aneurysm is most likely to have ruptured

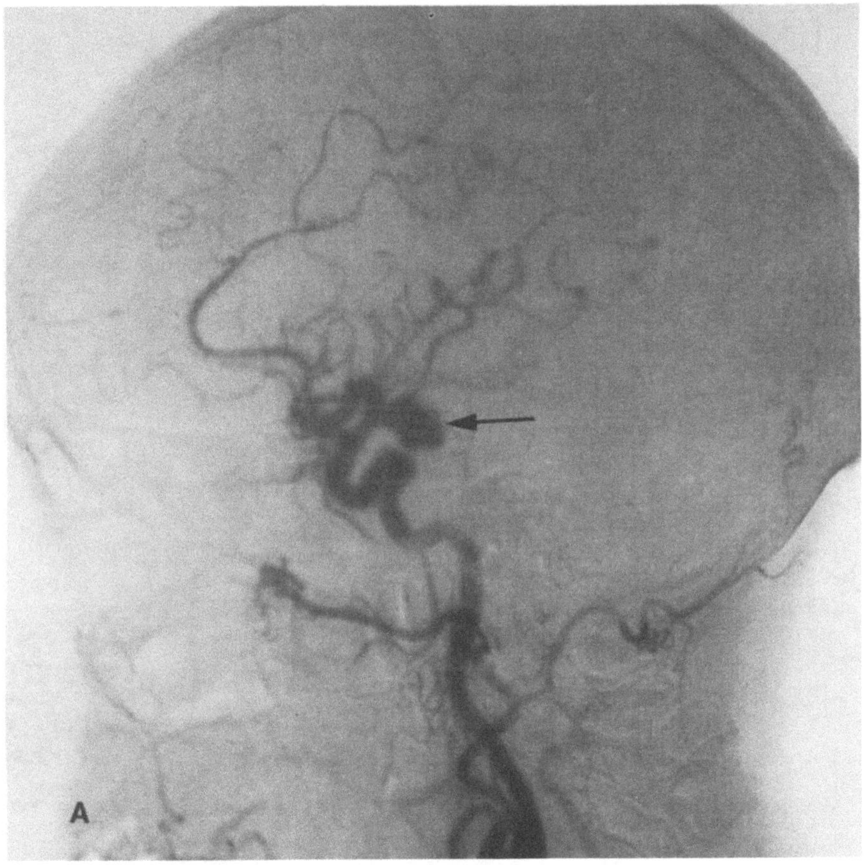

FIGURE 2–9A

FIGURE 2–9. Demonstration of the neck of a large aneurysm by dynamic CT scanning. A 1 × 1.6-cm aneurysm arising at the origin of the posterior communicating artery on the right side in a 63-year-old female. (A) Right carotid angiogram, lateral view. Note the lobulated aneurysm (arrow). In order to make a neurosurgical decision whether to clip the neck of the aneurysm or achieve its thrombosis by balloon occlusion of the internal carotid artery, it was necessary to demonstrate whether or not the aneurysmal neck was buried within the thrombosed portion of the lesion. A dynamic CT was done to obtain this information. (B) Serial plain CT images to show the thrombosed aneurysm (arrow) appearing as a mass lesion. (C) Dynamic contrast CT done at the level of the posterior communicating artery. This level was determined by looking at plain CT images and choosing the most appropriate image. 30 cc of contrast medium was injected as a bolus through an 18-gauge needle. Nine images were obtained. Scan time for each image was 4.8 seconds and intersection delay was 1.2 seconds. Note that the aneurysmal neck (arrow) can be clearly defined and that it is not enveloped by the thrombus. Reformatted sagittal and coronal imaging may provide further information on whether or not the neck is incorporated in the thrombus. MRI alone cannot provide this critical information.

(85% reliability); b) If there are two aneurysms of equal size, the one that shows most irregularity and locularity or a nipplelike protrusion at its dome is most likely to have rupture; and c) If neither of the above two criteria help, then the presence of focal vasospasm suggests the ruptured aneurysm. If the patient has survived a rebleed and if a subsequent angiogram is obtained, the aneurysm that has enlarged in the interim or shown some other change in its morphology is the one most likely to have ruptured. Extravasation of contrast material from the aneurysm is of course pathognomonic of a ruptured aneurysm, but this frightening phenomenon, fortunately for the angiographer, is only rarely observed. When observed it has perhaps no relation with angiography per se. If none of the above criteria help, then perhaps the aneurysm that is most likely to rupture stastically is chosen for surgical clipping. It should be remembered that the size of the ruptured aneurysm at angiography is less than actual, because the aneurysmal lumen becomes partially collapsed after rupture, or contains clot, or both [44].

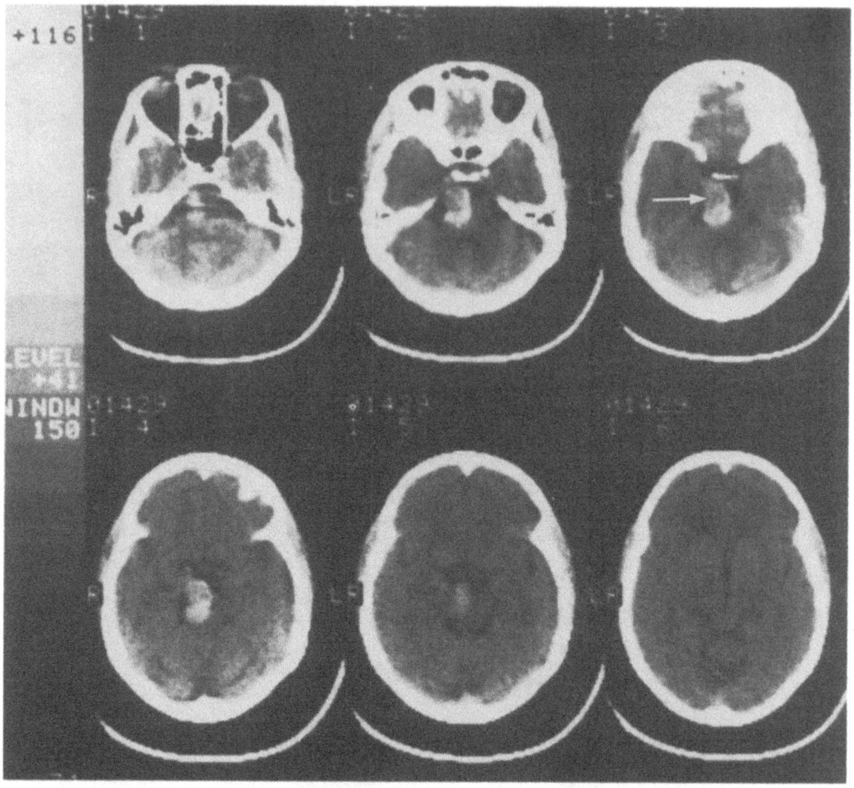

FIGURE 2–9B

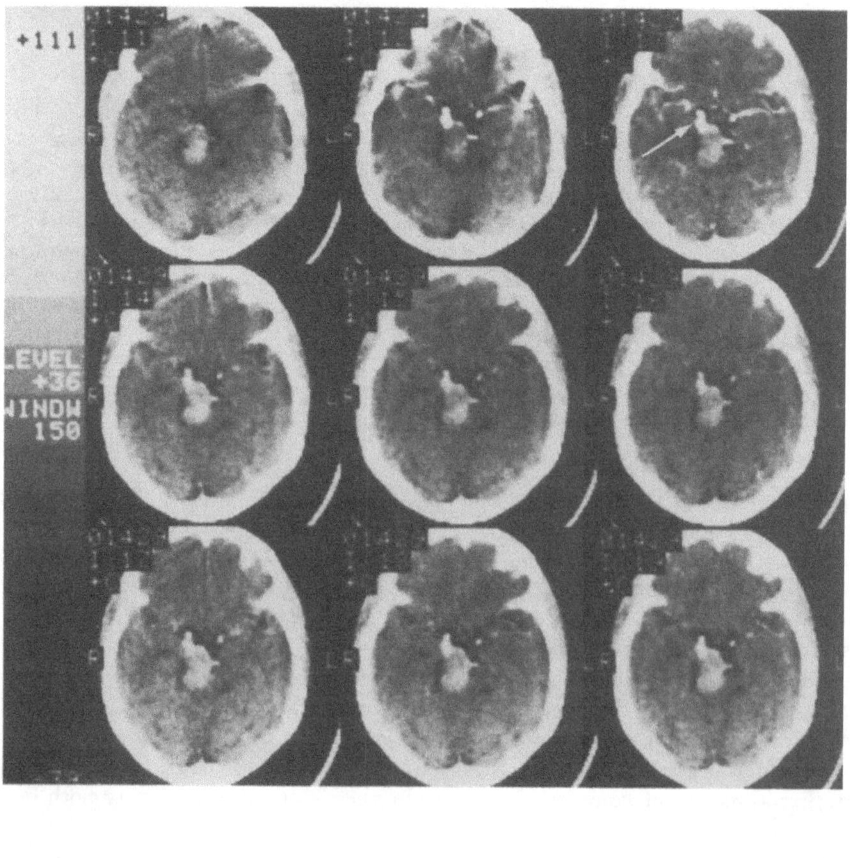

FIGURE 2–9C

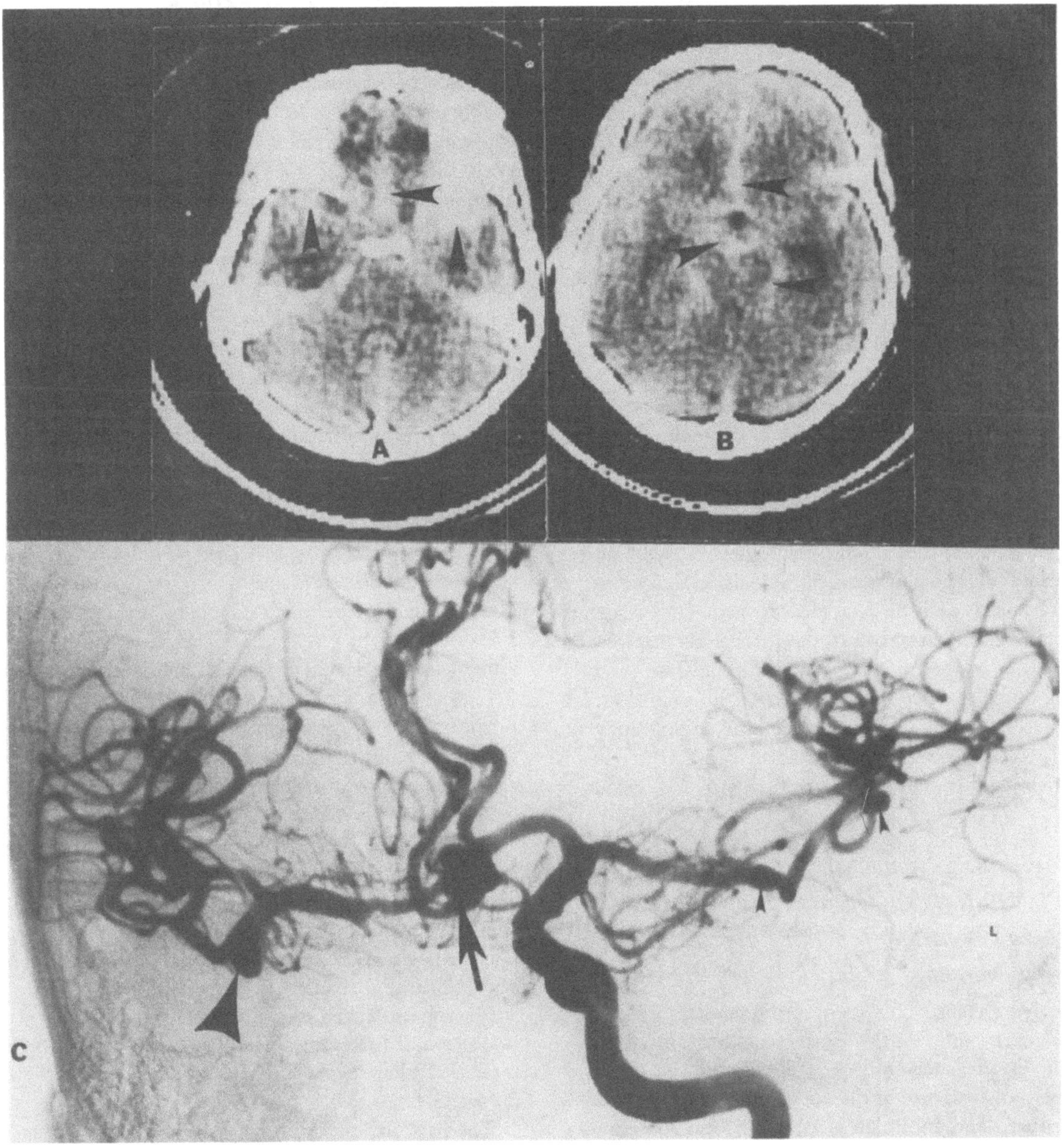

FIGURE 2–10. Usefulness of angiogram in detecting the ruptured aneurysm in the presence of multiple aneurysms. (A, B) Plain CT scan. There is only subarachnoid hemorrhage (arrowheads) without any parenchymal component. From CT it would be difficult to ascertain the culprit aneurysm. (C) Left carotid angiogram with compression of the right carotid artery. There are two aneurysms, one at the right middle cerebral artery bifurcation (large arrowhead) and the other at the anterior communicating artery (arrow). The larger size and irregularity of the anterior communicating artery aneurysm suggested that this was the symptomatic, ruptured aneurysm. (The small arrowheads on the left side are at arterial bends rather than aneurysms).

Timing of Cerebral Angiography

The timing of cerebral angiography should be done in consultation with the neurosurgeon. Most neurosurgeons wish to clip the aneurysm as soon as possible after the first episode of SAH, if the patient's clinical status permits. Using the grading system of Hunt and Hess [45] (table 2–1), most neurosurgeons would operate immediately if the patient's grade is I or II. That situation therefore calls for cerebral angiography as soon as possible. In grade III, the neurosurgeon may wish to obtain blood flow studies before

TABLE 2–1. Hunt and Hess Grading System [45]

Grade I	Minimal headache and slight neck stiffness.
Grade II	Moderate to severe headache, neck stiffness, no focal neurological deficit other than cranial nerve palsy.
Grade III	Drowsiness, confusion, mild focal neurological deficit.
Grade IV	Stupor, moderate to severe hemiparesis, possibly early decerebrate rigidity.
Grave V	Deep coma, decerebrate rigidity, moribund appearance.

requesting immediate cerebral angiography, since decreased cerebral blood flow is more likely to be attended with increased surgical morbidity or mortality. In grades IV and V there is certainly no need for immediate cerebral angiography; its timing depends upon the rate of neurological recovery of the patient and the neurosurgeon's decision to operate, if such a decision is made at all. If considerable spasm is present, such that the arterial lumen has diminished to 50% or more, it is wise to discontinue cerebral angiography since such a degree of luminal decrease denotes cerebral ischemia. Cerebral angiography or surgery under such circumstances carries the risk of causing additional neurological deficit.

Nonidentification of an Aneurysm in the Initial Angiogram in the Presence of SAH Well Substantiated by Clinical Symptomatology and by Lumbar Puncture

In the presence of a technically satisfactory angiogram, the following may account for nonopacification of an aneurysm if indeed it was the cause of SAH: a) small size of the aneurysm, which may have ruptured and annihilated itself, b) local spasm, disallowing good contrast entry into the aneurysm lumen, and c) spontaneous thrombosis, a rare phenomenon. The view of this author is that if the cerebral angiogram is of high quality and if there is no spasm, the chance of finding an aneurysm on a follow-up angiogram done two to three weeks after initial SAH is less than 1%. However, a follow-up angiogram is recommended for medicolegal reasons and for possibly identifying a nonaneurysmal cause of SAH (e.g., small arteriovenous malformation, arteritis) that may not have been manifest on the initial angiogram. Brismar et al. [46] have suggested that if no cause of SAH is detected on an initial angiogram, a good prognosis for the patient is likely.

Magnetic Resonance Imaging of Intracranial Aneurysms

The role of magnetic resonance imaging (MRI) in the investigation of intracranial aneurysms is still evolving. Many of these patients are quite sick and need intensive care that necessitates placement of various monitoring devices. Such circumstances usually preclude optimal conditions for MRI. Nor does it appear necessary that these patients be examined by MRI after sustaining acute SAH; CT has proved its ability to reveal SAH and parenchymal clot quite effectively. Because of the presence of deoxyhemoglobin in the intact RBCs in acute SAH, the blood may be isointense with CSF and thus elude recognition [47]. However, after several days it will become hyperintense, due mainly to the presence of paramagnetic methemoglobin in the clot [48]. The appearance of the parenchymal clot has been described by Zimmerman in chapter 1 of this volume.

MRI provides exquisite contrast. On thin slices obtained at the level of the Circle of Willis, MRI has the capability to reveal arterial anatomy quite well. Aneurysms 5–8 mm in size can be identified (figure 2–11). MRI may therefore be used to follow up unruptured incidental intracranial aneurysms less than 10 mm in size that have a very low probability of subsequent rupture [13]. A clot within an aneurysm that may cause embolic stroke (figure 2–12) appears as an area of hyperintensity surrounded by low-intensity flowing blood [49]. In a situation where there is no clinical or CT clue to the ruptured aneurysm in the presence of multiple aneurysms, one is then left to rely on angiographic criteria that are not infallible in order to recognize the offending aneurysm. In such a circumstance, the high-contrast resolution of MRI can detect a small clot as a hyperintense signal near the ruptured aneurysm [50] (figure 2–13).

The partially thrombosed giant aneurysm displays a variegated MRI morphology mirroring its complex pathology (figure 2–14). Pathologically, the giant aneurysm contains a lumen enveloped by a thrombus at various stages of evolution containing degradation products of RBCs. The thrombus contains tiny and fragile capillaries as part of the phagocytic process. They can rupture and cause repeated small hemorrhages within the mural thrombus. The lumen and the thrombus complex of the giant aneurysm is bordered by edema and/or gliosis. Calcification, usually rimlike, may be present [51]. These pathologic features can be readily seen on MRI. The parent vessel and the residual aneurysmal lumen of the partially thrombosed aneurysm show signal void on

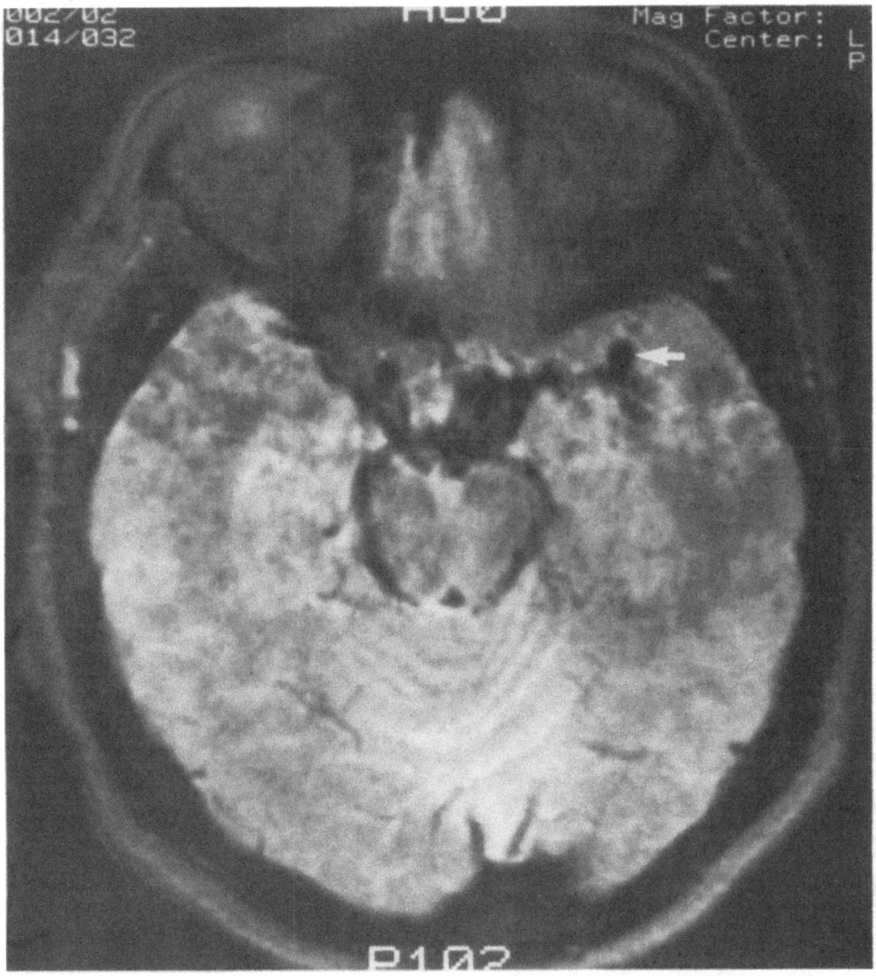

FIGURE 2–11. Asymptomatic aneurysm. A 64-year-old female with an asymptomatic 4 × 6-mm-size aneurysm (arrow) at left middle cerebral artery bifurcation. T2-weighted axial image (TR 2000 msec, TE 80 msec).

spin-echo imaging since the protons do not remain within a particular section for both the 90° and 180° pulses. The turbulent flow within the giant aneurysmal sac exhibits signal (figure 2–15) (table 2–2). Occasionally the lumen of the giant aneurysm on T1-weighted images may be isointense with the brain (figure 2–16). The rim that is interposed between the lumen and the thrombus shows high intensity on both T1- and T2- weighted images due to the enzymatic oxidation of the deoxyhemoglobin to methemoglobin, since this region lies in close proximity to the oxygenated, flowing blood [52]. The multilaminated mural thrombus contains varying components of methemoglobin, hemosiderin, and deoxyhemoglobin in intact RBCs [51, 52, 53], accounting for a varying combination of low- and high-intensity signals. A discrete peripheral rim of low intensity may represent predominantly hemosiderin-laden macrophages, deoxyhemoglobin in intact RBCs at the most peripheral part of the laminated mural thrombus [48, 53], calcification, or a combination of all these. The parenchymal edema that surrounds the giant aneurysm appears as a high-signal area. The gliosis also appears as a high-signal area due to the presence of tiny fluid-containing spaces that are part of the morphology of gliosis. Perianeurysmal small hemorrhages appear as high-intensity areas. In the commonly employed spin-echo technique of MRI, the symmetric echos of a multi-echo train may display a high signal within the giant aneurysm due to slow laminar flow [54]. This phenomenon is much reduced or totally absent for asymmetric echoes. It is also reduced when motion artifact suppression technique (MAST) is used.

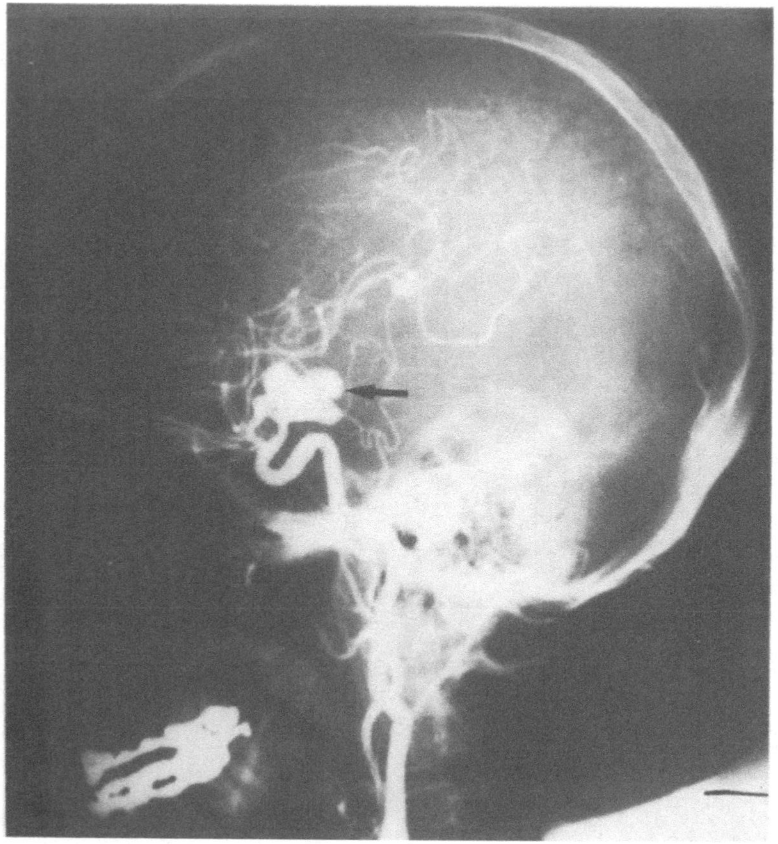

FIGURE 2–12A

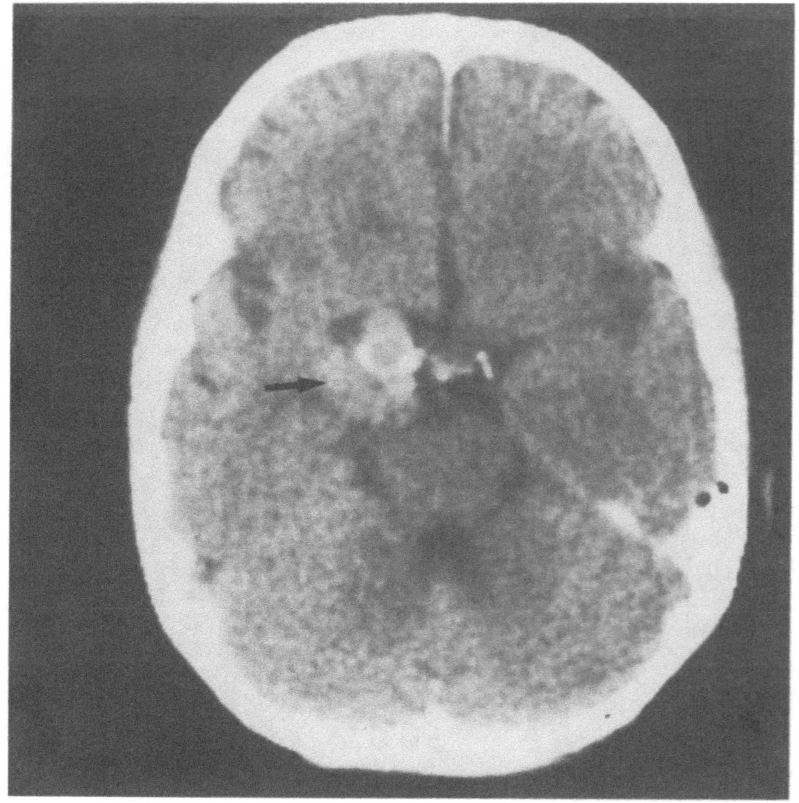

FIGURE 2–12B

FIGURE 2–12. Giant aneurysm. A 63-year-old female presented with decreased vision on the right side. Partly thrombosed 2 × 2.5-cm aneurysm at origin of the anterior choroidal artery. (A) Right carotid angiogram, lateral view. Note the multilocular aneurysm (arrow). (B) Plain CT. Partly calcified and thrombosed aneurysm appears as an isodense mass lesion (arrow). (C) Contrast CT. The aneurysmal lumen opacifies (arrowhead). (D) Sagittal reformatted contrast CT. The aneurysmal lumen (arrow) is buried in the thrombus. Oblique and coronal reformatted images (not shown) provided the same information. This critical information led the neurosurgeon to treat the aneurysm by balloon occulusion of the right internal carotid artery, since he could possibly not clip the aneurysmal neck that was incorporated in the thrombosed portion of the aneurysm. (E) Intermediate-weighted (TR 2000 msec, TE 30 msec) coronal MRI. The patent aneurysmal lumen appears as signal void (arrow) and the thrombosed portion (arrowhead) as high signal. (F, G) T2-weighted (TR 2000 msec, TE 67 msec) coronal MRI. Note that there is no apparent change in signal in the patent lumen (arrow) and the thrombosed portion (arrowhead) of the aneurysm. However, on this pulse sequence the gliosis and/or edema in the neural tissue that surrounds the aneurysm appears as an area of high signal intensity (open arrow). (H) Intermediate-weighted (TR 2000 msec, TE 30 msec) axial images at the level of the centrum semiovale. Note the multiple small hyperintense areas (arrows), probably representing infarctions resulting from emboli dislodged from the aneurysm.

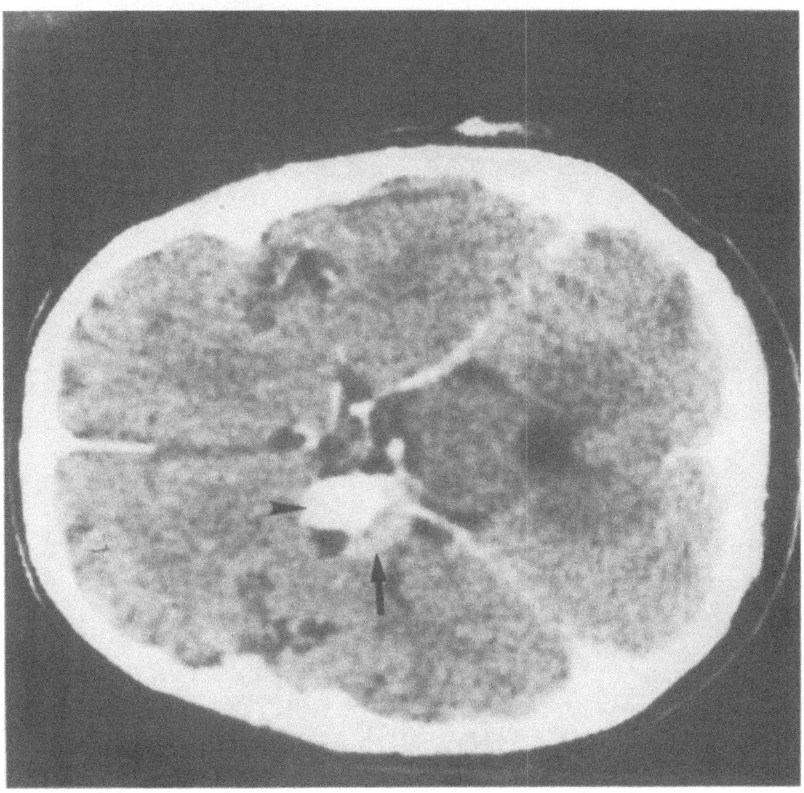

FIGURE 2–12C

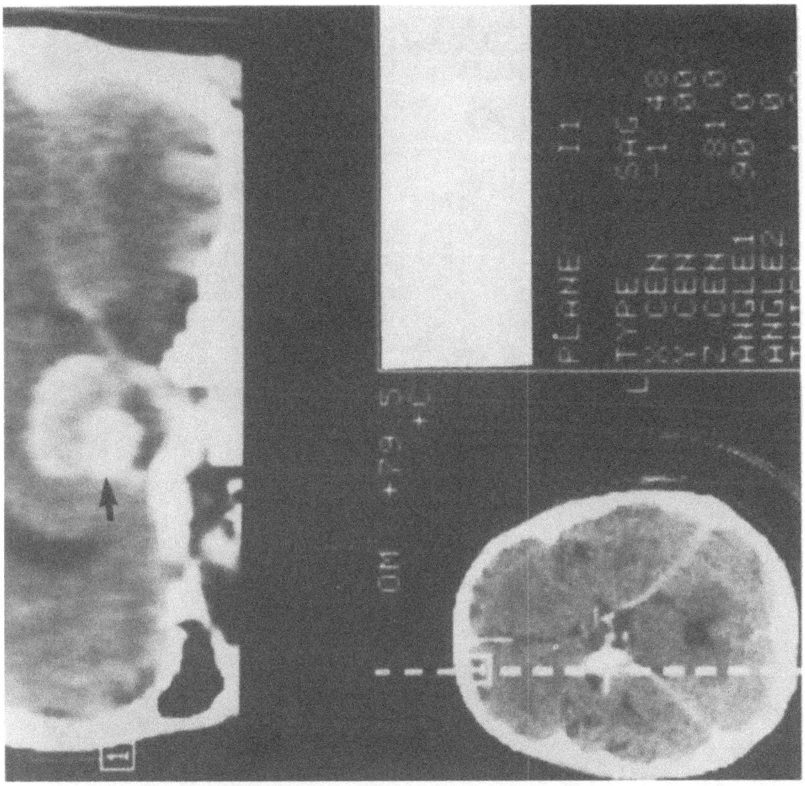

FIGURE 2–12D

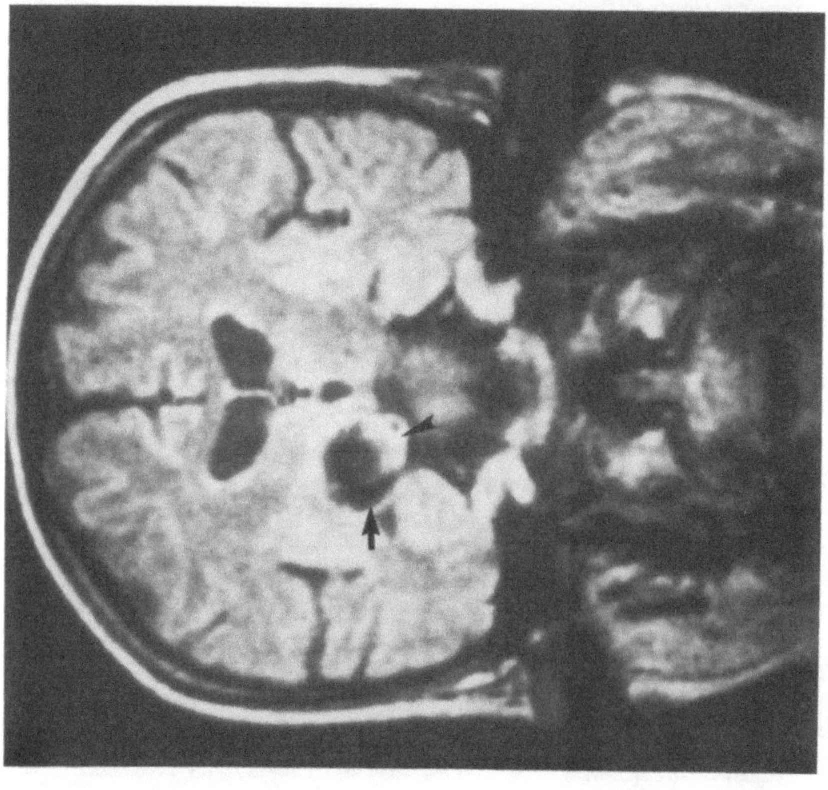

FIGURE 2–12E

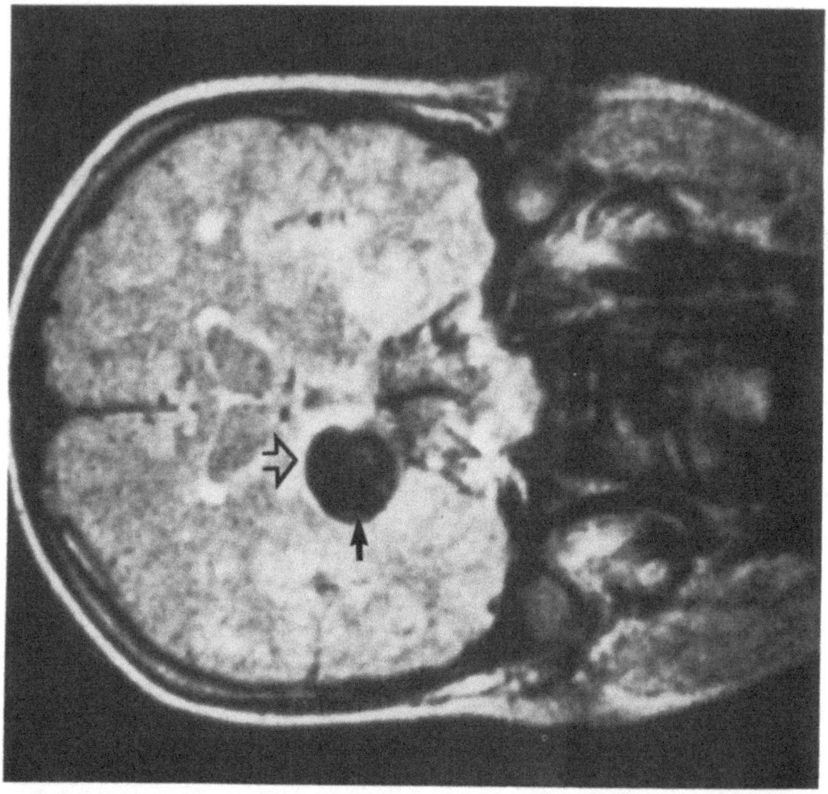

FIGURE 2–12F

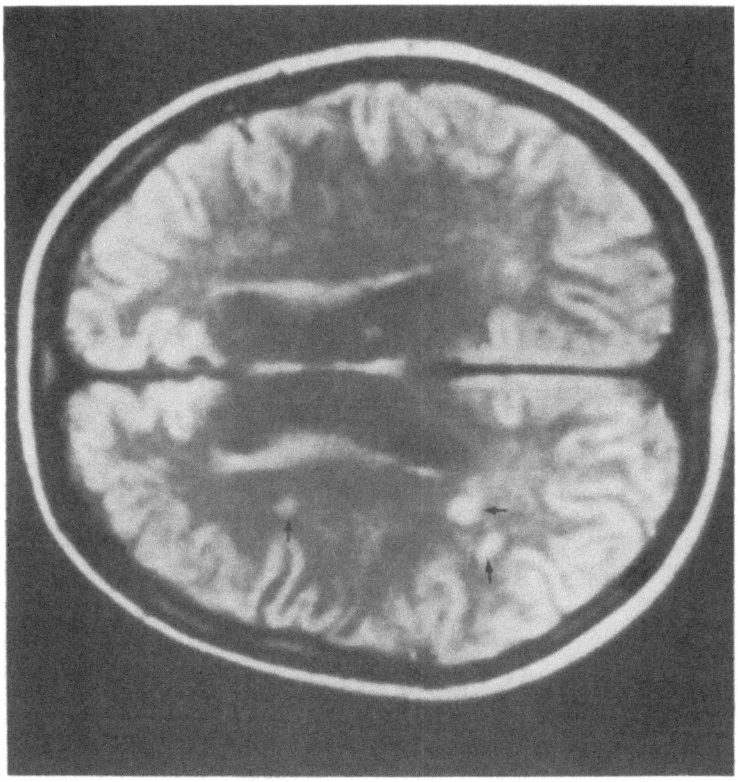

FIGURE 2–12H

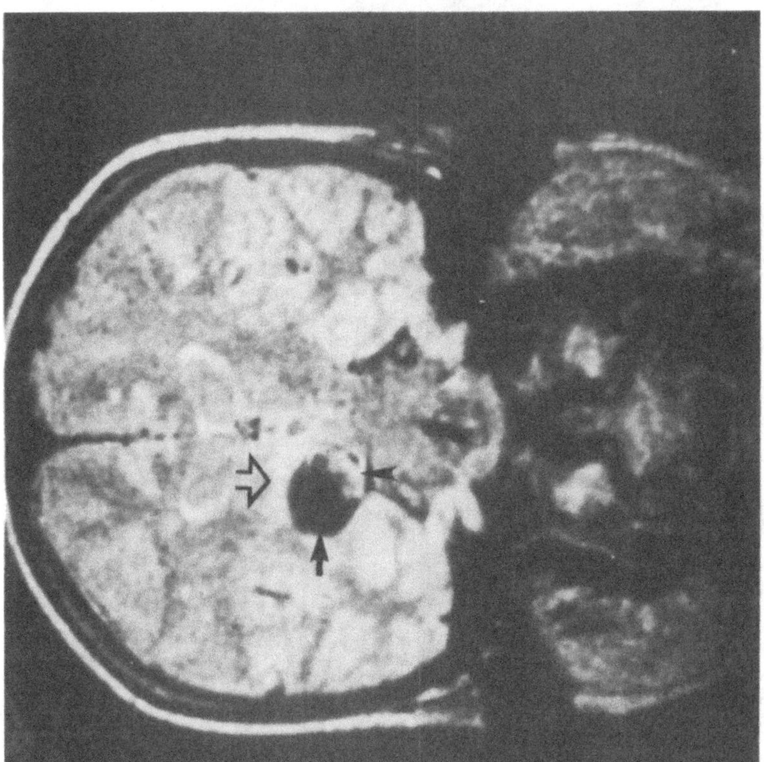

FIGURE 2–12G

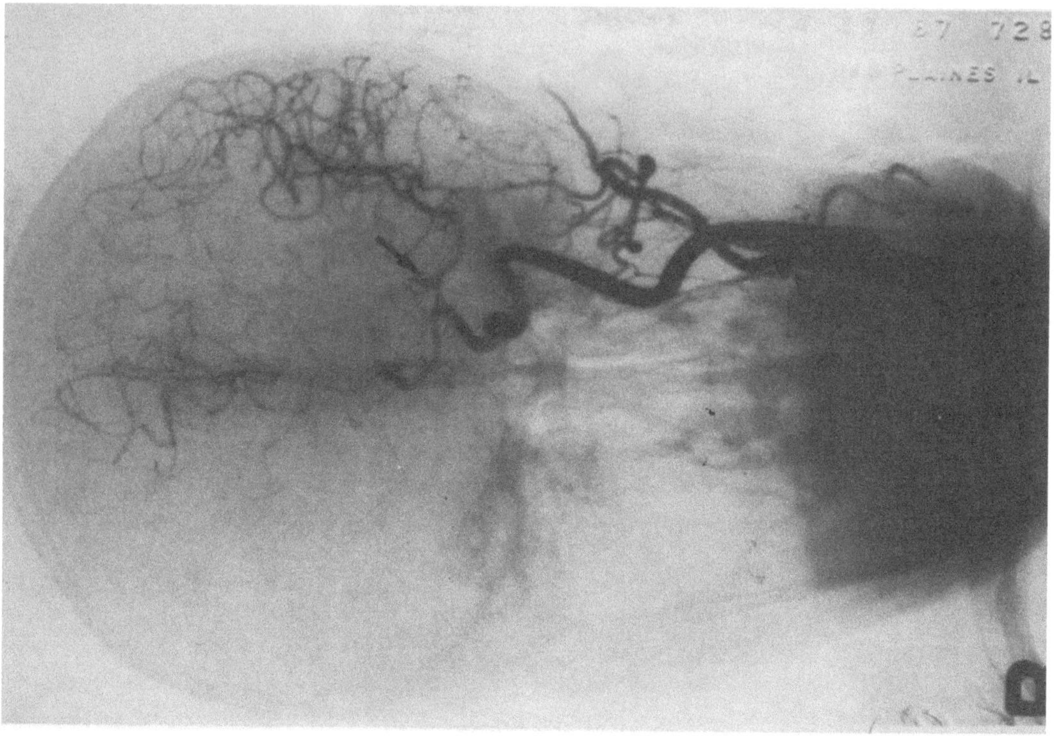

FIGURE 2–13B

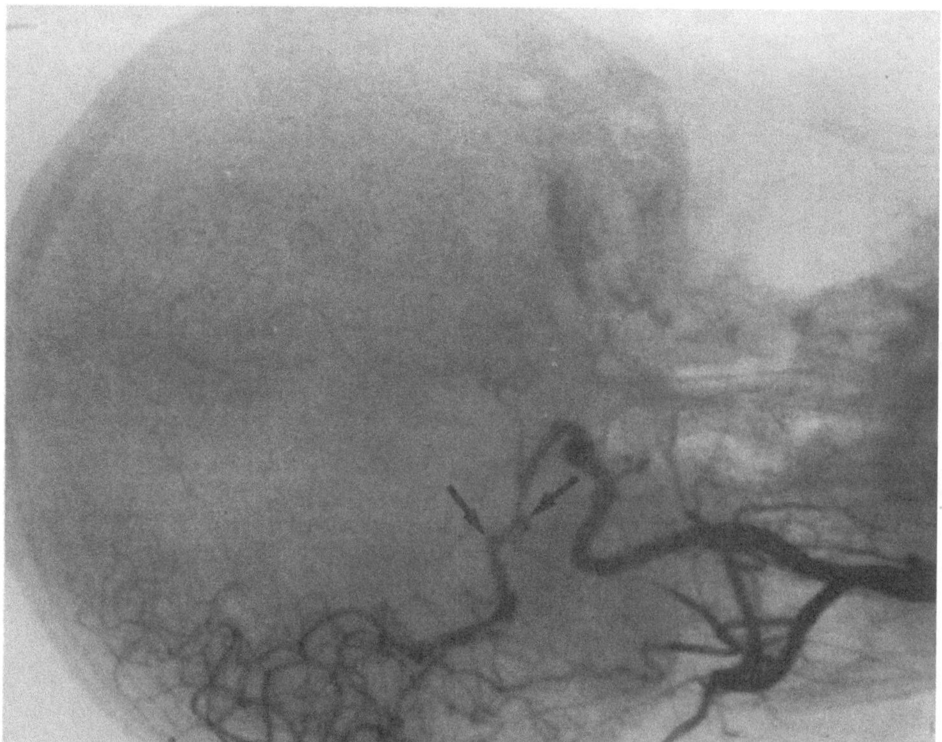

FIGURE 2–13A

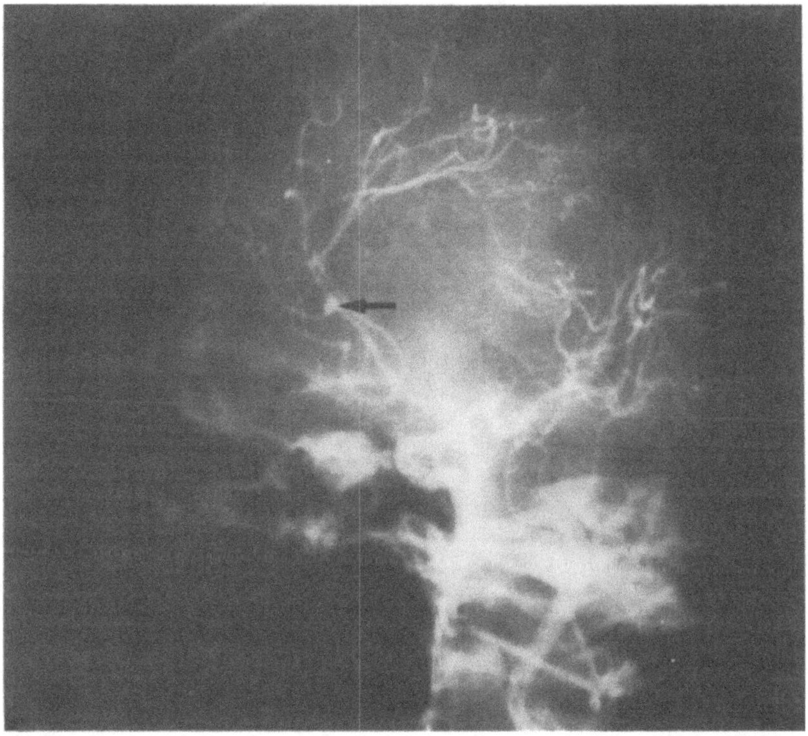

FIGURE 2–13C

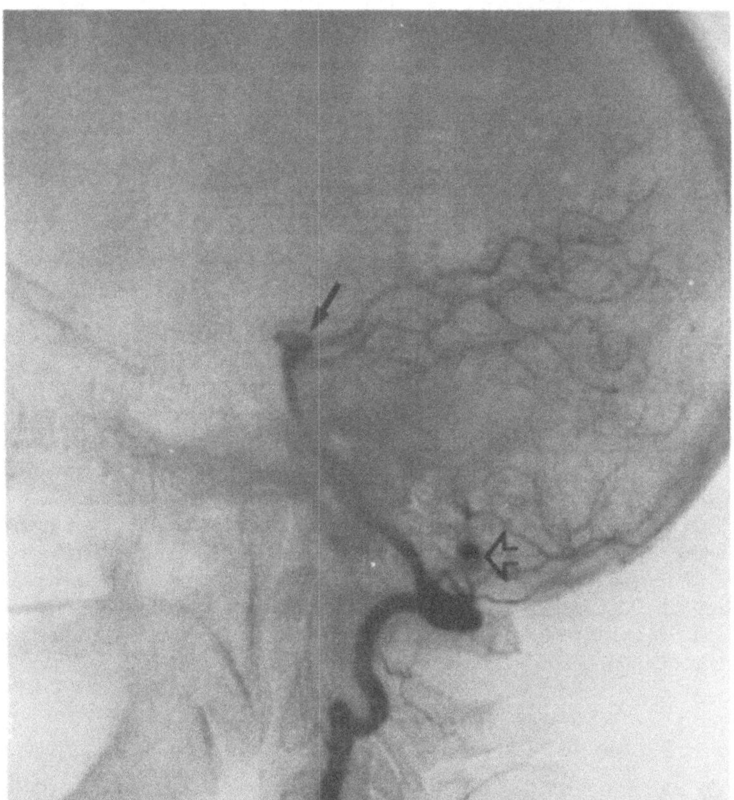

FIGURE 2–13D

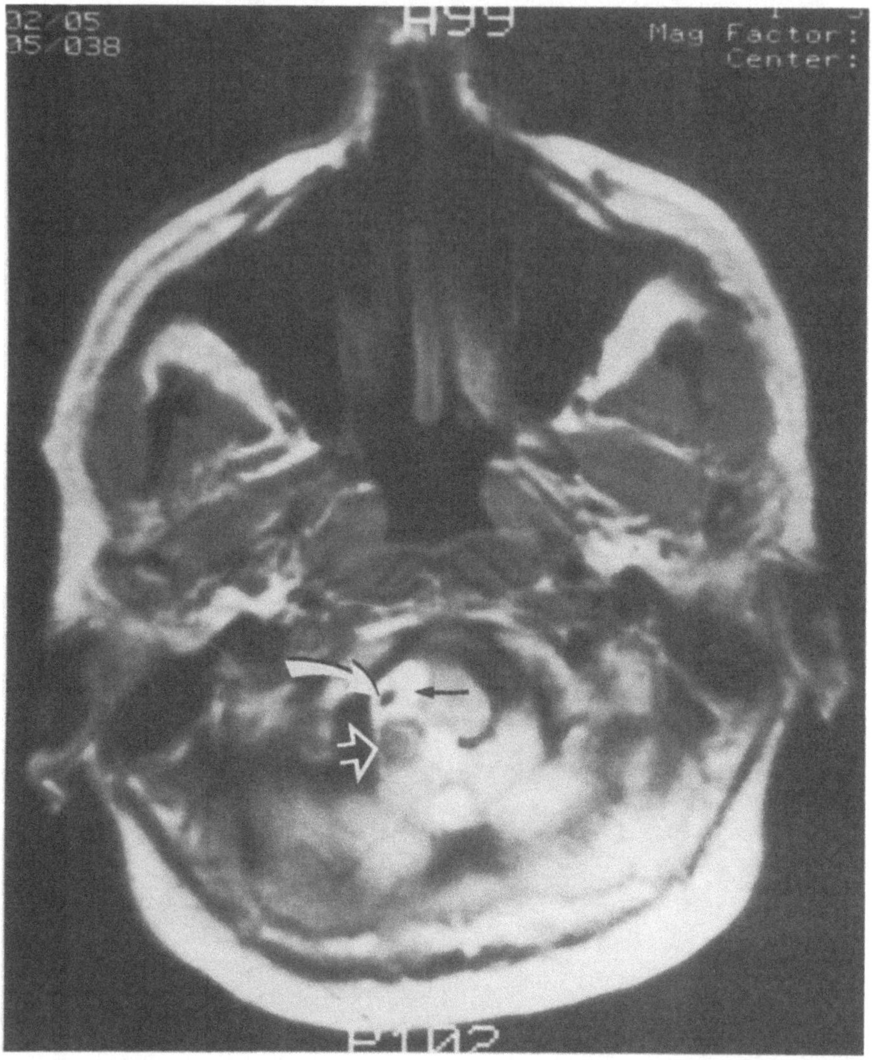

FIGURE 2–13E

FIGURE 2–13. Role of MRI in revealing the culprit aneurysm in the presence of multiple aneurysms. This 40-year-old female presented with sudden severe occipital headache and neck stiffness. A CT scan done on the second after ictus revealed no subarachnoid hemorrhage (SAH). Lumbar puncture confirmed SAH. Bilateral carotid and right vertebral angiograms ((A) right carotid, (B) left carotid, (C) oblique view of the left carotid, and (D) right vertebral) demonstrated multiple aneurysms (arrows). Localized high-intensity (due to methemoglobin) clot (arrow in (E)) on intermediate-weighted (TR 2000 msec, TE 20 msec) MRI (E) was seen near the unusual aneursym that arose from the right posterior inferior cerebellar artery (open arrow in (D)). This finding suggested that this was the offending aneurysm. This observation was confirmed surgically. This aneurysm measured 6 × 8 mm on the angiogram and 10 mm on MRI (open arrow). This discrepancy in size is explained by two factors: (a) one sees only the lumen of the aneurysm on the angiogram, and (b) the thrombosed part of the aneurysm contributes to the larger size of the aneurysm seen on MRI. Note that, as expected, the largest aneurysm had bled. The curved arrow in (E) is at the posterior inferior cerebellar artery trunk.

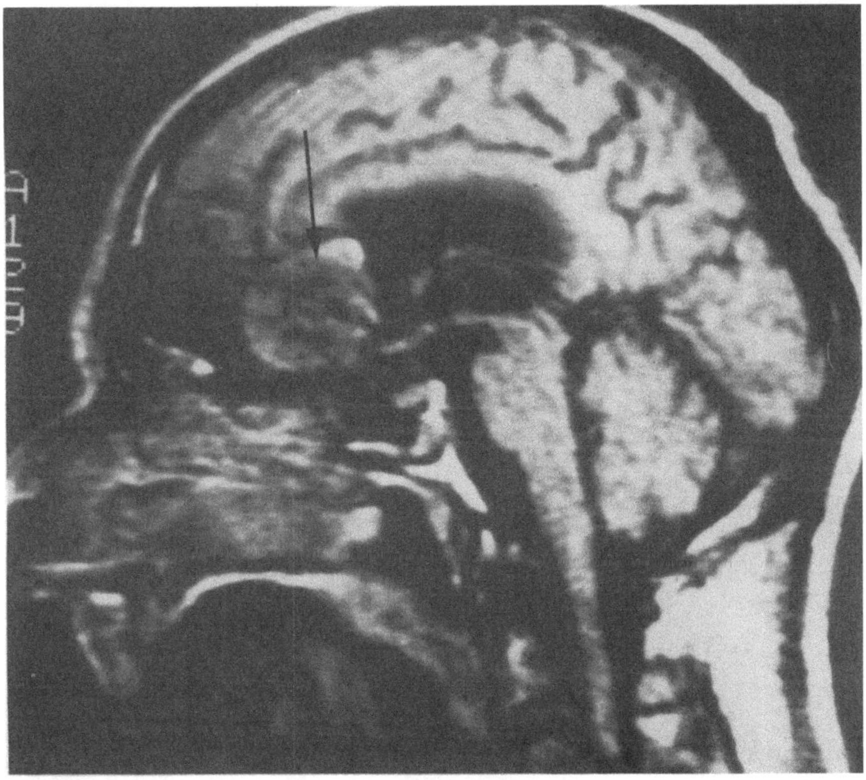

FIGURE 2–14. Giant thrombosed aneurysm of anterior communicating artery. T1-weighted (TR 500 msec, TE 25 msec) sagittal MRI showing a variegated morphology of the aneurysm (arrow). The laminated mixed signal represents a combination of thrombosis, small hemorrhages of different ages, hemosiderin, and calcium.

TABLE 2–2. Magnetic Resonance Imaging Morphology of Giant Intracranial Aneurysms

Component of partially thrombosed giant aneurysm	MR signal on spin-echo technique	
	T1-weighted	T2-weighted
1. Parent lumen	Signal void.	Same.
2. Aneurysm lumen	Predominantly decreased signal or mixed signal due to turbulent flow.	Same.
3. Area interposed between patent lumen and mural thrombus	High signal because of methemoglobin.	Signal may become more intense.
4. Mural thrombus	Layers of low-, and high-intensity signal due to intermingled methemoglobin and hemosiderin.	Signal from methemoglobin may become more intense.
5. Rim of aneurysm	Low signal due to hemosiderin in macrophages, deoxyhemoglobin in intact RBCs, calcium, or combination of these.	Same.
6. Surrounding brain parenchyma	Edema causes high signal. Gliosis also causes high signal due to presence of tiny cystic spaces within it.	Signal becomes more intense.

Large calcified lesions close to the expected location of intracranial aneurysms can appear as signal-void areas, thus mimicking giant aneurysms (figure 2–17). This mistake can be corrected by obtaining

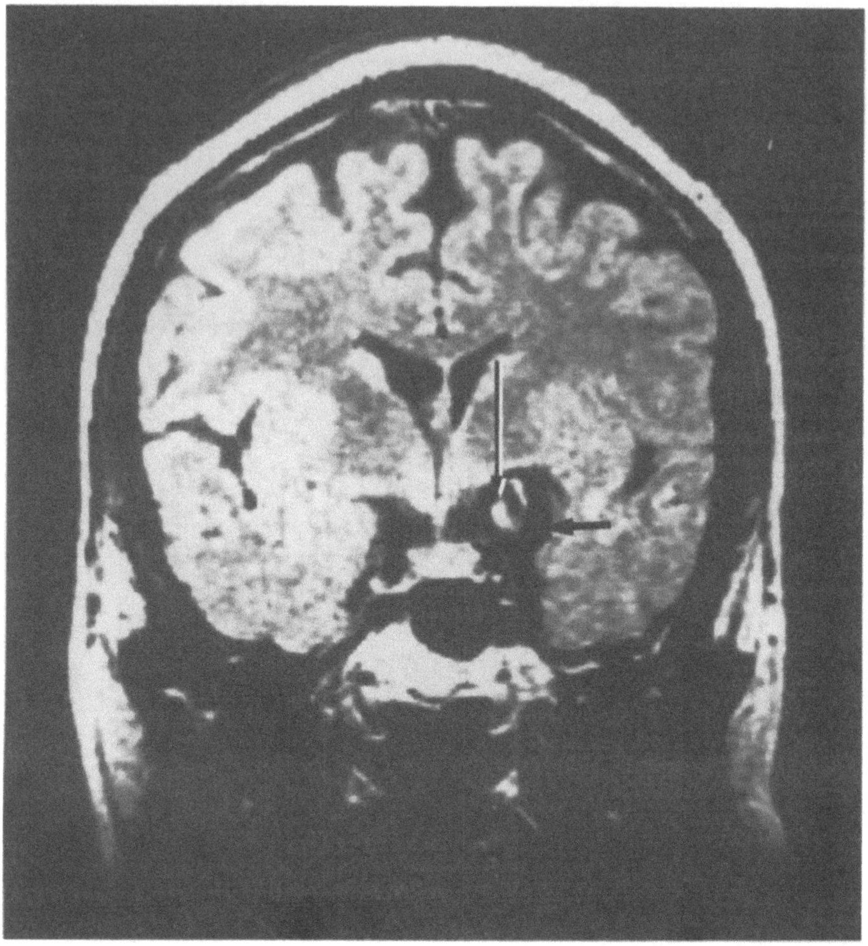

FIGURE 2–15A

FIGURE 2–15. Hyperintense core of nonthrombosed giant aneurysm on MRI. A giant aneurysm at early bifurcation of the left middle cerebral artery. (A) Intermediate-weighted (TR 2000, TE 20 msec) coronal image. The patent lumen (thick arrow) of the aneurysm appears as signal void and the swirling blood (thin white–black arrow) within the lumen appears as a mixed intensity signal. (B) Plain CT. Note the slightly hyperdense lesion (arrow). (C) Contrast CT. The entire lesion opacifies (arrow). No thrombosed portion is noted.

GRASS (gradient recalled acquisition at steady state), which is a fast pulse sequence (TR 24/TE 12) that depicts flowing blood as a hyperintense signal.

The tortuous course of dolichoectatic vessels is difficult to demonstrate on CT even by obtaining reformatted sagittal or coronal images. Before MRI, it was therefore necessary to resort to angiography. This situation no longer exists. MRI can very effectively demonstrate the entire complex morphology of the dolichoectatic vessels (figure 2–18).

Hydrocephalus that follows SAH and may require shunting gives an abnormal high-intensity periventricular signal [55] due to transependymal escape of water from CSF.

Conclusion

At present, CT and angiography remain the primary modalities for clinical management of patients with aneurysmal SAH. In a rare situation where the patient has multiple aneurysms and there are no clinical, CT, or angiographic clues to the ruptured aneurysm, MRI may provide some useful information because of its high-contrast ability. The effective role of MRI in depicting the complex morphology of partially thrombosed giant aneurysms is now established. In the future, when MRI becomes more readily available and nonmagnetic devices are de-

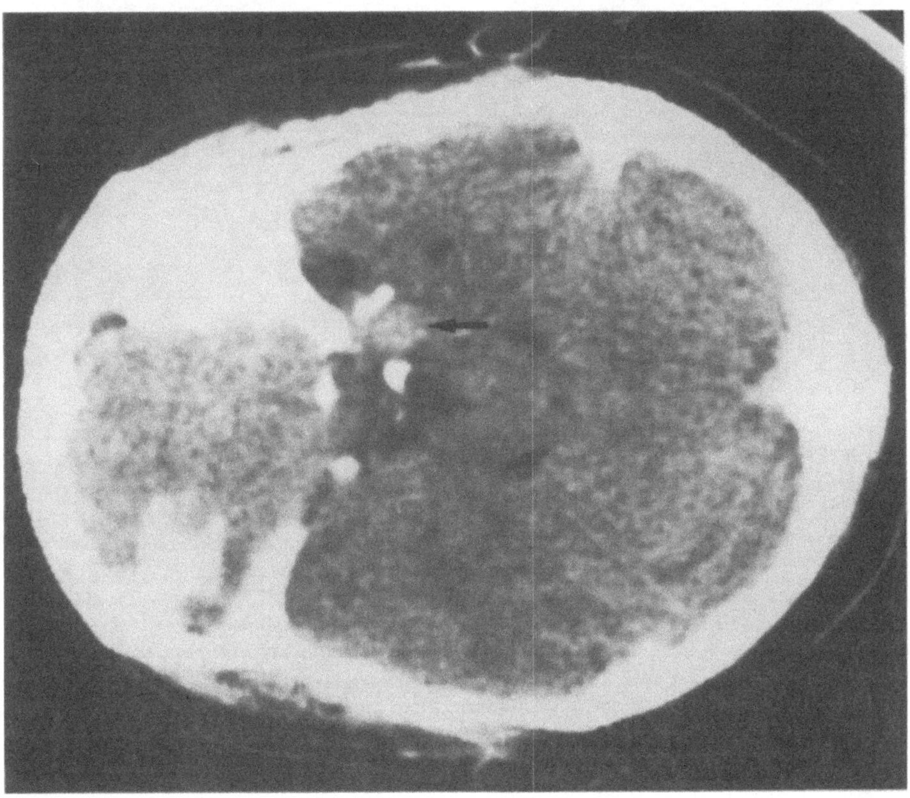

FIGURE 2–15B

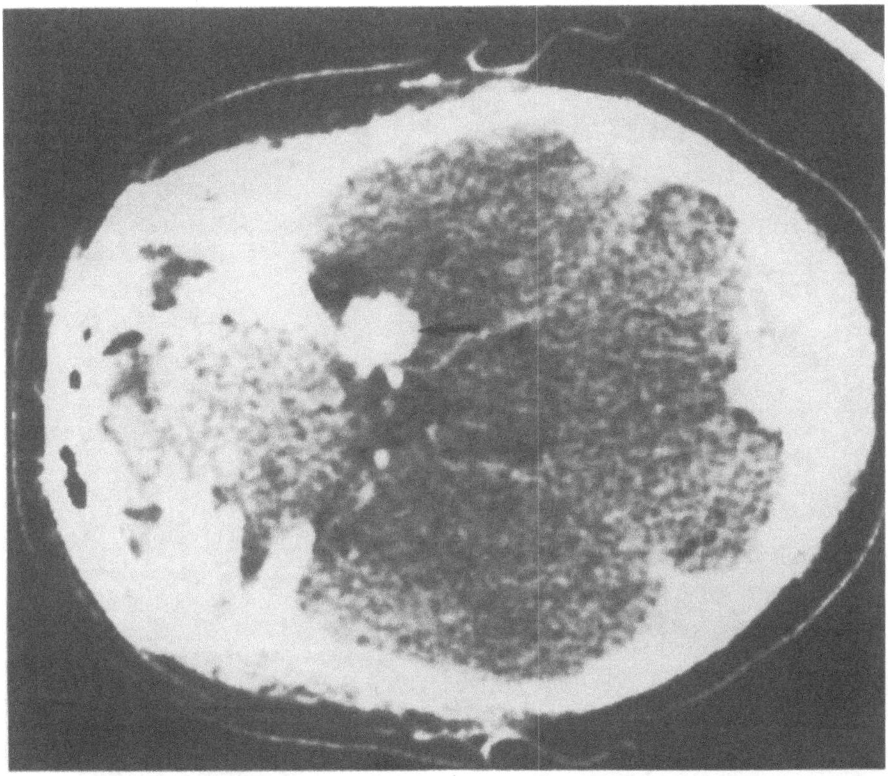

FIGURE 2–15C

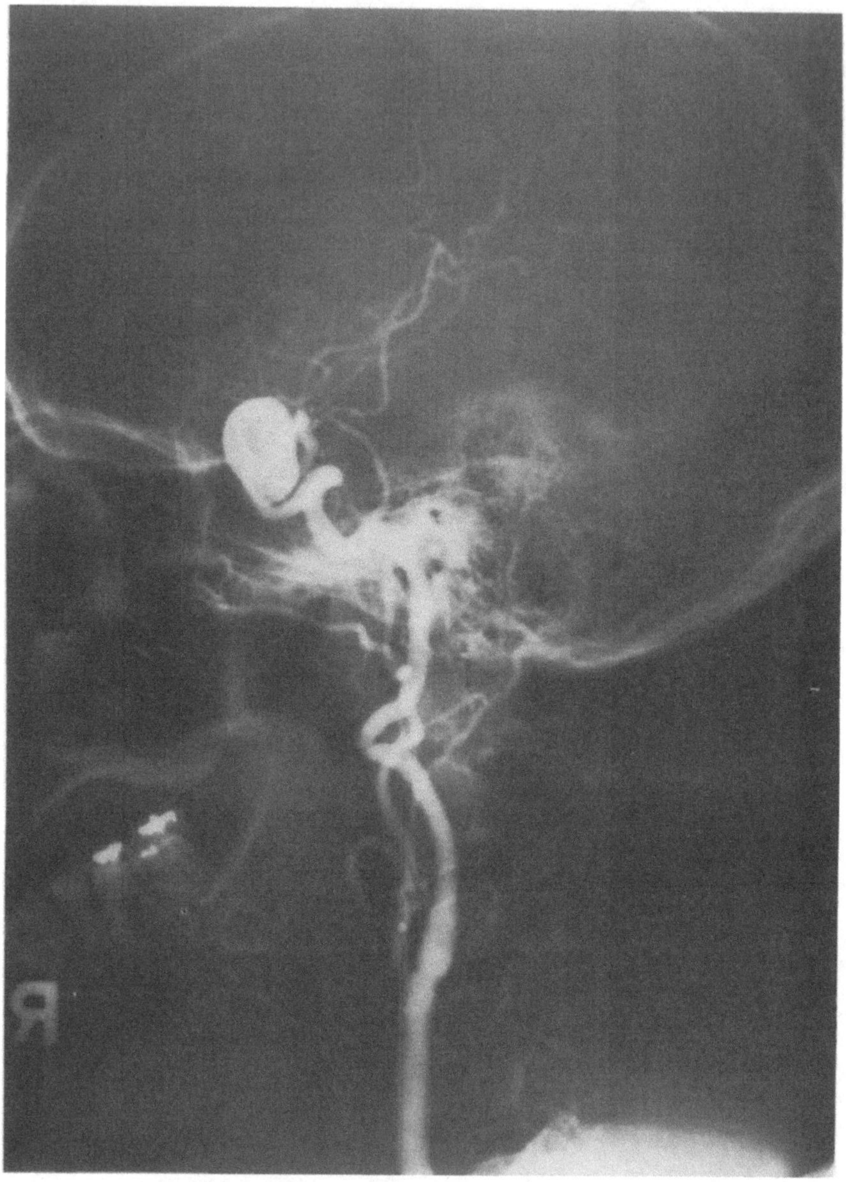

FIGURE 2–16A

FIGURE 2–16. Giant aneurysm. A 54-year-old female presented with almost total blindness in the right eye. Giant 1.8 × 2.5 cm aneurysm at the origin of the right ophthalmic artery. (A) Right carotid angiogram, lateral view. Note the giant aneurysm. (B, C) Partial saturation T1-weighted (TR 1000 msec, TE 25 msec) coronal images. There is an isointense lesion in the right cavernous sinus and the neighboring suprasellar cistern (arrow). In the image (C) that is 10 mm posterior to (B), the carotid artery (arrowhead) is seen as a signal-void structure. On this pulse sequence, a reliable diagnosis of an aneurysm cannot be made. A tumor such as meningioma could mimic this morphology. (D) Intermediate-weighted (TR 2000 msec, TE 20 msec) axial image. Note the giant aneurysm as a large signal-void structure (arrow). (E) T2-weighted (TR 2000 msec, TE 80 msec) axial image. The Circle of Willis and the aneurysm (arrow) are again clearly shown. (F) GRASS (Gradient-recalled acquistion in a study-state mode) axial images (TR 22 msec, TE 12 msec). The aneurysm appears as a hyperintense lesion (arrow). (G) Sagittal T1-weighted (TR 500 msec, TE 25 msec) image obtained after balloon occlusion of the right internal carotid artery to induce thrombosis within the aneurysmal lumen. Note the layering (arrow) within the aneurysm. This luminal layering is probably best explained by gravitational layering of the cellular elements of blood and slower flow in the dependent part of the giant aneurysm. No such layering was present on MRI done before balloon occlusion of the right internal carotid artery.

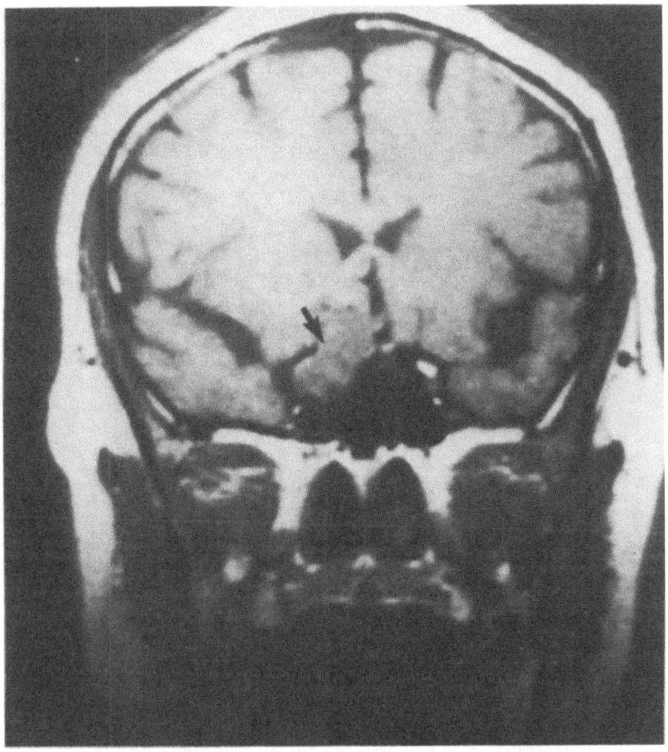

FIGURE 2–16B

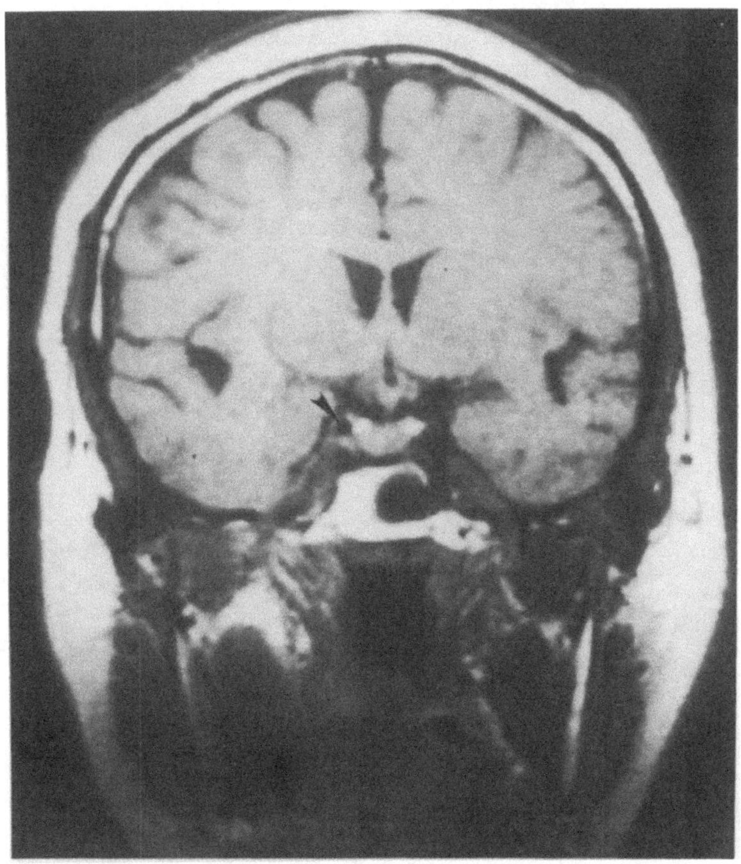

FIGURE 2–16C

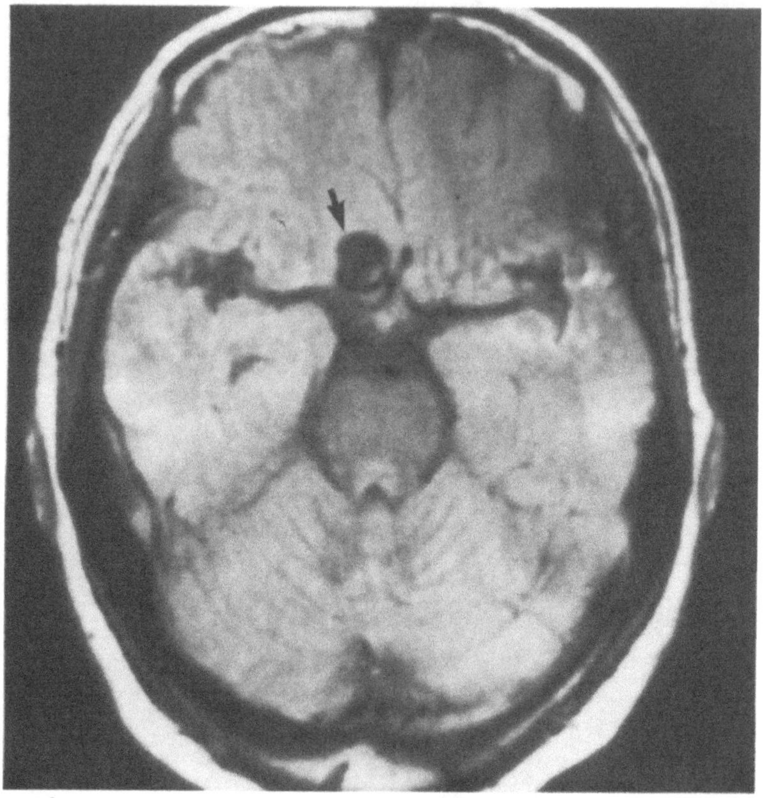

FIGURE 2–16D

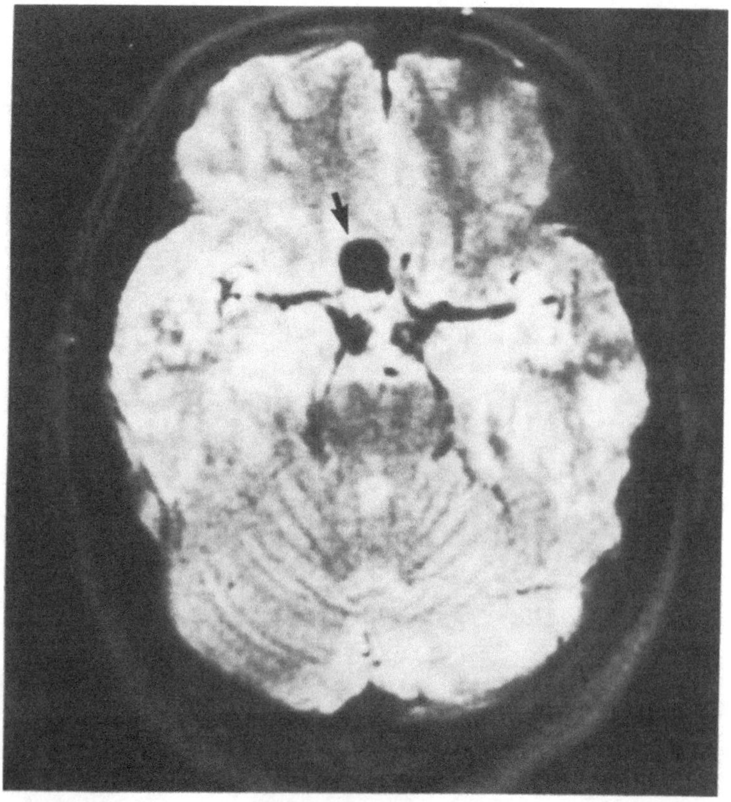

FIGURE 2–16E

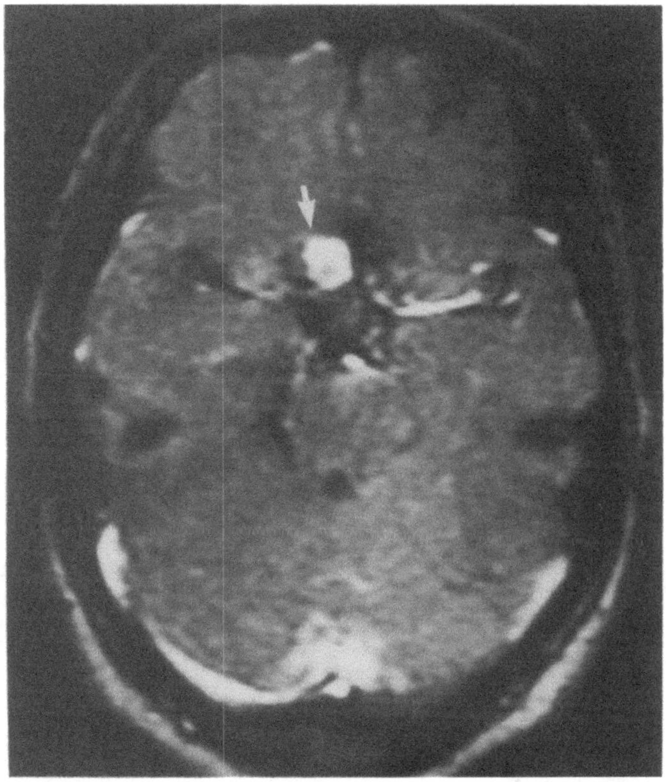

FIGURE 2–16F

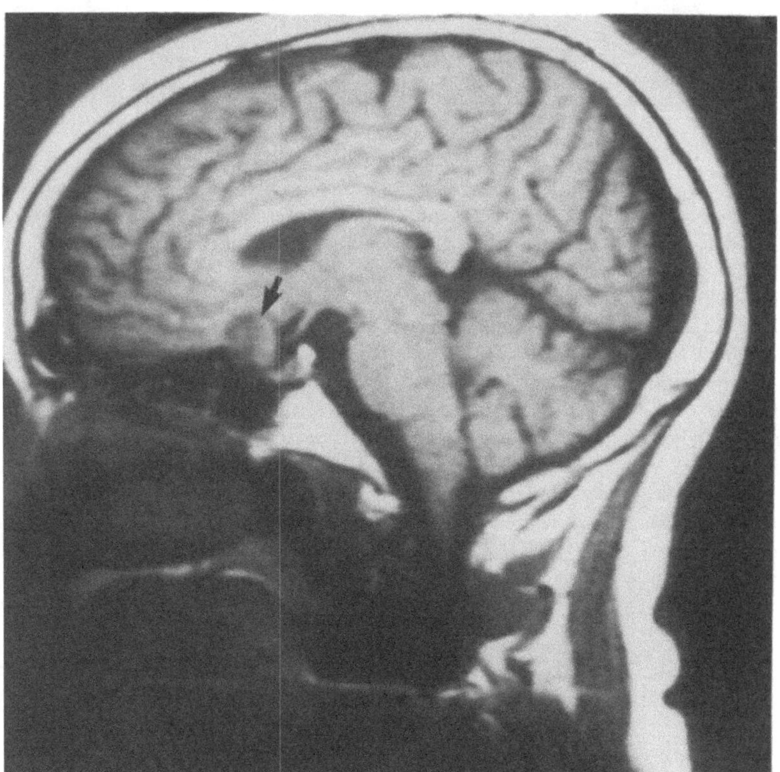

FIGURE 2–16G

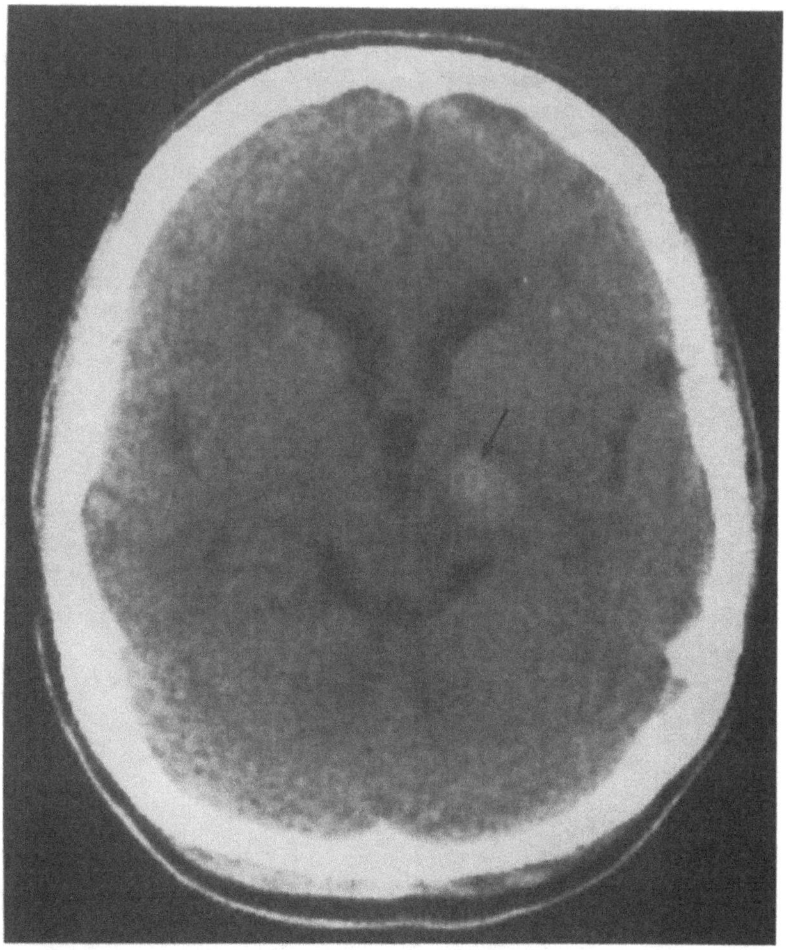

FIGURE 2–17A

FIGURE 2–17. A 17-year-old female with a thrombosed aneurysm arising from the lateral aspect of the horizontal portion of the left posterior cerebral artery. Its angiographically documented patent lumen was 5 mm in size. It progressively enlarged in size due to hemorrhages that underwent thrombosis and calcification. The case is being shown to demonstrate the usefulness of GRASS (gradient recalled acquisition at study state) pulse sequence (TR 21/TE 12) in the investigation of vascular lesions. (A) Plain CT of October, 1986. 1.8-cm-size mass lesion is present anterior to left cerebral peduncle (arrow). (B) Plain CT of August, 1987. The lesion has grown to a size of 2.5 cm and has shown calcification anteriorly and at the periphery. (C) Proton density (TR 2000/TE 20) and (D), T2-weighted (TR 2000/TE 80) axial images (August 1987). The peripheral low intensity (arrows) is mineralization (hemosiderin in macrophages and calcification). The slightly hyperintense component of the lesion seen on proton-weighted images (arrow) has shown a slight decrease in signal on T2-weighted images, indicating that it is mainly intracellular methemoglobin. The anterior irregularly circular area of hypointensity (open arrow), without the knowledge of its CT character, could either represent mineralization or lumen of the aneurysm. (E) GRASS sequence. Note that the rounded low-intensity area (arrow) anteriorly remains hypointense, indicating that there is no flowing blood (which would have appeared hyperintense on GRASS) in it; hence it is not an aneurysmal lumen. (F) Proton density (TR 2000/TE 20) axial image. The aneurysm is shown as a low-intensity signal (arrow). The adjoining cerebral peduncle is severely distorted and shows gliosis as a hyperintense area.

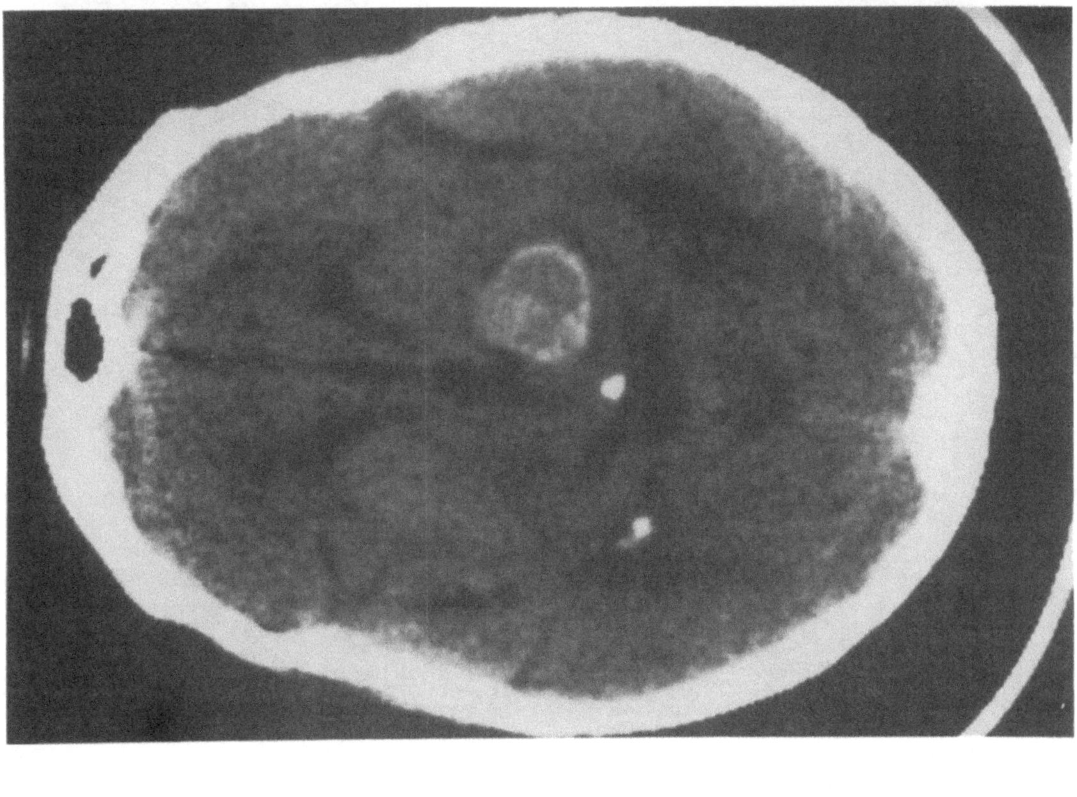

FIGURE 2–17B

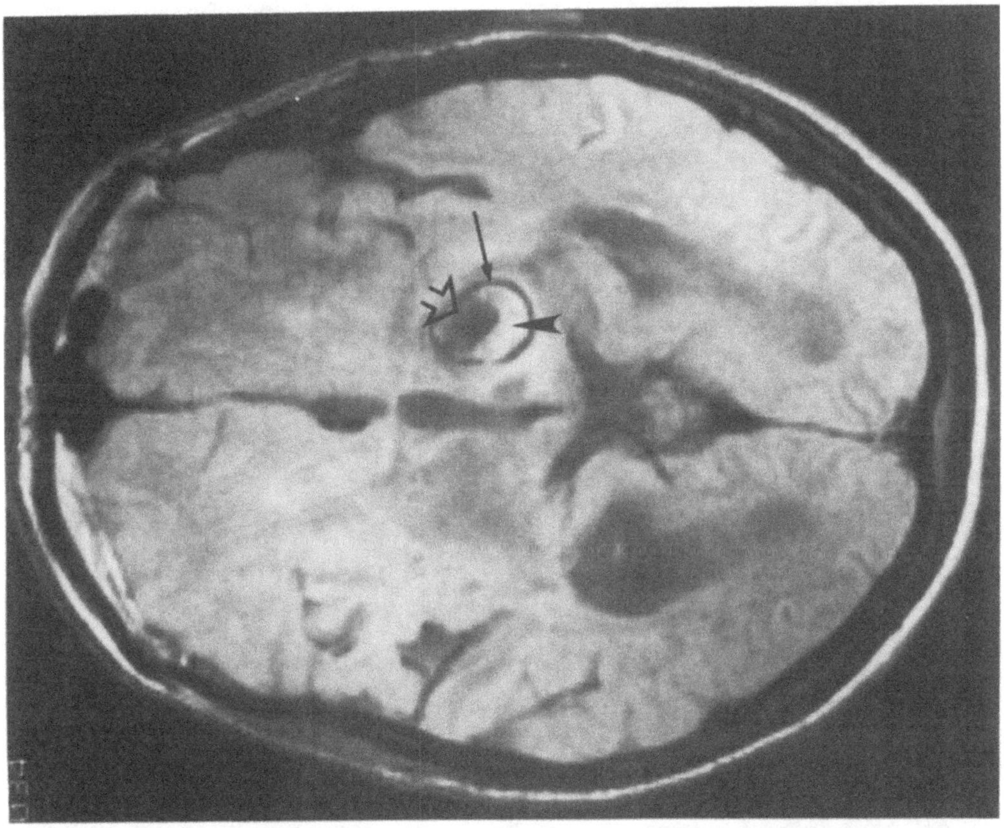

FIGURE 2–17C

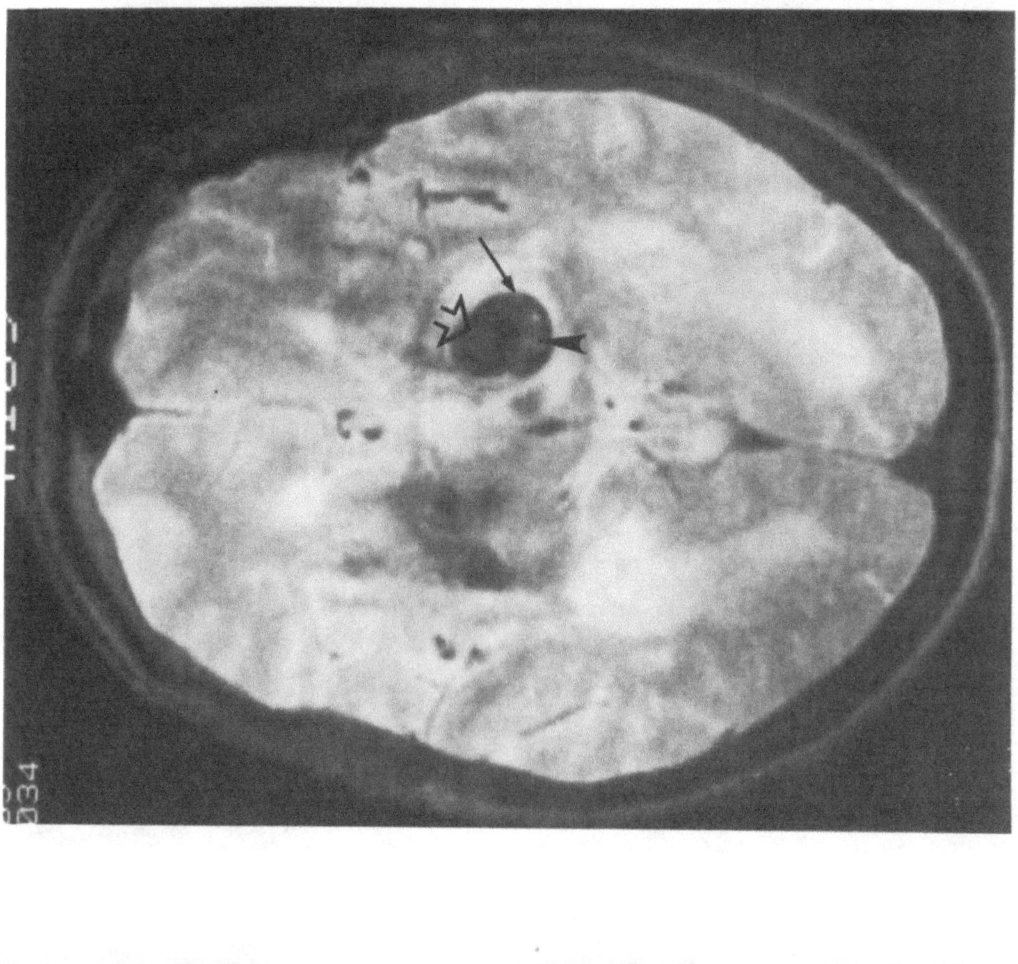

FIGURE 2–17D

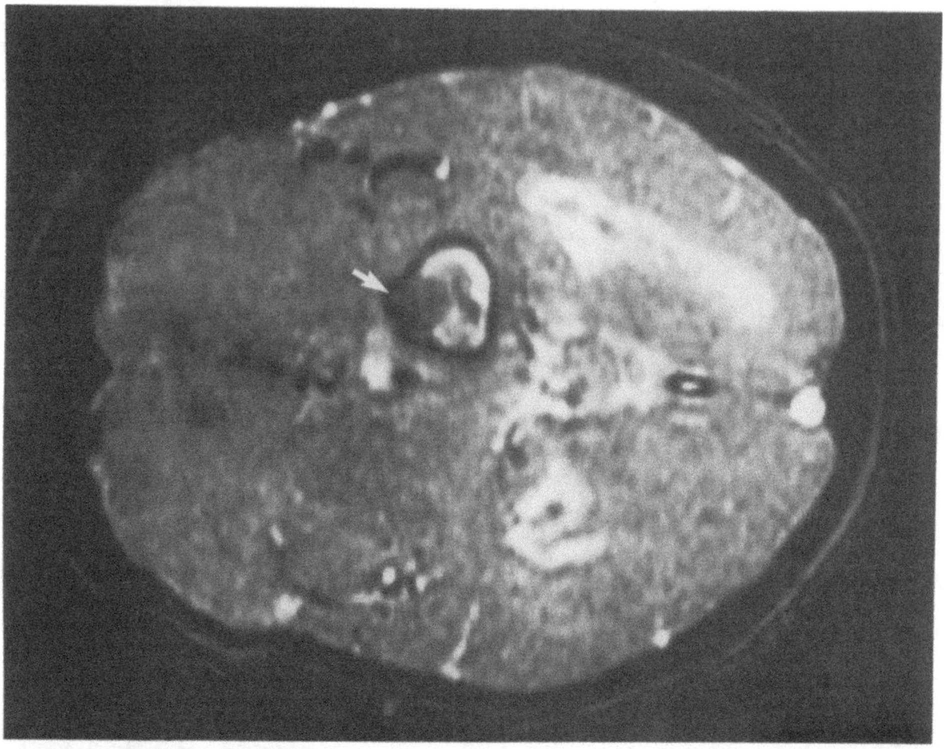

FIGURE 2–17E

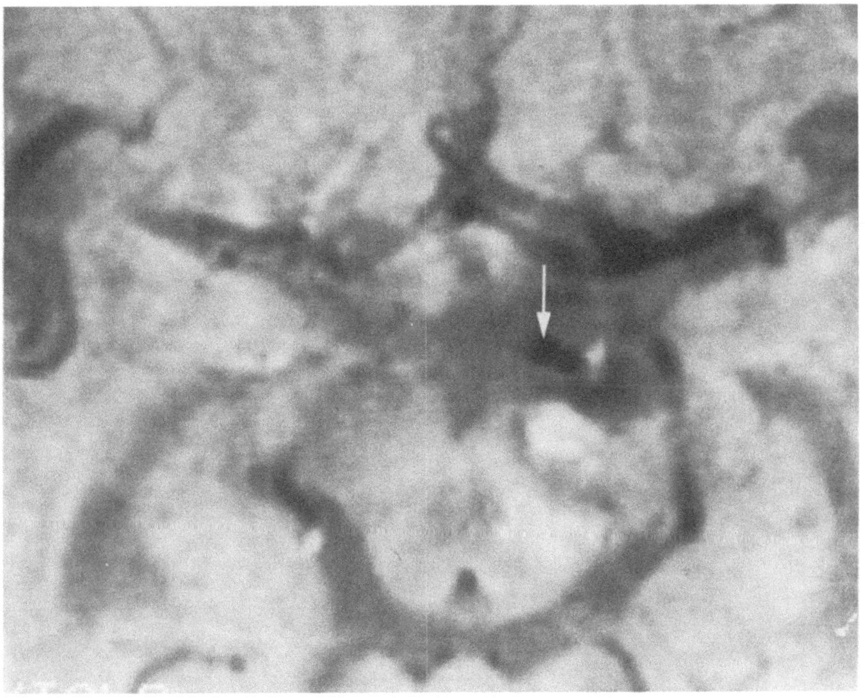

FIGURE 2–17F

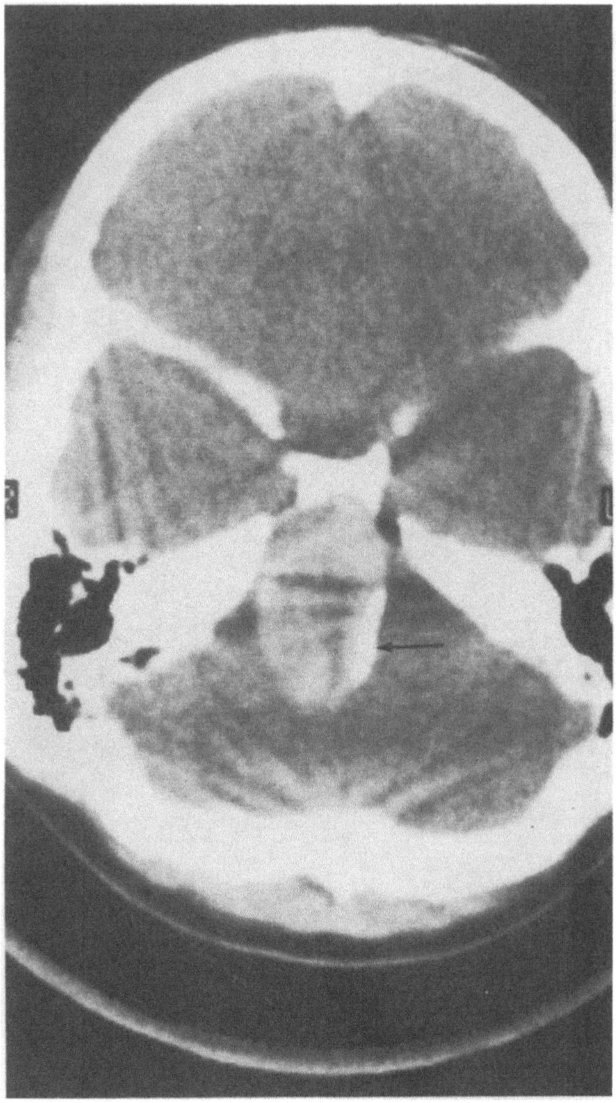

FIGURE 2–18A

FIGURE 2–18. Basilar artery dolichoectasia. A 20-year-old male presented with headaches associated with blurred vision and episodes of tonic–clonic movements of his body of several months duration. (A) Plain CT scan at a level just below the fourth ventricle. There is a large high-density extraaxial lesion with partly calcified rim (arrow). (B) Contrast CT at the same level as (A). Note that part of the lesion posteriorly (arrow) shows contrast enhancement. This is the patent lumen of the lower part of the dolichoectatic basilar artery. The enhancing rim of the lesion perhaps represents secondary tissue reaction (granulation tissue). (C) Reformatted sagittal image of a contrast CT. The entire lesion is shown. The patent lumen (arrows) of the dolichoectatic basilar artery is seen posteriorly. The compressed and posteriorly dislocated fourth ventricle is difficult to visualize. Note again the enhancing rim at the posterior margin of the lesion. (D) Partial saturation T1-weighted (TR 530 msec, TE 30 msec) sagittal MRI. The thrombosed part (arrow) of the lesion appears as a hyperintense area. The patent lumen of the dolichoectatic basilar artery is partly seen posteriorly (open arrow). The entire brainstem is humped over the lesion. The fourth ventricle and acqueduct are compressed. There is no hydrocephalus. (E) Intermediate-weighted (TR 2240 msec, TE 30 msec) coronal MRI. The lumen of the dolichoectatic basilar artery (arrow) appears as a tubular signal-void area surrounded by the hyperintense thrombosed component of the lesion. (F) Left vertebral angiogram, lateral view. The lumen of the dolichoectatic basilar artery (arrow) is irregular due to the surrounding thrombosis. The abnormal distance between the clivus and the basilar artery is caused by the thrombosed component of the lesion.

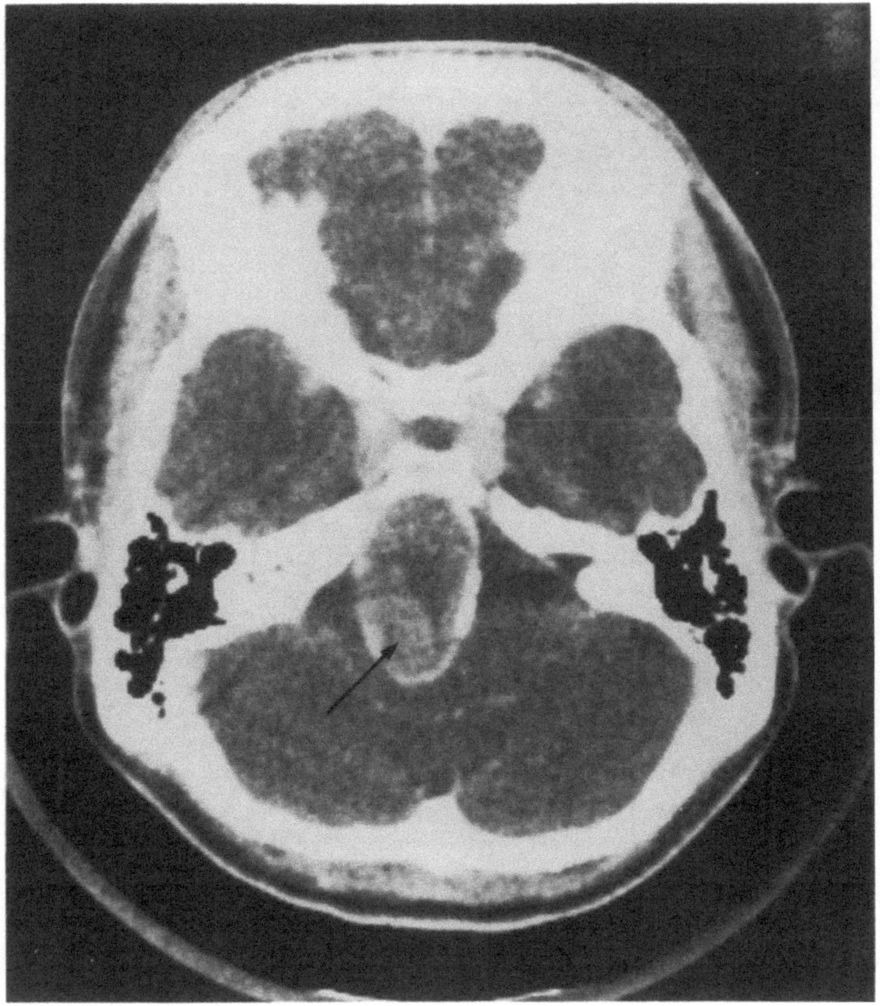

FIGURE 2–18B

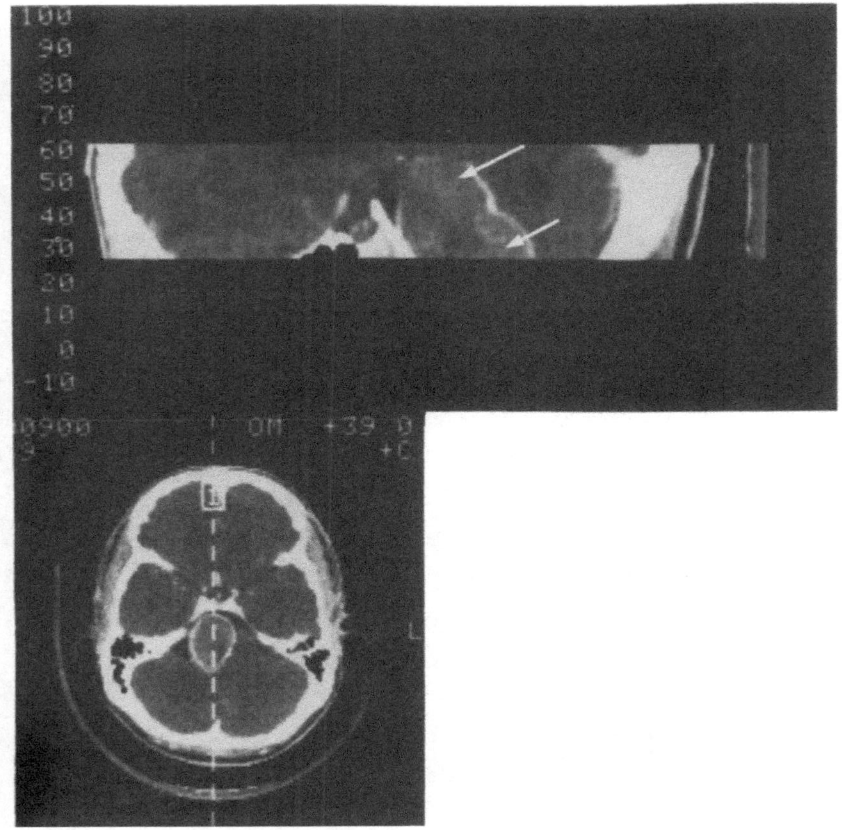

FIGURE 2–18C

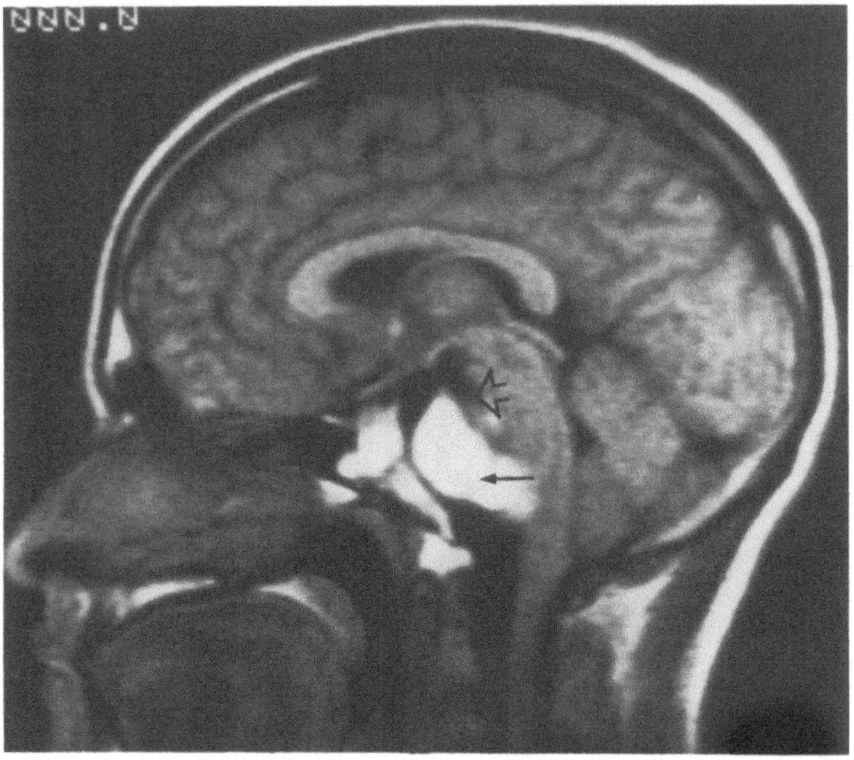

FIGURE 2–18D

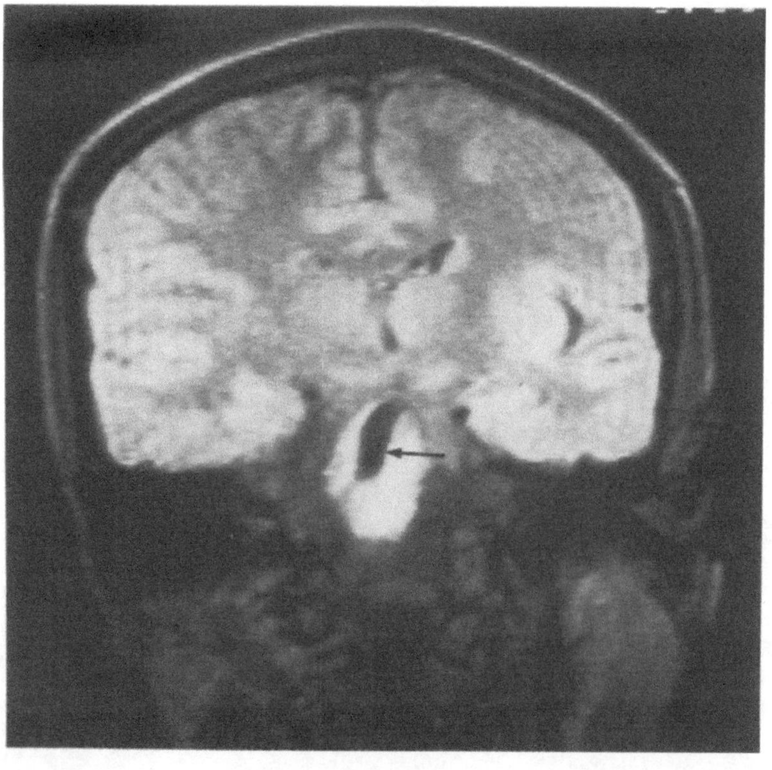

FIGURE 2–18E

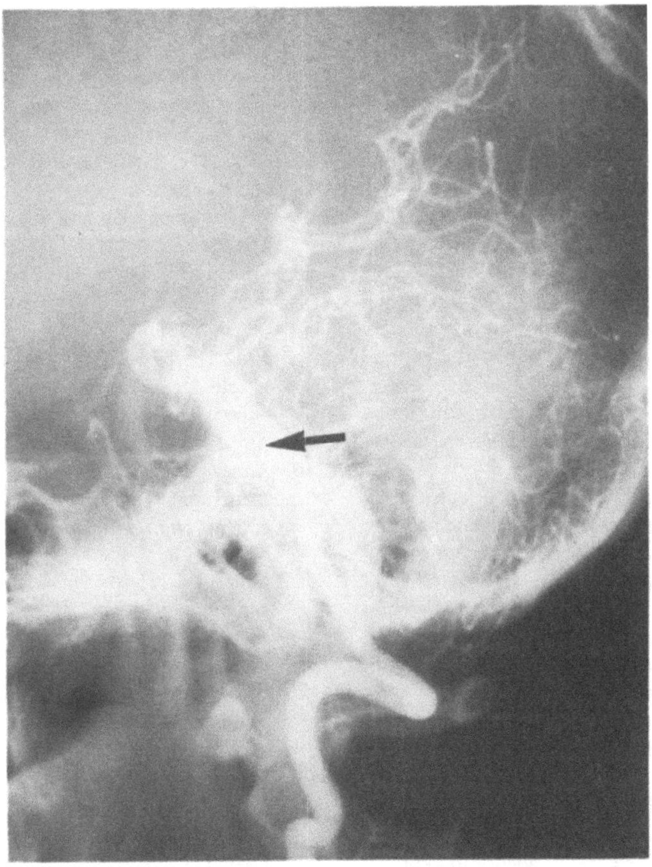

FIGURE 2–18F

signed to monitor the sick patients, MRI may assume a more prominent role in the investigation of aneurysmal SAH.

References

1. Crawford T. Some observations on the pathogenesis and natural history of intracranial aneurysm. *J Neurol Neurosurg Psychiat* 1959;22:259–266.
2. Crompton MR. Mechanism of growth and rupture in cerebral berry aneurysms. *Br Med J* 1966;1:1138–1142.
3. Forbus WD. On the origin of miliary aneurysm of the superficial cerebral arteries. *Bull Johns Hopkins Hosp* 1930;47:239–284.
4. Turnbull HM. Intracranial aneurysms. *Brain* 1918; 41:50–56.
5. Nystrom SHM. Development of intracranial aneurysms as revealed by electron microscopy. *J Neurosurg* 1973;20:329–337.
6. Cajander S, Hassler O. Enzymatic destruction of the elastic lamella at the mouth of cerebral berry aneurysms? An ultrastructural study with special regard to the elastic tissue. *Acta Neurol Scan* 1976; 53:171–181.
7. Crompton MR. The pathology of subarachnoid hemorrhage. *J R Coll Physicians Lond* 1973;7:235–237.
8. McCormick WF. The natural history of intracranial saccular aneurysms. Weekly Update. *Neurol Neurosurg* 1978;1:1–8.
9. McKissock W, Richardson A, Walsh L, et al. Multiple intracranial aneurysms. *Lancet* 1964;1:623–626.
10. Andrews RJ, Spiegel PK. Intracranial aneurysm. Age, sex, blood pressure, and multiplicity in an unselected series of patients. *J Neurosurg* 1979;51:27–32.
11. Nehls DG, Flom RA, Carter LP, Spetzler RF. Multiple intracranial aneurysms: determining the site of rupture. *J Neurosurg* 1985;63:342–348.
12. Rhoton AL, Jackson FE, Gleave J, et al. Congenital and traumatic intracranial aneurysms. *Clinical Symp* 1977;29:2–40.
13. Wiebers DO, Whisnant JP, Sundt TM Jr, et al. The significance of unruptured intracranial saccular aneurysm. *J Neurosurg* 1987;66:23–29.
14. Jane JA, Kassell NF, Torner JC, et al. The natural history of aneurysms and arteriovenous malformations. *J Neurosurg* 1985;62:321–323.

15. Ojemann RG, Crowell RM. *Surgical Management of Cerebrovascular Disease*. Baltimore/London: William and Wilkins, 1983: 134.
16. Heros RC, Zervas NT, Negoro M. Cerebral vasospasm. *Surg Neurol* 1976;5:354–362.
17. Simeone FA, Trepper PJ, Brown DJ. Cerebral blood flow evaluation of prolonged experimental vasospasm. *J Neurosurg* 1972;37:302–311.
18. Petruk KC, Weir BKA, Marriott MR, et al. Clinical grade, regional cerebral blood flow and angiographic spasm in the monkey after subarachnoid hemorrhage and subdural hemorrhage. *Stroke* 1973;4:431–445.
19. Kwak R, Niizuma H, Ohi J, et al. Angiographic study of cerebral vasospasm following rupture of intracranial aneurysms. Part 1. Time of the appearance. *Surg Neurol* 1979;11:257–262.
20. Du Boulay G. Distribution of spasm in the intracranial arteries after subarachnoid hemorrhage. *Acta Radiol* {Diagn} 1963;1:257–266.
21. Grubb RL Jr, Raichle ME, Eichling JO, et al. Effects of subarachnoid hemorrhage on cerebral blood volume, blood flow, and oxygen utilization in humans. *J Neurosurg* 1977;46:446–453.
22. Pitts LH, Macpherson P, Wyper DJ, et al. Cerebral blood flow, angiographic cerebral vasospasm, and subarachnoid hemorrhage. *Acta Neurol Scand* (Suppl 64) 1977;56:334–335.
23. Astrup J, Siesjo BK, Symon L. Threshold in cerebral ischemia—ischemic penumbra. *Stroke* 1981;12:723–725.
24. Knuckey NW, Fox RA, Surveyor I, et al. Early cerebral blood flow and computerized tomography in predicting ischemia after cerebral aneurysm rupture. *J Neurosurg* 1985;62:850–855.
25. Post KD, Flamm ES, Goodgold A, et al. Ruptured intracranial aneurysms. Case morbidity and mortality. *J Neurosurg* 1977;46:290–295.
26. Duncan AW, Rumbaugh CL, Caplan L. Cerebral embolic disease. A complication of carotid aneurysms. *Radiology* 1979;133:379–384.
27. Antunes JL, Correll JW. Cerebral emboli from intracranial aneurysms. *Surg Neurol* 1976;6:7–10.
28. Stewart RM, Samson D, Diehl J, et al. Unruptured cerebral aneurysms presenting as recurrent transient neurologic deficits. *Neurology* 1980;30:47–51.
29. Galera R, Frietz T. Hydrocephalus in the adult secondary to the rupture of intracranial aneurysm. *J Neurosurg* 1970;32:634–641.
30. Van Gijn J, Hijdra A, Wijdicks FM, et al. Acute hydrocephalus after aneurysmal subarachnoid hemorrhage. *J Neurosurg* 1985;63:355–362.
31. Black P McL. Hydrocephalus and vasospasm after subarachnoid hemorrhage from ruptured intracranial aneurysm. *Neurosurgery* 1986;18:12–16.
32. Suwanwela C, Suwanwela N, Charuchinda S, et al. Intracranial mycotic aneurysms of extravascular origin. *J Neurosurg* 1971;36:552–559.
33. Morley TP, Barr HWK. Giant intracranial aneurysms: diagnosis, course, and management. *Clin Neurosurg* 1969;16:73–94.
34. Bull JWD. Massive aneurysm at the base of the brain. *Brain* 1969;92:535–570.
35. Sarwar M, Batnitzky S, Schechter MM. Tumorous aneurysms. *Neuroradiology* 1976;12:79–97.
36. Sarwar M, Batnitzky S, Schechter MM, et al. Growing intracranial aneurysms. *Radiology* 1976;120:603–607.
37. Sacks JG, Lindenberg R. Dolicho-ectasia of intracranial arteries: symptomatology and pathogenesis of arterial elongation and distension. *Johns Hopkins Med J* 1969;125:95–106.
38. Corkill G, Sarwar M, Virapongse C. Evolution of dolichoectasia of the vertebrobasilar system as evidenced by serial computed tomography. *Surg Neurol* 1982;18:262–266.
39. Hassler O, Saltzman GF. Histologic changes in infundibular widening of the posterior communicating artery. *Acta Pathol Microbiol Scand* 1959;46:305–312.
40. Almaani WS, Richardson AF. Multiple intracranial aneurysms: identifying the ruptured aneurysm in patients with multiple intracranial aneurysms. *Surg Neurol* 1978;9:303–305.
41. Morttila I, Heiskanen O. Value of neurological and angiographic signs as indicators of the ruptured aneurysm in patients with multiple intracranial aneurysms. *Acta Neurochir* 1970;23:95–102.
42. Ghoshhajra K, Scotti L, Marasco J, et al. CT detection of intracranial aneurysm in subarachnoid hemorrhage. *AJR* 1979;132:613–616.
43. Wood EH. Angiographic identification of the ruptured lesion in patients with multiple cerebral aneurysms. *J Neurosurg* 1964;21:182–198.
44. Kassell NF, Torner JC. Size of intracranial aneurysms. *Neurosurgery* 1983;12:291–297.
45. Hunt WE, Hess RM. Surgical risk as related to time of intervention in the repair of intracranial aneurysms. *J Neurosurg* 1968;23:14–20.
46. Brismar J, Sundbarg G. Subarachnoid hemorrhage of unknown origin: prognosis and prognostic factors. *J Neurosurg* 1985;63:349–354.
47. Bradley WG, Schmidt PG. Effect of methemoglobin formation on the MR appearance of subarachnoid hemorrhage. *Radiology* 1985;156:99–103.
48. Gomori JM, Grossman RI, Goldberg HI, et al. Intracranial hematomas: imaging by high field MR. *Radiology* 1985;157:87–93.
49. Eller TW. MRI demonstration of clot in a small unruptured aneurysm causing stroke. Case report. *J Neurosurg* 1986;65:411–412.
50. Hackney DB, Lesnick JE, Zimmerman RA, et al. MR identification of bleeding site in subarachnoid hemorrhage with multiple intracranial aneurysms. Case report. *J Comput Assist Tomogr* 1986;10:878–880.
51. Olsen WL, Brant-Zawadzki M, Hodes J, et al. Giant intracranial aneurysms: MR imaging. *Radiology* 1987;163:431–435.
52. Atlas SW, Grossman RI, Goldberg HI, et al. Partially thrombosed giant intracranial aneurysms: correlation of MR and pathologic findings. *Radiology* 1987;162:111–114.

53. Hahn FJ, Ong E, McComb R, et al. Peripheral signal void ring in giant vertebral aneurysm: MR and pathologic findings. *J Comput Assist Tomogr* 1986; 10:1036–1038.
54. Alvarez O, Hyman O. Even-echo MR rephasing in the diagnosis of giant intracranial aneurysms. *J Comput Assist Tomogr* 1986;10(4):699–701.
55. Zimmerman RD, Fleming CA, Lee BC, et al. Periventricular hyperintensity as seen by magnetic resonance: prevalence and significance. *AJNR* 1986; 7:13–20.

3. INTRACRANIAL VASCULAR MALFORMATIONS

Roger A. Hyman

Steven F. Sands

Histological varieties of intracranial vascular malformations include arteriovenous malformations (AVMs), venous malformations, cavernous malformations, and capillary telangiectasia. They constitute a significant clinical problem. According to one estimate there are approximately 350,000 individuals in the United States with clinically evident or occult intracranial arteriovenous malformations [1]. Arteriovenous malformations are the lesions most familiar to practicing physicians because of their protean clinical manifestations. Autopsy series indicate that venous angiomas may actually have a higher prevalence, but these are frequently asymptomatic throughout life and of no clinical consequence. In a prospective autopsy series evaluating 4069 consecutive brains [2], 165 had one or more vascular malformations. Of these only 24 were AVMs while 105 were venous angiomas.

Hemorrhage is the most common mode of presentation of intracranial vascular malformations. Approximately 50% of all AVMs will present clinically with an intracerebral hemorrhage: two thirds of these will be predominantly parenchymal and one third predominantly subarachnoid. Intraventricular extension of hemorrhage occurs in 6% [3]. The largest clinical study of AVMs to date included 545 cases in a cooperative study of 6368 patients with subarachnoid hemorrhage [3]. In this study, approximately 8.6% of patients presenting with subarachnoid hemorrhage had an underlying AVM. The majority of the remaining patients in this series demonstrated one or more aneurysms at angiography. The incidence of AVMs as a cause of subarachnoid hemorrhage rises dramatically during pregnancy, with some series reporting an occurrence rate as high as 50% [4,5].

Other modes of clinical presentation include seizure, headaches, and manifestations of ischemia. Hemorrhage is more frequent than seizure as the initial symptom in most large series [6]. In larger AVMs, the incidence of hemorrhage and seizure is almost equal; in small AVMs, hemorrhage occurs more frequently. The death rate from hemorrhage associated with an AVM is lower than from hemorrhage associated with an aneurysm. Approximately 10% of patients will die from their first hemorrhage associated with an AVM, while 30% of patient will die from their first hemorrhage associated with an aneurysm [7]. This occurs because the bleeding may originate from the venous side of the shunt in an AVM, which is a somewhat lower pressure system. In addition, blood-induced vasospasm occurs infrequently with AVMs because the lesion is not usually at the skull base and the volume of subarachnoid blood is frequently less than with subarachnoid hemorrhage associated with an aneurysm. Once an AVM has hemorrhaged, the risk of recurrent bleeding is approximately 6% during the first year and 2–3% per year thereafter [7].

Although AVMs are congenital in origin, they rarely demonstrate clinical manifestations until adulthood. Symptoms most commonly have their onset in the third or fourth decades of life. Of 545 AVMs in the Cooperative Study, only 5% presented before age 11 [3].

In the Cooperative Study, 28% of patients with AVMs presented initially with seizures. Occasionally, seizures are the only manifestation of an AVM throughout a patient's lifetime. Five patients demonstrated seizures only, after a period of 10 to 40 years, in another long-term follow-up study of 70 patients with AVMs [8].

Certain hemodynamic aspects of AVMs are important in understanding their natural history and in making decisions with respect to therapy. The assumption that AVMs usually enlarge over time was confirmed in a study of 21 patients who underwent serial angiography. Twelve of these patients showed a verified increase in size over approximately four years in a group of patients whose average age at first angiography was 26 [9].

The increasing size of an AVM over time and the increasing size of the arteries feeding these lesions give rise to a phenomenon referred to as a cerebrovascular steal syndrome, with underperfusion of normal brain parenchyma in the distribution of the major parent arteries feeding the AVM. Steal syndromes associated with AVMs can lead to progressive neurological and mental deterioration [10], and may be a cause of seizures or transient ischemic attacks. In attempting to treat an AVM, whether by surgical extirpation, arterial embolization, or a combination of the two, the steal phenomena can lead to potential complications. Steal-induced ischemia of the tissues surrounding an AVM can lead to an impairment of normal autoregulation of the vessels in the region. When the AVM is then resected or embolized, high arterial pressure is now delivered to the vessels in the region which have been accustomed to low arterial pressure, and the result may be edema or hemorrhage into the surrounding tissues following treatment [11]. With large AVMs, this problem can be addressed by sequentially ligating or embolizing feeding vessels at separate sittings over time [12].

Pathological Classification

There are distinct biological and histological differences between the types of vascular malformations categorized here. Each has a characteristic natural history. The four categories include 1) arteriovenous malformation, 2) venous malformation, 3) cavernous malformation, and 4) telangiectasia [13].

ARTERIOVENOUS MALFORMATIONS

Of the vascular malformations, arteriovenous malformation is the most frequently symptomatic and, therefore, the most frequently encountered clinically. Angiography demonstrates shunts from the arterial to the venous circulation that are not usually seen in the other categories of vascular malformations. Histologically, AVMs consist of masses of abnormal arteries connected to abnormal veins without a normal intervening capillary network (figure 3–1a). The larger vessels are predominantly venous. Microscopically, areas of hemorrhage are frequently seen. Mineralization is not uncommon. Abnormal gliotic parenchyma is seen between the component vessels. These lesions are frequently encountered in the middle cerebral arterial territory and are much more common above the tentorium.

VENOUS MALFORMATIONS

Although venous malformations are the most common vascular malformations found at autopsy [2, 13], these lesions are rarely of clinical significance. They are not infrequently encountered as incidental findings on angiograms, computerized tomography (CT), and magnetic resonance (MR) studies, and therefore are of considerable interest radiologically. They are composed of anomalous veins, frequently arising in the white matter, that are separated by normal parenchyma. A collection of enlarged medullary veins usually converge to a single transcerebral or transcerebellar vein. This vein then courses through the white matter to end in a cortical vein, a dural sinus, or less frequently the deep venous system. The vessel walls may be thickened and hyalinized (figure 3–1b). Mineralization is rare but does occur. There is no arterial component. In rare instances these lesions may hemorrhage, particularly those located in the cerebellum [14].

CAVERNOUS MALFORMATIONS

Cavernous malformations are relatively uncommon lesions consisting of a contiguous grouping of sinusoidal-type vessels with no normal intervening brain parenchyma. The walls of the sinusoidal spaces are thin, without evidence of smooth muscle or elastic tissue. Mineralization and even ossification can occur, particularly in larger lesions. Frequently, old hemorrhage and gliosis are adjacent to the regions of sinusoidal thickening (figure 3–1c).

TELANGIECTASIA

Telangiectasia are composed of small capillary-type vessels (figure 3–1d). They are most commonly seen in the pons as an incidental finding at autopsy. Gliosis and hemorrhage are rare. These lesions are not detected by angiography and only rarely by CT and MR. Patients with Rendu–Osler–Weber syndrome may demonstrate these lesions throughout the brain.

CRYPTIC OR OCCULT MALFORMATIONS

By definition, cryptic or occult malformations are lesions with normal angiograms. Cryptic vascular malformations (CVMs) or occult vascular malformations (OVMs) may be of any of the four histological

FIGURE 3–1. Pathological classification of intracranial vascular malformations. (A) Arteriovenous malformation. Thick-walled arterialized veins are seen against a background of gliotic brain tissue. (B) Venous malformation. An abnormal collection of dilated slightly thick-walled veins is noted in the deep white matter. The intervening parenchyma is normal. (C) Cavernous malformation. Closely clustered dilated blood vessels of variable diameter crowd out the intervening brain parenchyma, which shows gliosis and hemosiderin deposition. (D) Capillary telangiectasia. Thin-walled dilated capillaries are present in a section through the pons. The intervening parenchyma is essentially normal.

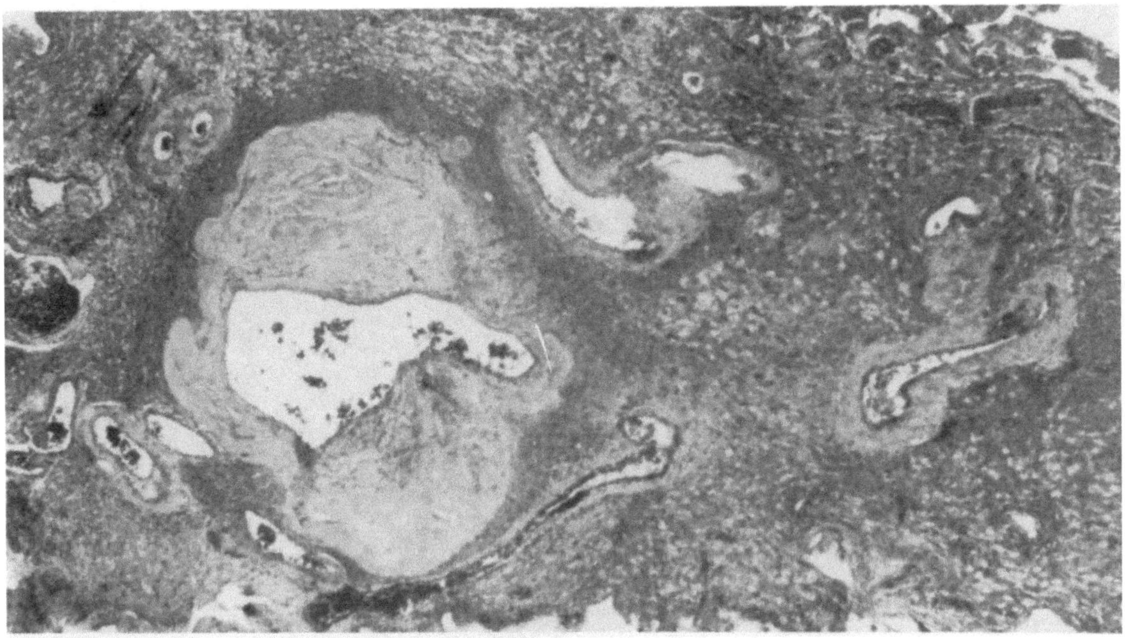

FIGURE 3–1A

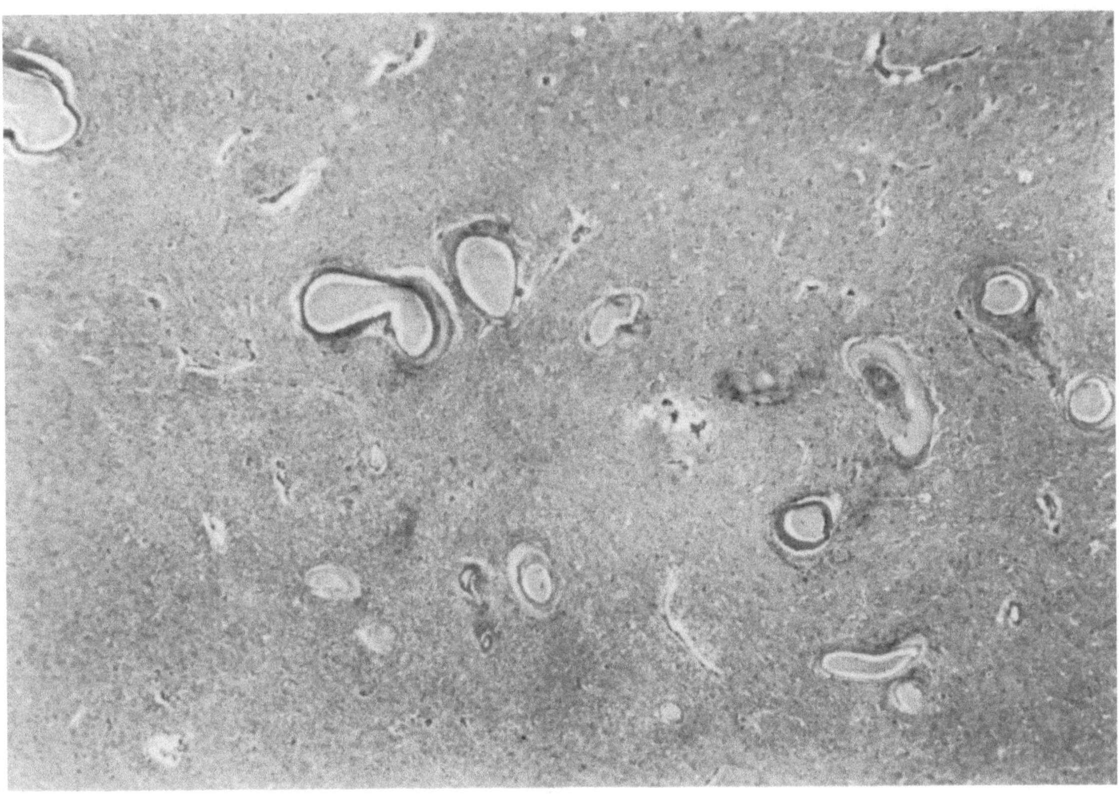

FIGURE 3–1B

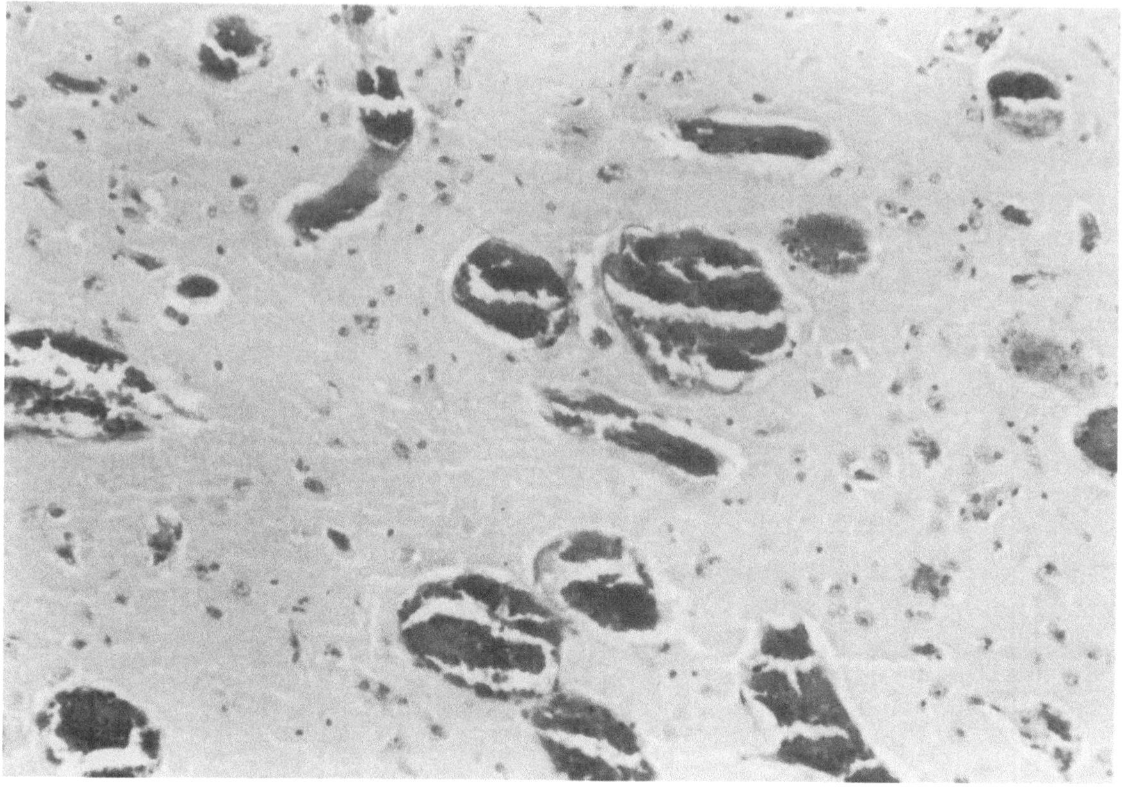

FIGURE 3–1C

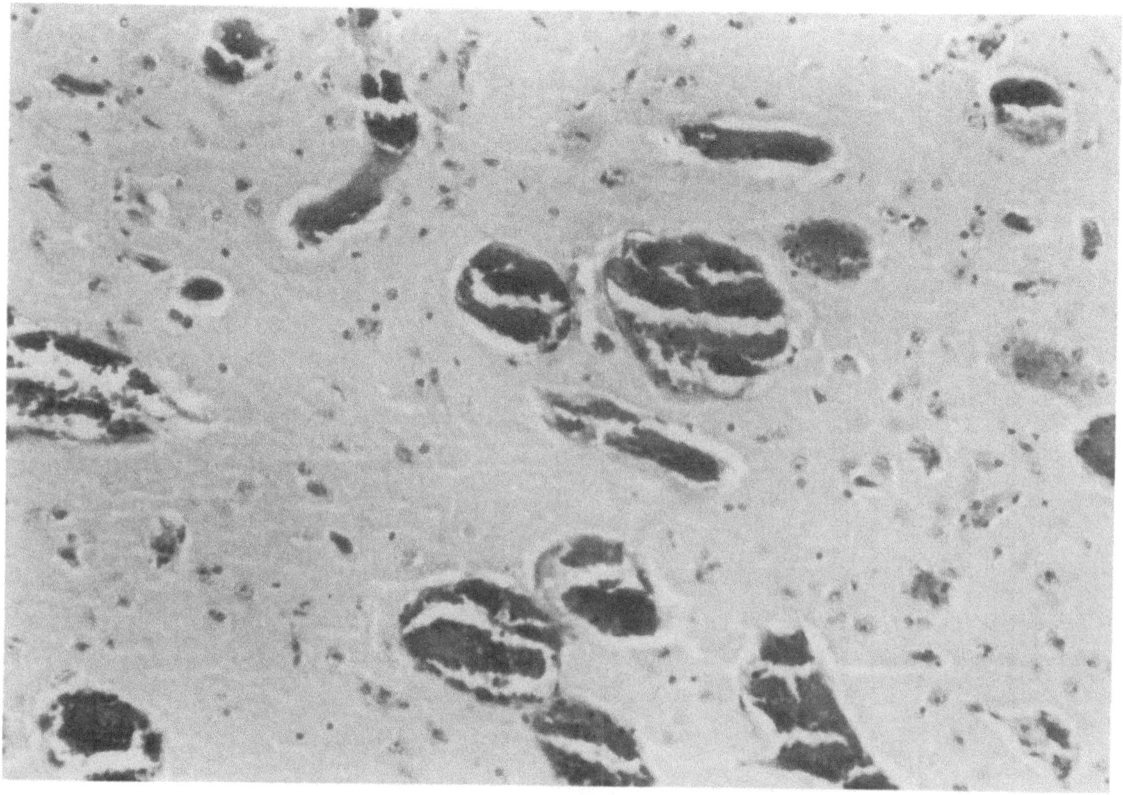

FIGURE 3–1D

types described or may be of transitional forms. Arteriovenous malformations that have thrombosed or bled and obliterated themselves may present as CVMs. In the 1950s, before CT and MR, the term was originally applied to lesions that had hemorrhaged spontaneously and were angiographically occult [15], only to be identified at surgery or autopsy. The interest in occult lesions has grown with the increasing use of CT and MR. Most CVMs identified currently are diagnosed by CT and MR and are not infrequently asymptomatic at the time of their detection.

Arteriovenous Malformations

A retrospective review of our own radiographic material revealed 85 intracranial vascular malformations diagnosed during the period 1974 through 1986 (table 3–1).

Sixty-eight percent of these lesions were demonstrated at angiography to be AVM's. This frequency is considerably higher than that reported in autopsy series and reflects the fact that AVMs demonstrate clinical manifestations far more frequently than any of the other histological types of vascular malformations of the brain.

CT SCANNING

During the past 15 years, the majority of newly diagnosed AVMs were first suspected following CT scans performed to evaluate a wide variety of symptoms. However, therapeutic decisions are rarely made on the basis of CT findings alone. Angiography is still required to confirm the findings suggested by CT and to further identify and characterize the arterial feeding vessels and the venous drainage. To determine which lesions will require surgery, embolization, a combination of the two, or proton beam irradiation, full angiographic examination is still required even when complete MR studies are available.

The CT findings in AVMs will depend on the manner of presentation. If the patient presents with an intracranial hemorrhage, the major findings will be related to the hemorrhage itself and will be seen on a noncontrast examination. Arteriovenous malformations that have not ruptured are not reliably demonstrated unless intravenous contrast material has been administered. Some authors report a high incidence of abnormal noncontrast CT scans in AVMs. One study indicated that 81% of precontrast scans were abnormal [16], and another that 43% of these lesions in children were calcified [17]. However, false negative noncontrast studies clearly occur. If hemorrhage is not present and if CT is to be used alone in routine screening for AVMs, postcontrast studies must be performed.

TABLE 3–1. Intracranial Vascular Malformations Diagnosed Radiographically 1974–1986

Arteriovenous malformations	58
Venous malformations	12
Cavernous malformations	1
Telangiectasia	0
Cryptic malformations with unverified histology	14
Total	85

The classic CT findings in unruptured AVMs include either a normal noncontrast examination or a serpiginous region of slight hyperdensity. Following contrast administration, dense enhancement is usually seen outlining the venous and occasionally the arterial components of the malformation (figure 3–2). The classic findings on CT scanning may be modified by previous hemorrhage. Regions of porencephaly or volume loss in a hemisphere can occur. Dense calcification at the site of a prior hemorrhage is occasionally seen (figure 3–3).

Enhancement noted on CT scans of AVMs is due primarily to the blood-pool effect of the increased iodine concentration in the large vessels of the AVM. To a lesser extent, the defective blood–brain barrier characteristic of central nervous system lesions also allows for some leakage of contrast into the matrix of the AVM. This second mechanism also allows for the visualization of an AVM on a delayed postcontrast scan [16], favored by some for screening symptomatic individuals. Most AVMs, however, are best seen on scans performed in close proximity to the time of injection of intravenous contrast material.

If an AVM has hemorrhaged, the most common finding is an intracerebral hematoma [18]. The noncontrast study will reveal a relatively homogeneous high attenuation mass. Frequently, routine images without intravenous contrast enhancement will demonstrate little to identify the etiology of the hemorrhage (figure 3–4). Therefore, in nonhypertensive younger patients further evaluation is warranted [19]. In this setting, a postcontrast CT scan may be helpful, particularly prior to emergency operative intervention for hematoma. Parenchymal hemorrhages associated with AVM will behave over time in a manner similar to hematomas of other etiologies. They will gradually lose density and eventually develop a peripheral region of ring enhancement [20]. They may resolve with little volume loss within

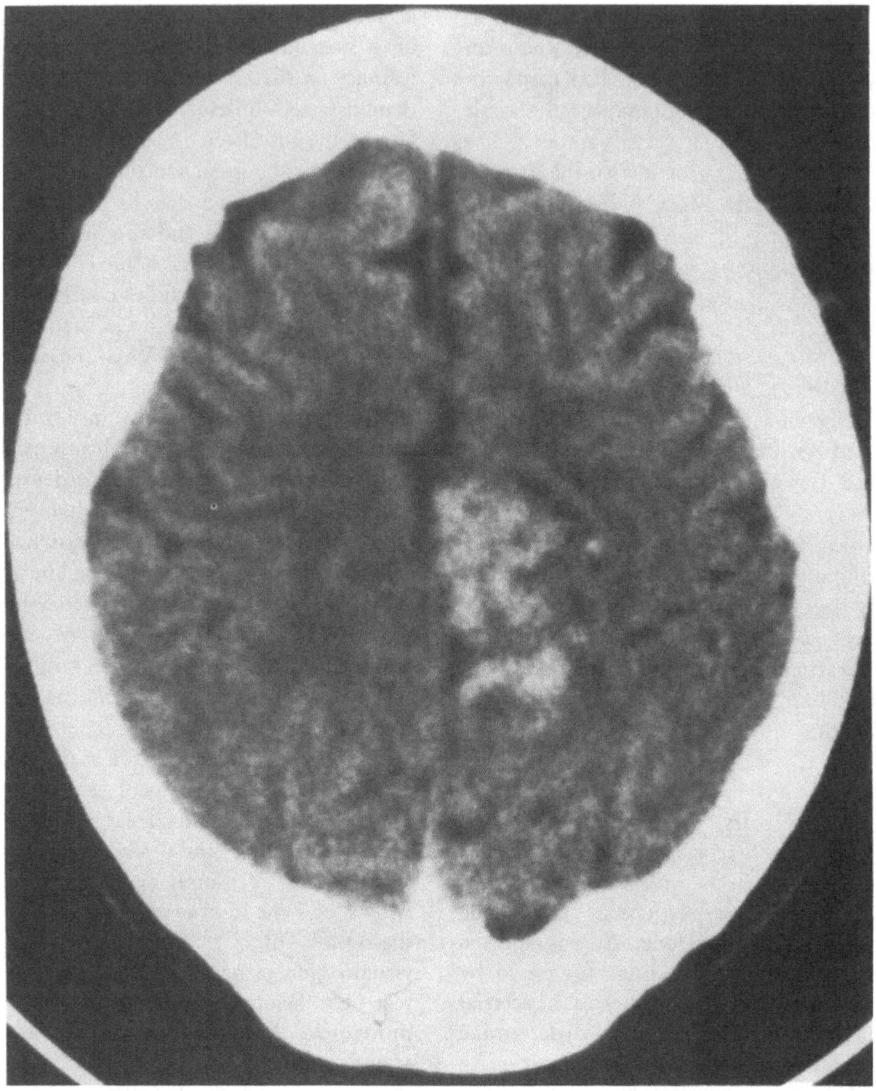

FIGURE 3–2A

FIGURE 3–2. Unruptured AVM. (A) An examination performed prior to contrast administration in this 22-year-old male with seizures demonstrates an abnormal region of increased density in the left frontoparietal region near the midline. Prominent vessels in an AVM are frequently slightly hyperdense to the brain even in the absence of calcification. (B) After intravenous contrast, prominent enhancement is seen. (C) A left internal carotid angiogram demonstrates both anterior (closed arrow) and middle (open arrows) cerebral arterial supply to this AVM and confirms the suggested CT diagnosis.

the hemisphere, or, if large, porencephaly may result. Calcification in the region can occur.

When the parenchymal hematoma is large, intraventricular extension may occur (figure 3–5). Occasionally, an AVM projects into a ventricle and an associated hemorrhage may outline the ventricular system. Patients with arteriovenous malformations located in or adjacent to the subarachnoid space may present with subarachnoid hemorrhage (figure 3–6). More commonly, parenchymal hematomas extend into the subarachnoid space and give rise to subarachnoid hemorrhage in combination with a hemispheric clot. Where parenchymal hematomas are identified, differential diagnostic possibilities include hypertensive hemorrhages, aneurysms, AVMs, hemorrhagic infarcts, and occasionally neoplasms. The patient's age, the history, and the location of the hemorrhage are helpful in the differential diagnosis. Hypertensive

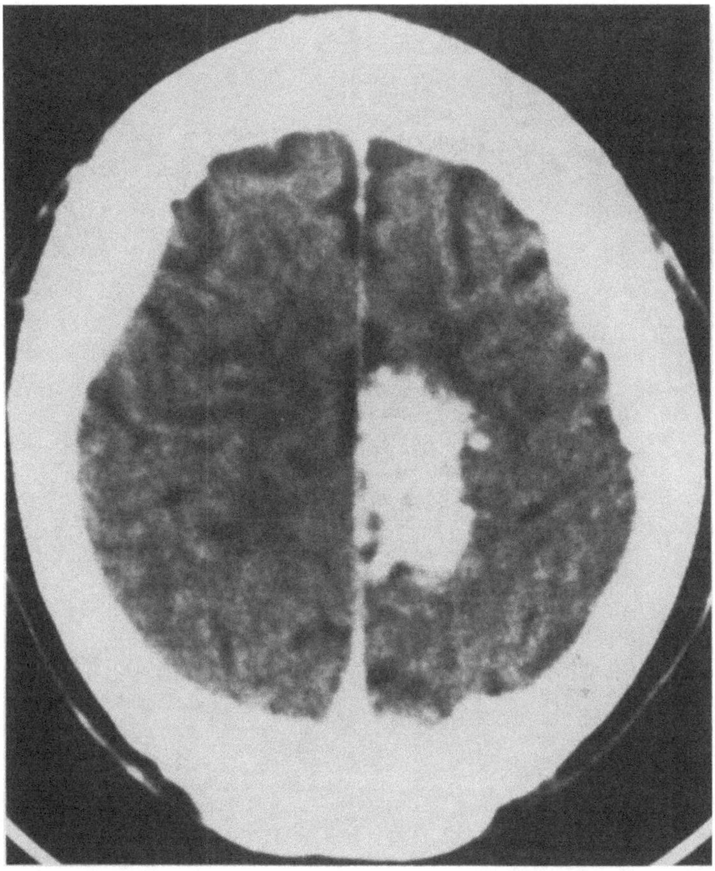

FIGURE 3–2B

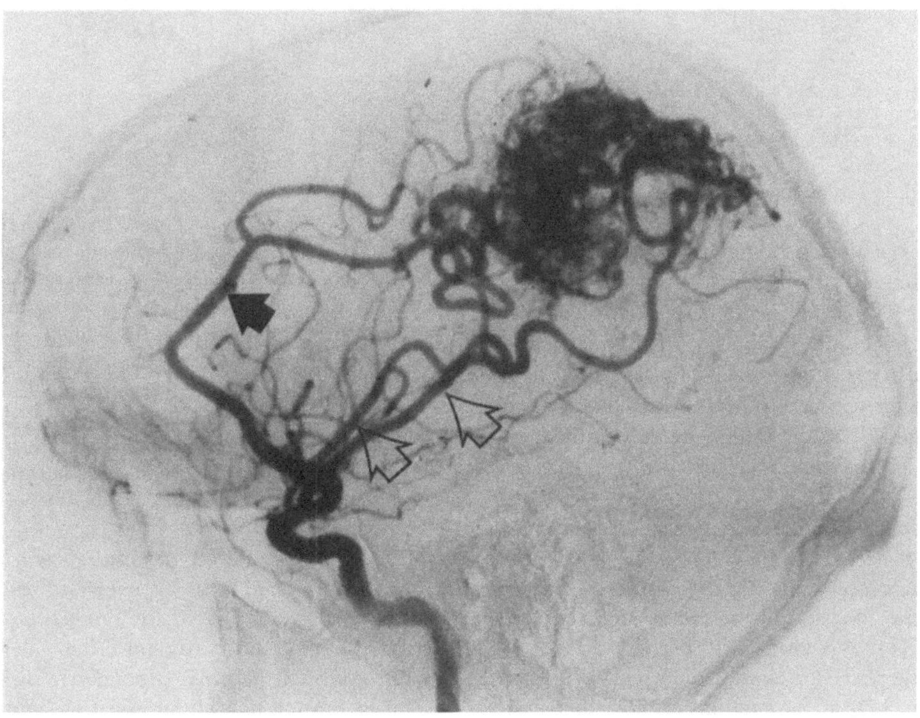

FIGURE 3–2C

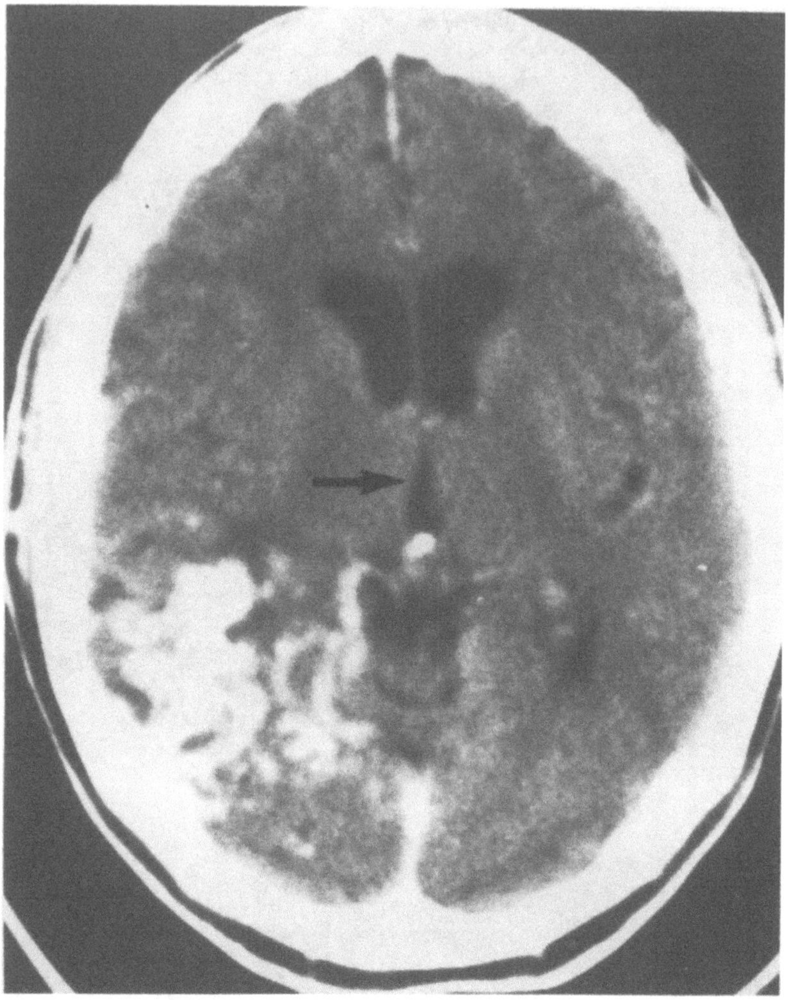

hemorrhages frequently occur within the basal ganglia and cerebellum of older patients. Parenchymal hemorrhages from AVMs are usually hemispheric and occur in a younger age group. Occasionally, a postcontrast study reveals enhancing vessels adjacent to a hematoma characteristic of an AVM. If a diagnosis on a postcontrast CT scan cannot be determined, angiographic evaluation is required where clinically appropriate.

Arteriovenous malformations are not commonly seen below the tentorium in adults. The Cooperative Study reported only 7% of their clinically apparent lesions to be located in the brain stem and the cerebellum [3]. A significantly higher incidence of lesions in the posterior fossa is reported in children. In one study [17], 40% of lesions in the pediatric age group were reported in an infratentorial location.

FIGURE 3–3. Ruptured AVM with late calcification. (A) A postcontrast study in a 65-year-old male demonstrates the tortuous configuration of vessels associated with an AVM in the right parietal region. The third ventricle (arrow) is displaced to the left. (B) A higher section reveals ring enhancement (arrows) around a central dense lesion consistent with a resolving hematoma associated with the AVM. (C) Two years later, a follow-up study demonstrates dense calcification in the old hematoma site and porencephalic dilatation of the right occipital horn on the noncontrast study.

Of particular interest in children is a rare lesion commonly referred to as an aneurysm of the vein of Galen. This entity actually represents an AVM, shunting blood primarily into the galenic venous system, which becomes secondarily dilated. The AVM is most commonly supplied by branches of the posterior cerebral arteries although other arterial

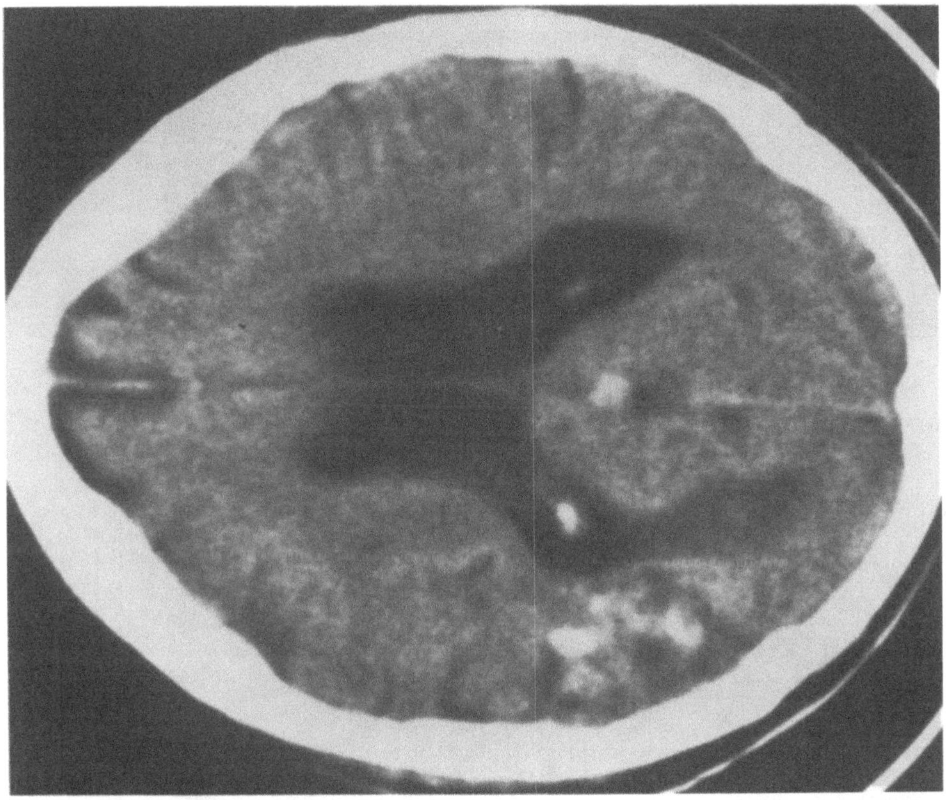

FIGURE 3–3C

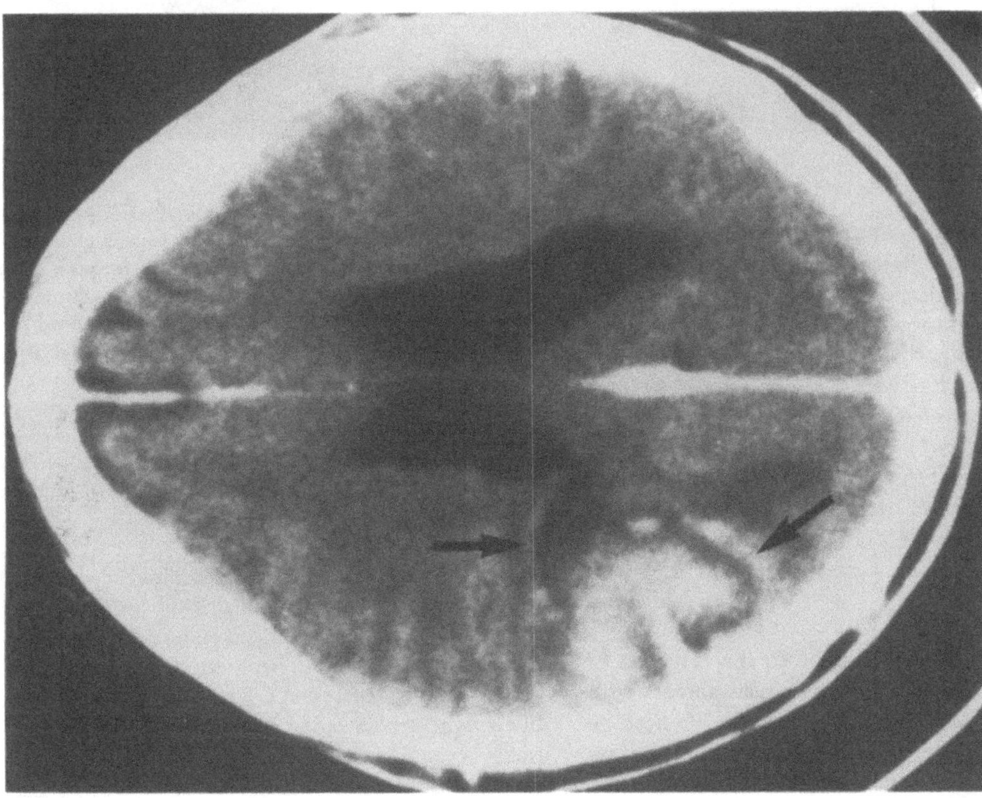

FIGURE 3–3B

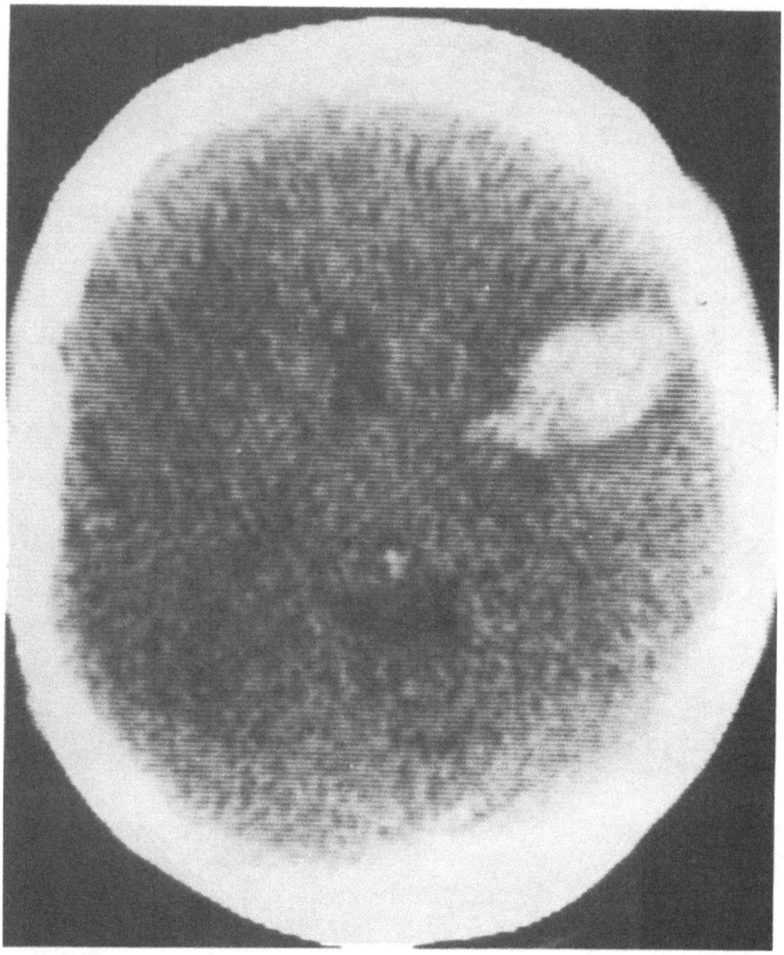

FIGURE 3–4A

FIGURE 3–4. Parenchymal hematoma associated with AVM. (A) A noncontrast study in a 34-year-old female with the sudden onset of right hemiparesis reveals a dense lesion in the left frontal lobe with displacement of the anterior horns to the right. This is consistent with a parenchymal hemorrhage. The AVM is not identified on the noncontrast study. (B) An angiogram reveals a region of increased vascularity in the left frontal lobe with an avascular mass posterior and inferior to the lesion corresponding to the hematoma. (C) The AVM drains superiorly into the superior sagittal sinus and inferiorly into the lateral sinus.

feeders may exist as well. The presenting symptoms are usually cardiac failure in the neonatal period or an abnormal increase in head circumference [21]. Noncontrast CT scans show the dilated vein of Galen as a homogeneous, slightly hyperdense mass in the region of the posterior third ventricle. This could be confused with a pineal region neoplasm; however, the post-contrast study will frequently demonstrate the malformation itself (figure 3–7). Angiographic evaluation is definitive. Recently, MR findings have been decribed as well [22]. The prognosis when the entity is identified in the neonatal period is grave.

MR IMAGING

MR imaging is highly sensitive to the diagnosis of intracranial AVMs and offers certain advantages with respect to CT [23, 24, 25]. There is no need to administer intravenous contrast material, and no ionizing radiation is utilized. Furthermore, small lesions in the posterior fossa and in the temporal lobes within the middle cranial fossa are not subject to the beam-hardening artifacts induced by surrounding bone on CT scans. These lesions are occasionally obliterated by artifacts on CT scans and would be less likely to be overlooked on MR. Pure dural malformations, because of their frequently flattened configuration against the calvarium, may be missed by both

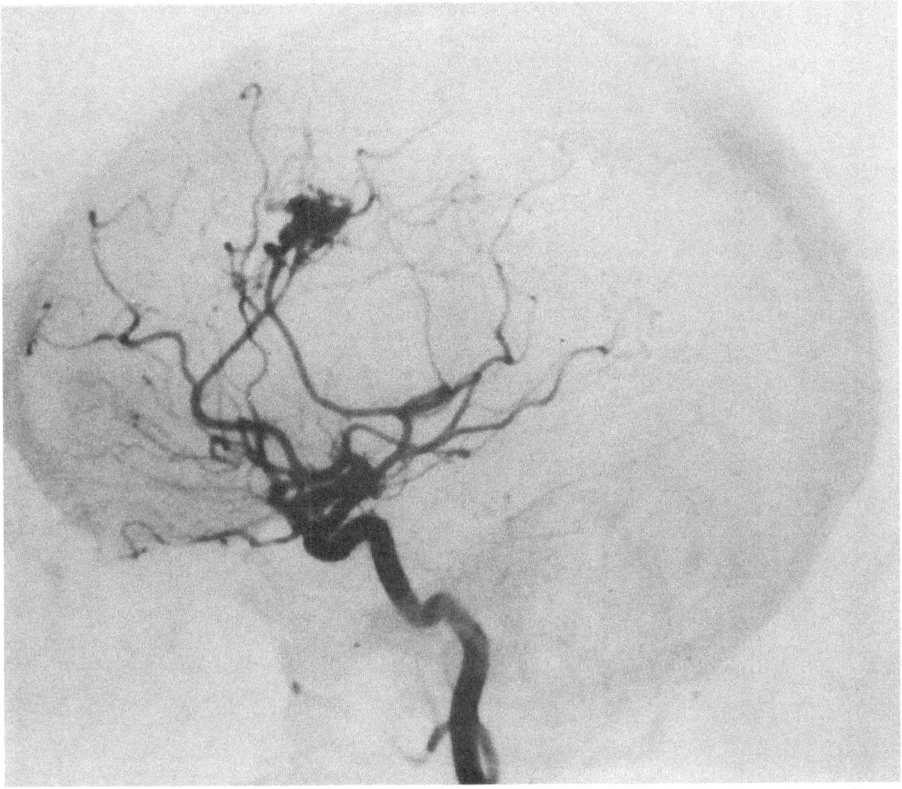

FIGURE 3–4B

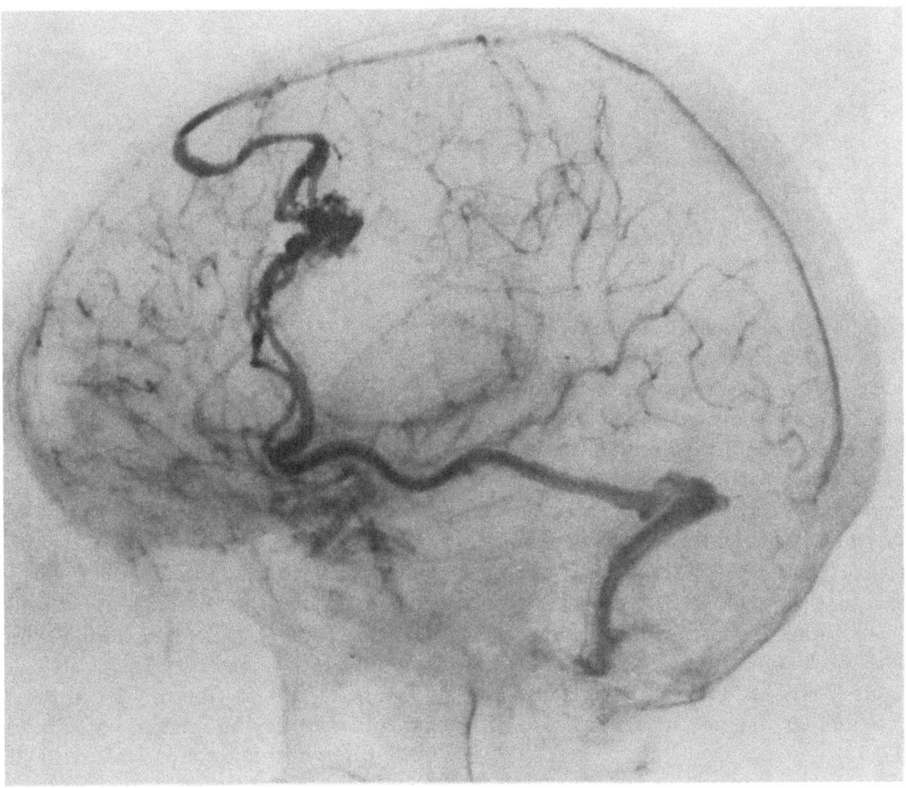

FIGURE 3–4C

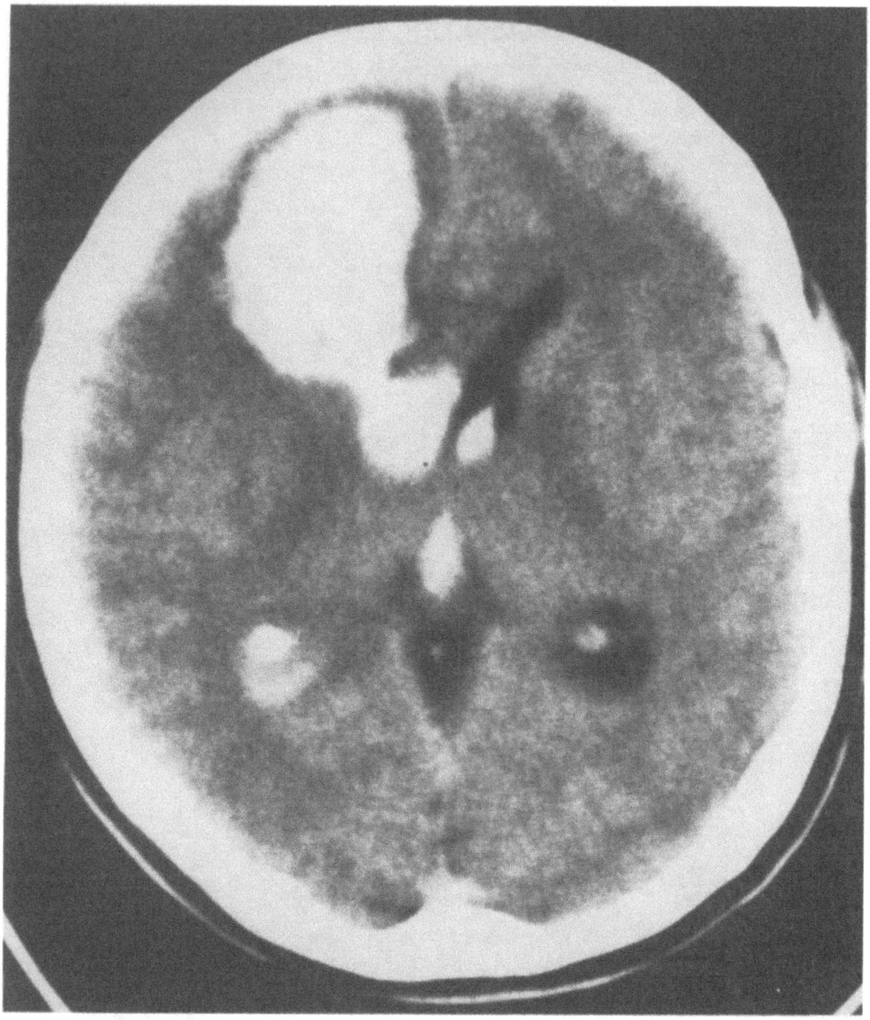

modalities [24]. MR has certain disadvantages with respect to CT as well. Calcium gives rise to no signal. Therefore, small lesions that present with minor calcifications only could be missed by MR, unless the calcification is large enough to present as an area of signal void. This same relative insensitivity of MR to calcification provides an instance where MR can be complementary to CT. Hemorrhage and calcification both appear hyperdense on CT scans and are both frequently seen in AVMs. On MR, hemorrhage and calcification are quite distinct, with hemorrhage appearing with high-signal intensity on both T1- and T2-weighted images after approximately 24 hours, and calcium giving rise to a signal void if visible at all.

MR is quite sensitive to the presence of parenchymal hemorrhage after 24 hours. Small amounts of

FIGURE 3–5. Intraventricular extension of hematoma with AVM. A 39-year-old male with recent ictus demonstrates a right frontal hematoma, with significant mass effect, that has extended into the lateral and third ventricles. Approximately 6% of hemorrhages from AVMs extend into a ventricle.

blood can be more easily identified by MR than CT [24]. However, during the first 24 hours after ictus and in the subarachnoid and intraventricular compartments, CT has the advantage in visualizing blood. Therefore, most patients entering hospitals with suspected intracranial hemorrhage are first studied by CT. The rapid scan time of CT compared to MR for patients that may be agitated presents advantages as well. Elective studies of patients with chronic headaches, seizures, or ischemic signs have a higher yield when performed with MR.

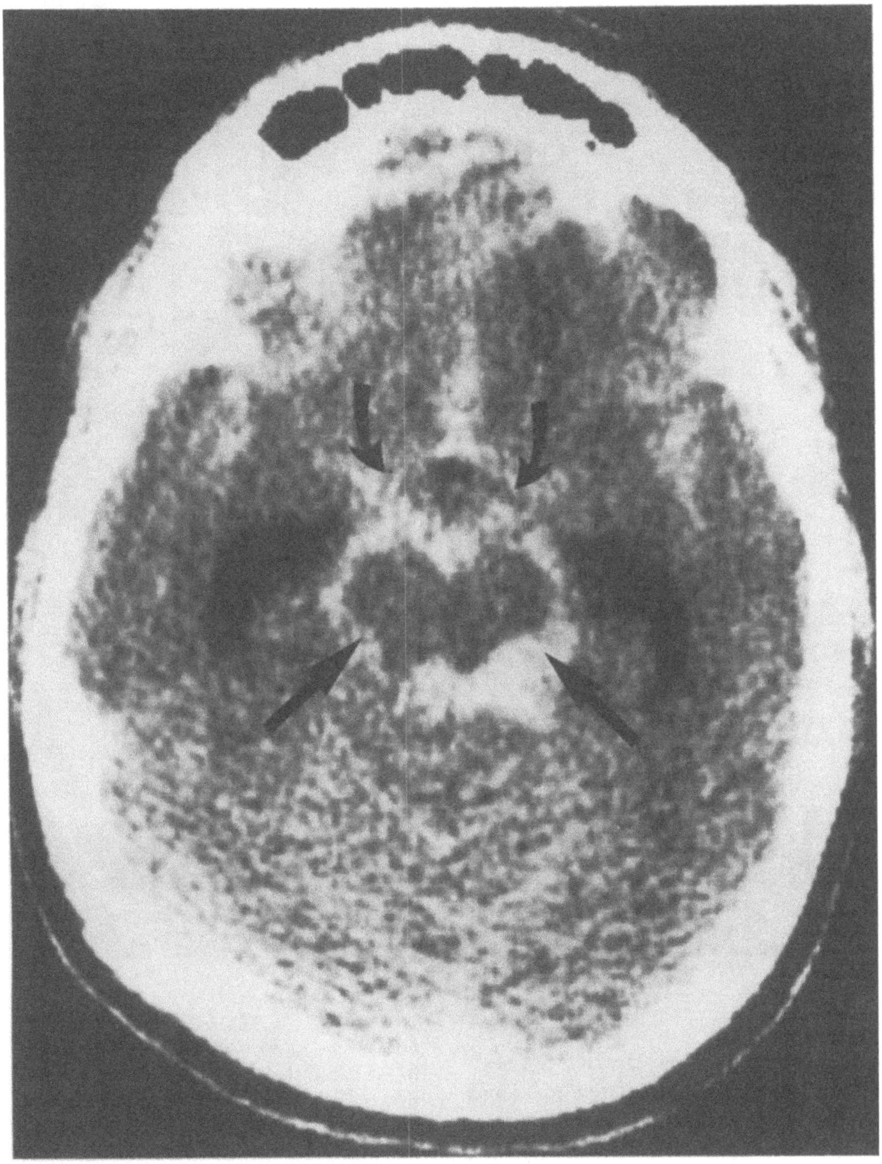

FIGURE 3–6. Subarachoid hemorrhage in AVM. The suprasellar (curved arrows) and perimesencephalic (straight arrows) cisterns are opacified by subarachnoid blood turning these usually black structures white. The bleeding arose from an AVM in the superior cerebellar vermis.

Arteriovenous malformations that have not thrombosed or been previously treated are high-flow shunts with blood passing rapidly from arteries to veins. This results in the classic flow-void effect demonstrated on MR (figure 3–8). Arteries and veins are identified as serpiginous structures with a lack of signal on both T1- and T2-weighted images.

MR is extremely sensitive to subtle changes in tissues adjacent to AVMs. Because of the steal phenomena frequently encountered with AVMs, cerebral tissue between and adjacent to vessels of the AVM and its nidus may be affected by anoxic change, edema, or gliosis. These changes are seen on MR images as areas of increased signal intensity on T2-weighted images (figure 3–9), and as normal or low-signal regions on T1-weighted images. These changes are readily differentiated from hemorrhage, which demonstrates increase in signal intensity on

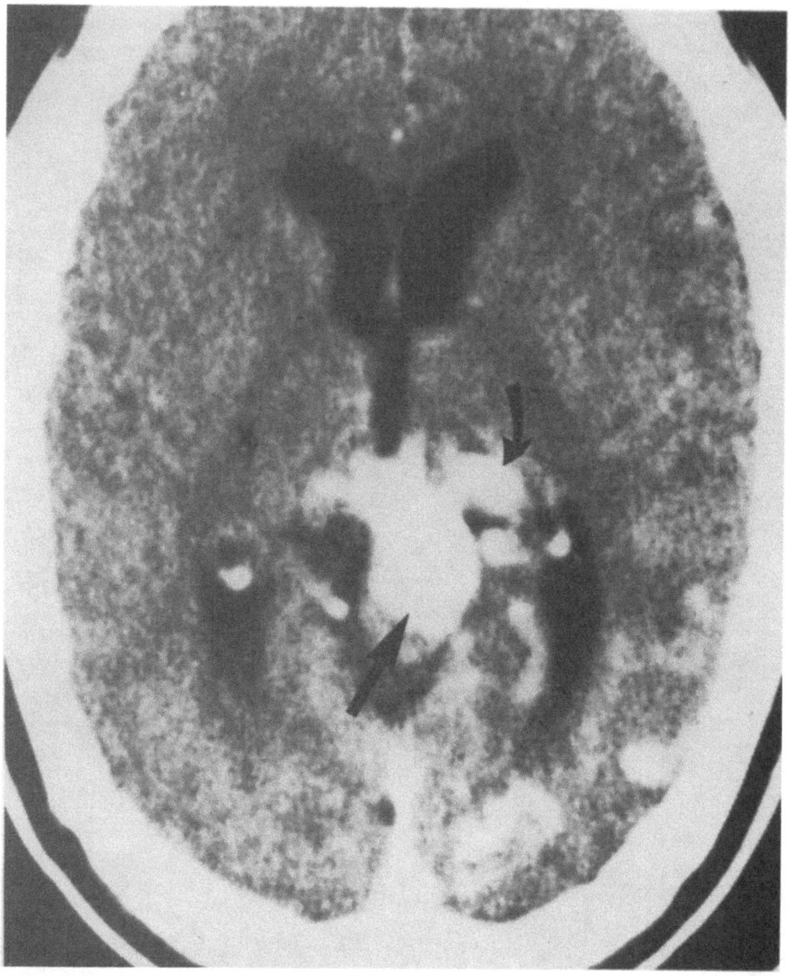

FIGURE 3–7A

FIGURE 3–7. Vein of Galen malformation. (A) CT scan following intravenous contrast administration shows the markedly enlarged vein of Galen (straight arrow) and a portion of the malformation in the left posterior thalamic region (curved arrow). (B) A T1-weighted MR image in the transaxial plane shows absence of signal (flow void) in the region of the vein of Galen (arrow) and within the more anterior malformation. (C) Coronal T2-weighted image shows flow void within the vein of Galen (G) as well as within dilated vessels along the tentorium (arrows). (D) Dilated left and right posterior cerebral arteries supply the malformation as demonstrated at angiography. (Illustrations courtesy of Michael J. Brantley, M.D. Walter Reed Army Medical Center, Washington, D.C.)

both T1- and T2-weighted images after approximately 24 hours. Subtle changes related to anoxia may not be appreciated on CT scans.

MR can identify individual vessels related to the AVM nidus; however, one cannot always ascertain whether these are arteries or veins. The sagittal projection can be helpful in this regard and may identify arteries arising from the anterior, middle, or posterior cerebral arteries, or more commonly may demonstrate large draining veins. The even-echo rephasing phenomena may be helpful in distinguishing the slower-flowing blood in veins from the more rapidly flowing blood in arteries when a multiecho series is utilized [26, 27]. An AVM that is not obscured by beam-hardening artifacts on CT and that does not demonstrate subtle calcifications, hemorrhage, or evidence of previous ischemia will demonstrate similar findings on MR and CT (figure 3–10). It should be considered, however, that similar findings are demonstrated at the cost of an injection of iodinated contrast material as well as a distinct, albeit minor, radiation dose for CT.

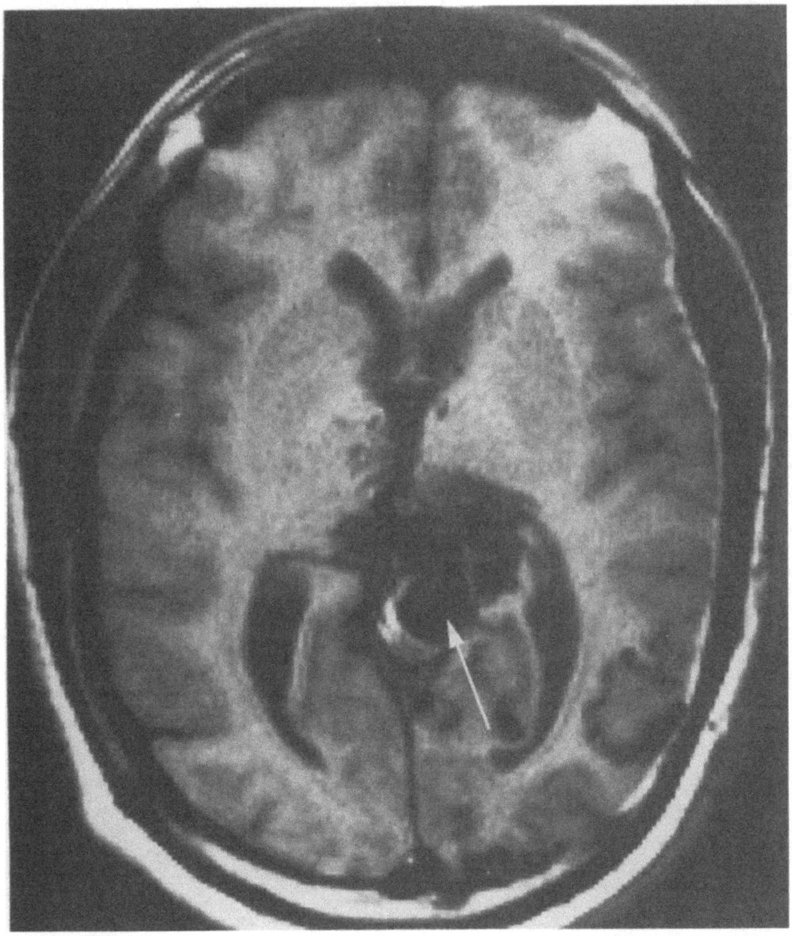

FIGURE 3–7B

ANGIOGRAPHIC FINDINGS

Despite the advances made over the past 15 years with CT and MR, angiography remains the gold standard for the complete evaluation of AVMs. Arteriovenous malformations have been classified angiographically on the basis of their arterial blood supply into pial, dural, and mixed pial–dural types. Lesions that are purely pial are intraaxial with no meningeal arterial contribution; they are supplied only by cerebral or cerebellar arteries. Pure dural AVMs are supplied by meningeal arteries, which may be branches of the internal or external carotid artery or of the vertebral artery. The anterior falx artery and the tentorial branch of the meningohypophyseal trunk (also referred to as the artery of Bernasconi and Cassinari) are examples of branches of the internal carotid artery that supply dural structures and may represent the sole blood supply to pure dural AVMs.

Lesions that are mixed pial–dural AVMs demonstrate a blood supply both from the meningeal circulation and from branches to the cerebrum or cerebellum (figure 3–11). The relative occurrence of pial, dural, and mixed AVMs in one reported series [28] is summarized in table 3–2.

TABLE 3–2. Occurrence of Pial, Dural and Mixed AVMs in 129 Patients

	Supratentorial	Infratentorial
Pure pial	81	13
Mixed pial–dural	16	4
Pure dural	6	9
Total	103	26

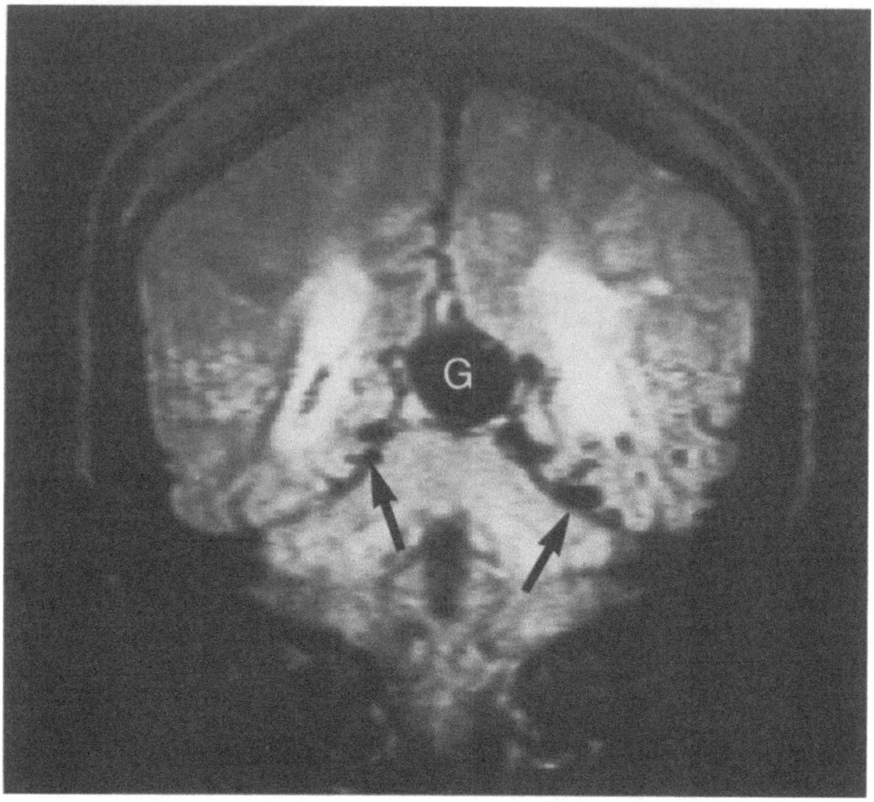

FIGURE 3–7C

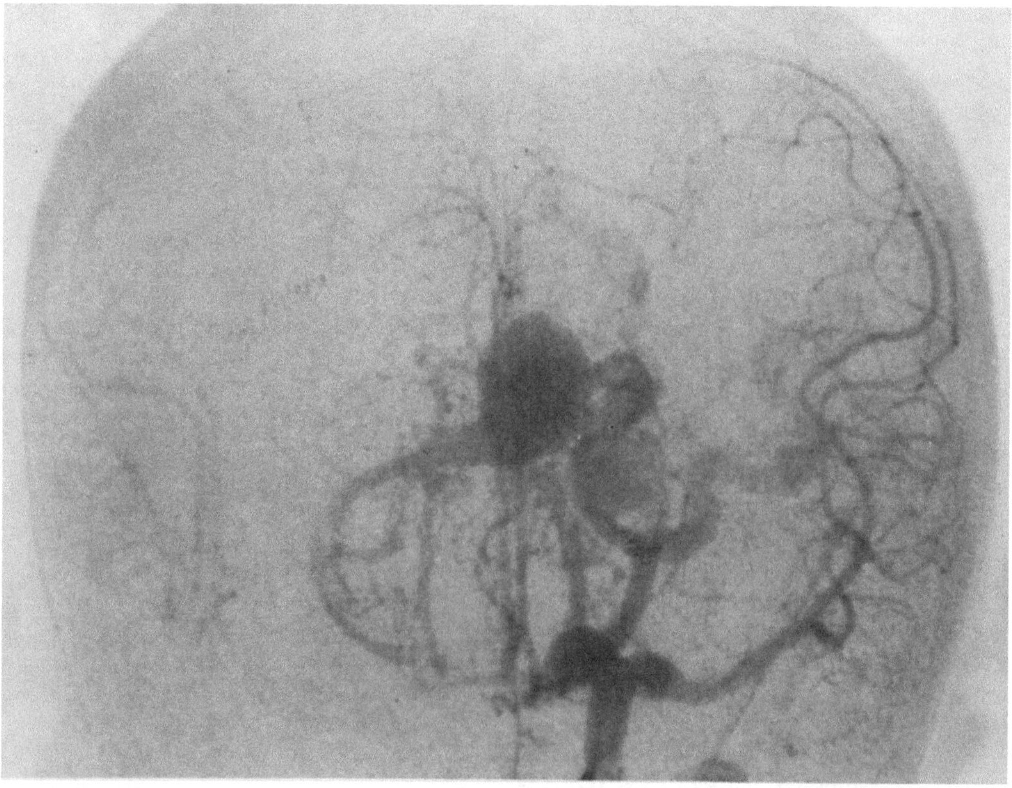

FIGURE 3–7D

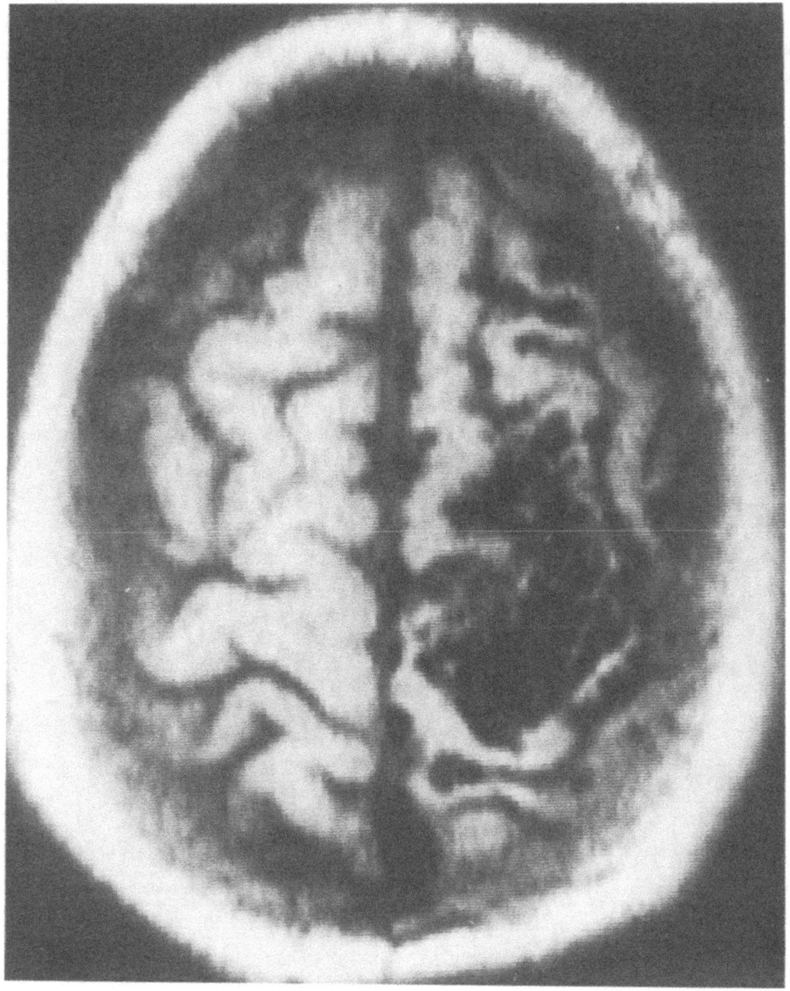

FIGURE 3–8A

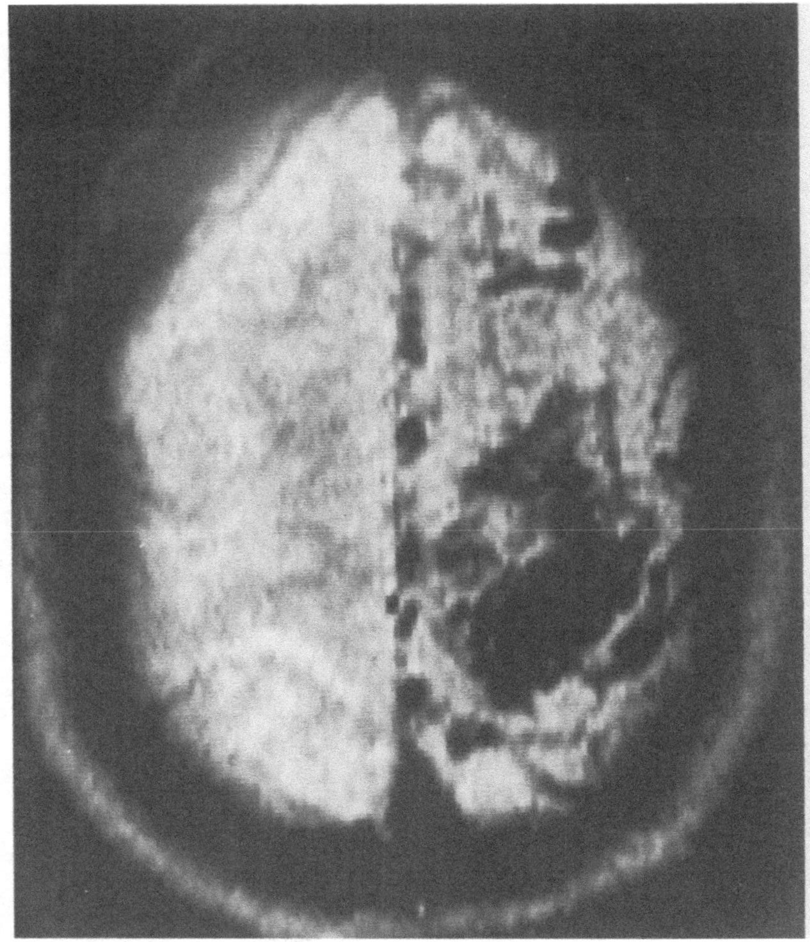

FIGURE 3–8B

FIGURE 3–8. MR findings in unruptured AVM. (A) TR = 500 msec., TE = 30 msec. T1-weighted image demonstrating a serpiginous area of absent signal in the left frontoparietal region in this 30-year-old male who presented with headaches only. (B) TR = 2000 msec., TE = 90 msec. On this T2-weighted image, the cerebrospinal fluid has increased signal and the AVM is now more dramatically outlined in its full extent. The flow-void phenomenon is accentuated by the increased cerebrospinal fluid signal surrounding it.

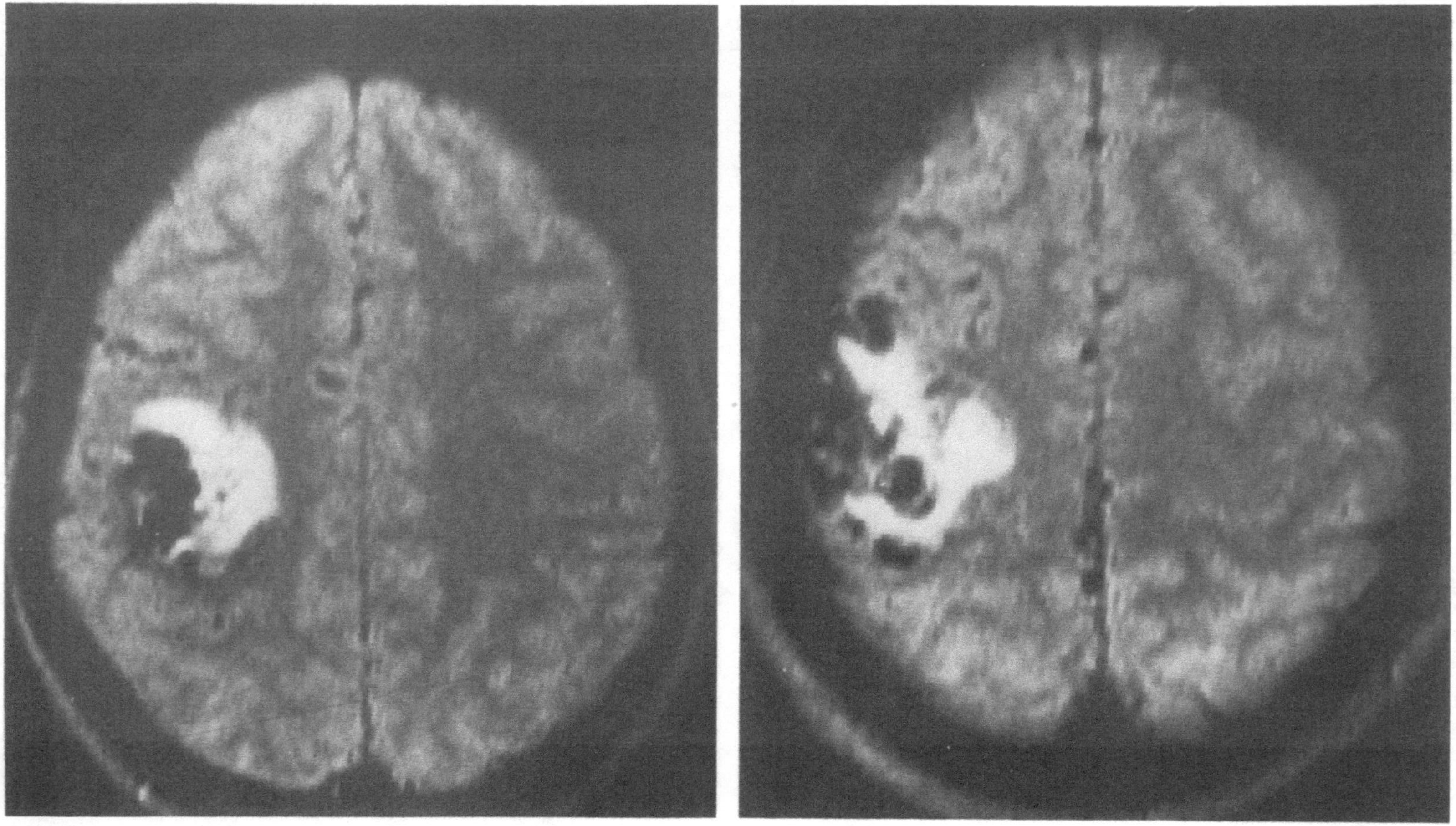

FIGURE 3–9B

FIGURE 3–9A

FIGURE 3–9. Anoxic change on MR imaging. (A) TR = 2000 msec., TE = 90 msec. This 28-year-old female with seizures demonstrates a low-signal region extending into the white matter in the right frontoparietal region with a medial area of increased signal intensity. (B) The same pulse sequence at a higher level confirms the serpiginous appearance of an AVM The medial high-signal region was not seen on T1-weighted images and corresponds to a region of anoxia probably related to the steal phenomena.

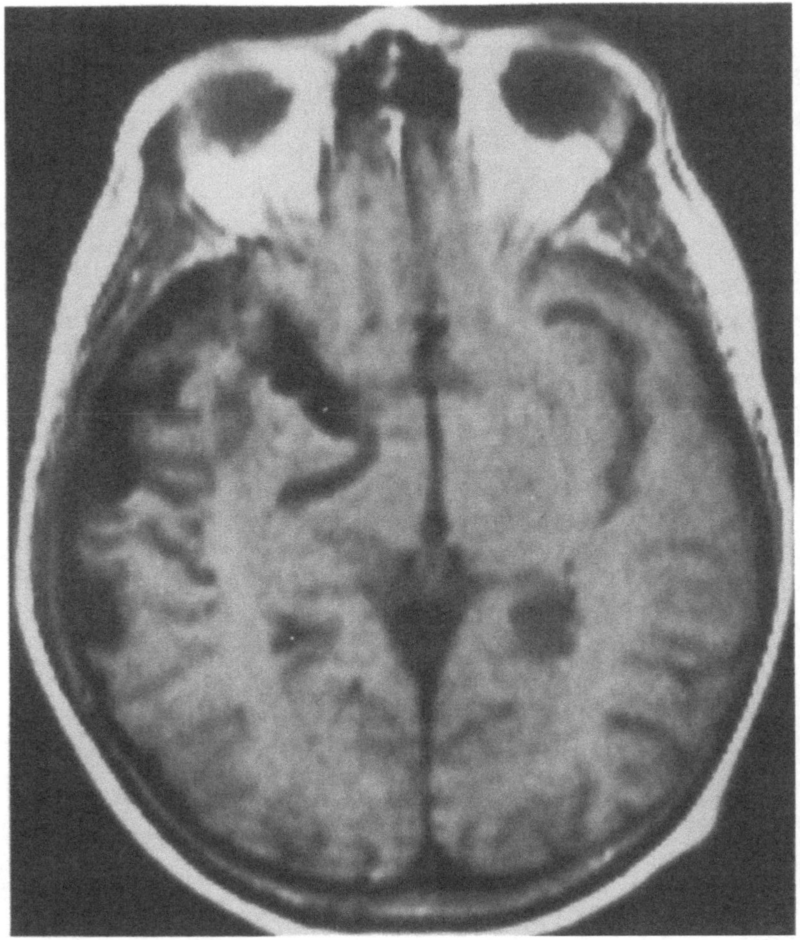

FIGURE 3–10A

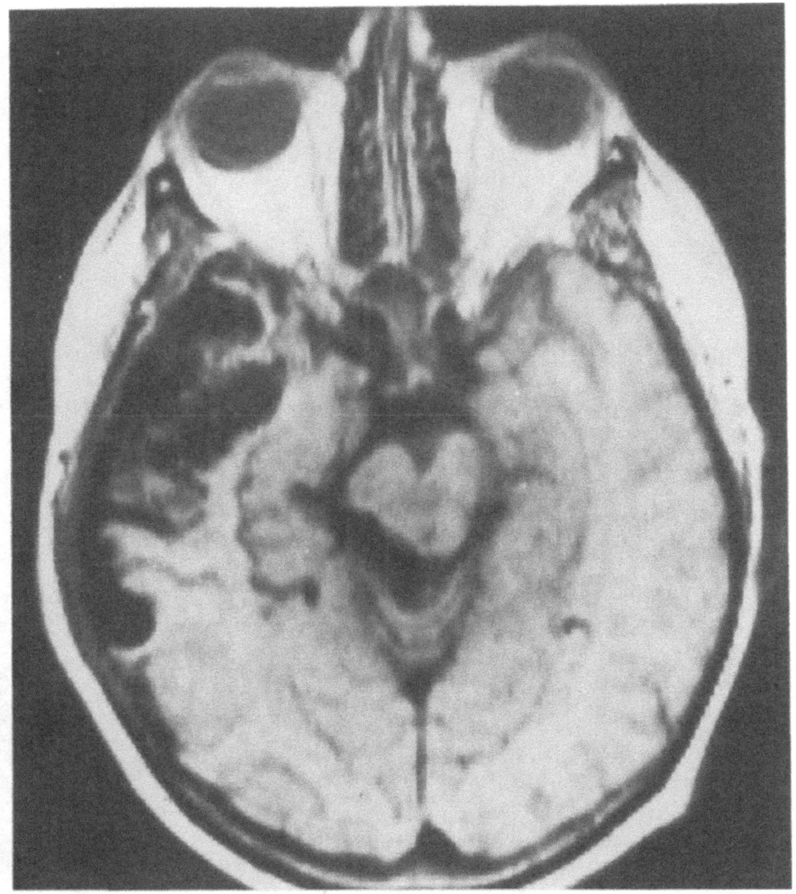

FIGURE 3–10B

FIGURE 3–10. Comparable MR and CT findings in AVM. (A & B) TR = 525 msec., TE = 32 msec. There is a large AVM in the right temporal lobe with draining veins demonstrated. Note the rotation of the brainstem and displacement of the third ventricle to the left, indicative of mass effect without evidence of hemorrhage—an uncommon, but reported, finding with AVM. (C & D) Postinfusion CT scan. The findings demonstrated are essentially identical to MR. However, the noncontrast examination provided far less detail with respect to the AVM. Intravenous contrast is necessary for an equivalent examination.

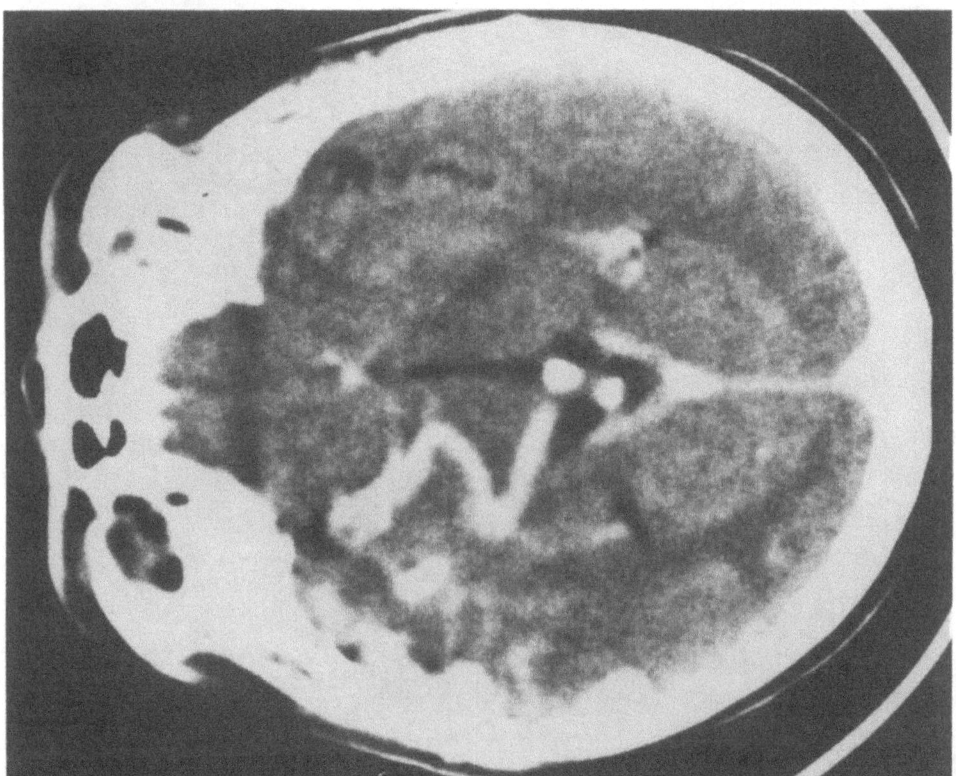

FIGURE 3–10C

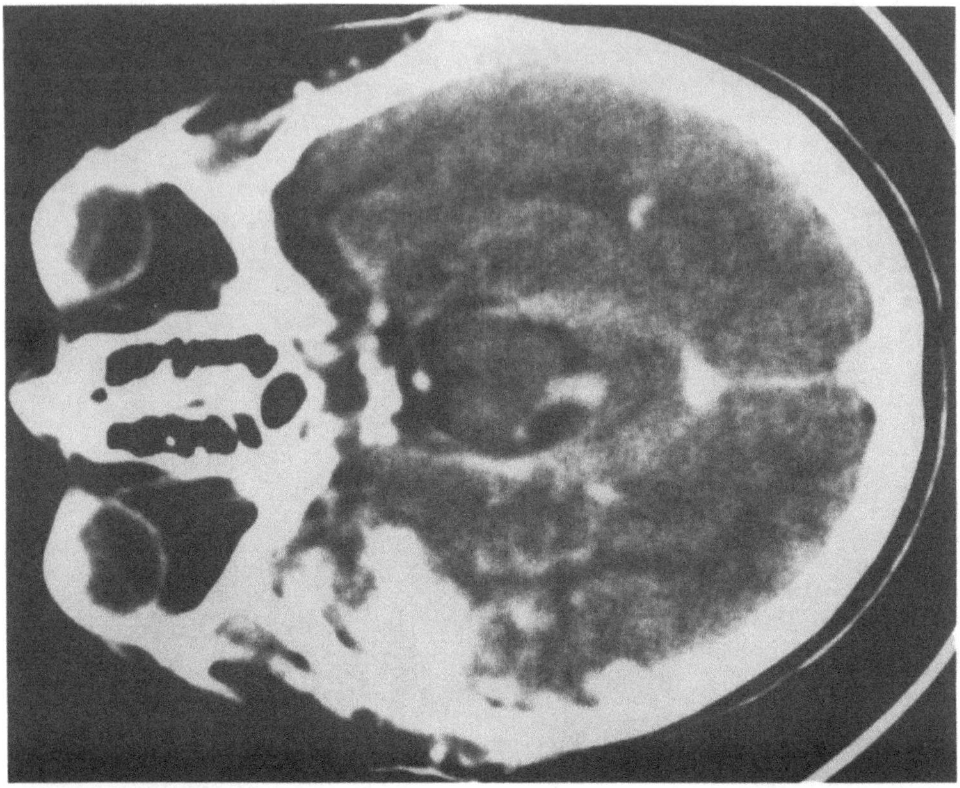

FIGURE 3–10D

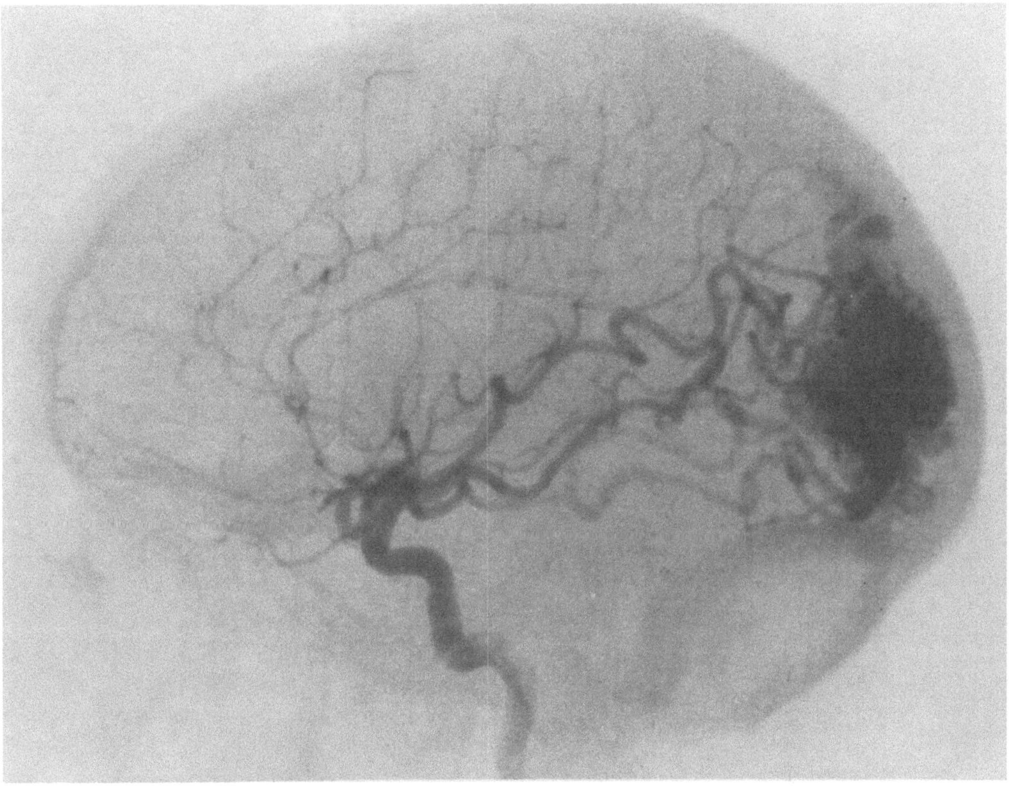

FIGURE 3–11A

FIGURE 3–11. Mixed pial–dural AVM. This AVM demonstrated both pial and dural arterial supply from both the anterior and posterior circulations. (A) Large middle cerebral arterial branches are seen to supply the nidus. (B) A meningeal branch of the external carotid artery contributes blood supply as well. (C) In the posterior fossa, posterior cerebral branches (curved arrows) of the basilar artery as well as the posterior meningeal branch (straight arrow) of the vertebral artery supply the malformation.

Supratentorial pial AVMs frequently are supplied by more than one vessel. A review of angiograms performed in 98 patients with AVMs, in one series [29], revealed 88% to be located in the supratentorial compartment. Two thirds of these were in or near a watershed area of the brain and were supplied by two or more major cerebral arteries. If an AVM is in the distribution of a single vessel, the vessel is most commonly the middle cerebral artery. Most of the pial lesions not in the supratentorial compartment are located in the cerebellar vermis.

Pure dural AVMs are relatively uncommon [30]. They account for 2–12% of all AVMs, depending on the series [28–33]. Most arise within or adjacent to a major venous sinus in the posterior fossa and have predominantly dural venous drainage. Because they are located outside of the brain, these lesions are usually occult to CT and MR unless they have hemorrhaged, which is rare. Although most dural AVMs are asymptomatic and present as incidental findings encountered during angiography of unrelated lesions, they can be associated with buzzing sounds and subarachnoid hemorrhage on rare occasions [32].

Because the blood supply to an AVM can encompass virtually any intracranial vessel, a complete angiographic evaluation would require injection of both internal and both external carotid arteries as well as both vertebral arteries. However, since most AVMs encountered currently have been well studied by CT and/or MR prior to angiography, if the AVMs are located close to the surface of one hemisphere and well away from a potential supplying artery, some investigators prefer to omit injections of vessels remote from the lesion. Unlike cavernous malformations and telangiectasia, AVMs are multifocal only very rarely [34].

Cerebral ischemia can be caused by AVMs due to the steal phenomena. Occasional patients may present with transient ischemic attacks. One author has suggested that seizures associated with AVMs may be

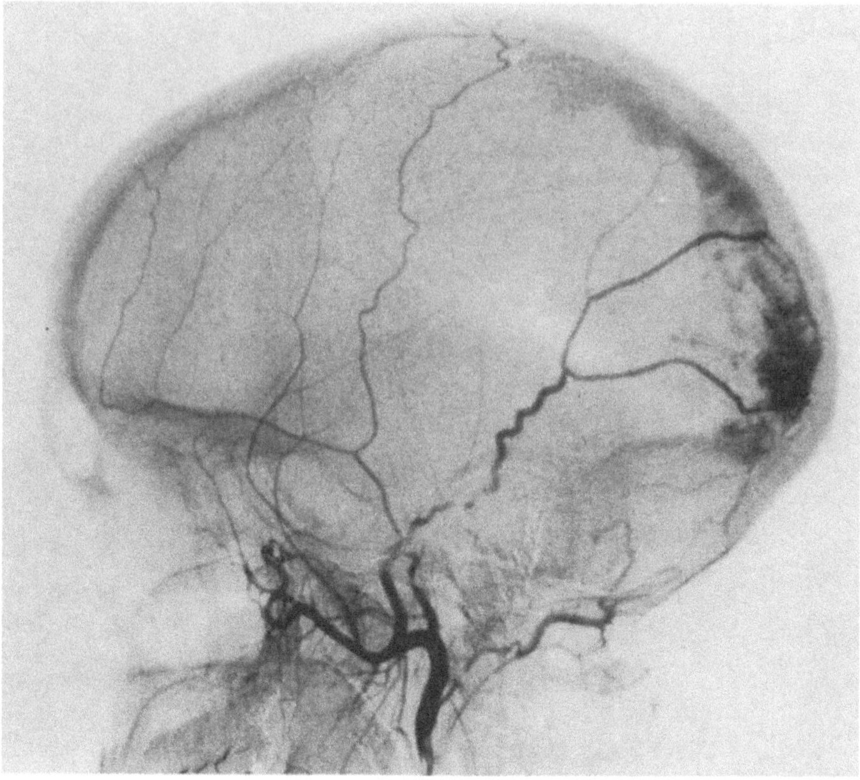

FIGURE 3–11B

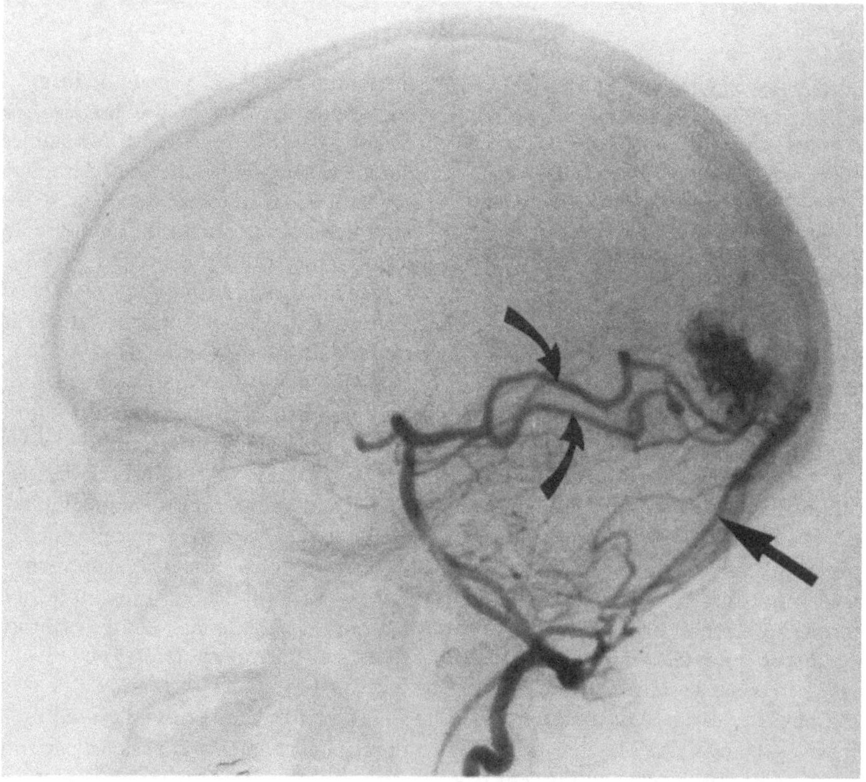

FIGURE 3–11C

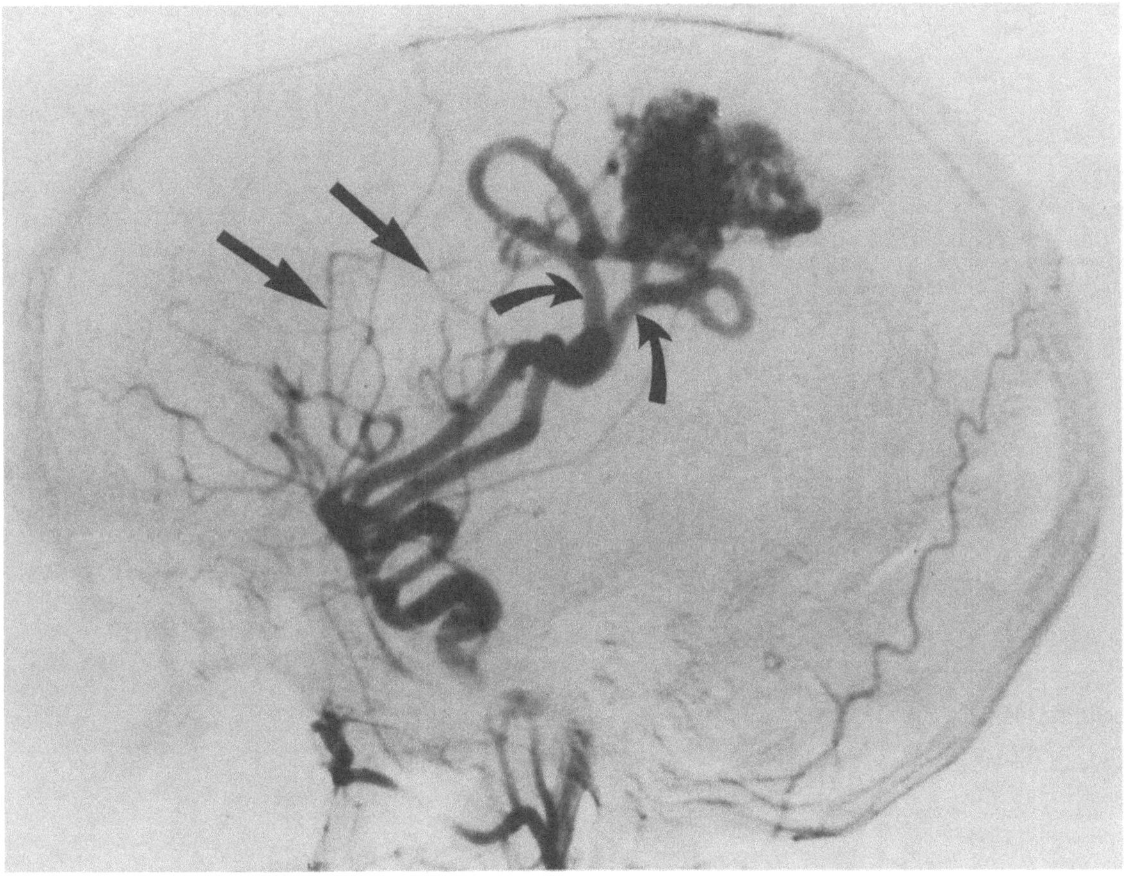

FIGURE 3–12. Steal phenomena demonstrated at angiography. A typical AVM in the posterior frontoparietal region. The posterior branches of the middle cerebral artery are markedly dilated (curved arrows). The anterior middle cerebral arterial branches appear attenuated and underperfused (straight arrows). The steal phenomena can give rise to ischemia and perhaps seizures. Rapid reversal after treatment of the lesion can cause edema and hemorrhage in the previously underperfused region.

related to ischemia as well [1]. Angiographic correlates of the steal phenomena are demonstrable. The enlarged arteries with decreased peripheral resistance that supply the AVM will decrease the perfusion and the perfusion pressure within other arteries with the same vascular territory (figure 3–12). In a similar fashion, the decreased peripheral resistance within the nidus of the AVM creates a sump effect that can diminsh the perfusion even to regions relatively remote to the AVM (figure 3–13).

One rare occasions, an AVM that has not been treated may manifest partial regression when serial angiographic studies are preformed [9]. There are, however, very few reports of complete spontaneous regression [35]. One patient in our own series demonstrated total obliteration of his AVM on serial angiographic studies with no evidence of parenchymal hemorrhage on CT scans performed concurrently with the angiographic studies (figure 3–14). A new hemorrhage, however, may substantially alter the angiographic appearance of an AVM. Hemorrhages that are extensive may totally obliterate the angiographic findings associated with an AVM. Lesser degrees of compression by hematoma can alter the usual angiographic findings as well (figure 3–15).

The association of intracranial aneurysms with AVMs is well documented and is of significant interest. The reported incidence varies from 3–9% depending on the series [36, 37]. In the Cooperative Study, 37 of 490 AVM patients demonstrated an aneurysm for an incidence of 7.6%. In our own series, three of 58 patients or 5.2% of the patients with AVMs had aneurysms. These rates are all considerably higher than the incidence rates of aneurysm

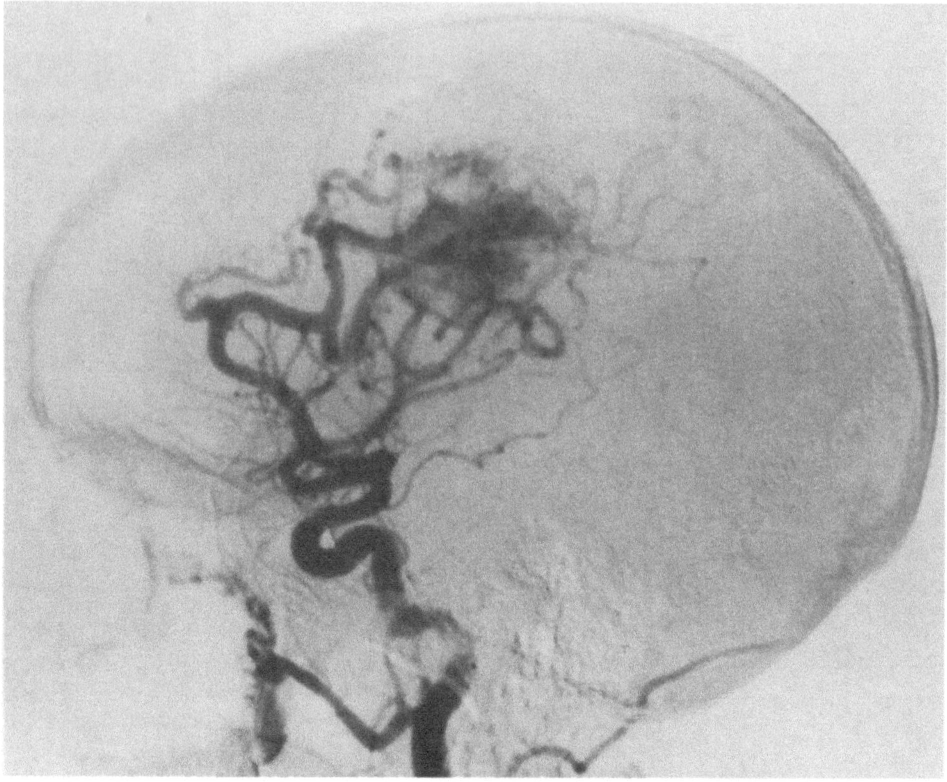

FIGURE 3–13A

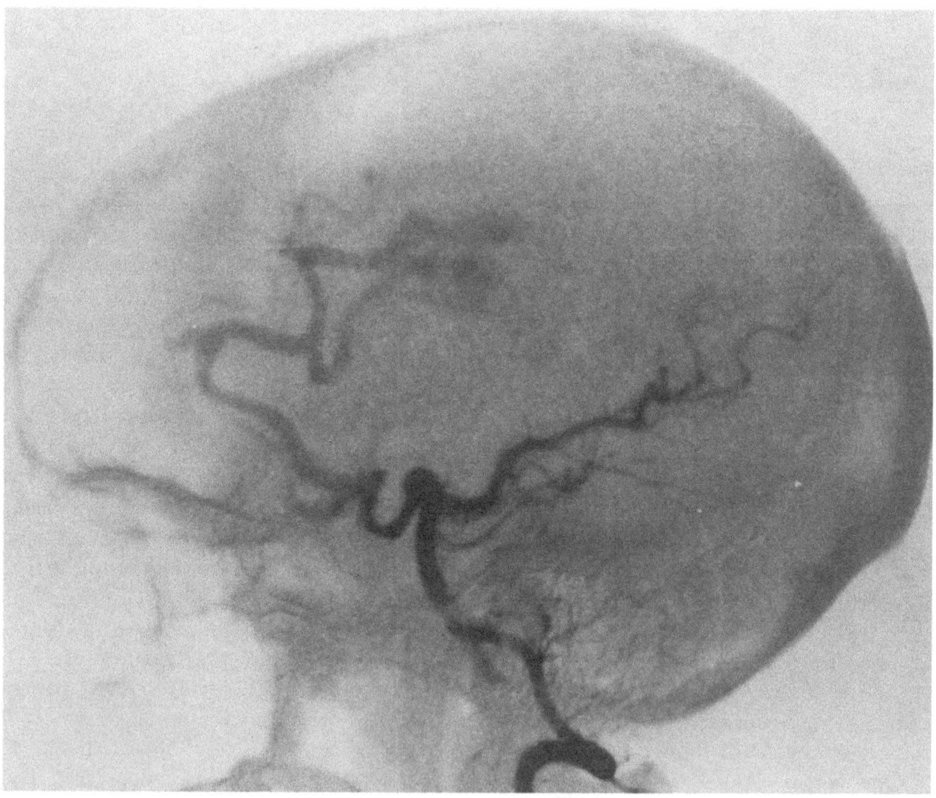

FIGURE 3–13B

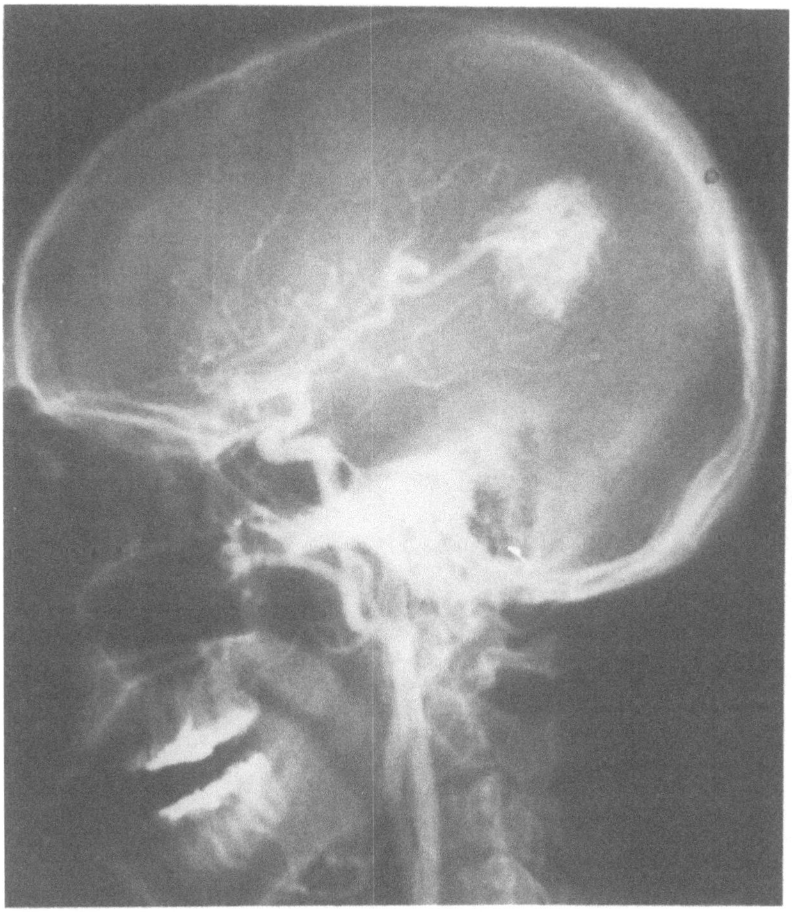

FIGURE 3–14A

FIGURE 3–14. Spontaneous thrombosis of an AVM. (A) Right parietal AVM. (B) Two years later, with no intervening treatment, the lesion is no longer visualized at angiography. Note the same arterial configuration as two years previously, and the more peripheral arterial perfusion of normal structures (arrows).

FIGURE 3–13. Sump effect of a large AVM. (A) A carotid injection demonstrates predominantly anterior cerebral artery but also middle cerebral artery feeders to this large AVM. (B) Vertebral injection reveals contrast filling the pericallosal and callosomarginal arteries through flow in the posterior communicating artery due to the sump effect of the AVM. Perfusion in the posterior fossa will be decreased.

in the population at large. Several theories have been promulgated to explain this association. The most enticing of the explanations relates to the increased flow through the arteries supplying the AVM. This theory states that the increased hemodynamic stresses in the enlarged arteries result in an increased propensity for congenital weaknesses at bifurcation points in the arterial walls to become aneurysms [38]. This suggestion is reinforced by the much higher than normal incidence of aneurysms in arteries supplying AVMs [37] and their frequent occurrence in locations that would otherwise be unusual for congenital aneurysms (figure 3–16).

The angiographic appearance of an AVM is usually diagnostic. Occasionally, the differential diagnostic possibility of a primary or secondary tumor must be entertained. This can particularly be the case if the AVM is small or compressed by hematoma and the characteristic draining veins are not readily apparent. An AVM commonly shows considerably more dilata-

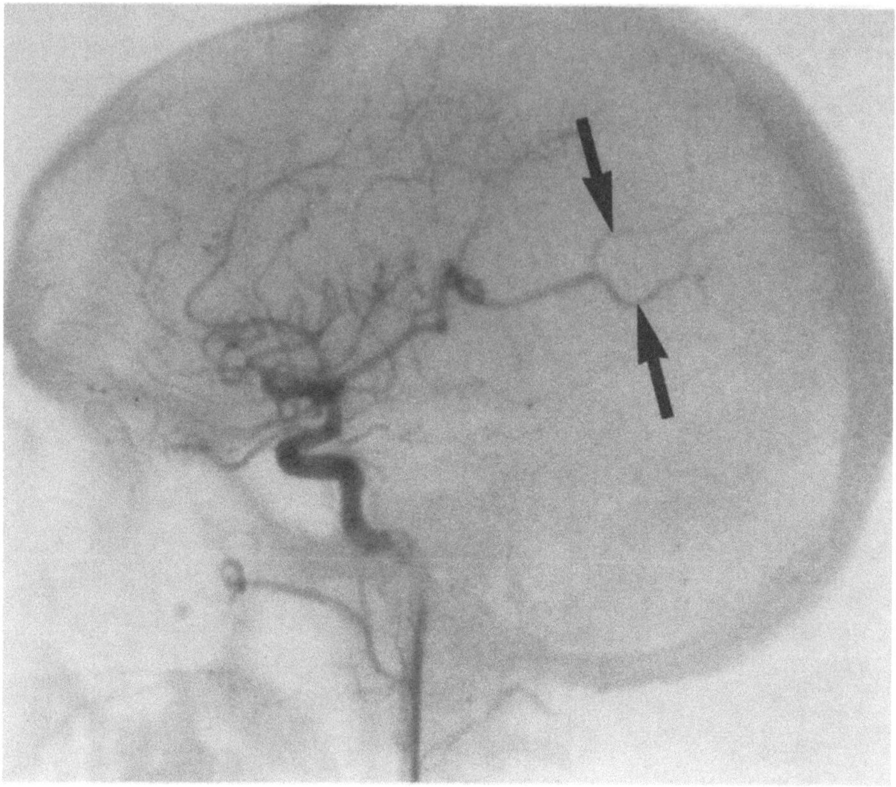

FIGURE 3–14B

tion and undulation of its afferent vessels than does a neoplasm [39]. Furthermore, veins draining neoplasms are rarely large. The abnormal vessels or neovascularity encountered in malignant intraaxial tumors such as glioblastoma frequently demonstrate a widely scattered distribution with intervening avascular zones. The nidus of an AVM is usually quite compact with little if any empty space, particularly if all supplying vessels are opacified. When the CT and/or MR findings are considered as well, this issue is usually not difficult to resolve.

Venous Malformations

Venous malformations (also referred to as venous angiomas) are frequently described as being rare [40, 41, 42]; however, they are the vascular malformation most often encountered in some series at autopsy [2, 13]. In our own series, 12 of 85 vascular malformations encountered in the brain were venous malformations, for an incidence of 14% (table 3-1). In the vast majority of these patients, the lesions were incidentally found and not related to the symptoms for which the patient sought medical attention. This lack of correlation of the presenting symptoms with the location of the venous malformation has been confirmed by others [40]. One type of symptomatology has been established, however. Any of the four major classifications of vascular malformations can bleed. Most commonly this is associated with an AVM. In a literature review, Rothfus [14] found 20 cerebellar venous malformations that had bled out of 56 reported, a relatively high hemorrhage rate of 36%. Hemorrhage in venous malformations above the tentorium is rare. Cerebellar venous malformations may be expected to bleed at a considerably higher rate than lesions encountered in the cerebral hemispheres.

CT SCANNING

The hallmark of the CT findings in a venous malformation is the identification of the transcerebral or transcerebellar draining vein [43, 44]. Because of the normal variations encountered in the venous drainage

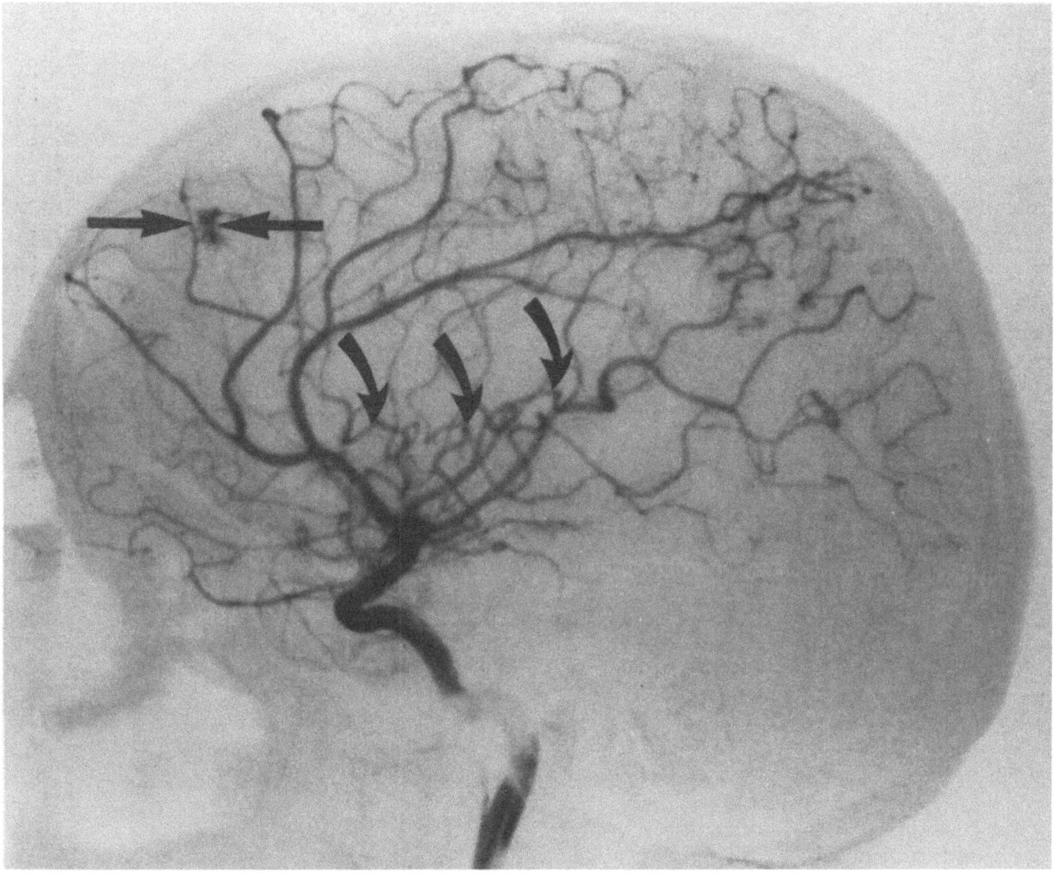

FIGURE 3–15. Alteration of angiographic findings by hematoma. The large hematoma in this patient compressed the malformation, resulting in only a small collection of vessels being seen at angiography in the right frontal lobe (straight arrows). Note the downward displacement of the sylvian triangle (curved arrows) caused by the hematoma. This patient's concurrent CT scan is illustrated in figure 3–5.

in many areas of the body, it is sometimes difficult to determine if a venous structure represents a normal variant or is draining a small venous malformation or an AVM. Normal large veins do not cross white-matter tracts; therefore, if a large vein is encountered in a white-matter location, one can assume that it is associated with a venous malformation or an AVM.

CT scanning performed following contrast administration will usually demonstrate a curvilinear enchancing structure representing the associated anomalous venous channel (figure 3–17). The findings on CT scanning are suggestive, but not pathognomonic of a venous malformation. In order to correctly make the diagnosis on the basis of the CT scan alone, one must identify the dilated medullary veins of the venous malformation feeding the ectopic transcerebral or transcerebellar vein. This is frequently not possible because the dilated medullary veins will be below the resolution of the CT scanning process. In some cases, however, the diagnosis can be established with a high degree of certainly by CT scanning alone [43]. In one of our patients who presented with right cerebral dysfunction, the incidental diagnosis of a left frontal venous malformation could be made on the basis of the CT scan alone (figure 3–18). Visualization of the perpendicular linear enhancing structures entering the draining vein is the CT equivalent of the *caput medusa* appearance characteristic of the lesion at angiography.

Occasionally, calcification within venous malformations can be demonstrated on CT scans [44] (figure 3–19). One must then consider the possibility of a low-grade neoplasm or an old inflammatory, hemorrhagic, or ischemic process. In these instances, further

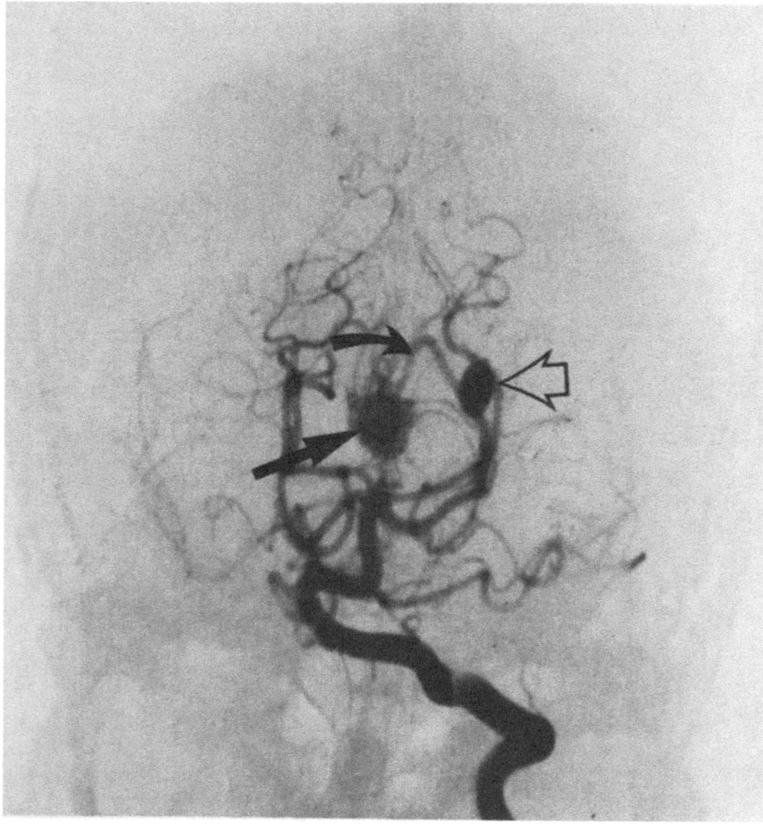

FIGURE 3–16A

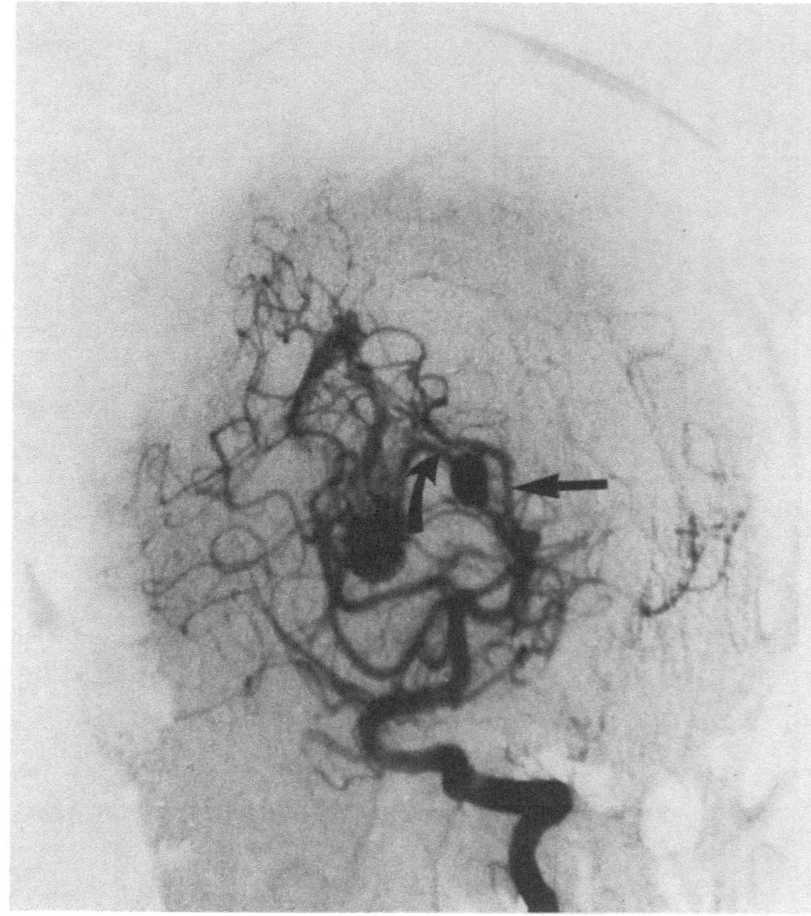

FIGURE 3–16B

FIGURE 3–16. Aneurysm associated with AVM. (A) Left vertebral injection shows an AVM (straight closed arrow) supplied by the superior vermian artery (curved closed arrow), a branch of the superior cerebellar artery. An aneurysm is identified (open arrow), but in this projection it is not clear if it is arising from the superior cerebellar artery or the posterior cerebral artery. (B) Because the origin of the aneurysm was equivocal on the lateral view as well, an oblique view was performed that showed the aneurysm to project free of the posterior cerebral artery (straight arrow) and to be associated with the superior cerebellar artery (curved arrow) supplying the AVM.

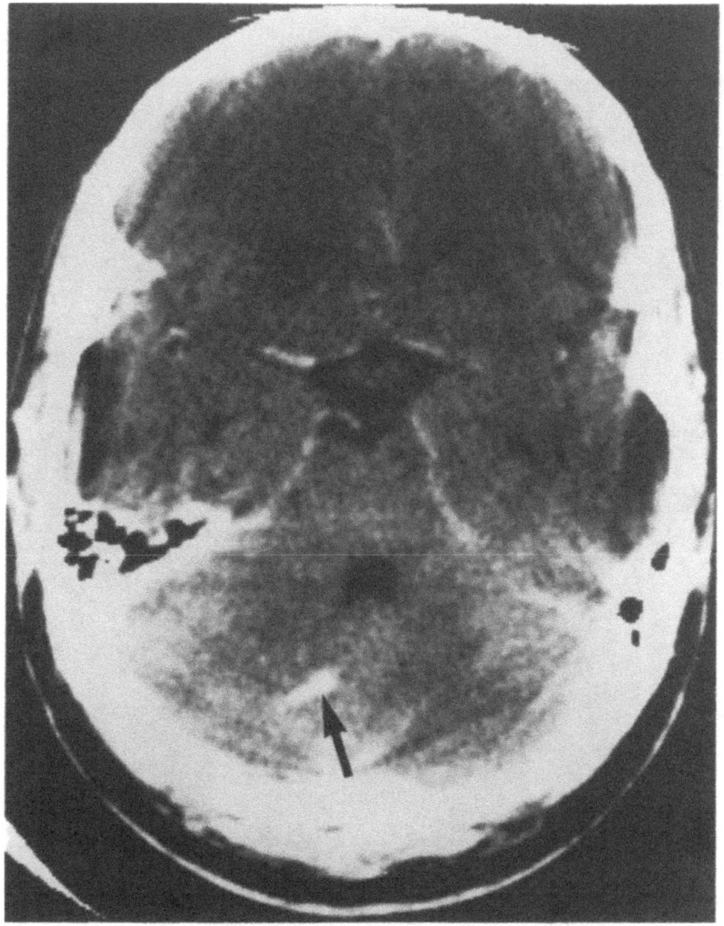

FIGURE 3–17A

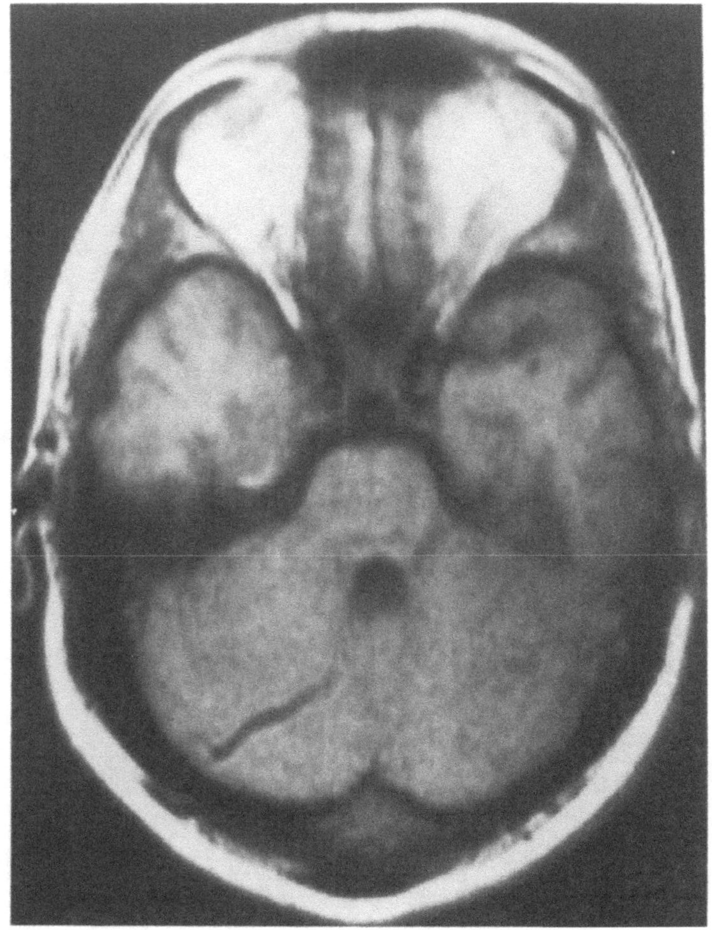

FIGURE 3–17B

FIGURE 3–17. Cerebellar venous malformation. (A) A postinfusion CT scan demonstrated a curvilinear enhancing structure coursing through the white matter of the right cerebellar hemisphere (arrow). This is suggestive of the typical transcerebellar vein draining a venous malformation. (B) TR = 525 msec., TE = 32 msec. A T1-weighted MR study shows flow void (absent signal) in the same region. (C) A section obtained slightly below that shown in (B) with the same pulse sequence demonstrates a region of increased signal in the inferomedial aspect of the right cerebellar hemisphere. This was seen also as a high-signal region on a comparable T2-weighted image and is consistent with a hemorrhage. (D) An angiogram shows the dilated medullary veins (small arrows) of the cerebellar venous malformation draining into the transcerebellar vein (large arrow) demonstrated on CT and MR. This is the typical *caput medusa* seen at angiography.

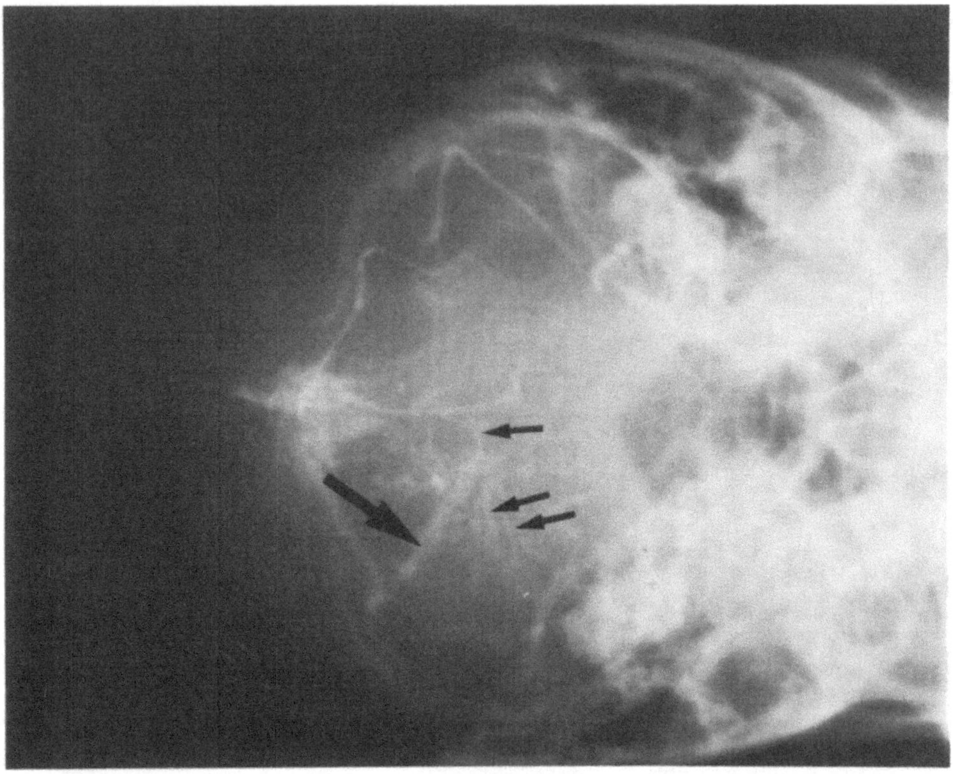

FIGURE 3–17D

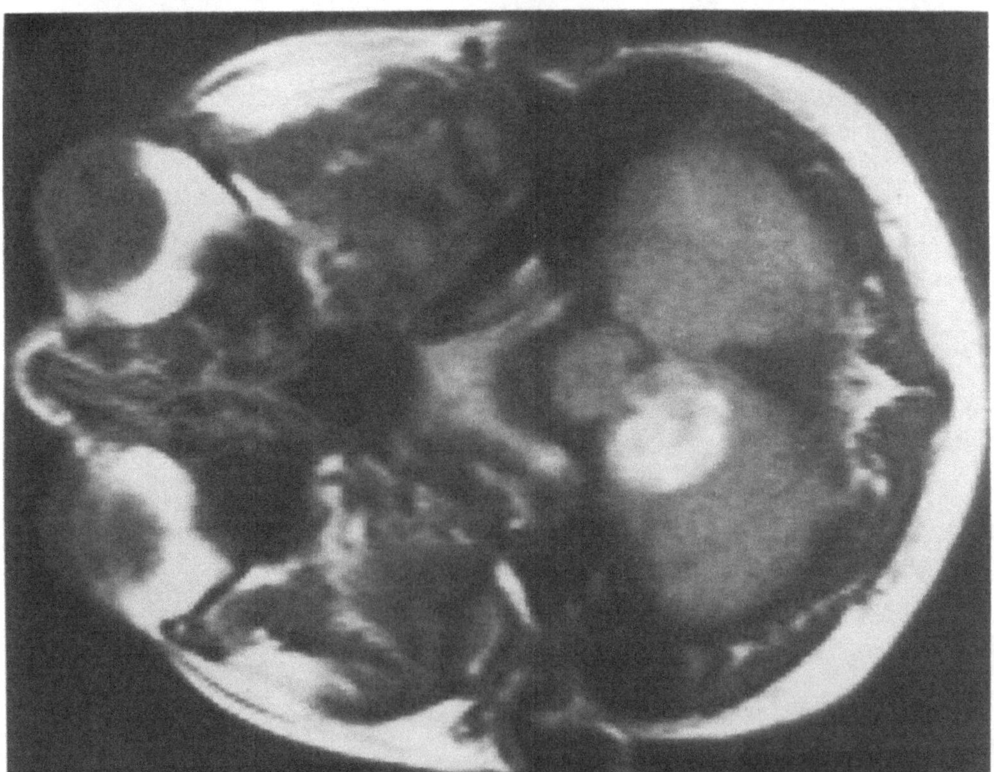

FIGURE 3–17C

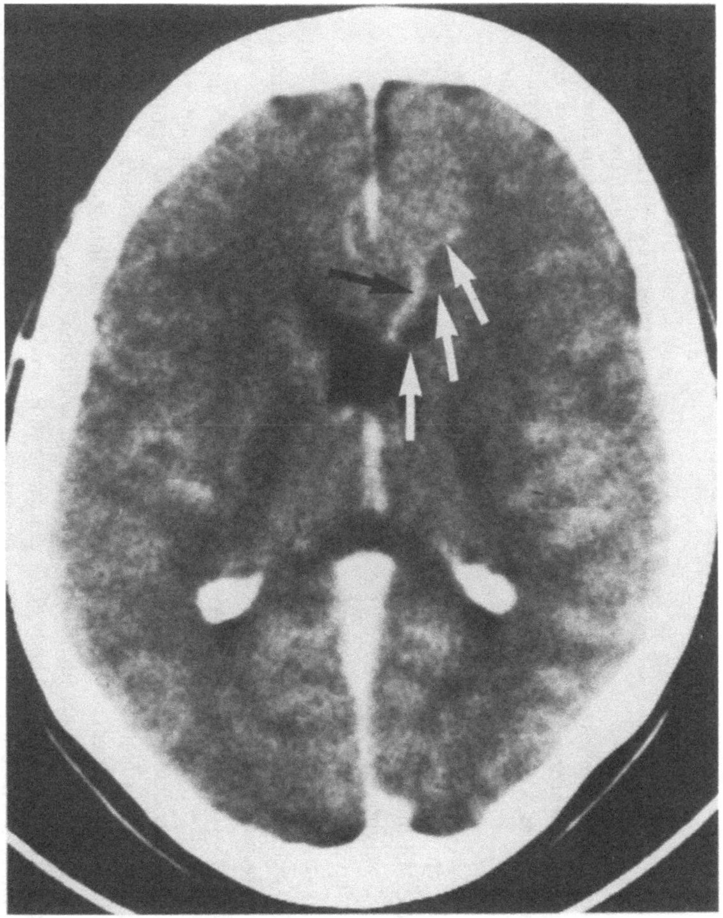

FIGURE 3–18A

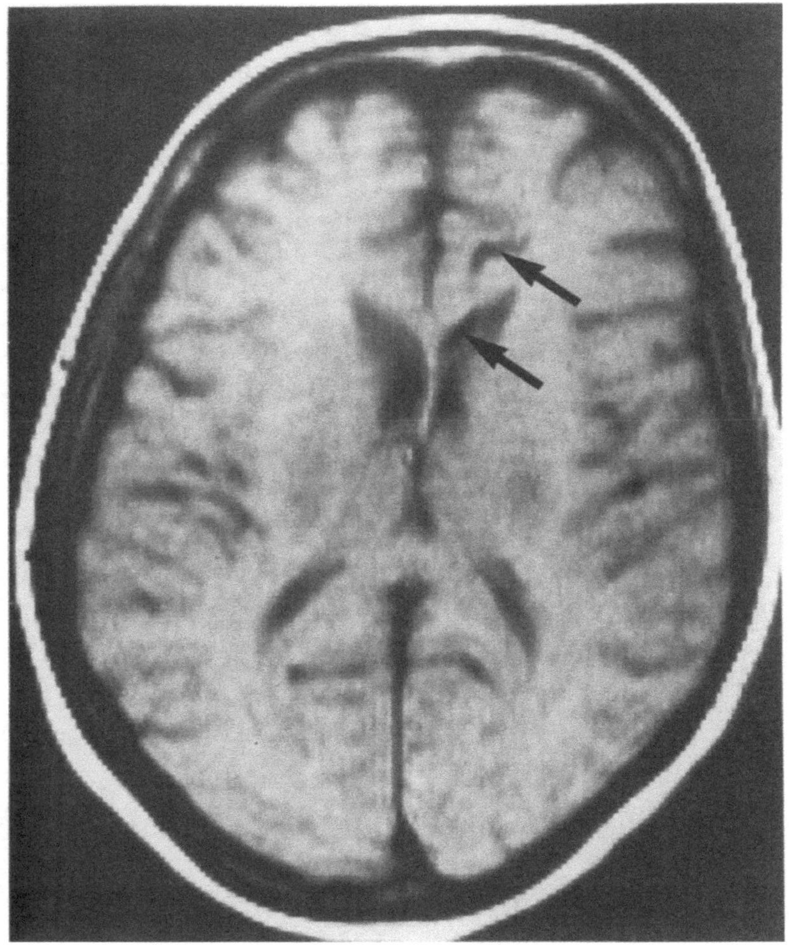

FIGURE 3–18B

FIGURE 3–18. Frontal lobe venous malformation. (A) The diagnosis of venous malformation can rarely be virtually pathognomonic on CT alone. This postinfusion study demonstrates a large enhancing linear structure (black arrow) medial to the left frontal horn which is associated with multiple perpendicular enhancing structures (white arrows) consistent with dilated medullary veins. (B) T1-weighted MR demonstrates flow void in the region of the draining vein (arrows). (C) An angiogram confirms the diagnosis. Medullary veins (straight arrows) drain into an enlarged vein (curved arrow) that enters the internal cerebral vein (open arrow). The frontal view showed the transcerebral vein to be lateral to the septal vein as suggested on CT.

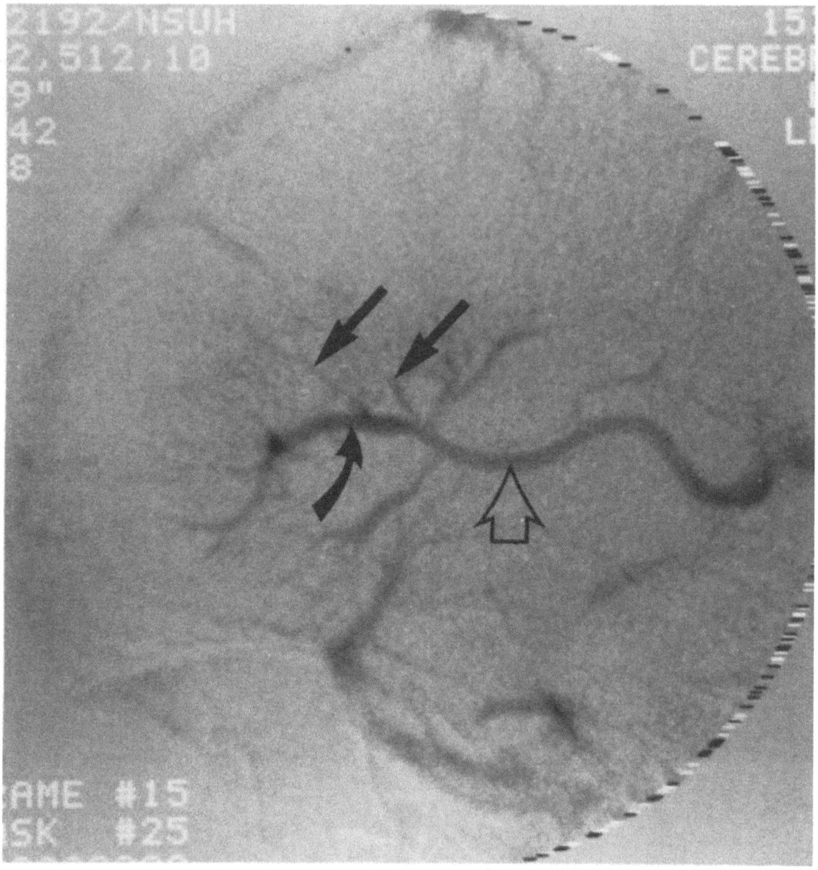

FIGURE 3–18C

FIGURE 3–19. Calcification in a venous malformation. (A) A noncontrast CT study demonstrates a focus of calcification in the right frontal lobe adjacent to the anterior horn. (B) Following intravenous contrast administration, enhancement involving the subependymal region is seen, but no mass effect on the adjacent lateral ventricle is apparent. (C) Angiography demonstrates the dilated medullary veins in the subependymal region draining into a large transcerebral vein (arrow) coursing to the superior sagittal sinus.

evaluation is warranted to confirm the diagnosis. Mass effect is almost always absent in venous malformations unless hemorrhage has occured.

MR IMAGING

Venous malformations are frequently well documented on T1-weighted MR images (figures 3–17b, 3–18b) as well as on T2-weighted images. The most characteristic feature observed is a curvilinear or round region of decreased signal intensity, usually more clearly defined on T2-weighted images. On T2-weighted images, cerebrospinal fluid demonstrates increased signal, adding contrast against the flow-void phenomena usually seen within the veins of a venous malformation. The individual medullary veins making up the body of the venous malformation are usually beyond the resolving capabilities of MR imaging. However, a region of increased signal intensity on T2-weighted images corresponding to the body of the venous malformation is often seen [45]. Occasionally, it may be difficult to distinguish an artery from a vein or a vein draining a venous malformation from a vein draining an AVM on MR imaging. Even-echo rephasing may be helpful in distinguishing arteries from veins [26]. This phenomenon occurs in vascular structures that exhibit slow laminar flow, such as large veins and dural sinuses. It is characterized by increased signal intensity on the second, fourth, and additional even echoes of a multiecho chain when flow void (absent signal) is present within the vascular structure on odd echoes (figure 3–20).

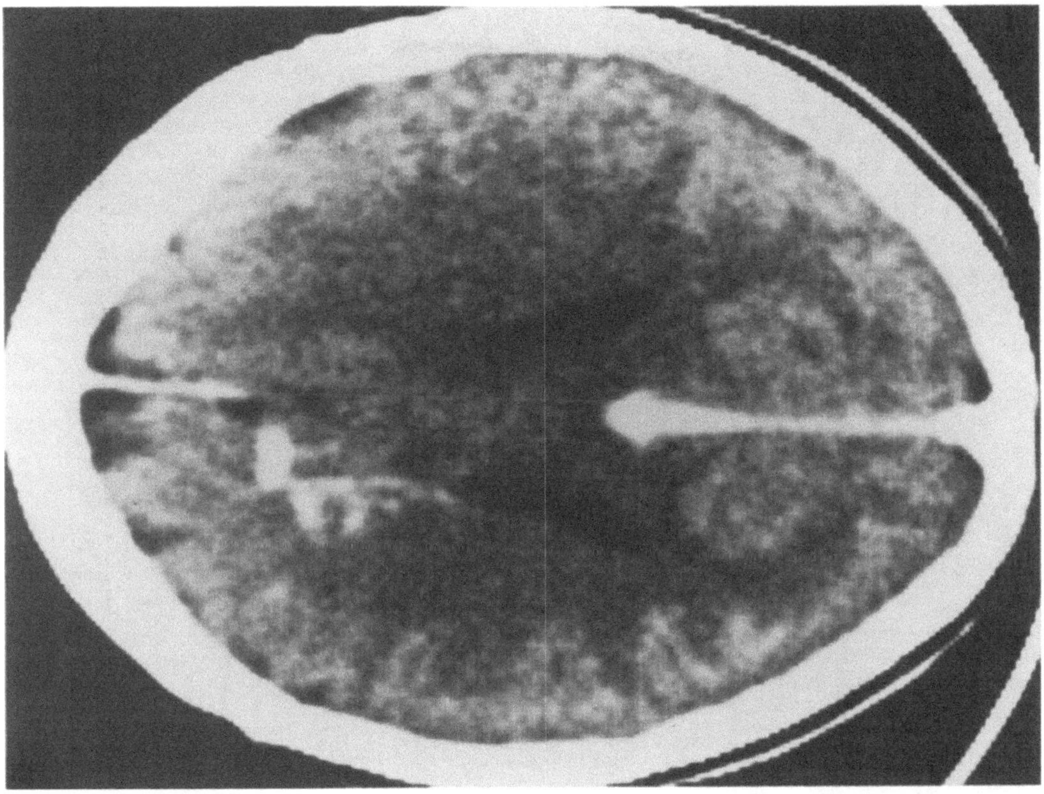

FIGURE 3–19B

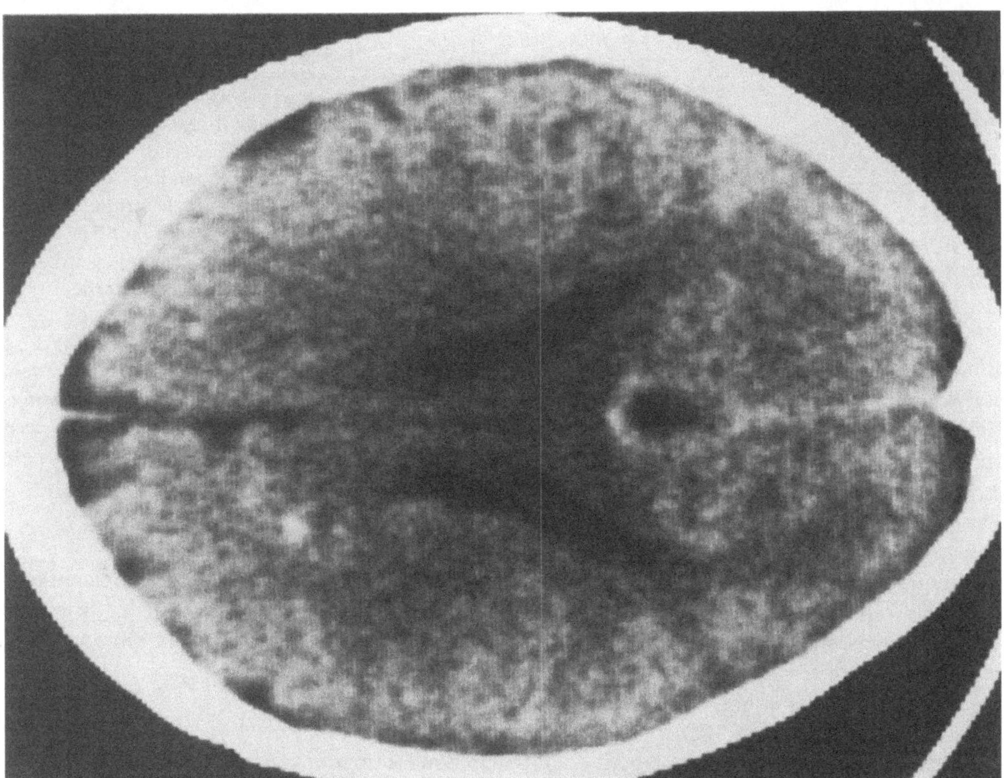

FIGURE 3–19A

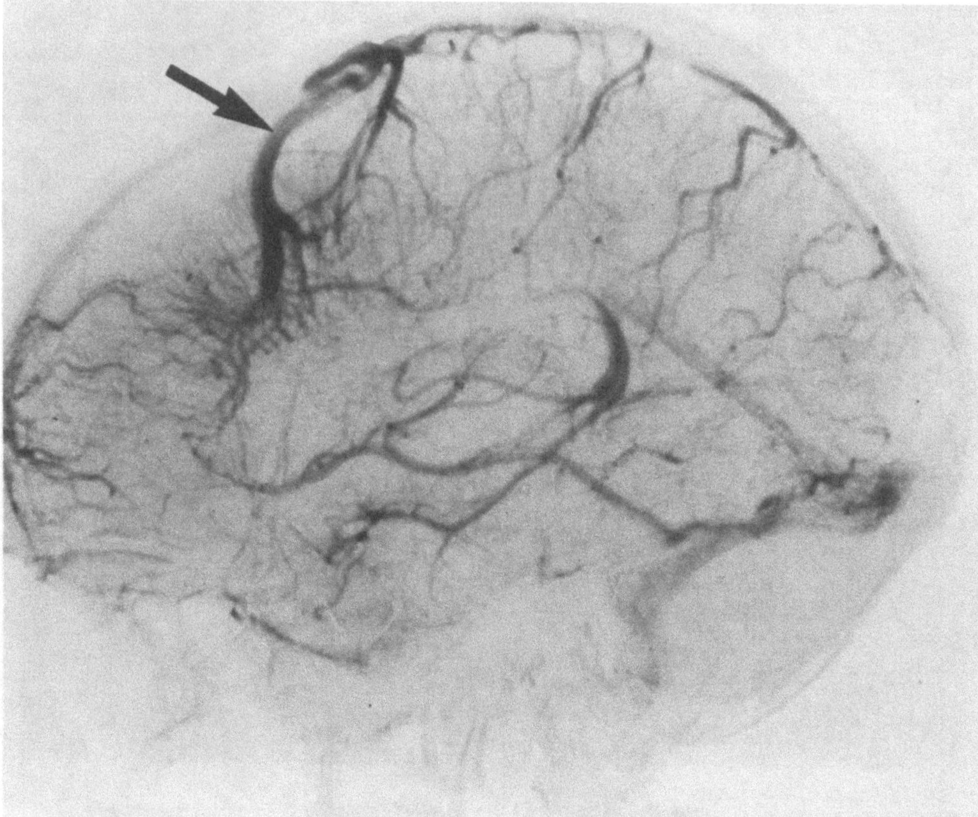

FIGURE 3–19C

FIGURE 3–20. Even-echo rephasing in venous malformation. Venous malformation in an 18-year-old male. (A) TR = 2000 msec., TE = 32 msec. (First echo). A curvilinear region of decreased signal is seen extending to the ependymal surface of the right lateral ventricle (arrow). (B) TR = 2000 msec., TE = 64 msec. (second echo). The draining vein demonstrates increased signal, rendering it barely visible due to even-echo rephasing. The body of the venous malformation abutting the ventricular surface is of mixed signal. (C) TR = 2000 msec., TE = 96 msec. (third echo, most T2-weighted image shown). The draining vein is again clearly seen in the absence of even-echo rephasing. The body of the venous malformation is now clearly seen as a region of increased signal (arrow) on the heavily T2-weighted image.

ANGIOGRAPHIC FINDINGS

Angiography remains the most accurate modality for the detection and characterization of venous malformations. CT and MR can occasionally be diagnostic as well, but not infrequently some ambiguity as to the exact diagnosis exists prior to the performance of an angiogram.

The arterial phase of an angiogram is almost invariably normal, but occasionally abnormalities such as a poorly circumscribed blush or slightly enlarged arterial branches have been reported [42]. The circulation time is normal, with the malformation appearing in the normal course of the venous phase [46]. The classic angiographic abnormalities include the presence of numerous enlarged medullary veins, frequently draining in a perpendicular fashion into a prominent transcerebellar (figure 3–17d) or transcerebral vein. The transcerebral vein will usually drain through the white matter into a superficial sinus (figure 3–19c), but less commonly may drain into the deep venous system (figure 3–18c).

When the characteristic findings of a venous malformation are identified on a high-resolution CT or MR study and the patient's symptoms are related to a region remote from the lesion, or are nonspecific, angiography is probably no longer warranted.

Cavernous Malformations

Cavernous malformations (also referred to as cavernous angiomas) usually become symptomatic between the

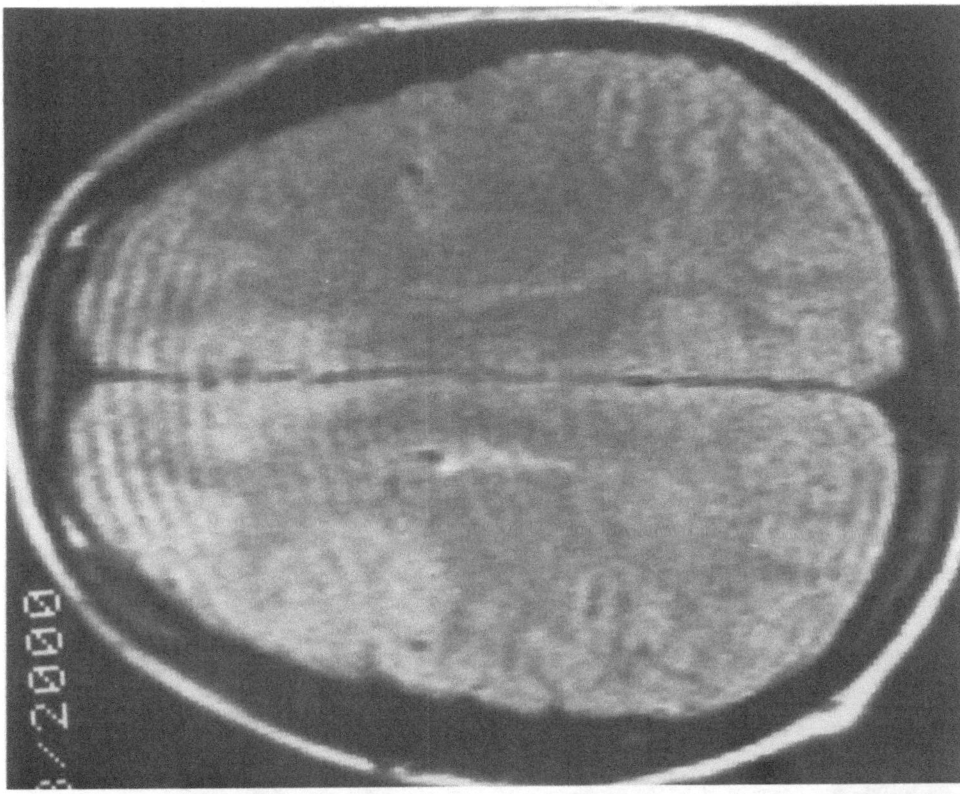

FIGURE 3–20B

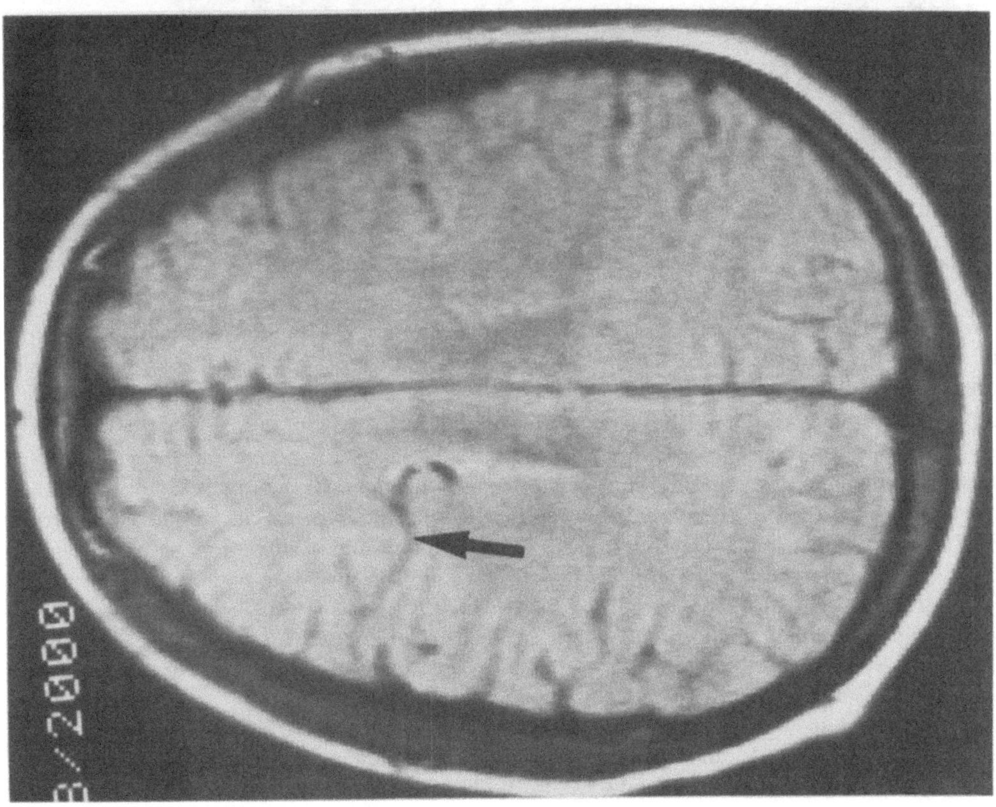

FIGURE 3–20A

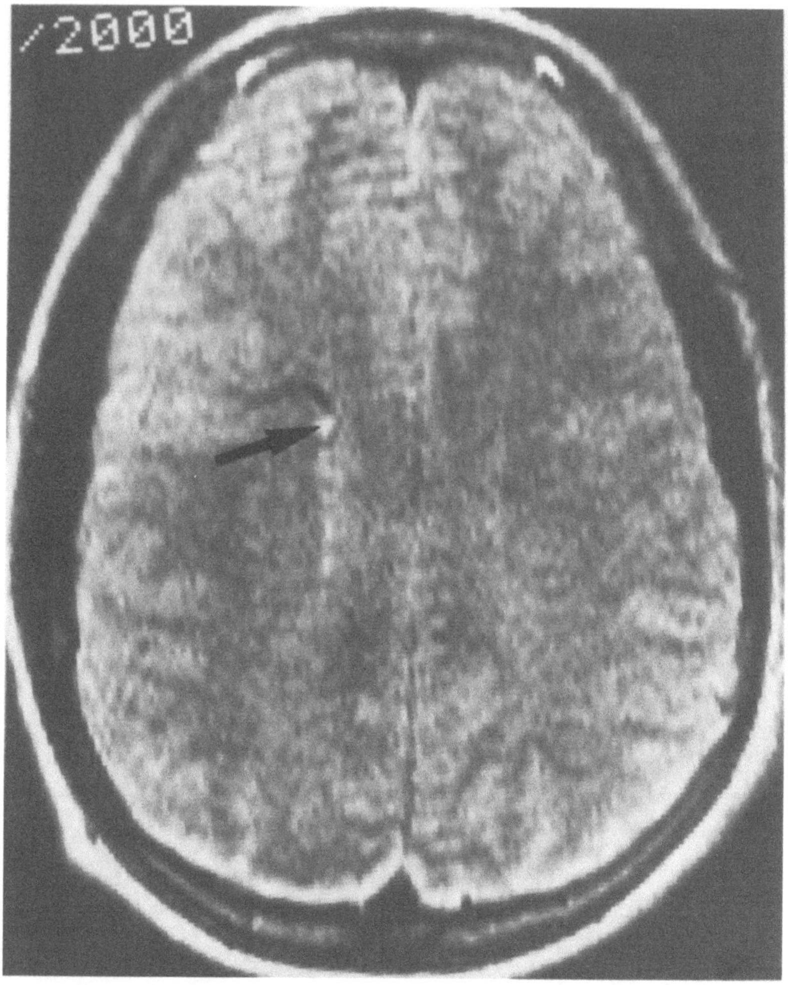

FIGURE 3–20C

ages of 20 and 50 and have a slightly higher incidence in men. Approximately 7% of vascular malformations seen in a large autopsy series were cavernous malformations [2]. However, these lesions are seen with less frequency clinically. This is indicative of the fact that cavernous malformations are considerably less biologically aggressive than AVMs. These lesions are most frequent above the tentorium and are not infrequently identified in the temporal lobes (figure 3–21), where they may give rise to seizure activity. The second most common location is the pons. They are occasionally seen in the cerebellum. When symptomatic, cavernous malformations present with signs and symptoms of hemorrhage, seizures, or headaches. The findings identified on CT scanning are characteristic, but they are not pathognomonic [47, 48]. Unlike AVMs and venous malformations, cavernous malformations cannot be diagnosed specifically without histological confirmation. The lesions are well circumscribed. They are made up of a network of sinusoidal spaces that are closely packed together and have no normal intervening neural tissue. Calcification is frequently present. Gliosis and hemosiderin deposition are frequently seen, as is recent and remote hemorrhage. There are no feeding arteries or draining veins of significant size [13]. Because of their well-circumscribed nature and lack of prominent afferent or efferent vessels, they are quite susceptible to neurosurgical intervention, which is frequently curative.

The characteristic CT findings relate to the well-circumscribed nature of the lesions as well as to the frequently present calcification. Noncontrast studies reveal hyperdense lesions that may demonstrate mild enhancement following contrast administration (figure 3–22). Mass effect is usually mild in the absence

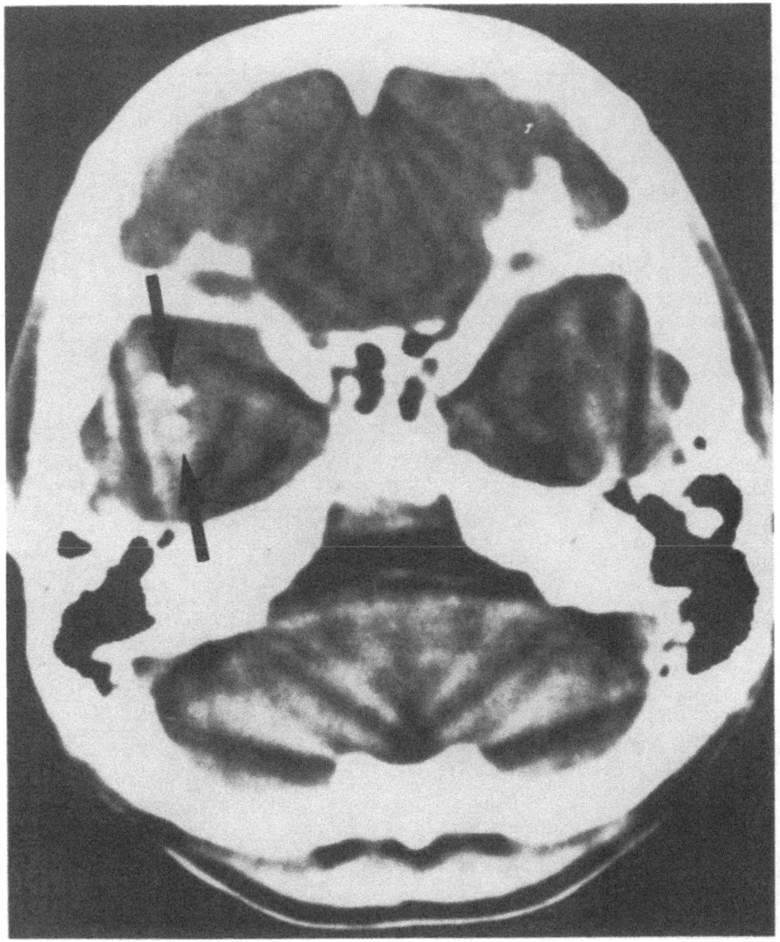

FIGURE 3–21A

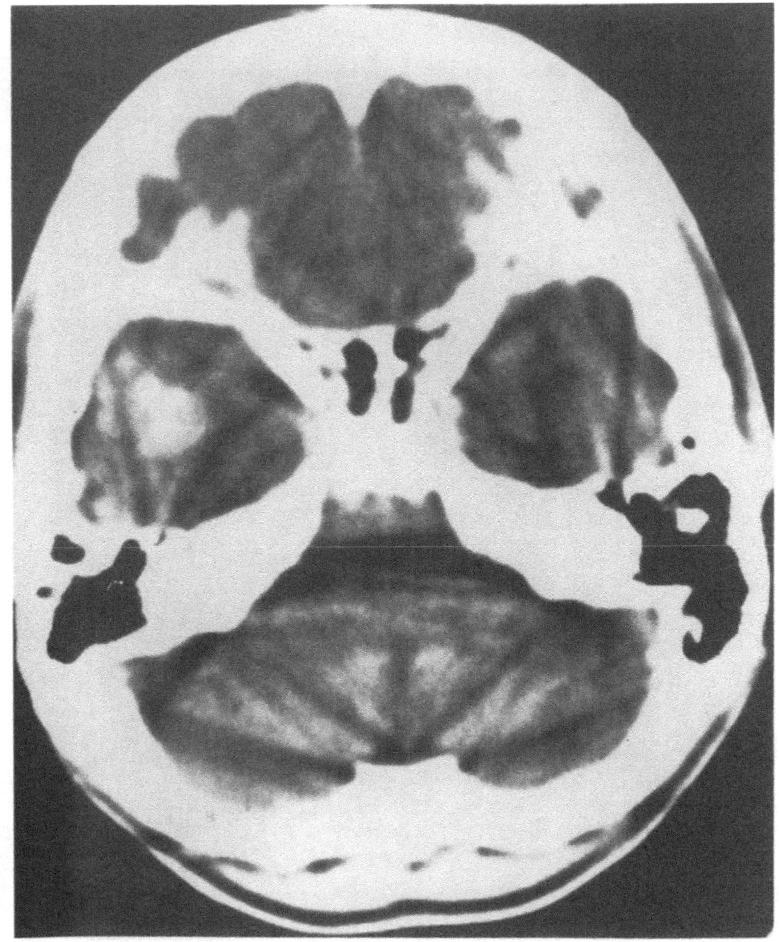

FIGURE 3–21B

FIGURE 3–21. Cavernous malformation. A 40-year-old female presented with seizures. (A) The noncontrast CT scan shows a calcified lesion in the right temporal lobe (arrows). The lesion is difficult to visualize because of the streak artifacts emanating from the bony structures surrounding the middle fossa. (B) Mild enhancement around the calcified zone is seen following contrast administration. (C) TR = 525 msec., TE = 32 msec. An isodense to slightly hyperdense well-circumscribed lesion is seen in the same region on a T1-weighted MR scan. Note the clear visualization of the lesion without the artifacts in this region frequently seen on CT. (D) TR = 2218 msec., TE = 64 msec. The lesion is well circumscribed and of high signal on this T2-weighted image. The surrounding border of decreased signal is related to calcium or hemosiderin.

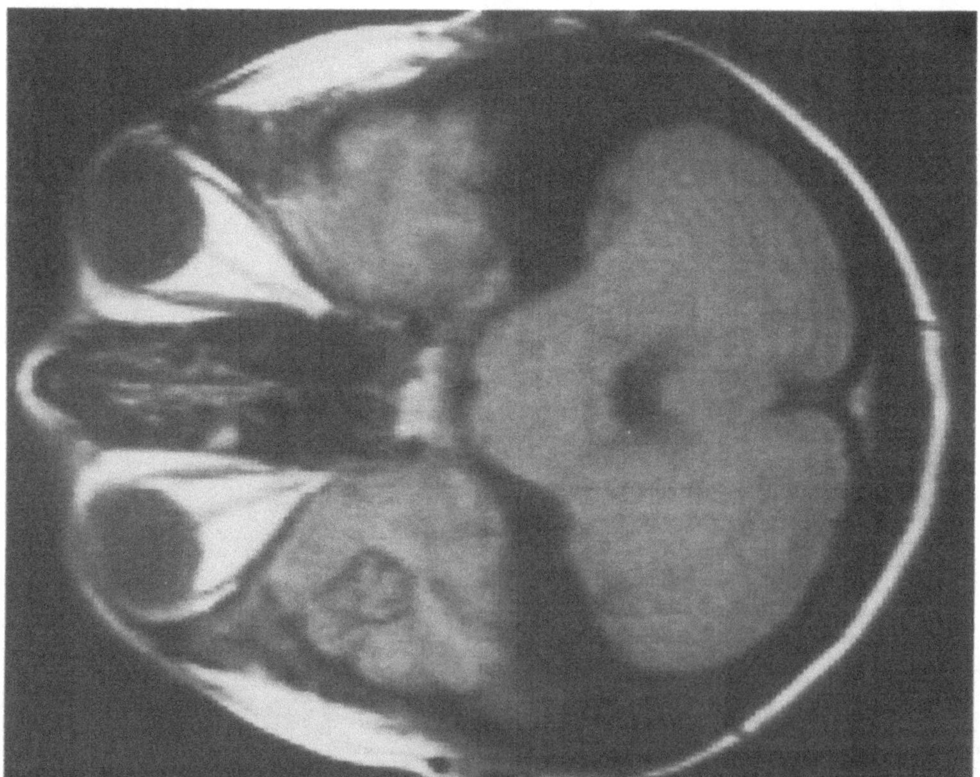

FIGURE 3–21C

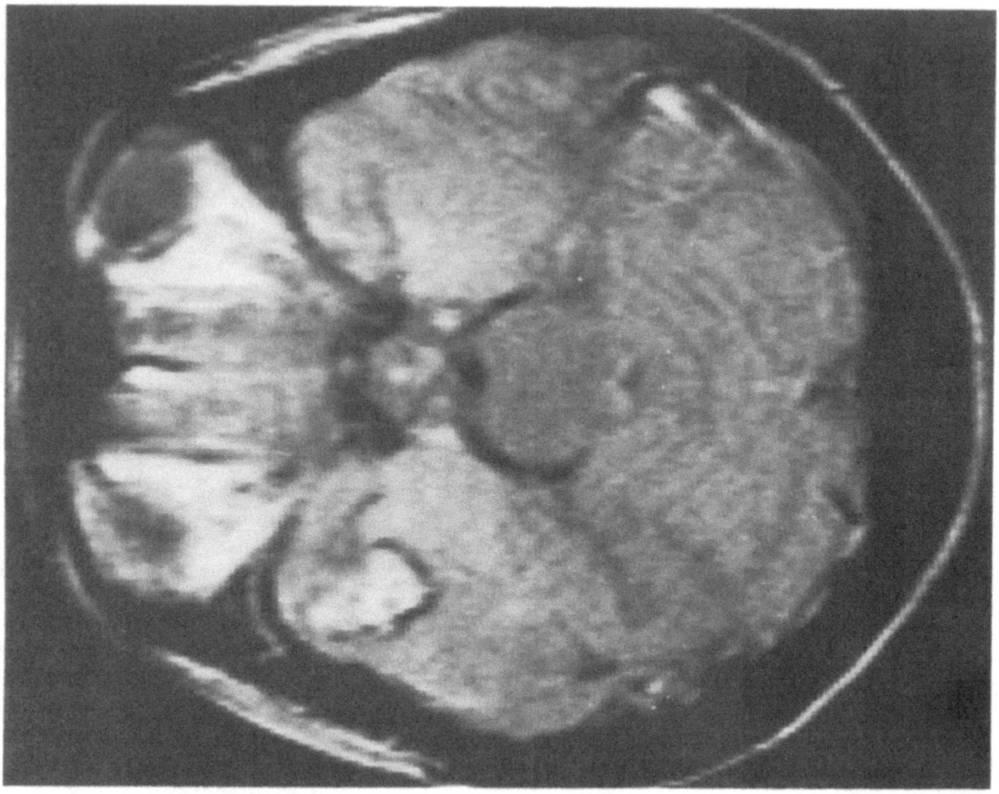

FIGURE 3–21D

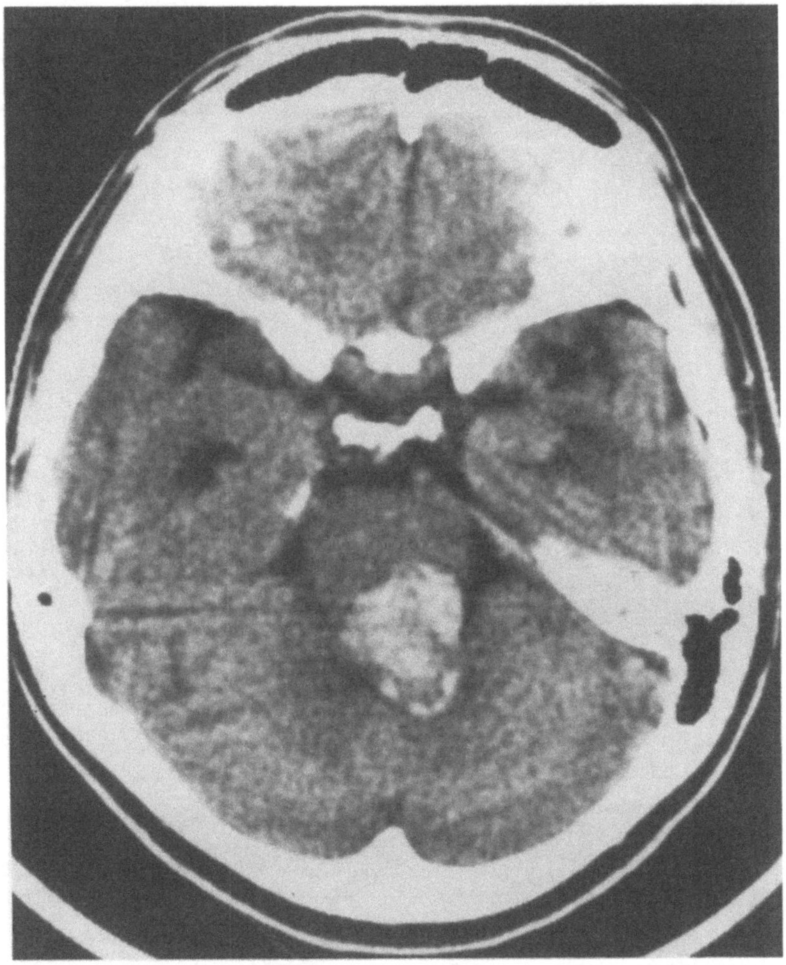

FIGURE 3–22A

FIGURE 3–22. Cavernous malformation. (A) Noncontrast CT scan reveals a hyperdense lesion in the pons and mesencephalon. The posterior margin is calcified. The lesion is sharply circumscribed. (B) Following intravenous contrast administration, there is minimal enhancement noted in the anterior aspect of the lesion. These findings are characteristic of cavernous malformations, but may be seen in other entities, such as low-grade gliomas. (C) T1-weighted MR scan reveals the lesion to be of increased signal intensity, highly suggestive of hemorrhage or thrombosis. Histologically verified cavernous malformation.

of hemorrhage. While these findings are characteristic, other entities, such as low-grade glioma, prior hemorrhage, and old inflammatory lesions can give rise to similar CT appearance.

The addition of MR to the diagnostic armamentarium of the neuroradiologist has made the overall constellation of radiographic findings of cavernous malformations somewhat more suggestive. Cavernous malformations usually demonstrate increased signal intensity on both T1- and T2-weighted images (figures 3–21, 3–22), consistent with previous hemorrhage or thrombosis, as do other angiographically occult vascular malformations, such as thrombosed AVMs [49]. In fact, cavernous malformations account for a substantial portion of all cryptic vascular malformations (CVMs) [50]. Although cavernous malformations are most frequently occult to angiography, occasional nonspecific angiographic abnormalities are demonstrated [51, 52]. These usually consist of a mild stain or blush and occasionally a small early draining vein. Cavernous malformations are multiple in 5–15% of patients depending on the series, a much higher rate than with AVMs.

Telangiectasia

Telangiectases are abnormal collections of capillary vessels with normal intervening neural parenchyma

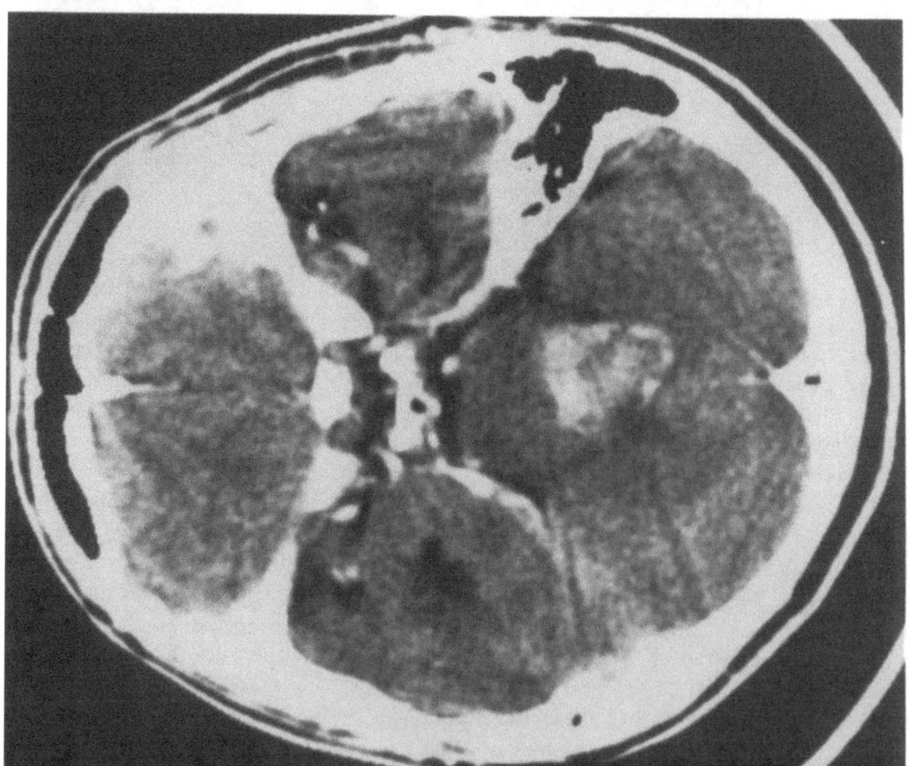

FIGURE 3–22B

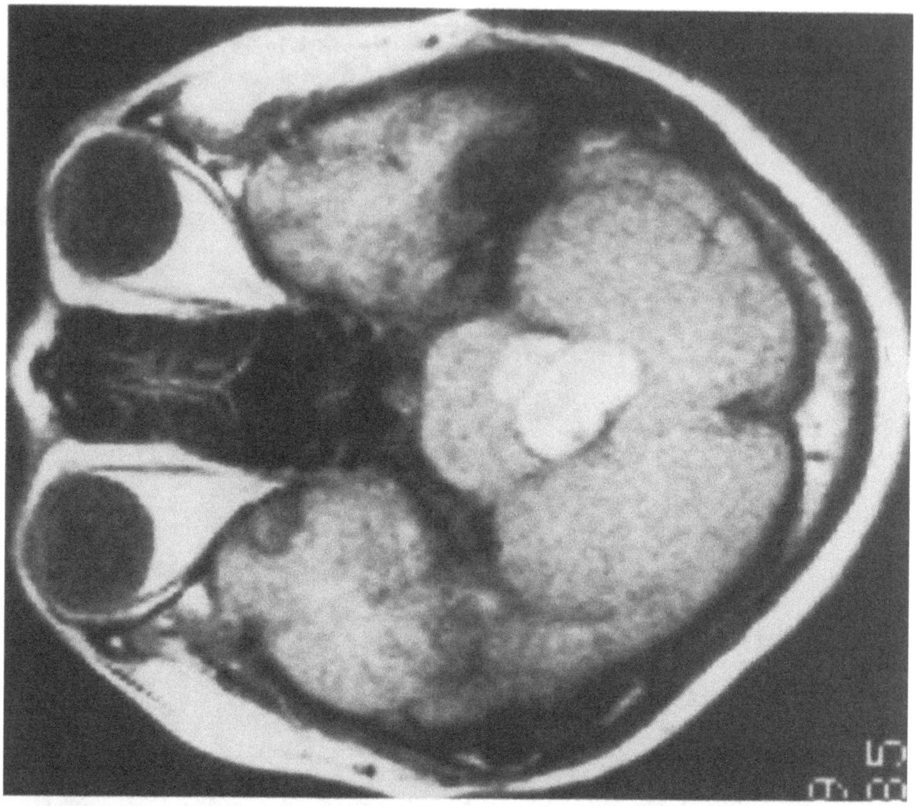

FIGURE 3–22C

which are usually incidental autopsy findings. They are most frequently encountered in the pons, but may be seen anywhere in the brain [53]. Hemorrhage and calcification are rare, but can occur [13]. These lesions are angiographically occult and because of their small size and usual pontine location are infrequently identified by CT [18]. They may be identified as small CVMs on MR studies [27].

Cryptic Vascular Malformations

Cryptic vascular malformations may be thrombosed AVMs, cavernous malformations, venous malformations, telangiectases, or transitional forms. The advent of MR has led to an increased interest in CVMs, a subject about which a great deal of interest had already existed [9, 15, 54–60]. Prior to CT and MR, this entity was described by Crawford and Russell [15], who reported 20 cases with spontaneous cerebral hemorrhages, unremarkable angiograms, and vascular malformations identified at surgery or autopsy. The description of radiographic findings in CVMs were further refined with CT. Numerous studies were published describing the characteristic findings [54–59]. Almost invariably, these authors reported that the lesions demonstrated increased density on the noncontrast CT scan, were relatively well circumscribed, and usually demonstrated relatively mild enhancement, if any, following intravenous contrast administration. When histological evaluations were available, the majority of these lesions were AVMs that had undergone thrombosis [55, 56]; however, the percentage of AVMs was considerably lower than that seen in usual clinical practice. As a corollary, the incidence of cavernous malformations and venous malformations were considerably higher than normally seen.

Typical CT scans of patients with CVMs are seen in figures 3–21 and 3–22, which are histologically verified cavernous malformations, and in figure 3–23, which is a histologically unverified lesion with radiographic findings characteristic of a CVM. MR examinations reveal sharply demarcated lesions of mixed but predominantly high signal on both T1- and T2-weighted images. Areas of decreased signal represent calcium and/or hemosiderin within and surrounding the lesions. MR studies of CVMs show high detection rates, even higher than with CT [60]. The findings can be quite specific as well. The relatively high rate of multiplicity of occult lesions has also been demonstrated. In one series performed at a high field strength, 47% of patients (9 of 19) demonstrated two or more CVMs [60]. Two CVMs of considerable size variance were encountered in one of our patients as well (figure 3–24).

It is clear that MR can identify small CVMs that are occult to both angiography and CT (figure 3–25). These lesions may remain small and asymptomatic throughout a patient's lifetime [59]. Deeply located vascular malformations can be readily identified by MR, and if the characteristic appearance is present, the need for angiography may be obviated in selected cases.

A summary of the pertinent CT, MR, and angiographic findings demonstrated by the four major histological categories of vascular malformations, as well as by CVMs, is presented in table 3–3.

Syndromes Associated with Vascular Malformations

STURGE–WEBER SYNDROME

Encephalotrigeminal angiomatosis is a syndrome with typical clinical and radiographic findings. Patients with this neurocutaneous disorder may present with a port-wine stain of the facial skin (nevus flammeus), seizures, hemiatrophy of the brain with contralateral hemiparesis, and glaucoma. Radiographic findings include calcifications and skull asymmetry that may be apparent on plain films [61]. CT demonstrates the calcifications effectively and at an earlier stage than plain radiographs (figure 3–26). Commonly the calcifications are unilateral and are noted to conform to the cerebral convolutions in the parietal and/or occipital regions. These calcifications lie in the superficial layers of the cerebral cortex. Angiographic abnormalities related to the leptomeningeal angiomatosis have been reported as well [62]. Occasionally, superficial enhancement has been noted on CT scans. Recently, angiomatous malformations of the choroid plexus have been demonstrated on MR examinations [63].

RENDU–OSLER–WEBER SYNDROME

Hereditary hemorrhagic telangiectasia is a dominantly inherited disease with associated arteriovenous anomalies of the lung, gastrointestinal tract, skin, and other viscera. Although most commonly this syndrome effects the central nervous system because of pulmonary shunts and resulting brain abscesses, primary brain AVMs have been reported [64, 65]. One patient in our series with Rendu–Osler–Weber syndrome presented with a large right parietal AVM. Many patients with this syndrome have asymptomatic telangiectasia discovered at autopsy.

WYBURN–MASON SYNDROME

In its complete form Wyburn–Mason syndrome presents with cerebral, retinal, facial, and cutaneous AVMs [66]. A characteristic AVM of the retina

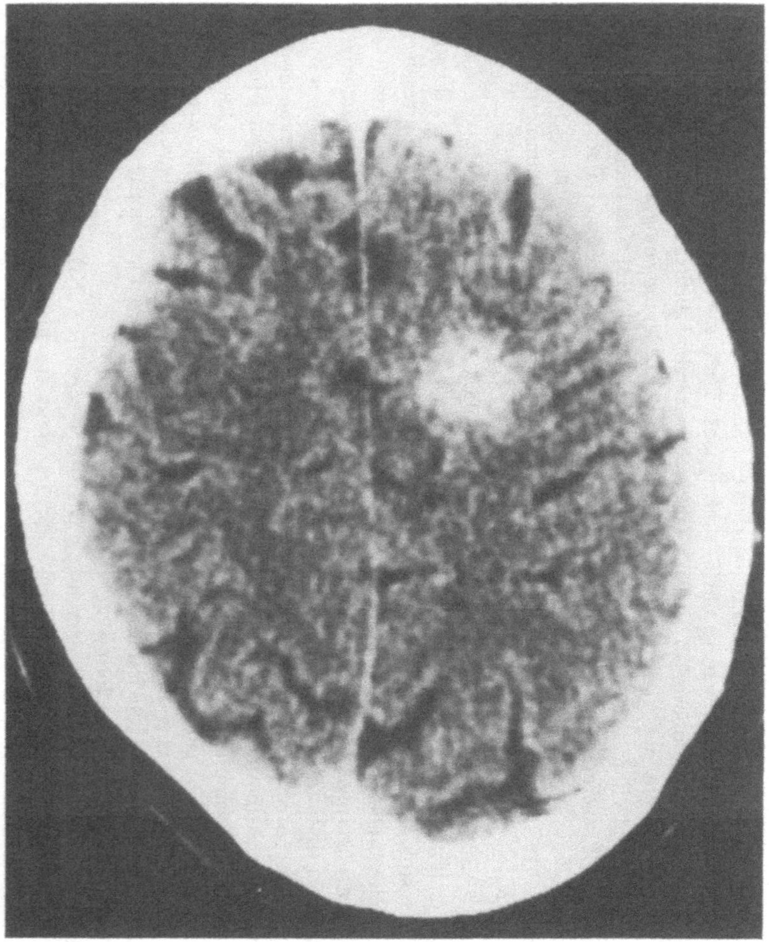

FIGURE 3–23A

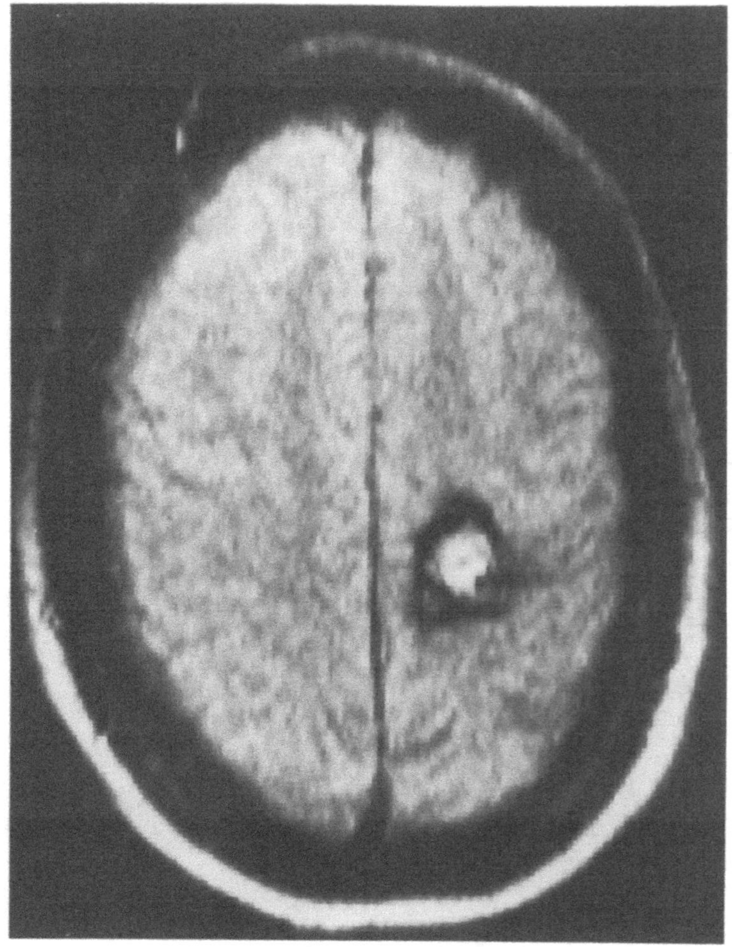

FIGURE 3–23B

FIGURE 3–23. Cryptic vascular malformation. (A) Non-contrast CT scan demonstrates a well-circumscribed dense lesion in the left frontal lobe. No enhancement was seen on a postcontrast CT study. (B) T2-weighted MR, TR = 2128 msec., TE = 64 msec. A hypersignal-intense lesion with a surrounding low-signal zone (thought to represent calcium or hemosiderin) is seen. The lesion was of increased signal on T1-weighted images. Angiography was within normal limits. The apparently more anterior position of the lesion on CT relates to standard angulation techniques utilized for CT and MR studies.

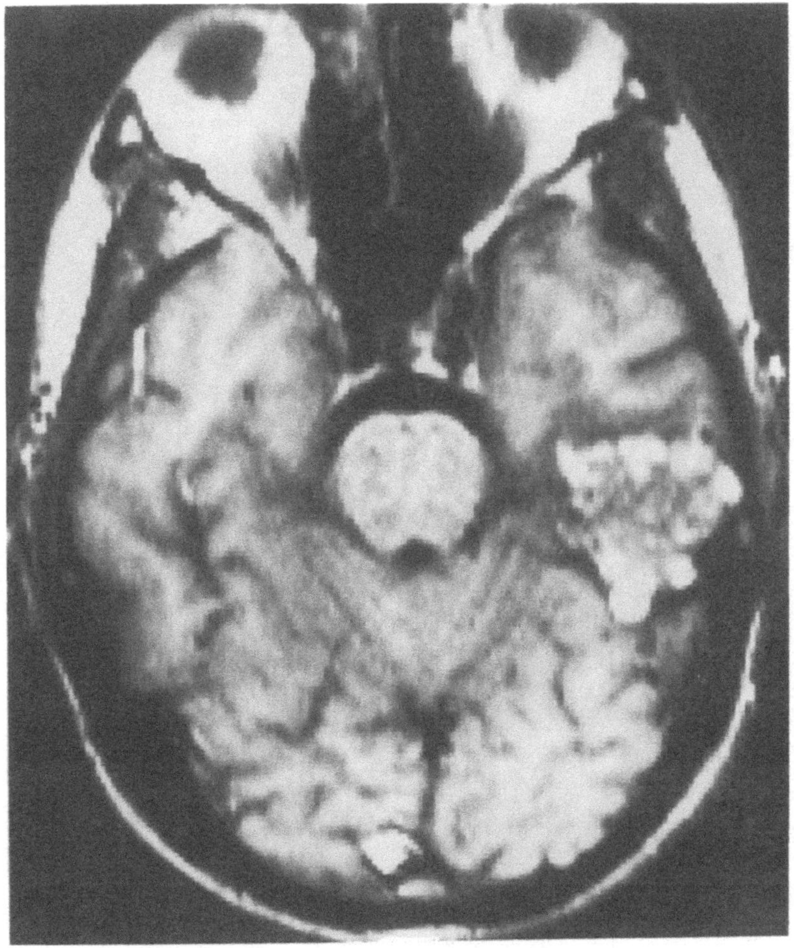

FIGURE 3–24A

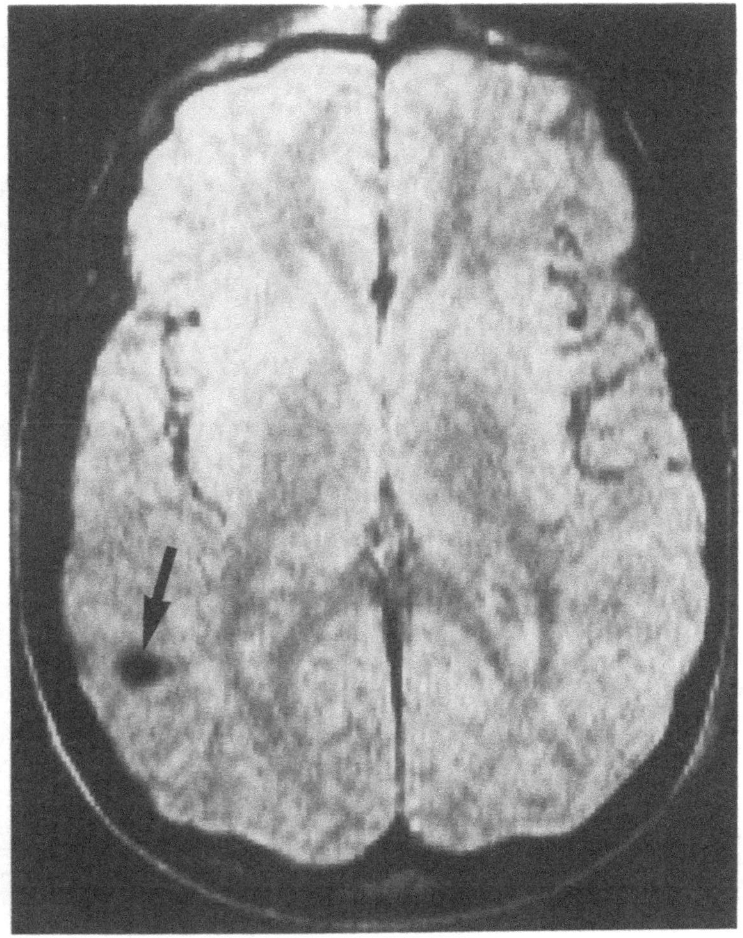

FIGURE 3–24B

FIGURE 3–24. Multiple CVMs (A) TR = 525 msec., TE = 32 msec. A large inhomogeneous but predominantly high-signal-intensity, well-circumscribed lesion is seen in the left temporal lobe on this T1-weighted image. The lesion was hypersignal-intense on T2-weighted images as well. (B) TR = 2128 msec., TE = 64 msec. A second lesion in the right parietal area is seen at a higher level (arrow). The lesion is of low signal intensity, reflecting hemosiderin deposition, since it was not seen on CT as an enhancing vessel or a region of calcification.

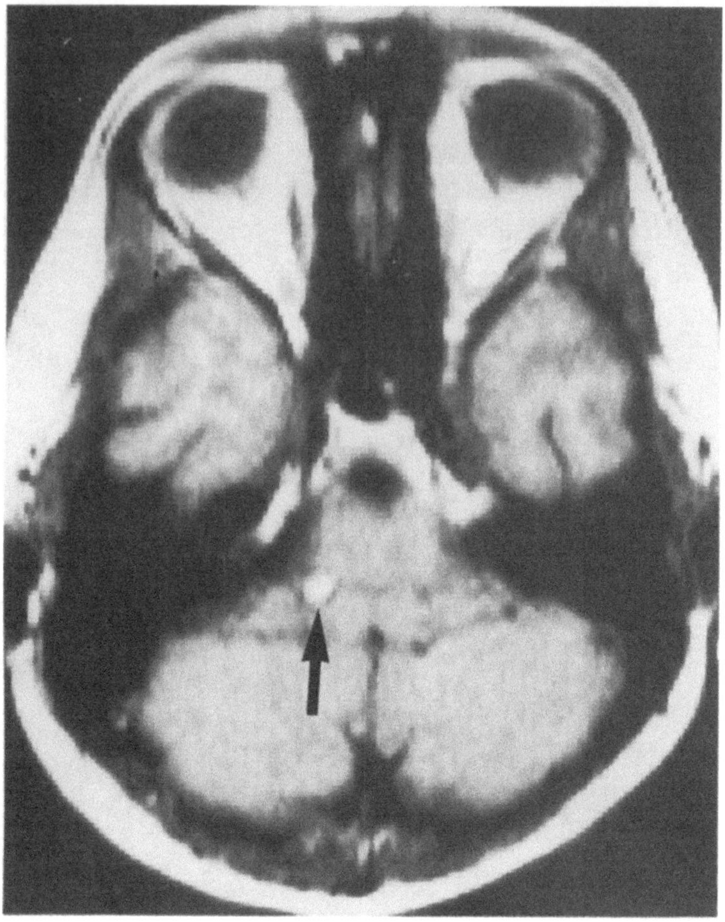

FIGURE 3–25A

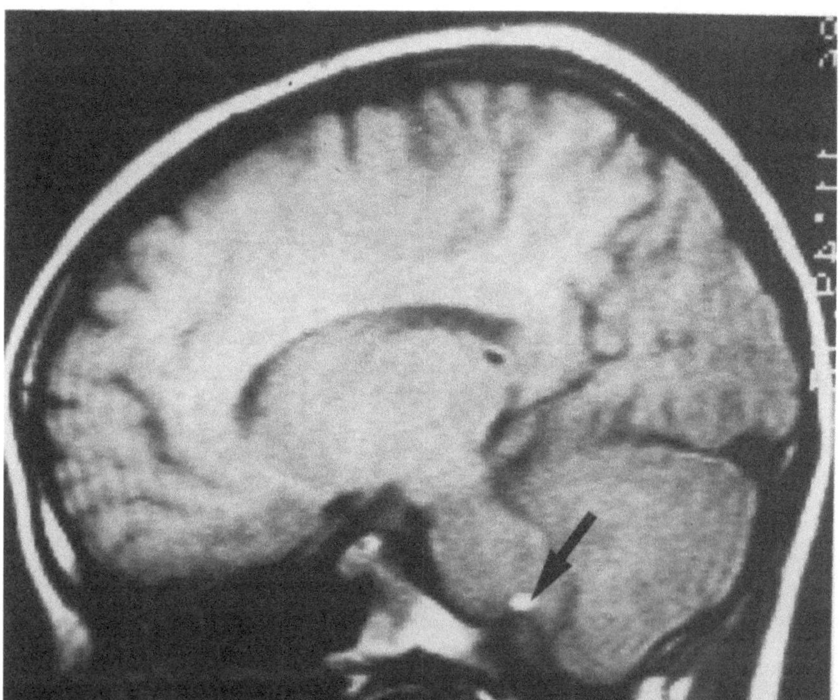

FIGURE 3–25B

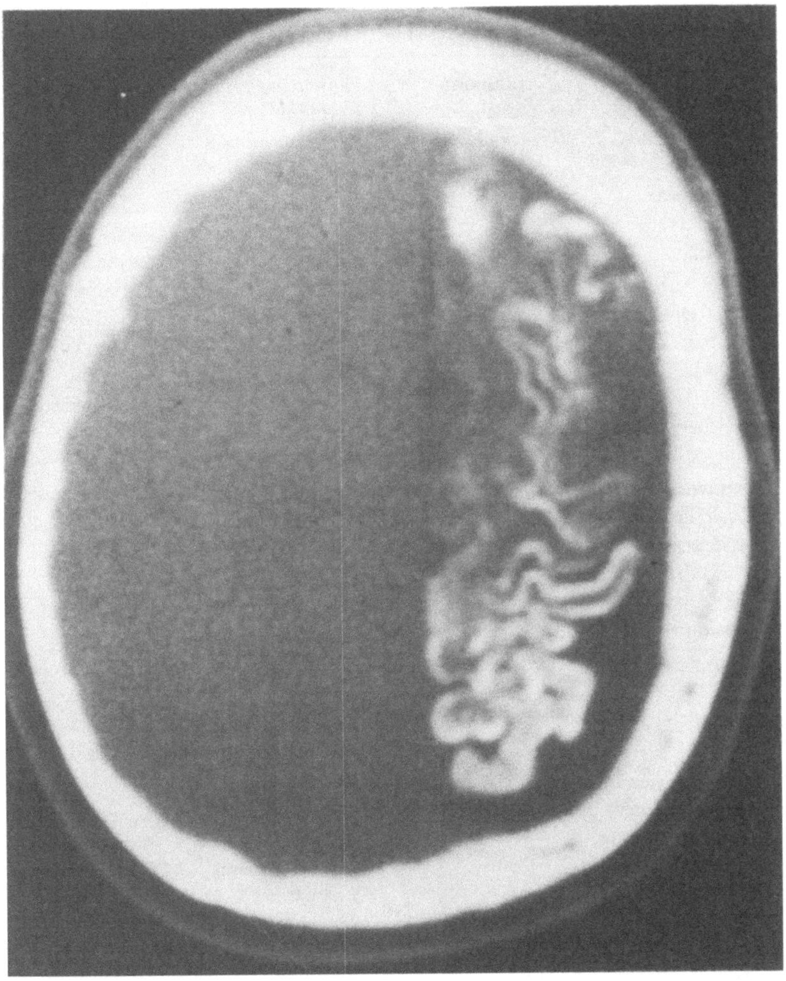

FIGURE 3–26. Sturge-Weber syndrome. CT scan with bone window settings demonstrates calcifications conforming to the cerebral convolutions in the left fronto-parietal and occipital regions. The calvarium on the left is thickened and there is hemiatrophy of the left cerebral hemisphere.

FIGURE 3–25. Cryptic vascular malformation occult to angiography and CT. (A) TR = 525 msec., TE = 32 msec. A tiny region of increased signal intensity is seen in the region of the right pontomedullary junction (arrow). A similar region of increased signal was seen on the T2-weighted images. (B) The same pulse sequence, sagittal plane. The lesion is clearly identified within the brain stem (arrow).

which may be seen on fundoscopic examination suggests the diagnosis.

BLUE–RUBBER–BLEB–NEVUS SYNDROME

This is a rare neurocutaneous disorder where characteristic blue nevi occur on the body surface. Vascular malformations associated with this syndrome are occasionally reported [67].

Recommended Protocol for Radiological Evaluation of Intracanial Vascular Malformations

A protocol for radiological evaluation of vascular malformations must recognize that patients do not present to physicians with vascular malformations, tumors, inflammatory diseases, or congenital abnor-

TABLE 3–3. Summary of Radiographic Findings in Intracranial Vascular Malformations

	Noncontrast CT	Postcontrast CT	MR: T1-weighted image	MR: T2-weighted image	Angiography
Arteriovenous malformations	Hemmorrhage. Calcification. Serpiginous areas of increased density.	Serpiginous enhancing vessels.	Hemorrhage. Flow void in vasculature.	Ischemic regions. Flow void in vasculature.	Typical rapid AV shunts, large afferent and efferent vessels, nidus.
Venous malformations	May show hemorrhage particularly in cerebellum, rarely calcification. Otherwise negative.	Curvilinear region of enhancement coursing through white matter. May see medullary veins rarely.	Hemorrhage in cerebellum. Flow void in curvilinear or round fashion.	Flow void in curvilinear or round fashion. Even echo rephasing. Increased signal from body of lesion.	Caput medusa. Transcerebral or transcerebellar vein.
Cavernous Malformations	Frequently calcified, well-circumscribed and slightly dense lesion. May be hemorrhagic.	May demonstrate slight enhancement.	Usually increased signal. Usually sharply demarcated.	Usually increased signal. Usually sharply demarcated.	Usually negative. May show blush in some cases.
Telangiectasia	Usually negative.	Usually negative.	Usually negative. May show as region of increased signal.	Usually negative. May show as region of increased signal.	Usually negative.
Cryptic vascular malformations	May show variable calcification, hemorrhage, or thrombosis. Usually sharply demarcated.	May show slight enhancement.	Increased signal, mixed signal regions. Usually shaply demarcated.	Increased signal, mixed signal regions. Usually sharply demarcated.	Negative.

malities, but rather with the associated symptoms and signs, which may be similar for many disease entities. Therefore, an appropriate imaging algorithm for the utilization of CT, MR, and angiography must consider the manner in which these lesions present. For intracranial vascular malformations, this includes acute neurological deficits most commonly caused by intracranial hemorrhage, seizure, headaches, and other progressive or intermittent neurological deficits.

Patients with acute neurological deficits (stroke) are best studied initially by CT. CT will clearly distinguish between a parenchymal or intraventricular hemorrhage and an ischemic event. MR is less successful in making this distinction during the first 24 hours following ictus. Subarachnoid hemorrhage, which may present with an acute neurological event or sudden onset of severe headaches, will also be more effectively visualized with CT. Once an intracerebral, intraventricular, or subarachnoid hemorrhage has been demonstrated by CT, further evaluation will depend on clinical circumstances. An elderly hypertensive patient with a hematoma in the basal ganglia or cerebellum will probably require only follow-up CT studies to see that the lesion resolves. A young patient with an intracerebral hematoma who requires immediate surgical decompression should usually undergo a postcontrast CT study, if time permits, in an attempt to visualize the lesion prior to surgery. A patient with a hematoma that is not in need of immediate decompression, but that because of the patient's age and the location of the clot is suspected to be associated with a vascular malformation or aneurysm, should undergo postcontrast CT and angiographic evaluation. Similarly, patients with subarachnoid hemorrhage require complete angio-

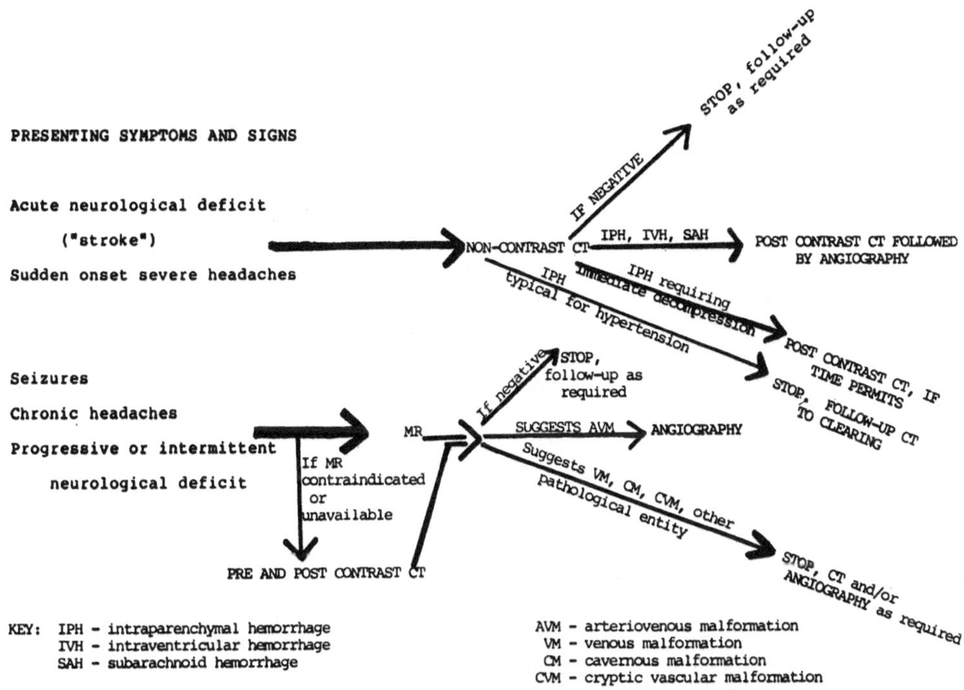

FIGURE 3–27. Suggested imaging algorithm for patients with suspected intracranial vascular malformations.

graphic evaluation to exclude aneurysm or vascular malformation after pre- and postcontrast CT scanning.

Patients presenting with seizures, chronic headaches, or other progressive or intermittent neurological deficits are best studied initially by MR. Magnetic resonance has a number of advantages over CT in elective circumstances where acute hemorrhage is not a clinical consideration. Lesions in the posterior fossa and in the temporal lobes are visualized without the potentially obscuring effects of beam-hardening artifacts produced by bone on CT. This can be of particular significance in identifying vascular malformations in the temporal lobes and in identifying small CVMs in the posterior fossa and brainstem. Cryptic vascular malformations identified in these deep locations may be quite characteristic on MR studies, and the need for angiography may be obviated. Furthermore, small CVMs identified by MR can be occult not only to angiography but also occasionally to CT.

When an AVM is demonstrated by MR, angiographic evaluation is warranted prior to undertaking therapy, whether it be surgery, embolization, or proton beam irradiation. When a venous malformation is suggested but not confirmed by MR, we perform a pre- and postcontrast CT scan in an attempt to establish the diagnosis by these two modalities in concert. If a question still remains as to the possible presence of an AVM, angiography is performed for diagnostic purposes. Our approach to lesions that may be cavernous malformations or low-grade gliomas is similar: we perform pre- and post-contrast CT following MR and, if necessary, angiography. With some CVMs the MR diagnosis is so definitive that further imaging studies are not warranted.

While we prefer MR to CT as the initial imaging modality under almost all elective circumstances, certain contraindications remain. Patients with potentially magnetic intracranial aneurysm clips and with cardiac pacemakers should not undergo MR imaging. In these instances, pre- and post contrast CT can be utilized instead. Logistical and economic considerations are encountered as well in some cases. When MR is unavailable, CT with and without contrast can be substituted as the initial imaging modality. In many areas, MR studies cost approximately twice as much as CT at this time, a factor worth considering in certain circumstances. A diagrammatic representation of a suggested imaging algorithm for patients with suspected intracranial malformations is given in figure 3–27.

References

1. Michelson WJ. Natural history and pathophysiology of arteriovenous malformations. *Clin Neurosurg* 1979; 26:307–313.
2. Sarwar M, Mc Cormick WF. Intracerebral venous angioma. *Arch Neurol* 1978;35:323–325.
3. Perret G, Nishioka H. Report on the cooperative study of intracranial aneurysms and subarachnoid hemorrhage: Section 6, arteriovenous malformations: An analysis of 545 cases of craniocerebral arteriovenous malformations and fistulae reported to the cooperative study. *J Neurosurg* 1966;25:467–490.
4. Robinson JL, Hall CJ, Sedzimir CB. Subarachnoid hemorrhage in pregnancy. *J Neurosurg* 1972;36: 27–33.
5. Robinson JL, Hall CJ, Sedzimir CB. Arteriovenous malformations, aneurysms and pregnancy. *J Neurosurg* 1974;41:63–70.
6. Waltimo O. The relationship of size, density and localization of intracranial arteriovenous malformations to the type of initial symptom. *J Neurol Sci* 1973;19:13–19.
7. Graf CJ, Perret GE, Torner JC. Bleeding from cerebral arteriovenous malformations as part of their natural history. *J. Neurosurg* 1983;58:331–337.
8. Henderson WR, Gomez RRL. Natural history of cerebral angiomas. *Br Med J* 1967;4:571–574.
9. Waltimo O. The change in size of intracranial arteriovenous malformations. *J Neurol Sci* 1973;19:21–27.
10. Nornes H, Grip A. Hemodynamic aspects of cerebral arteriovenous malformations. *J Neurosurg* 1980;53: 456–464.
11. Spetzler RF, Wilson CB, Weinstein P, et al. Normal perfusion pressure breakthrough theory. *Clin Neurosurg* 1978;25:651–672.
12. Kvam DA, Michelson WJ, Quest DO. Intracerebral hemorrhage as a complication of arterial embolization. *Neurosurgery* 1980;7:491–494.
13. Mc Cormick WF. Pathology of vascular malformations of the brain. In:Wilson CB, Stein BM (eds). *Intracranial Arteriovenous Malformations*. Baltimore, MD: Williams and Wilkins, 1984:44–63.
14. Rothfus WE, Albright AL, Casey KF, et al. Cerebellar venous angioma: "benign" entity? *AJNR* 1984;5: 61–66.
15. Crawford JV, Russell DS. Cryptic arteriovenous and venous hematomas of the brain. *J Neurol Neurosurg Psychiat* 1956;19:1–11.
16. Hayman LA, Fox AJ, Evans RA. Effectiveness of contrast regimens in CT detection of vascular malformations of the brain. *AJNR* 1981;2:421–425.
17. Brunelle FUS, Harwood-Nash DCF, Fitz CR, Chuang SH. Intracranial vascular malformations in children: Computed tomographic and angiographic evaluation. *Radiology* 1983;149:455–461.
18. Haughton V. Vascular diseases. In: Williams AL, Haughton VM (eds). *Cranial Computed Tomography*. New York: CV Mosby Co, 1985:88–147.
19. Leblanc R, Ethier R, Little JR. Computerized tomography findings in arteriovenous malformations of the brain. *J Neurosurg* 1979;51:765–772.
20. Enzmann DR, Britt RH, Lyons BE, et al. Natural history of experimental intracerebral hemorrhage: Sonography, computed tomography and neuropathology. *AJNR* 1981;2:517–526.
21. Diebler C, Dulac O, Renier D, et al. Aneurysms of the vein of Galen in infants aged 2 to 15 months. *Neuroradiology* 1981;21:185–197.
22. Roosen N, Schirmer M, Lins E, et al. MRI of an aneurysm of the vein of Galen. *AJNR* 1986;7:733–735.
23. Lee BCP, Herzberg L, Zimmerman RD, Deck MDF. MR imaging of cerebral vascular malformations. *AJNR* 1985;6:863–870.
24. Kucharczyk W, Lemme-Pleghos L, Uske A, et al. Intracranial vascular malformations: MR and CT imaging. *Radiology* 1985;156:383–389.
25. Schorner W, Bradac GB, Treisch J, et al. MRI in the diagnosis of cerebral arteriovenous angiomas. *Neuroradiology* 1986;28:313–318.
26. Alvarez OA, Hyman RA. Even echo MR rephasing in the diagnosis of giant intracranial aneurysm. *J Comput Assist Tomogr* 1986;10:699–701.
27. Norman D. Vascular disease: Hemorrhage. In: Brant-Zawadzki M, Norman D (eds). *Magnetic Resonance Imaging of the Central Nervous System*. New York: Raven Press, 1987;209–220.
28. Newton TH, Crouqist S. Involvement of dural arteries in intracranial arteriovenous malformations. *Radiology* 1969;93:1071–1078.
29. Houser OW, Baker HL Jr, Svien HJ, Okazaki H. Arteriovenous malformations of the parenchyma of the brain. *Radiology* 1973;109:83–90.
30. Houser OW, Baker HL Jr, Rhoton AL Jr, Okazaki H. Intracranial dural arteriovenous malformations. *Radiology* 1972;105:55–64.
31. Kosnik EJ, Hunt WE, Miller CA. Dural arteriovenous malformations. *J Neurosurg* 1974;40:322–329.
32. Newton TH, Weidner W, Greitz T. Dural arteriovenous malformations in the posterior fossa. *Radiology* 1968;90:27–35.
33. Ito J, Imamura H, Kobayashi K, et al. Dural arteriovenous malformations of the base of the anterior cranial fossa. *Neuroradiology* 1983;24:149–154.
34. Schlachter LB, Fleischer AS, Faria MA Jr, et al. Multifocal intracranial arteriovenous malformations. *Neurosurgery* 1980;7:440–444.
35. Sartor K. Spontaneous closure of cerebral arteriovenous malformation demonstrated by angiography and computed tomography. *Neuroradiology* 1978;15:95–98.
36. Suzuki J, Onuma T. Intracranial aneurysms associated with arteriovenous malformations. *J Neurosurg* 1979; 50:742–746.
37. Miyasaka K, Wolpert SM, Prager RJ. The association of cerebral aneurysms, infundibula and intracranial arteriovenous malformations. *Stroke* 1982;13:196–203.
38. Hayashi S, Arimoto T, Itakura T, et al. The

association of intracranial aneurysms and arteriovenous malformations of the brain. *J Neurosurg* 1981;55: 971–975.
39. Goree JA, Dukes HT. The angiographic differential diagnosis between the vascularized malignant glioma and the intracranial arteriovenous malformations. *AJR* 1963;90:512–521.
40. Saito Y, Kobayashi N. Cerebral venous angiomas: Clinical evaluation and possible etiology. *Radiology* 1981;139:87–94.
41. Olson E, Gilmor RL, Richmond B. Cerebral venous angiomas. *Radiology* 1984;151:97–104.
42. Wendling LR, Moore JS, Kieffer SA, et al. Intracerebral venous angioma. *Radiology* 1976;119:141–147.
43. Lotz PR, Quisling RG. CT of venous angiomas of the brain. *AJNR* 1983;4:1124–1126.
44. Michels LG, Bentson JR, Winter J. Computed tomography of cerebral venous angiomas. *J Comput Assist Tomogr* 1977;1:149–154.
45. Augustyn GT, Scott JA, Olson E, Gilmore RL, Edwards MK. Cerebral venous angiomas: MR imaging. *Radiology* 1985;156:391–395.
46. Fierstein SB, Pribram HW, Hieshima G. Angiography and computed tomography in the evaluation of cerebral venous malformations. *Neuroradiology* 1979;17:137–148.
47. Bartlett JE, Kishore PRS. Intracranial cavernous angioma. *AJR* 1977;128:653–656.
48. Ahmadi J, Miller LA, Segall HD, et al. CT patterns in histopathologically complex cavernous hemangiomas. *AJNR* 1985;6:389–93.
49. Lemme-Pleghos L, Kucharczyk W, Brant-Zawadzki M, et al. MR imaging of angiographically occult vascular malformations. *AJNR* 1986;7:217–222.
50. Cohen HCM, Tucker WS, Humphreys RP, et al. Angiographically cryptic histologically verified cerebrovascular malformations. *Neurosurgery* 1982;10: 704–714.
51. Rao VRK, Pillai SM, Shenoy, et al. Hypervascular cavernous angiomas at angiography. *Neuroradiology* 1979;18:211–214.
52. Jonutis AJ, Sondheimer FK, Klein HL, et al. Intracerebral cavernous hemangioma with angiographically demonstrated pathologic vasculature. *Neuroradiology* 1971;3:57–63.
53. Stahl SM, Johnson KP, Malamud N. The clinical and pathological spectrum of brain stem vascular malformations. *Arch Neurol* 1980;37:25–29.
54. Bell BA, Kendall BE, Syman L. Angiographically occult arteriovenous malformations of the brain. *J Neurol Neurosurg Psychiat* 1978;41:1057–1064.
55. Becker DH, Townsend JJ, Kramer RA, et al. Occult cerebrovascular malformations: A series of 18 histologically verified cases with negative angiography. *Brain* 1979;102:249–287.
56. Wharen RE, Scheithauer BW, Laws ER Jr. Thrombosed arteriovenous malformations of the brain. *J Neurosurg* 1982;57:520–526.
57. Kramer RA, Wing SD. Computed tomography of angiographically occult cerebral vascular malformations. *Radiology* 1977;123:649–625.
58. Yeates A, Enzmann D. Cryptic vascular malformations involving the brain stem. *Radiology* 1983;146:71–75.
59. New PEJ, Ojemann RG, Davis KR, et al. MR and CT of occult vascular malformations of the brain. *AJNR* 1986;7:771–779.
60. Gomori JM, Grossman RI, Golderg HI, et al. Occult cerebral vascular malformations: High field MR imaging. *Radiology* 1986;158:707–713.
61. Di Chiro G, Lindgren E. Radiographic findings in 14 cases of Sturge–Weber syndrome. *Acta Radiol* 1951;35:387–399.
62. Bentson JR, Wilsom GH, Newton TH. Cerebral venous drainage pattern of the Sturge–Weber syndrome. *Radiology* 1971;101:111–118.
63. Stimac GK, Solomon MA, Newton TH. CT and MR of angiomatous malformations of the choroid plexus in patients with Sturge–Weber disease. *AJNR* 1986;7:623–627.
64. Reagan TJ, Bloom WH. The brain in hereditary hemorrhagic telangiectasia. *Stroke* 1971;2:361–368.
65. Tress BM, Moseley IF. Cleidocranial dysostosis, hereditary haemorrhagic telangiectasia and epilepsy: A rare association. *Neuroradiology* 1977;12:233–236.
66. Wyburn-Mason R. Arteriovenous aneurysm of midbrain and retina, facial naevi and mental changes. *Brain* 1943;66:163–203.
67. Satya-Murti S, Navada S, Eames F. Central nervous system involvement in Blue-Rubber-Bleb-Nevus syndrome. *Arch Neurol* 1986;43:1184–1186.

4. ATHEROSCLEROTIC EXTRACRANIAL OCCLUSIVE CEREBROVASCULAR DISEASE

Solomon Batnitzky
John H. McMillan
Glendon G. Cox
Stanton J. Rosenthal

The term *stroke* denotes acute vascular and brain changes leading to rapid development of neurologic deficits. This condition was first described by Hippocrates as apoplexy, a term still in use today. The original Greek word, *apoplexia*, means *to strike down*—hence, the term *stroke* [1]. Hippocrates thought that the reason for apoplexy was hot blood vessels filled with white phlegm and black bile. Since Hippocrates's time, a large body of knowledge about brain circulation, metabolism, and pathology has been acquired, especially over the last 30 years, and it is now well recognized that stroke may be caused by both intracranial and extracranial disease.

Stroke is the third leading cause of death and one of the leading causes of long-term disability in the United States [2]. About two million Americans are affected by stroke and about 200,000 die of it each year. As a result, stroke has serious medical, social, and economic consequences in our society and is a major public health problem.

Arterial occlusive disease, including thrombosis and embolism, is responsible for approximately 80% of strokes. More than 90% of these are caused by atheroclerosis of the extracranial vasculature and its embolic complications [3].

Other causes of stroke are intracerebral hemorrhage, aneurysm (hemorrhage or infarction secondary to spasm), arteritis, tumor, and infection. For a discussion of imaging of the brain in cerebrovascular disease, the reader is referred to *Neuroradiology of Cerebral Infarction* (chapter 6 of this volume).

In about 10% of cases, the source of the stroke is cardiac in origin. Cardiac emboli occur in several disease entities such as valvular heart disease, arrhythmias, myocardial infarction with mural thrombus, endocarditis, atrial myxoma, nonbacterial thrombotic endocarditis, and congenital heart disease, as well as in prosthetic heart valves [4].

In 1906, Chiari first suggested a relationship between ulceration of the extracranial carotid arteries and encephalomalacia [5]. In 1914, Ramsey Hunt [6] suggested that stenosis of a carotid artery in the neck could cause a stroke. It was not until 1951 that Fisher [7] drew widespread attention to the extracranial carotid arteries in the investigation of cerebrovascular insufficiency. Soon after the introduction of carotid endarterectomy in the early 1950s [8], there was a surge of interest in cerebrovascular disease, including an ever-increasing demand for precise radiographic evaluation of patients with symptomatic and asymptomatic cerebrovascular disease. In fact, lesions surgically accessible by present techniques are demonstated in one or more of the extracranial vessels in 40–50% of symptomatic patients.

Atherosclerosis

PATHOGENESIS

Atherosclerosis is a noninflammatory degenerative disease that can affect segments of almost any artery in the body. Atherosclerosis is characterized by the presence of nodular, irregularly distributed yellow, fatty plaques involving the intima of large and medium-sized arteries [9]. The term *atheroma*, which is synonymous with and is used interchangeably with atherosclerosis and arteriosclerosis, is derived from the Greek word for *mush* and designates the pultaceous material in advanced intimal plaques. Atherosclerosis is by far the most common vascular disorder and is the underlying cause of death in most persons over 50 years of age.

Atherosclerotic plaques in themselves produce no symptoms. Atherosclerotic plaques cause clinical disease by two mechanisms: (a) obstruction of blood flow and (b) ulceration on the surface of the plaque.

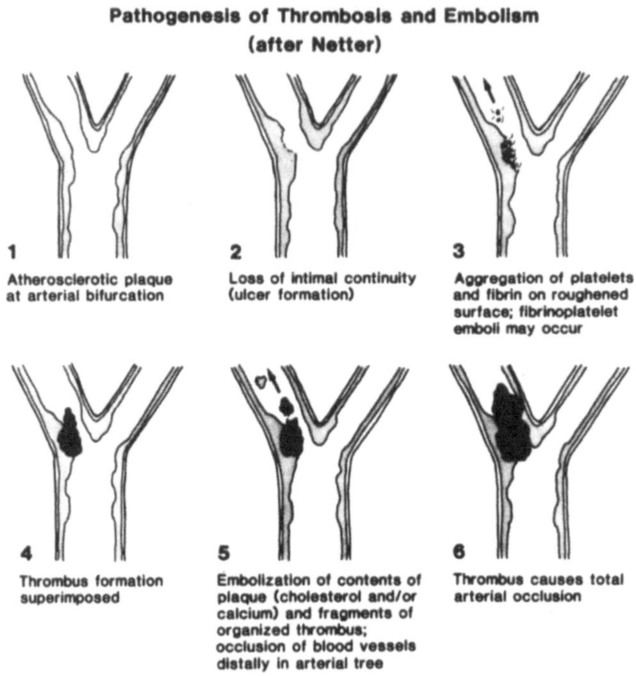

FIGURE 4–1. Diagramatic representation of the pathogenesis of atherosclerotic thrombosis and embolism.

Hemorrhage into the intima adjacent to the plaque may increase the size of the lesion and thereby narrow the arterial lumen [10]. Stenotic lesions account for only 10% of ischemic symptoms. The cerebral vasculature has a tremendous intrinsic ability to autoregulate and thereby maintain normal blood flow at a markedly reduced perfusion pressure. For a stenosis to be hemodynamically significant in lowering the ipsilateral arterial pressure, a luminal narrowing of at least 75% must be present [11, 12]. In such cases, slight degrees of blood pressure reduction may result in a markedly reduced cerebral blood flow, since the brain has already utilized its full autoregulatory ability. A far greater degree of reduction (90%) is necessary to reduce flow in normotensive or hypertensive patients [11].

Loss of continuity in the endothelial surface overlying the atheromatous plaque may result in discharge of cholesterol or calcific debris into the blood stream [10, 13] (figure 4–1). The presence of these emboli have been documented in the retinal arteries [14]. Loss of the endothelial surface also usually initiates the development of a thrombus within the arterial lumen. The collagen in the subendothelial layer, exposed to the blood by the loss of the intima, causes platelets to adhere to the ruptured surface. Such a thrombus may progress to occlusive thrombosis, or more commonly, it may fragment to produce thrombotic embolization. Most strokes are caused by emboli rather than inadequate hemodynamics [11].

Any branch of the aorta may be involved by the atheromatous process. Such lesions occur most often at sites of mechanical stress, particularly at points of major arterial bifurcations [15]. In most patients, multiple vessels are involved. The common carotid artery bifurcation, including the proximal 1–2 cm of the internal carotid artery, is the site in the extracranial circulation most frequently involved in the atherosclerotic process [1, 3, 11] (figure 4–2). It accounts for approximately one third of cases that have clinical evidence of cerebral ischemia [16]. The next most frequent site of atherosclerosis is the proximal segment of each vertebral artery. The proximal segment of the left subclavian artery is the third most frequent site of stenosis, with lesions occurring in about 12% of patients who have cerebral ischemia. The right subclavian is involved only 8% of the time,

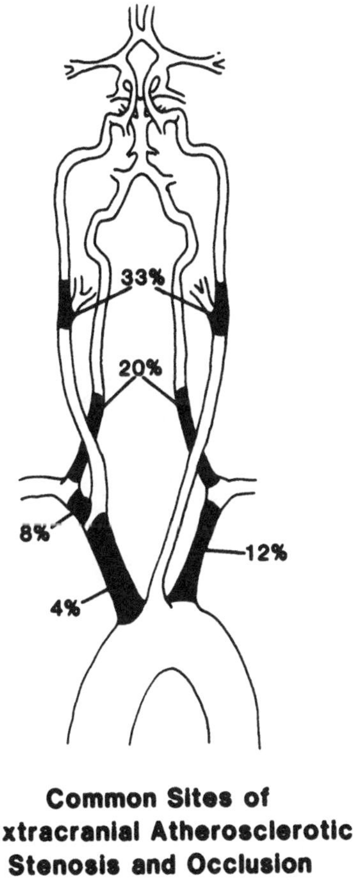

FIGURE 4–2. Diagramatic representation of the common sites of involvement in the extracranial vascular tree by atherosclerotic lesions.

but when stenosis of the brachiocephalic trunk (figure 4–3) and of the right subclavian artery are considered together, the frequency is approximately the same on both sides.

Ulcerative lesions most frequently involve the carotid artery bifurcation (figure 4–4) but have been described in other vessels. The ulcer is usually located in the posterior wall of the carotid bifurcation or at the origin of the internal carotid artery. Ulceration rarely occurs in the vertebrobasilar system.

The atherosclerotic process is unpredictable and variable [2]. The ultimate outcome, even with complete occlusion of a vessel, is influenced by a number of factors. These include effectiveness and adequacy of the collateral circulation, blood pressure and general cardiovascular efficiency, and associated lesions of the other internal carotid artery and of the vertebral arteries. The outcome ranges from permanent neurological deficits with frank infarction to no tissue damage with no clinical manifestations—e.g., occlusion of the internal carotid artery in one patient may be accompanied by only transitory neurologic deficits or no symptoms, whereas in another patient, neurologic dysfunction may be catastrophic.

NATURAL HISTORY OF TRANSIENT ISCHEMIC ATTACKS

Despite the accumulation of extensive literature on occlusive cerebrovascular disease, nothing has been discovered that will reduce the size of a cerebral infarct once it has occurred or that will promote regeneration of infarcted necrotic tissue. Prevention of stroke, therefore, has emerged as the only logical and sensible course. It has been towards these goals that the major efforts in the treatment and management of cerebrovascular disease have been directed.

A transient ischemic attack (TIA) is an episode of focal or temporary neurologic deficit that usually lasts from one to several minutes but no more than 24 hours [2]. Neurologic examination between attacks is generally normal. Symptoms of TIAs rarely arise from the aortic arch or its major vessels, although these vessels may be involved by atherosclerosis. The relative contribution of other embolic sources, particularly those that are cardiogenic in origin, to the production of TIAs is not certain.

In determining the potential benefits and risks of any form of therapy, one must be aware of the natural history of the disease process. Although the precise magnitude of risk is controversial and dependent on many factors, it is generally agreed that TIA is an important risk factor for stroke. In fact, TIA is the most important harbinger of stroke. However, not all patients with stroke are warned far in advance of potential infarction. In fact, less than 50% of patients with strokes will have a prior TIA [17].

The greatest risk for stroke is in the first year following the onset of TIA. Among patients with TIA who do not die of a cause other than stroke, about one third will suffer a stroke within five years [18, 19], more than 20% in the first month of the initial attack and 50% within the first year [18, 20]. After the first six months following the onset of TIA, the stroke occurrence rate is about 5% per year, which is approximately five times the expected rate for a normal population with a similar age distribution [17, 18, 21].

The number of TIAs and the average length of individual TIAs are less useful indicators of the risk of stroke [18, 22, 23].

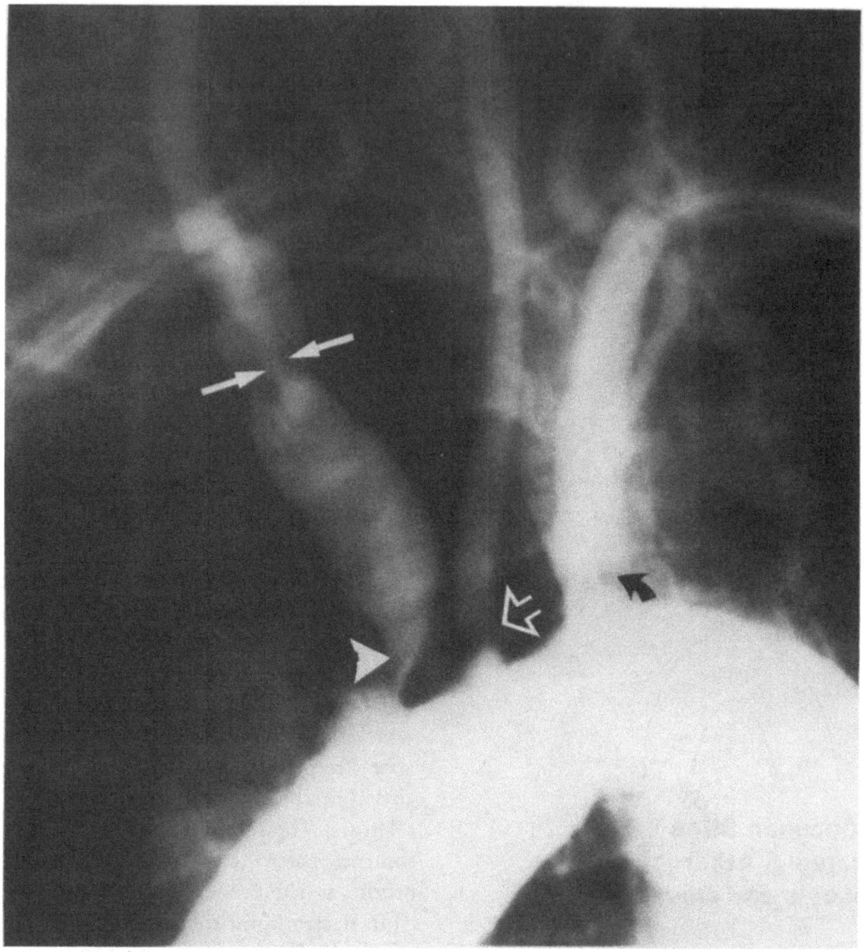

FIGURE 4–3. A 68-year-old female, 11 days status post-abdominal aortic aneurysm repair, who had an episode of right-sided amaurosis fugax. Right posterior oblique aortic arch study, performed via the left axillary artery route, demonstrates a greater than 90% stenosis of the origin of the innominate artery by a large atherosclerotic plaque (arrowhead). A high-grade stenosis of the right common carotid artery origin (arrows) is also identified. The origins of the left common carotid (open arrow) and left subclavian (curved arrow) arteries are also involved by the atheromatous process. Both common carotid bifurcations were unremarkable.

In another one third of patients, the TIAs will have ceased after three years, and in the remaining one third, the TIA will still be occurring.

CONTROVERSIES IN TREATMENT

The literature on the medical and surgical management of patients with TIAs is extensive, controversial, and often contradictory. Anticoagulant therapy, antiplatelet agents, and surgery have been advocated in an effort to improve the morbidity and mortality in patients with TIAs. Several reviews have cast doubt on the efficacy of each of these forms of therapy. At the present time, there is no conclusive evidence to suggest that any single mode of treatment of TIA is superior to another in preventing stroke [24].

The first results of carotid endarterectomy were reported over 30 years ago. Although the number of carotid endarterectomies performed in North America continues to climb [25], controversy still exists about the indications for and the risks and value of the procedure.

On the one extreme is the view that carotid endarterectomy is not justified for general use on the basis of the evidence presently available [26]. However, there are many experts who would recommend ca-

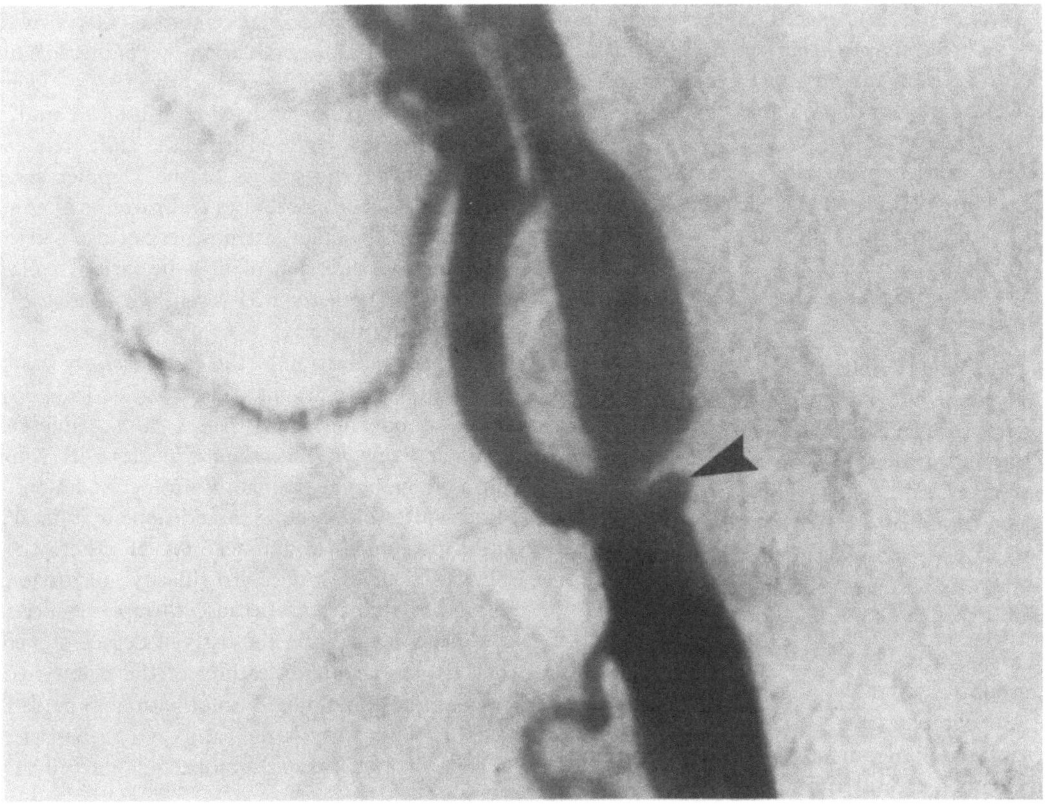

FIGURE 4–4. Atherosclerotic ulcerated plaque. Left lateral carotid angiogram reveals an atheromatous plaque with ulceration (arrowhead) on the posterior wall of the common carotid artery bifurcation. The plaque is causing a greater than 50% stenosis of the proximal internal carotid artery.

rotid endarterectomy to patients with ipsilateral carotid system TIAs with appropriately demonstrated lesions if the operation could be performed with a perioperative mortality and stroke morbidity of about 3% [11, 24, 27, 28]. In these carefully selected patients, there is evidence to suggest that surgery can reduce subsequent stroke morbidity.

Many believe that a "hemodynamically significant carotid stenosis" is a risk to the patient and should be treated by endarterectomy [29, 30]. Many surgeons recommend carotid endarterectomy for patients with an asymptomatic carotid bruit who are found on evaluation to have an operable lesion [24]. Others recommend evaluation of the carotid arteries in all patients who are to undergo a major surgical procedure, particularly coronary artery bypass surgery and peripheral vascular reconstruction [24]. This is based on the view that these patients are at a reltively higher risk for perioperative stroke and that carotid endarterectomy in patients with asymptomatic but operable lesions will reduce this risk.

The available data from a large number of publications [27, 28, 31–35] do not support a role for carotid endarterectomy in asymptomatic patients. This restraint applies equally to casually detected lesions, lesions found on the side opposite to symptomatic lesions, and lesions found in patients with only vertebrobasilar symptoms. An asymptomatic bruit is more a marker for the presence of atherosclerotic disease than a predictor of stroke. In fact, patients with an asymptomatic carotid bruit face a greater threat of death from heart disease than they do from stroke [11, 27, 36, 37].

Imaging Techniques

Imaging techniques play a major role in the diagnosis and management of vascular diseases involving the central nervous system. The approach to and choice of the imaging technique or techniques varies from one institution to another. It depends on regional philosophies of therapy in occlusive vascular disease and on the relative strengths of the different imaging procedures at each institution. The choice of the imaging technique also depends on how much information the treating physician requires to make a therapeutic decision and on his or her willingness to accept data inferior to that obtained by superior but riskier procedures.

There are a variety of imaging procedures available for the examination of patients suspected of having cerebrovascular disease. Procedures to find specific evidence of vascular disease may be divided into two categories: noninvasive and invasive.

NONINVASIVE TECHNIQUES

Although there are currently various noninvasive diagnostic tests available that can detect the presence of significant cervical carotid artery disease in a majority of cases, none is fully satisfactory either alone or in combination. At the present time, these techniques are generally regarded as having screening value only.

Noninvasive techniques include

1. Plain films
2. Ultrasonography
3. Dynamic computed tomography (CT) scanning of the common carotid artery bifurcation
4. Magnetic resonance imaging (MRI) of the common cartotid artery bifurcation

Plain Films. Since the advert of CT and MRI, the use of plain films of the head and neck has fallen into disfavor. However, this noninvasive modality can provide useful and important information in the evaluation of patients with cerebrovascular disease. According to Taveras [5], more than one half of patients who have calcifications 1 cm or more in length in the region of the common carotid artery bifurcation, as seen on soft-tissue films of the neck, have significant narrowing of the ICA at its origin. However, severe occlusive disease is frequently seen with minimal or no calicification.

The calcification of Monckeberg's sclerosis is noted in the tunica media of the vessel. Radiographically, it appears as a dense uniform band involving the entire circumference of the vessel, compared to the irregular and scattered nature of intimal calcification. The calcification of Monckeberg's sclerosis is not associated with stenosis of the artery.

Cervical spine films are also very useful in patients with vertebrobasilar TIAs. Compression of the vertebral artery can occur secondary to osteophytes impinging upon the foramen transversarium (see vertebral artery evaluation).

Cerebrovascular Ultrasonography. The superficial location of the extracranial cerebrovascular structures make them ideal for evaluation by noninvasive ultrasound techiques. Since the early 1970s, Doppler ultrasound instruments have been used to study the carotid arteries. By measuring the shift frequencies and analyzing the spectra of the Doppler spectra, information about local rates and patterns of flow can be obtained. Doppler instruments can also determine the rate and direction of flow in various collateral channels in patients with occlusive disease of the internal carotid artery.

Doppler evaluation of the carotid artery has been found to have an accuracy of 85–95% [38–43] for lesions producing more than a 50% reduction in luminal diameter. The major difficulty with Doppler ultrasonography is its inability to detect lesions that do not alter blood flow; in addition, it is incapable of displaying anatomic features of atherosclerotic plaques such as surface irregularity, ulceration, or thrombus formation. Because therapeutic decisions are made not only on the basis of degree of stenosis but also on anatomic features of the plaques themselves, the physiologic information provided by Doppler flow analysis represents only a partial evaluation of the patient's suspected cerebrovascular insufficiency.

Oculoplethysmorgraphy is one of a number of noninvasive tests that indirectly evaluate the state of the carotid vessels by examining the distal circulation for evidence of reduced flow [44, 45]. It uses the Doppler signal to monitor the flow in the supratrochlear and supraorbital branches of the ophthalmic artery. Alterations in pressure and flow at the common carotid bifurcation can be detected in these branches. However, lesions of the carotid system at any point between the common carotid artery and the ophthalmic artery can produce similar abnormalities in indirect testing.

To overcome the insensitivity of Doppler sonography to plaque producing less than 50% diminution in diameter, high-resolution real-time B-mode imaging is now routinely used as an adjunct to the Doppler examination. B-mode scanning can demonstrate the early changes of atherosclerosis long before a lesion produces alteration in blood flow (figures 4–5 to 4–7). Using B-mode instruments operating in the 7.5–10 mHz range, it is possible to demon-

FIGURE 4–6. Longitudinal B-mode ultrasound image showing calcific plaque (arrowheads) with posterior acoustic shadow (arrows) proximal to the common carotid bifurcation.

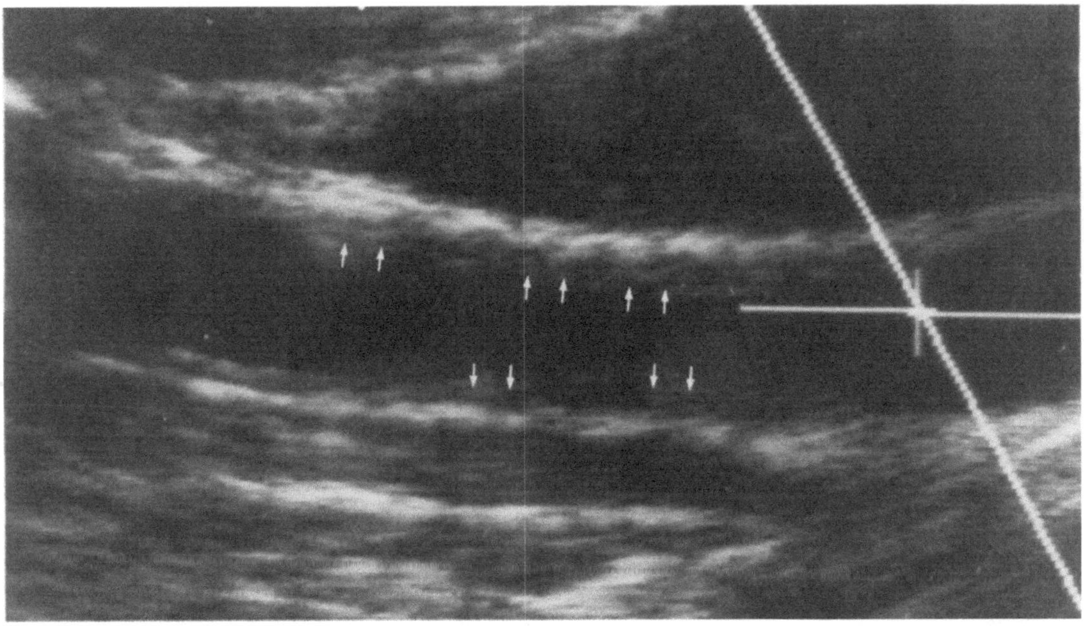

FIGURE 4–5. Longitudinal B-mode ultrasound image showing atheromatous plaque producing low-level echoes (arrows) along the anterior and posterior walls of the common carotid artery.

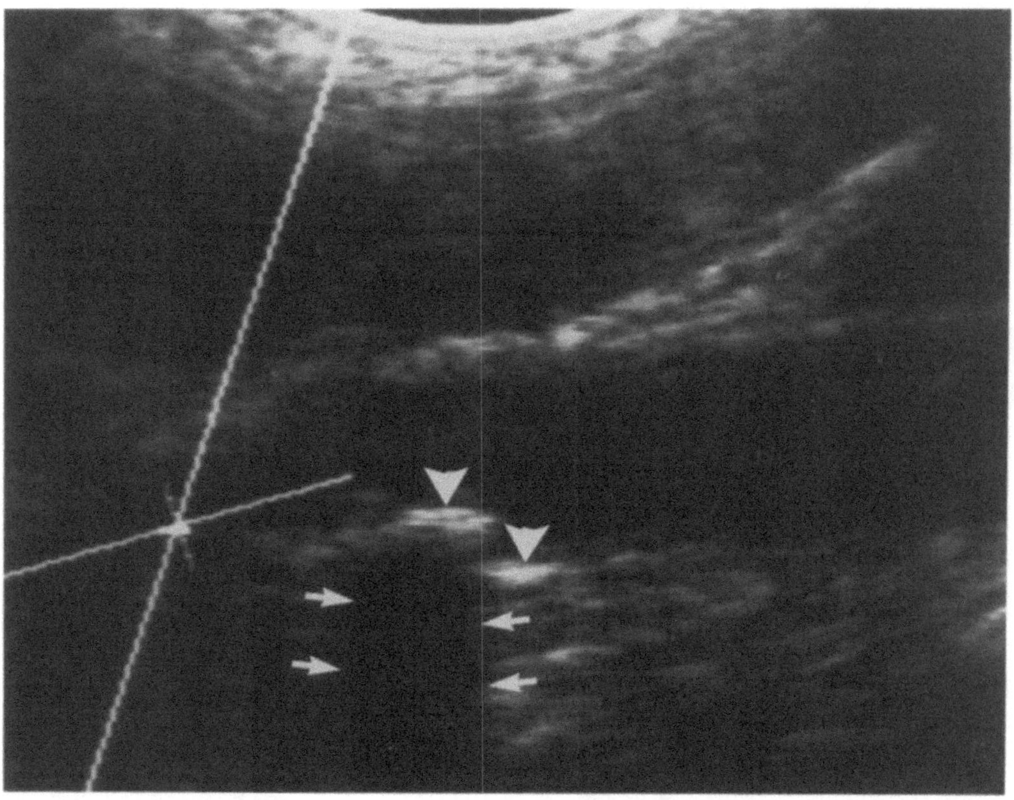

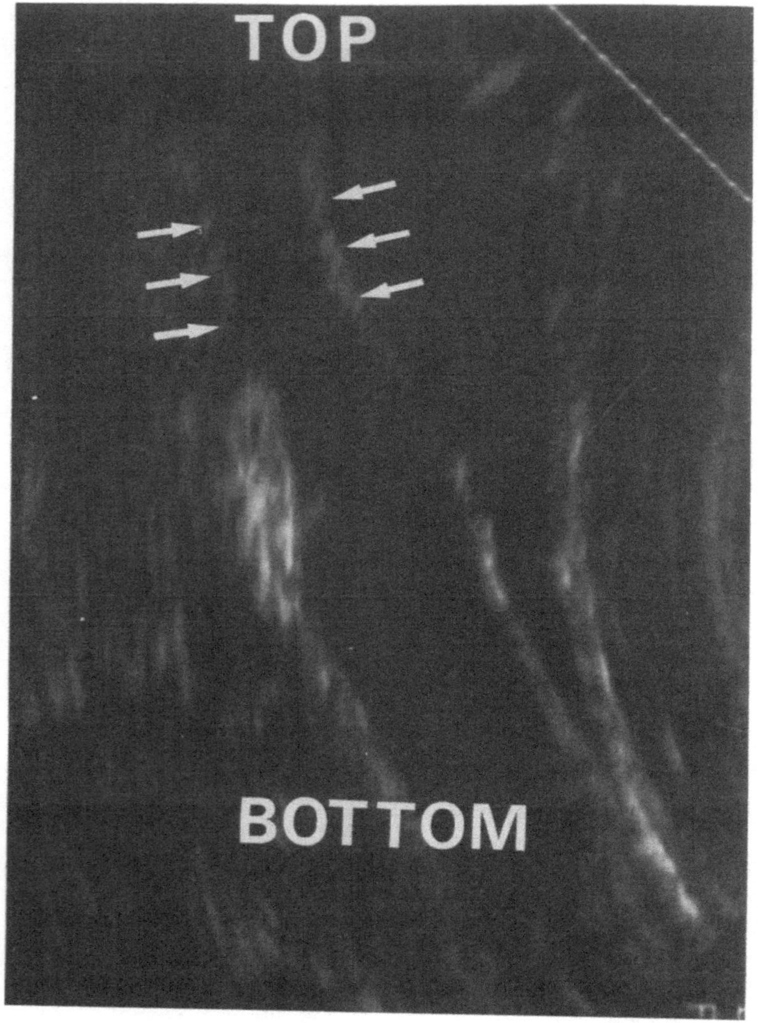

FIGURE 4–7A

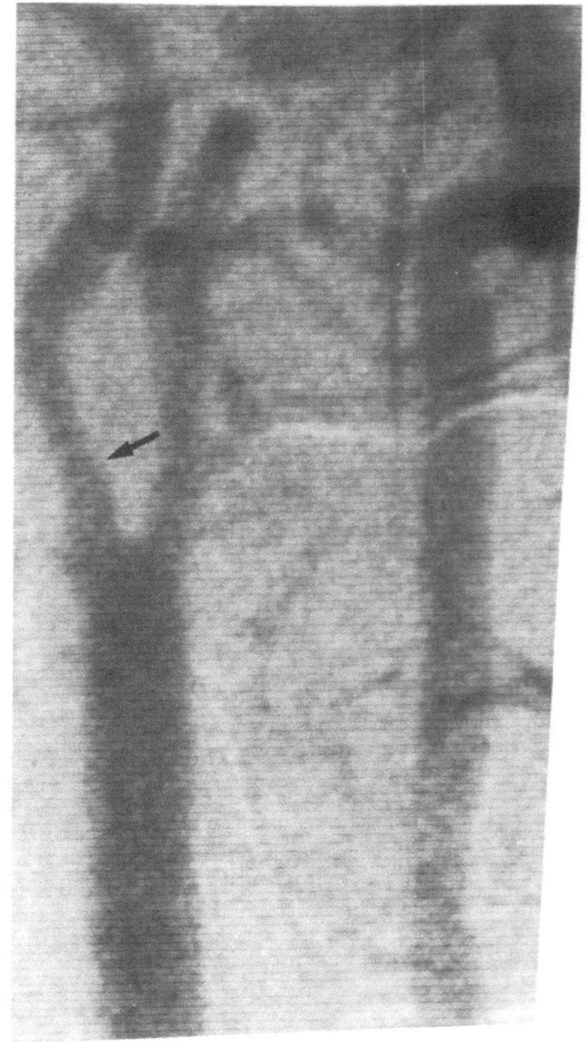

FIGURE 4–7B

FIGURE 4–7. (A) Longitudinal B-mode ultrasound image demonstrating moderate stenosis of the ICA (arrows). (B) IV-DSA confirms the moderate stenosis of the ICA (arrow).

strate ulceration or intraplaque hemorrhage in up to 90% of lesions showing such changes at surgery [46–48]. There are, however, significant difficulties with the use of B-mode imaging alone. The accuracy of B-mode imaging suffers as the severity of the disease increases [46, 49]. With real-time scanning, it is frequently impossible to distinguish a highly stenotic from a totally occluded vascular segment. B-mode scanning may underestimate the severity of disease due to incomplete visualization of the vessel as a result of plaque calcification. In addition, some plaques and thrombus may be overlooked unless careful attention is paid to subtle differences in echogenicity within the vessel lumina.

Given the accuracy of Doppler ultrasound for lesions producing more than 50% decrease in luminal diameter and the ability of B-mode scanning to demonstrate less severe lesion and lesion anatomy, the two modalities are now routinely combined for carotid artery evaluation. Such examinations are referred to as duplex ultrasonography and represent the state of the art in the noninvasive evaluation of the carotid arteries. Reported accuracy for duplex carotid sonography varies from 87–95% [50–53]. Although this technique suffers from an inability to evaluate the origins of the common carotid and vertebral arteries and the intracranial portions of the carotid arteries, it is now acceptable as a reliable method of screening for carotid atherosclerotic disease. Current opinion suggests that such procedures should be used in the initial evaluation of patients who are asymptomatic with a carotid bruit. Patients whose symptomatology is not localized to the carotid circulation nor definitive for cerebrovascular disease and patients in whom surgery would not be considered likely because of other serious medical problems or a probable nonsurgical cause for the TIA, such as a cardiac embolus.

CT Imaging of the Common Carotid Bifurcation. Imaging of the common carotid bifurcation by CT has been described as direct visualization of the arterial wall [54–56]. Following intravenous contrast administration, dynamic scanning utilizing multiple thin-section axial images is employed by this technique. Cross-sectional images are supplemented by various types of 3-D reformations. Preliminary reports suggest that CT scanning of the common carotid bifurcation is a potentially sensitive method for the detection of disease in the extracranial common carotid arteries. This technique can provide valuable and accurate anatomic information concerning the arterial wall but has yet to be widely used and accepted.

MRI. MRI has a unique ability for vascular study. Individual sagittal images can demonstrate the carotid artery bifurcation well (figure 4–8). However, the oblique and tortuous course of the carotid bifurcation as well as the eccentric location of the plaques make sagittal imaging difficult to interpret on a consistent basis. According to Goldberg [57], cross-sectional images are superior to sagittal images and permit appreciation to the best advantage of luminal occlusion by atheromatous plaques from any portion of the wall. In the future, with further improvement and refinement of the technology. MRI may well replace angiography in the investigation of patients with cerebrovascular disease.

INVASIVE TECHNIQUES

In spite of the advances made over the past 15 years using noninvasive techniques in the evaluation of cerebral vascular disease, angiography remains the gold standard and the definitive neuroimaging technique for the complete evaluation of cerebrovascular disease. Angiography is nearly always performed prior to a final decision for surgical intervention. Angiography should not be done unless the physician is prepared to advise surgical intervention should a surgically resectable lesion with appropriate symptoms be discovered.

The advent of cranial computed tomography (CT) and magnetic resonance imaging (MRI) have had a major impact in the management of the stroke patient, particularly in establishing the presence of an evolving or completed infarct (see Chapter 6). CT and MRI also serve to exclude other lesions that on occasion may clinically present similarly to a stroke. Every patient suspected of having cerebrovascular disease in whom angiography is to be performed should have a CT head scan or MRI of the head prior to angiography. Not infrequently, a patient with a subdural hematoma or a brain tumor, particularly a meningioma, may present with symptoms that mimic a TIA (figure 4–9). Other disease entities such as a berry aneurysm or an unsuspected AVM may be the cause of a patient's symptoms, rather than extracranial vascular occlusive disease.

Indications. Angiography is indicated when there is consideration of corrective vascular surgery, subarachnoid hemorrhage, or an uncertain diagnosis. Angiography is usually indicated in patients with TIAs and in patients who have had reversible ischemic attacks. (The neurological deficit clears completely within a few days to a week.) TIAs are the strongest indication for surgery when a surgically

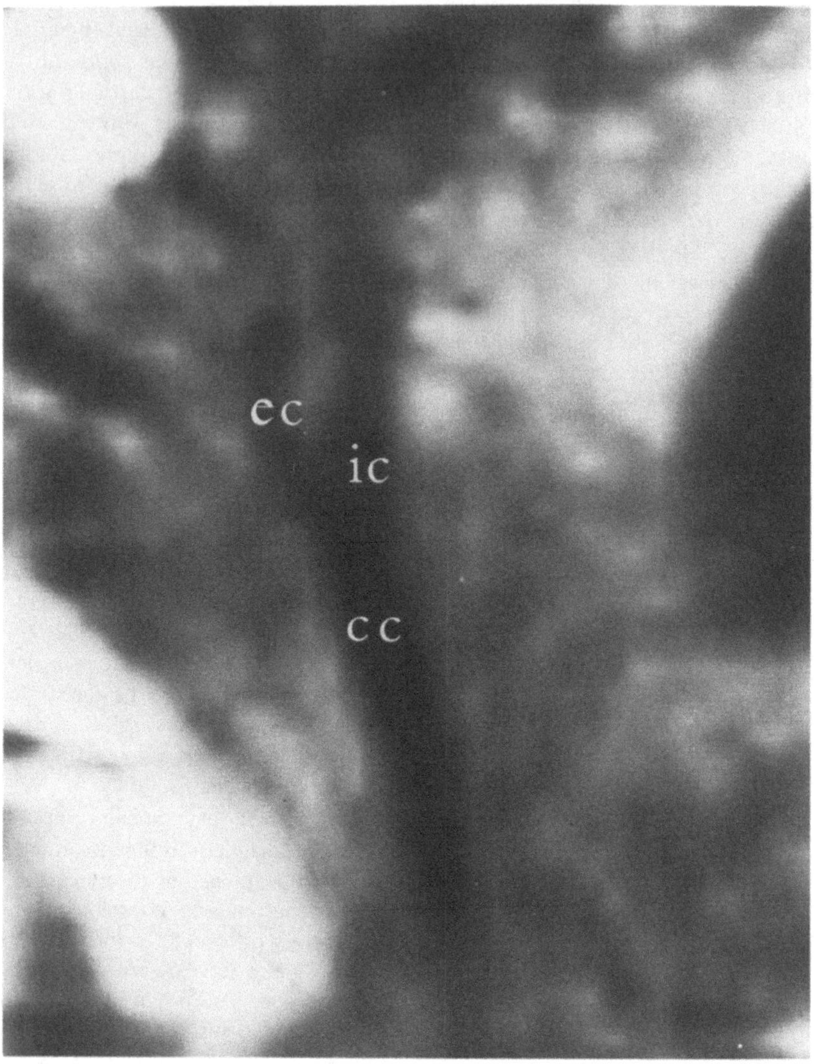

FIGURE 4–8. T1-weighted oblique sagittal MR image of the neck demonstrates a normal-appearing common carotid (cc) artery and its bifurcation into internal carotid (ic) and external carotid (ec) arteries.

accessible lesion in an artery supplying the appropriate area of the brain is identified.

Angiography and surgical intervention may also be indicated in patients with a fixed neurological deficit that may have initially improved but has subsequently worsened, and in patients where the neurological deficit progressively worsens. In stroke patients with subarachnoid hemorrhage, angiography should be performed if an aneurysm or arteriovenous malformation is considered a likely etiology. Finally, angiography should be performed in those petients in whom the neurological diagnosis has not been clearly established either by clinical or other diagnostic evaluation.

General Considerations. The ideal angiographic examination should fully visualize the entire cervicocranial vasculature from the aortic arch through its intracranial circulation, since atherosclerotic disease is frequently widespread and may involve both the extracranial and intracranial arteries.

Each exmination should be tailored to the individual patient's needs. However, certain principles should be followed. In order to define reliably and consistently the clinically significant but frequently radiographically subtle atheromatous lesions in the

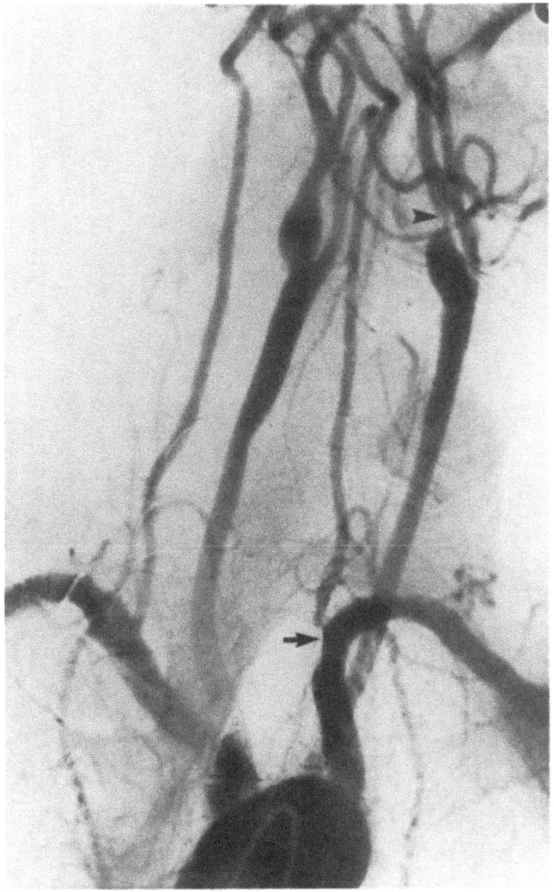

FIGURE 4–9A

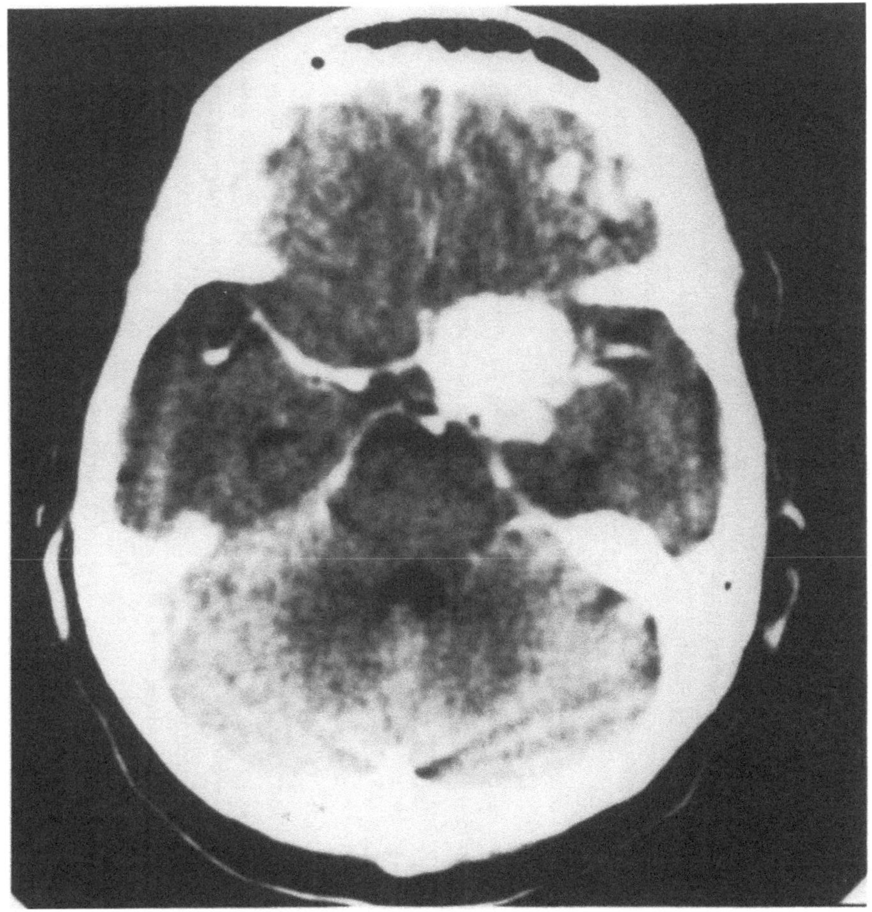

FIGURE 4–9B

FIGURE 4–9. 63-year-old female who complained of left hemispheric TIAs. Right posterior oblique arch aortogram (A) demonstrates an approximately 80% stenosis of the origin of the left ICA (arrowhead). An atheromatous plaque involving the right subclavian artery and not producing any significant hemodynamic stenosis is also identified immediately proximal to the origin of the left vertebral artery (arrow). The origin of the left vertebral artery is mildly stenotic. Mild but diffuse atherosclerotic changes are also identified in the left subclavian, innominate, right subclavian, and right common carotid arteries. The patient underwent a left carotid endarterectomy. Her symptoms did not improve but persisted. Postcontrast CT head scan (B) six weeks postoperatively demonstrates a large enhancing mass in the left medial sphenoid wing region which proved to be a meningioma. This was resected with resolution of the patient's symptoms.

carotid system extracranially, it is critical that high detailed visualization of this region be obtained. To accomplish these ends, a high-quality study including subtraction films should be obtained.

Each vessel must be examined in at least two different projections. Biplane studies are necessary for accurate evaluation, since in one plane the vessel may appear entirely or nearly normal. However, on the opposite view, a severe stenosis or ulcerative lesion may be demonstrated.

No angiographic study of cerebral vascular disease is complete without evaluation of the intracranial carotid circulation (see Chapter 6). The presence of a significant intracranial occlusive lesion in association with an extracranial stenosis might dictate a change in the management of the patient. In the case of tandem stenotic lesions—that is, one in a major intracranial artery and another in an extracranial vessel—simple repair of the latter may not improve the blood flow to the brain [58].

In addition, evaluation of the intracranial circulation occasionally reveals unsuspected nonocclusive vascular lesions that could also alter therapy (figure 4–9). Some, such as aneurysms, are encountered incidentally, while others, such as a subdural hematoma and neoplasm, are disease processes that may mimic cerebral ischemic event. Visualization of the intracranial circulation is also important to identify patterns of collateral blood flow.

The extracranial and intracranial vasculature can be studied by a number of different angiographic techniques and routines. Some of the more common methods are listed below:

1. Intravenous digital subtraction angiography (IV-DSA)
2. Intraarterial DSA (IA-DSA)
3. Conventional or film-screen angiography
   a. Retrograde right brachial angiography with direct percutaneous left carotid anigography
   b. Arch aortography
   c. Selective catheterization of carotid arteries
   d. Evaluation of vertebrobasilar system

Intravenous Digital Subtraction Angiography (IV-DSA). IV-DSA has been advocated both as a screening examination for cervical carotid disease and as a definitive diagnostic examination for patients with symptomatic cerebrovascular disease [59–64] (figure 4–10). Despite the enthusiam it generated when introduced in the mid-1970s and its early popularity, IV-DSA has fallen into considerable disfavor.

IV-DSA has several advantages over conventional angiography:

1. It has the potential for full examination of the cervicocranial vasculature
2. It is less invasive and safer than conventional angiography because it does not involve arterial catheterization
3. The cost of the examination is much less than for conventional angiography

IV-DSA has significant limitations that are mainly related to its inferior spatial resolution and motion artifacts (due to patient motion and swallowing). Low contrast density secondary to decreased cardiac output is another significant drawback. About 20–25% of examinations may be technically unsatisfactory due to these problems [65–68]. Furthermore, superimposition of opacified vessels in the intracranial circulation make the evaluation of these vessels difficult if not impossible. The large amount of contrast material used for this examination is also a disadvantage and may be a concern in patients with marginal renal, pulmonary, and cardiac states.

In our department, we reserve the use of IV-DSA mainly as a screening examination in the following clinical situations:

1. Patients with asymptomatic carotid bruits
2. Asymptomatic patients in whom cerebrovascular disease is suspected clinically
3. In the postoperative evaluation of patients who have undergone carotid endarterectomy

Intraarterial Digital Subtraction Angiography (IA-DSA). Intraarterial injection of contrast material provides superior and far more consistent visualization of the extracranial and intracranial arteries than do intravenous studies but with the added risks of arterial catheterization (figure 4–10B). Using IA-DSA, many of the shortcomings of IV-DSA are avoided while most of its advantages are preserved.

IA-DSA is of great value, especially in patients with renal or cardiac disease, since excellent opacification of extracranial and intracranial arterial systems can be obtained with this technique using much lower doses of iodinated contrast material than are used with conventional angiography and IV-DSA.

The resolution of IA-DSA is not quite that of conventional angiography (figures 4–10C, 4–13, 4–14). However, IA-DSA has disease demonstration capabilities almost equivalent to conventional angiography [4, 11] and can provide all the necessary information in the vast majority of cases.

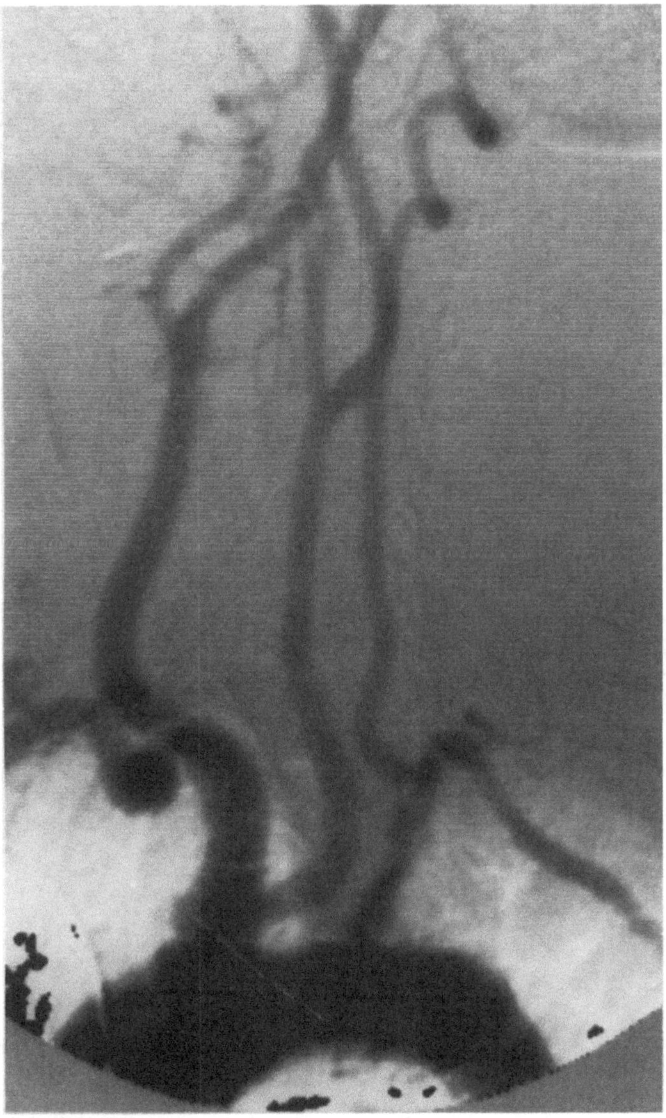

FIGURE 4–10A

FIGURE 4–10. Comparison of IV-DSA aortic arch, IA-DSA aortic arch and conventional selective right common carotid angiograms in a 72-year-old female who initially presented with bilateral asymptomatic carotid bruits and who then developed right hemispheric TIAs following coronary artery bypass surgery. (A) IV-DSA demonstrates normal common carotid birfurcations bilaterally. The origin of the brachiocephalic vessels are normal except for mild atherosclerotic changes in the left subclavian artery. The definition and resolution of both common carotid bifurcations on the IV-DSA compare favorably with the IA-DSA arch aortogram (B). The lateral view of the selective right common carotid angiogram (C) in the same patient is superior in resolution to the IV-DSA and IA-DSA studies but also does not demonstrate a significant abnormality.

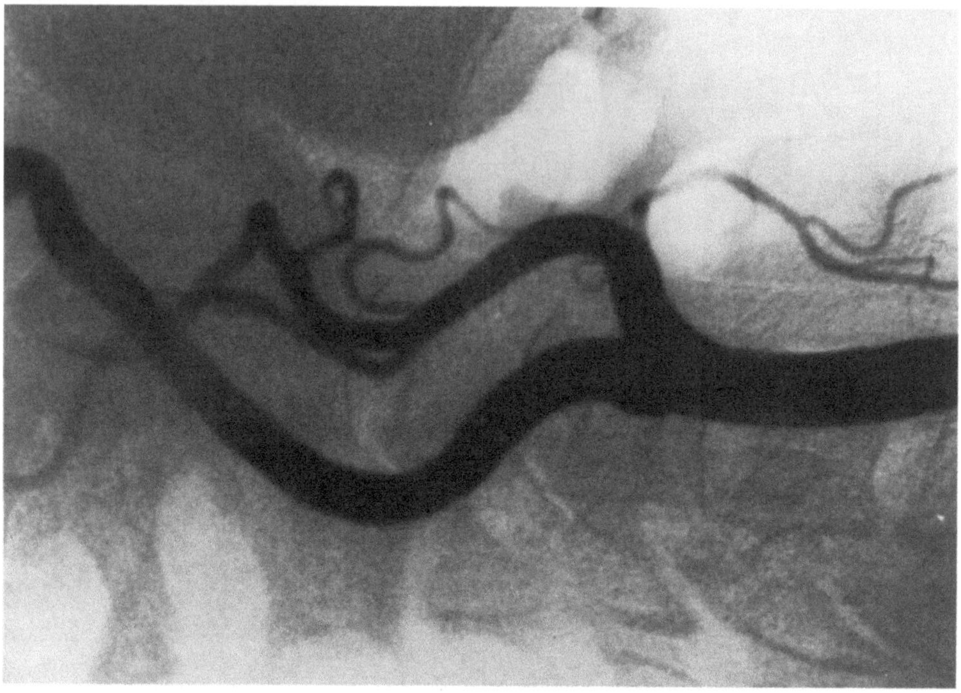

FIGURE 4–10C

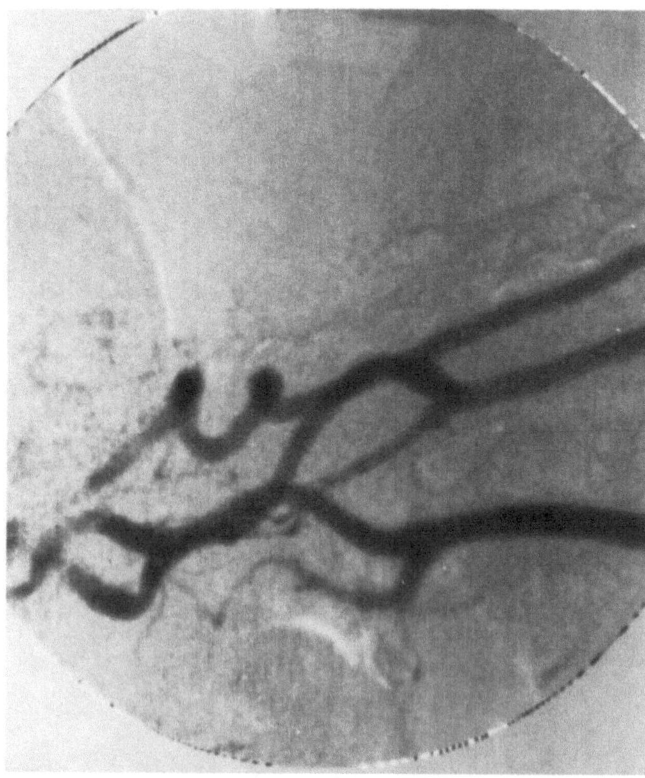

FIGURE 4–10B

Retrograde Brachial Angiography. In some instances where transfemoral arteriography is not technically feasible, retrograde brachial angiography may be performed (especially in patients with extensive peripheral vascular disease precluding the use of the femoral arteries and the Seldinger technique). Retrograde right brachial angiography is used to evaluate the antegrade flow through the right vertebral and the right carotid arteries. The left carotid circulation can be studied using a direct percutaneous left carotid approach. The left vertebral artery can be studied using a left brachial approach unless the left vertebral arises directly off the aortic arch.

Retrograde brachial angiography is a safe method for visualizing the cerebral vessels. It does have certain drawbacks. Right brachial angiography results in simultaneous opacification of the right carotid and vertebral arteries causing superimposition of these vessels on the AP radiographs. In general, the diagnostic quality of these examinations is poorer than with antegrade methods (selected catheterization and arch aortography).

Arch Aortography. The angiographic evaluation of cerebrovascular disease should not be limited to an aortic arch study. Although aortic arch injection provides adequate resolution of the extracranial arteries in most cases, it is not adequate for evaluation of the intracranial circulation. Furthermore, not infrequently the study of the extracranial vessels may be impaired by superimposition of vessels. The resolution is inferior to selective common carotid artery injection (see below) (figures 4–11, 4–12). Ulcerations may be difficult to identify, and evaluation of the degree of stenosis may be inaccurate [69]. A very high grade stenosis of the internal carotid artery may appear to represent a complete occlusion on aortic arch study [70, 71]—the so-called pseudo-occlusion of the internal carotid artery.

Selective Common Carotid Angiography. Angiography is best performed via the femoral approach using the Seldinger technique. This permits both aortic arch and selective common carotid artery injection. Selective common carotid injection provides superior anatomic detail concerning the nature and extent of vessel changes [5, 11, 12, 72] (figures 4–11, 4–12).

In high-risk patients with severe aortofemoral disease, which precludes this approach, right or left axillary artery catheterization can also achieve these objectives (figure 4–3).

In our department, we begin our angiographic investigation of patients with cerebrovascular disease with an IA-DSA aortic arch study (right posterior oblique projection) using a 5 French pigtail catheter. This is followed by selective catheterization of the common carotid artery on the side of interest after exchanging the pigtail catheter for another 5 French catheter.

Biplane filming of the common carotid artery bifurcation is performed using conventional cut film technique. All surfaces of the arterial wall may not be fully visualized even with biplane projections. If there is a suspicious change on the symptomatic side at the common carotid bifurcation or if the bifurcation is not adequately visualized, additional biplane oblique views should be obtained.

The opposite common carotid artery is then selectively catheterized and similar sets of radiographs are obtained.

If one has difficulty in selecting a particular vessel, the initial IA-DSA aortic arch study that mapped out the origin and course of the great vessels can be of great help.

At some institutions, the aortic arch study is performed after the selective common carotid artery studies, the rationale being that the aortic arch study requires the use of a number of 6 or 7 French catheters whereas a 5 French catheter is generally used for a selective common carotid artery catheterization. Bleeding would occur at the catheterization site if a smaller catheter was inserted after a larger one. This problem can easily be resolved by a number of methods. A femoral exchange sheath can be used to overcome this problem. Furthermore, the arch study can be performed, as is our custom, using a 5 French high-flow pigtail catheter so that the same size catheter is used for both the arch and selective common carotid artery studies.

The risk of neurologic complication is a factor that must be considered when angiography is contemplated. Since there is also an increased risk in the setting of acute cerebral infarction [73], angiography in the acute stroke setting is therefore performed only when the diagnosis of vascular occlusive disease is uncertain or when the procedure may elucidate an immediately treatable disease process such as arteritis or a surgically correctable vascular lesion.

Vertebral Artery Evaluation. The origin of the vertebral artery is a frequent location for occlusive vascular disease (figure 4–9). Clinical and angiographic studies demonstrate posterior circulation disease in 20–50% of all patients with cerebrovascular disease [74]. However, not all lesions are significant.

We usually evaluate the vertebral artery circula-

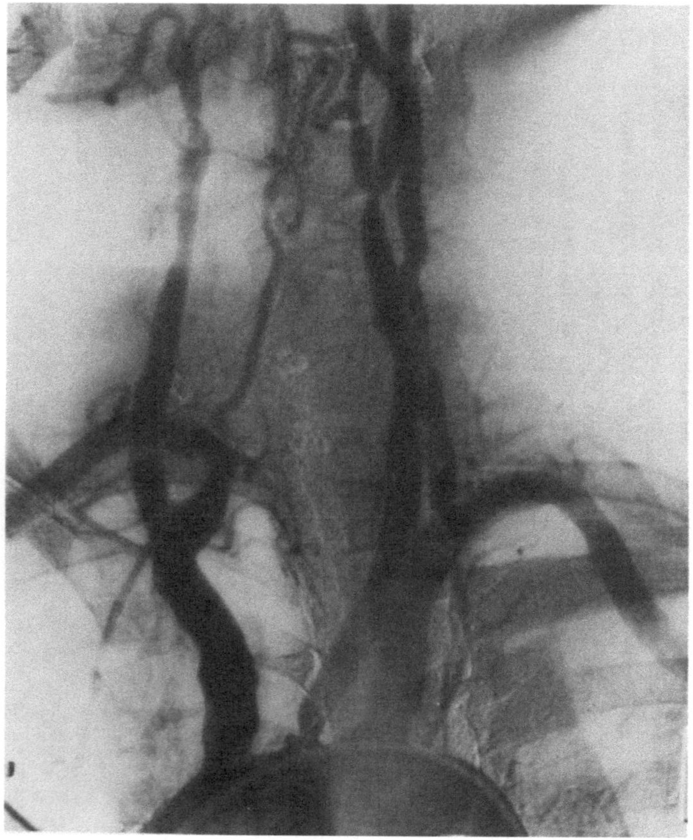

FIGURE 4–11A

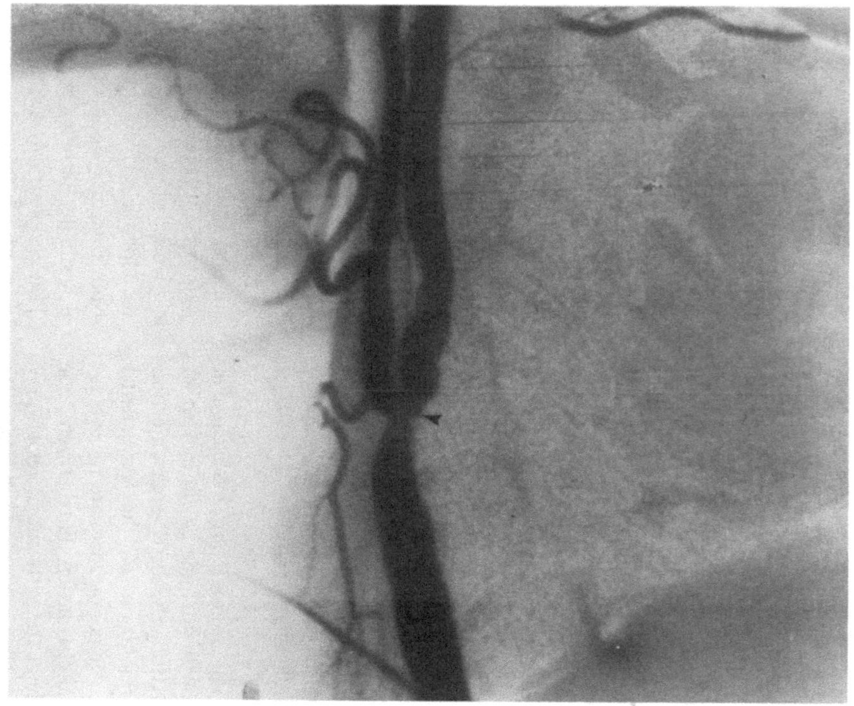

FIGURE 4–11B

FIGURE 4–11. A 60-year-old male with diffuse atherosclerotic disease of the extracranial vessels. (A) Right posterior oblique arch aortogram demonstrates extensive and diffuse atherosclerotic changes involving the innominate, right and left subclavian, and left vertebral arteries. There is a greater than 50% stenosis of the origin of the left common carotid artery. The right common carotid bifurcation is not well seen in this projection. (B) Lateral view of selective left common carotid angiogram demonstrates the atheromatous involvement of the common carotid bifurcation to a much better degree than the arch study. An 80% stenosis of the origin of the ICA is identified, as well as a small ulcer crater (arrowhead) that was confirmed at surgery.

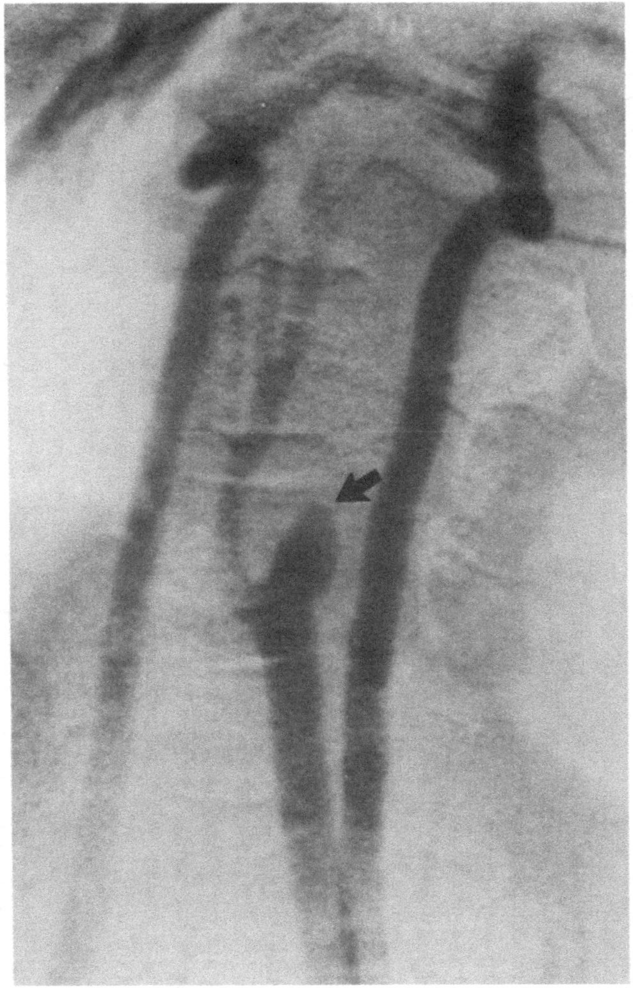

FIGURE 4–12A

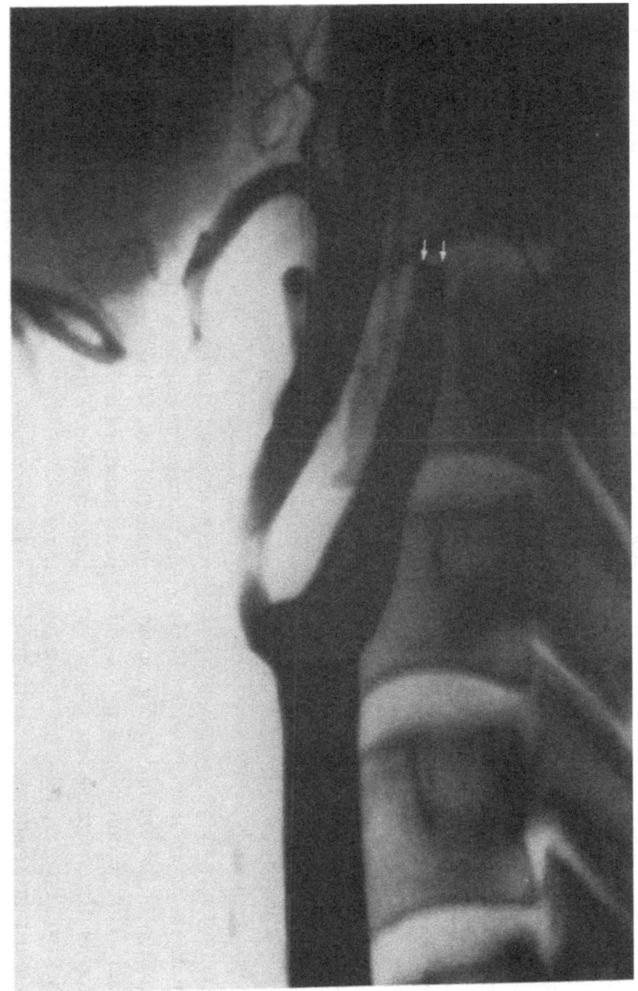

FIGURE 4–12B

FIGURE 4–12. (A) Right posterior oblique arch aortogram in a 63-year-old male demonstrates complete occlusion (arrow) of the left ICA 1.5 cm distal to its point of origin. (B) Lateral view of selective left common carotid angiogram demonstrates that the point of occlusion is in fact more distal than shown in the arch aortogram. Furthermore, the point of occlusion is not tapered, but a meniscus (arrows) of the proximal portion of the thrombus can be seen.

tion only on the aortic arch study unless specific symptoms dictate a more detailed examination. On an aortic arch injection, it may be difficult to visualize the origin of the vertebral arteries, because they arise posteriorly from the subclavian arteries. It may be necessary to profile the origin of the vertebral arteries by obtaining appropriate oblique views.

It is safer to study the vertebrobasilar circulation without selectively catheterizing the vertebral arteries. There is an increased risk of neurological complication with selective catherization of a diseased vertebral artery or when there are symptoms of vertebrobasilar insufficiency [72], either from dislodging atheromatous lesions or from occluding an already stenosed lumen with the catheter.

In patients with symptoms of vertebrobasilar insufficiency, if the aortic arch study is not satisfactory, injection of the contrast into the subclavian artery at or near the origin of the vertebral artery with blood-pressure-cuff occlusion of the brachial artery runoff may be performed. This technique provides good contrast resolution of the vertebral artery and its intracranial circulation. Bilateral selective common carotid artery examinations should also be performed to image the posterior cerebral arteries if these are not well seen on subclavian injections.

The vertebrobasilar circulation may also be compromised by intermittent compression of the cervical vertebral artery by osteophytes encroaching upon the artery in the foramen transversarium during head and neck turning. Cervical spine films and vertebral angiography with the head and neck turned to the symptomatic position will localize the side and level of the compression in such patients [75].

The vertebral circulation may be seriously affected by occlusion of the left subclavian artery in its first part, proximal to the left vertebral origin. When this occurs, the left vertebral artery acts as a collateral to the left subclavian artery (figure 4–13). Blood flows up the right vertebral artery and then in a retrograde fashion down the left vertebral artery to the left subclavian artery, where blood pressure has been lowered by the proximal subclavian block. This siphonage of blood from the vertebrobasilar system to the arm has been termed the *subclavian steal* syndrome. The diversion of blood from the brain may result in cerebral ischemia [2]. The abnormality is well demonstrated by arch aortography. Less commonly, siphonage of blood down the right vertebral artery may occur if there is an innominate or proximal right subclavian block (figure 4–14).

Angiographic Features of Atherosclerosis

Ulcerating and stenosing plaques most often arise from the posterior wall of the internal carotid artery at or near its origin (figure 4–4), but not infrequently they arise only from a side wall. An ulcerative plaque is seen as an irregularity of the vessel wall in mild cases (figures 4–15, 4–16) or as a niche or out-pouching in more advanced disease (figure 4–17); that is, the ulcer is seen in profile. In some cases, it may appear as a well-circumscribed double density superimposed on the artery, and this represents the ulcer seen *en face* rather than in profile.

Some discrepancies and inaccuracies exist between the angiographic findings and the surgical appearance of atheroma [76–78]. An ulcer crater covered by a thrombus at the time of angiography may lead to a false-negative diagnosis. In one series, only 60% of carotid artery bifurcations that showed ulceration at surgery were diagnosed as having ulcers using conventional angiography [76]. Half the remaining ulcers occurred in smooth, benign-appearing plaques. An incorrect angiographic diagnosis of ulceration was made in 17 of 50 carotid arteries, in most cases due to a subintimal hematoma or out-pouching secondary to destruction of the media with intact endothelium (figure 4–15).

In most cases, these subintimal hematomas have a typical angiographic appearance consisting of a sharply marginated, rounded, eccentric filling defect located near the extracranial carotid bifurcation. This characteristic radiographic appearance occurs in 50% of cases of subintimal hematomas. However, a smooth atheromatous plaque without hemorrhage may also have this appearance.

For diagnostic precision and accuracy, Kricheff [11] has recommended avoiding the use of the term *ulcer* and using descriptive terms such as *out-pouching*, *smooth plaque*, or *irregular plaque*.

A thrombus may occasionally be seen projecting into the arterial lumen (figures 4–18, 4–19).

As mentioned earlier, stenosis becomes hemodynamically significant when the cross-sectional area of the lumen is decreased by 75%. Angiographically, this is represented by a 50% decrease in the transverse diameter of the opacified lumen [11, 12]. A far greater degree of reduction (90%) is necessary to reduce flow [11]. There is no general agreement as to the best method for evaluating the degree of stenosis.

The distal cervical carotid segment diameter, not that of the bulb, should be used for determining the percent stenosis of a proximal internal carotid artery (ICA). Using the carotid bulb diameter as the normal ICA diameter significantly overestimates the degree of effective carotid stenosis [72]. Using the distal internal carotid diameter as the normal reference diameter, a 60% stenosis was found to produce a highly consistent change in ophthalmic artery dynamics as measured by oculoplethysmography.

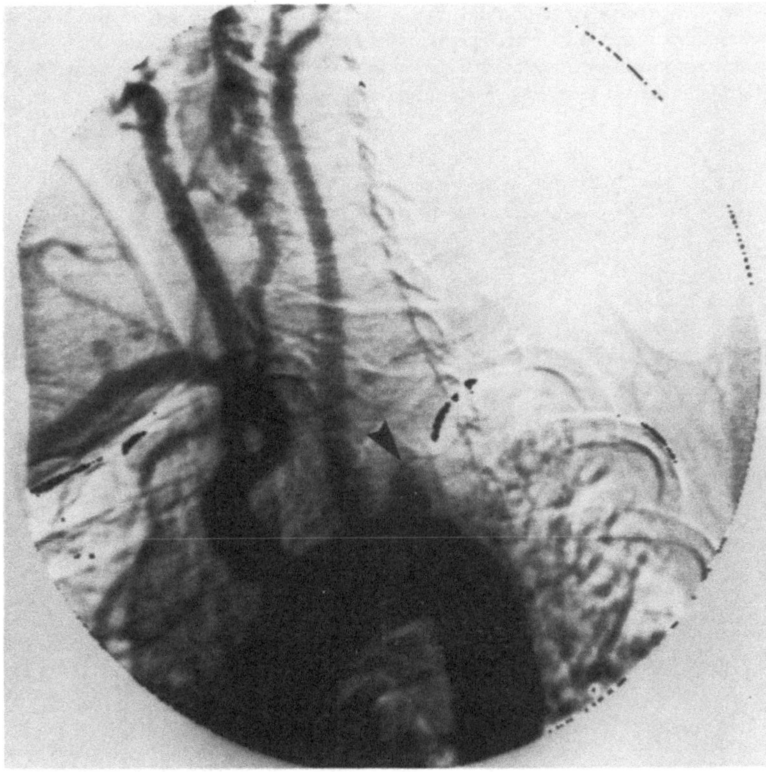

FIGURE 4–13A

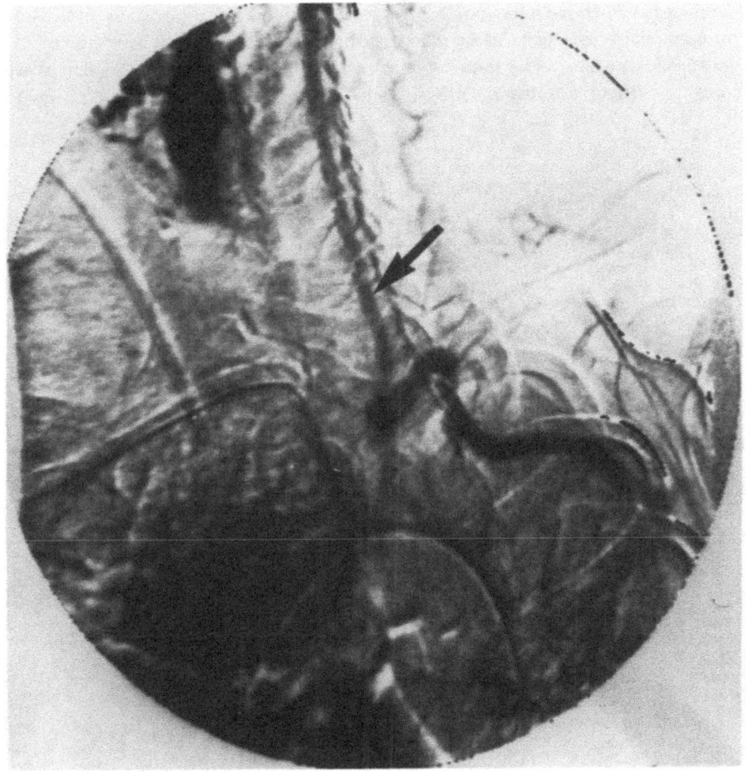

FIGURE 4–13B

FIGURE 4–13. Subclavian steal syndrome in a 66-year-old female. (A) IA-DAS arch aortogram demonstrates complete occlusion of the proximal portion of the left subclavian artery (arrowhead) 2 cm distal to its point of origin. (B) Film obtained several seconds after (A) demonstrates reconstitution of the distal portion of the left subclavian artery by retrograde filling of the left vertebral artery (arrow).

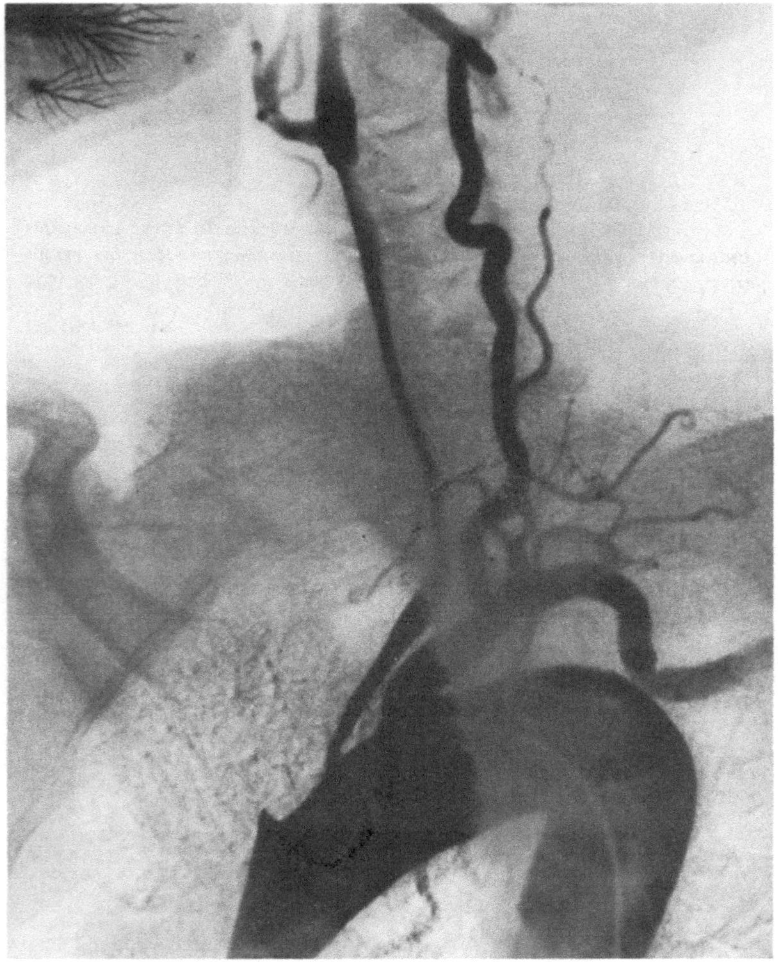

FIGURE 4–14A

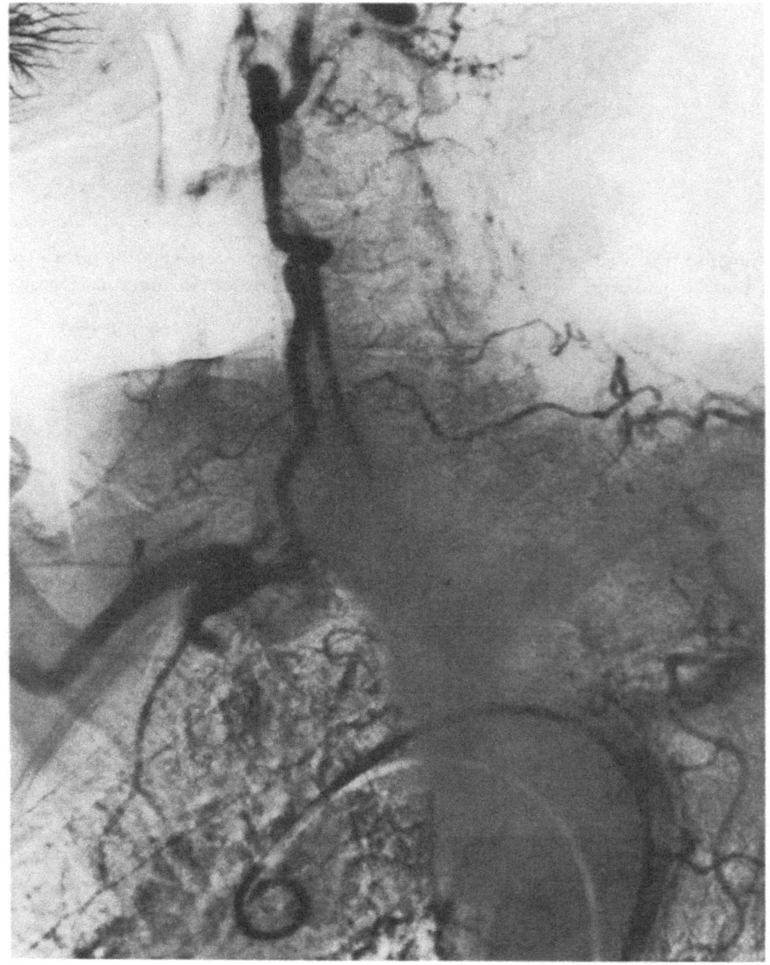

FIGURE 4–14B

FIGURE 4–14. Subclavian steal. (A) Note complete occlusion of the right innominate artery at its origin with nonfilling of the right common carotid, subclavian, and vertebral arteries. Also note the tapered and narrowed ascending aorta and left common carotid artery. (B) Film obtained several seconds after (A). Note retrograde opacification of the right vertebral artery and distal right subclavian artery. The underlying pathology in this case was a dissecting aneurysm of the aorta that occluded the innominate artery and produced a right-sided subclavian steal syndrome. Atherosclerosis causing occlusion of the innominate or proximal right subclavian artery can produce similar findings.

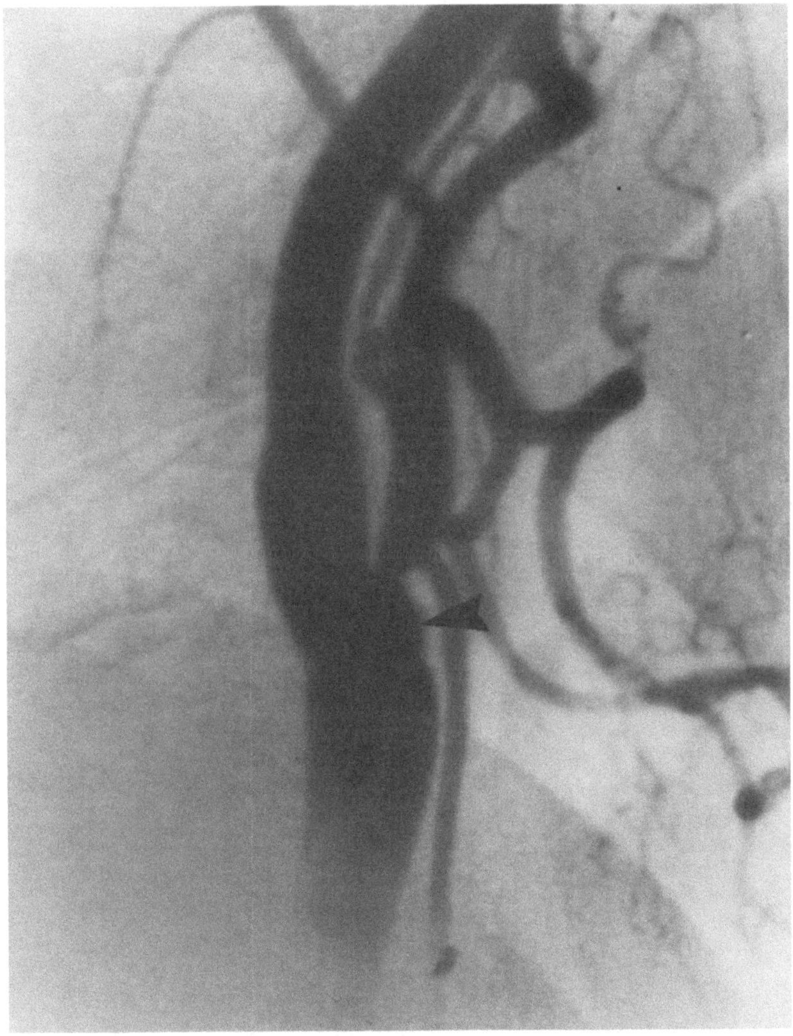

FIGURE 4–15. A 69-year-old female with right hemispheric TIAs. Lateral view of selective right common carotid angiogram reveals atheromatous plaque with an irregular out-pouching (arrowhead) involving the anterior wall of the common carotid bifurcation. The atheromatous plaque extends cephalad to produce an approximately 70% stenosis of the external carotid artery. Ulcers on the anterior wall of the artery are far less common than on the posterior wall. At surgery, this out-pouching proved to be due to a subintimal hematoma and not an ulcer.

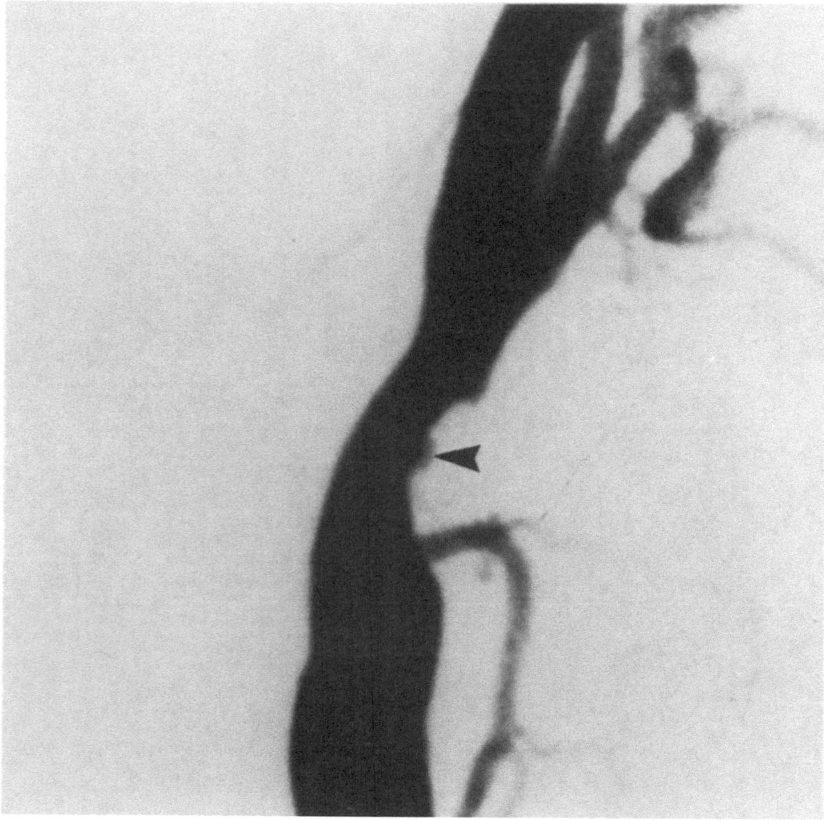

FIGURE 4–16A

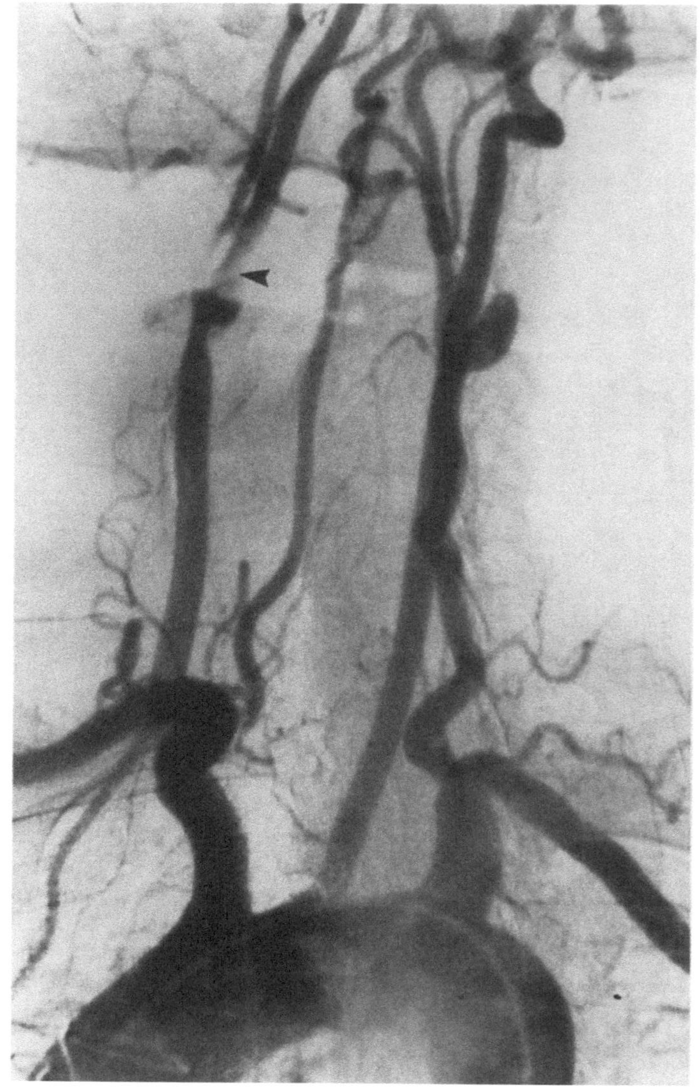

FIGURE 4–16B

FIGURE 4–16. (A) Selective right common carotid angiogram in a 54-year-old male with right hemispheric TIAs demonstrates an ulcerated plaque (arrowhead) involving the origin of the right ICA. This cannot be appreciated on the arch study. (B) (arrowhead). The left ICA is totally occluded.

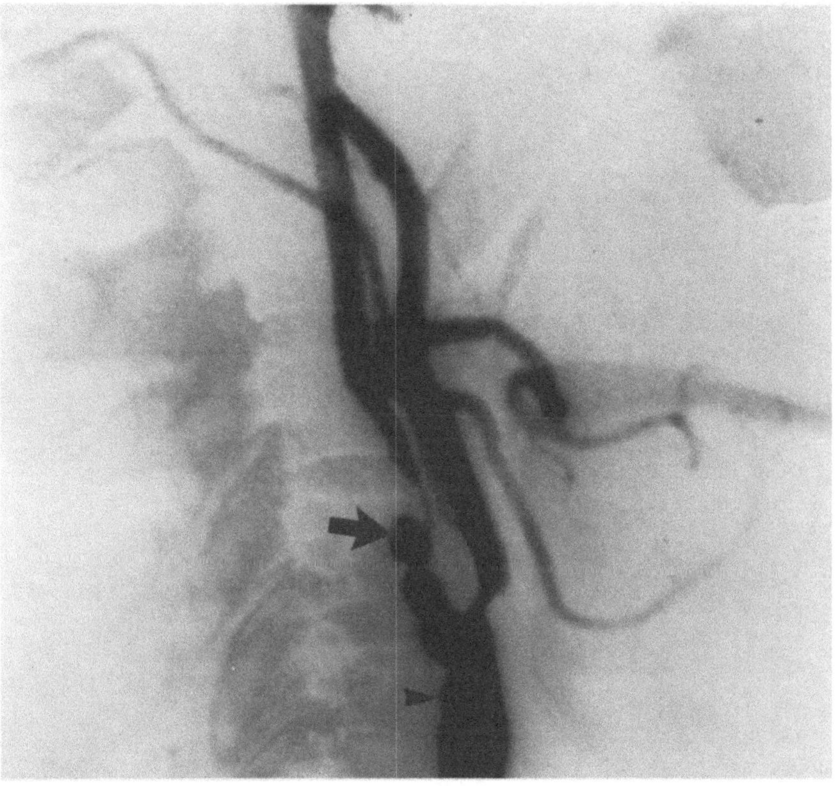

FIGURE 4–17. Extensive ulcerated atheromatous disease. Right lateral common carotid angiogram reveals extensive atheromatous disease involving the common carotid artery bifurcation with a small ulceration (arrowhead) on its posterior wall. A larger ulcerated plaque (arrow) is seen involving the proximal ICA, which is significantly narrowed by the atheromatous plaque. The origin of the external carotid artery is also narrowed.

Others have suggested that the most appropriate method for evaluating the hemodynamic significance of a stenosis is the determination of the minimal residual luminal diameter. A carotid artery with a residual diameter of less than 2 mm in absolute size after correction for magnification is considered significantly stenosed [11, 79].

The stenosis may extend over a long or short segment [81]. Both the length of the segment and its diameter are important. A long segment of stenosis may be more significant than a short segment with the same degree of narrowing [82].

As discussed earlier, if more than one stenotic lesion is present along a vascular distribution, the hemodynamic effect is not the cumulative effect of all the stenoses but is mainly determined by the most severe stenosis. Such tandem lesions are often found in the internal carotid artery, with one lesion at the common carotid artery bifurcation and the other in the carotid siphon. When the carotid siphon stenosis exceeds that of the proximal internal carotid artery, removal of the proximal lesion alone does not increase blood flow through the distal stenosis.

Stenosing plaques may be a smooth or rough surface. Plaques with irregular surfaces are often difficult to distinguish from atheromatous ulceration [11, 82].

Occlusion of the extracranial arteries is less common than stenosis. The end stage of atheromatous narrowing is total occlusion of the vessel. The distribution of arterial involvement is similar, since occlusion usually is sequel of advanced stenosis. Occlusion is recognized angiographically by the abrupt termination of flow of contrast medium and lack of filling of the artery distally.

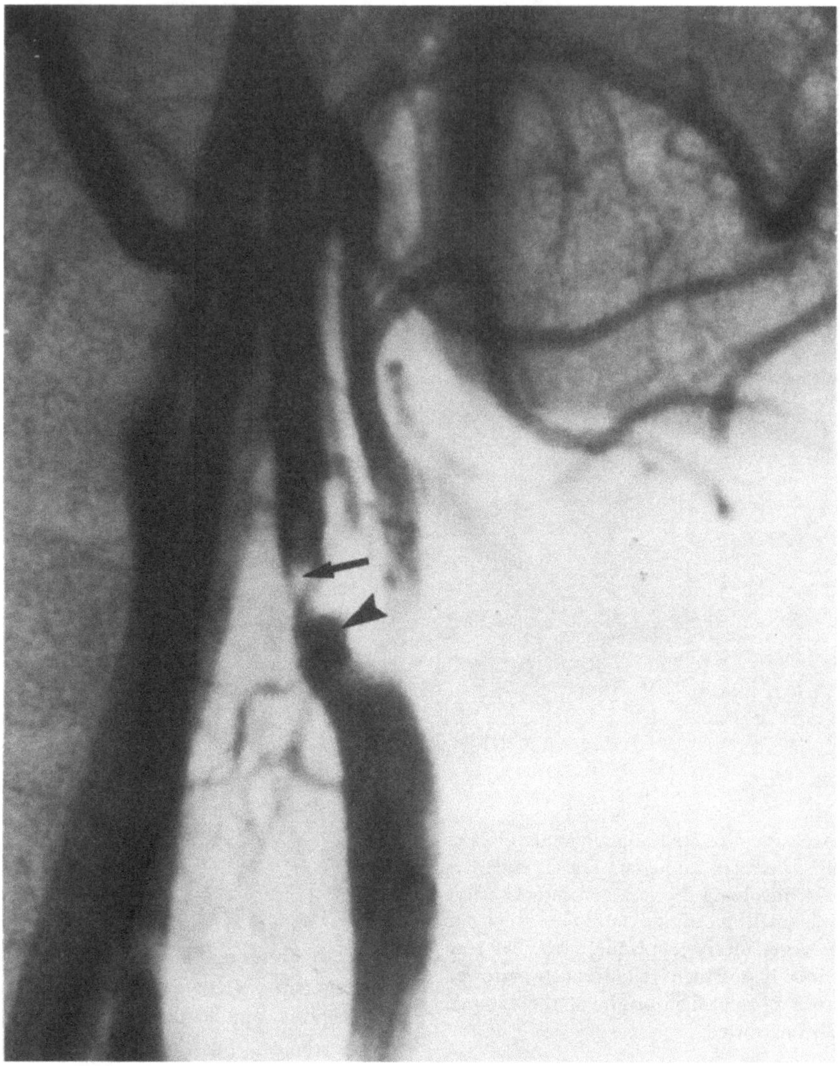

FIGURE 4–18. Ulcerated plaque with intramural thrombus. A 50-year-old female presented complaining of left hemispheric TIAs. Left posterior oblique arch aortogram demonstrates ulcerated plaque (arrowhead) involving the origin of the left ICA with a filling defect (arrow) distal to the plaque which represents an intramural thrombus.

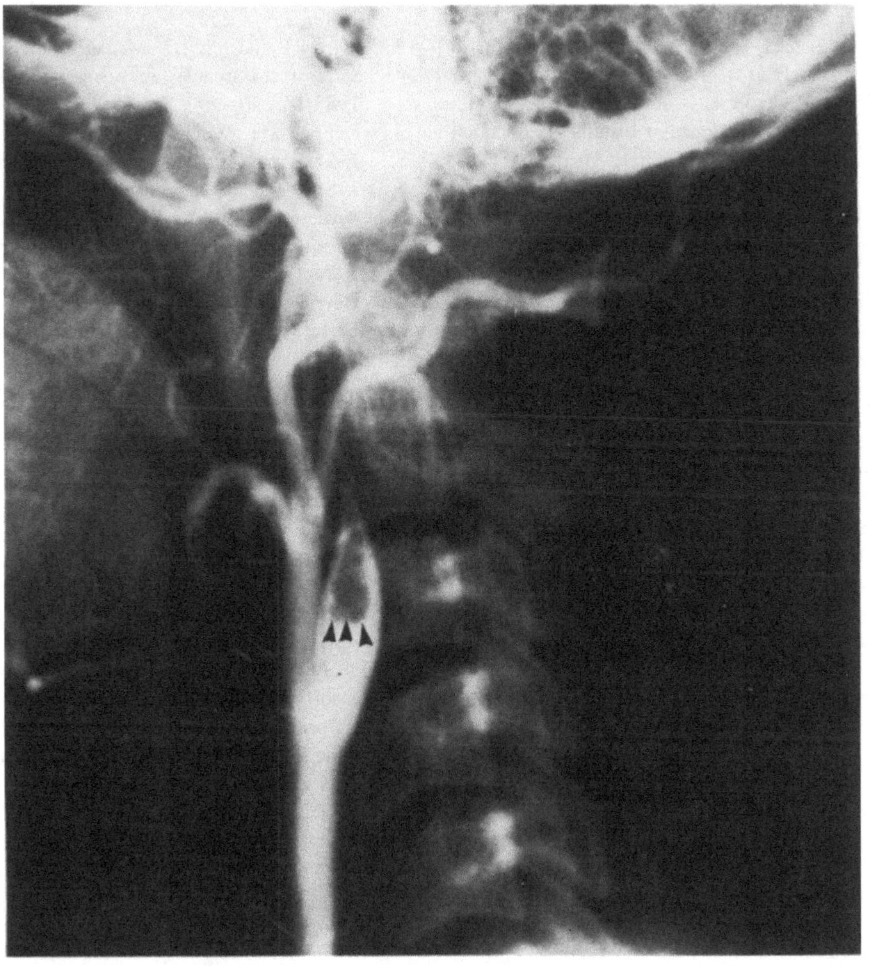

FIGURE 4–19. Intramural thrombus with total occlusion. A 63-year-old male with left hemispheric TIAs that had stopped three days prior to the angiogram. Lateral view of selective left common carotid angiogram demonstrates occlusion of the ICA 4 cm distal to its origin. A filling defect due to a thrombus (arrowheads) can be seen at the site of occlusion.

References

1. Fields WS. Aortocranial occlusive disease (Stroke). *Clin Symp* 1974;26:3–31.
2. Sarwar M. Cerebrovascular disease and nontraumatic intracerebral hemorrhage. In: Sarwar M. AzarKia B, Batnitzky S (eds). *Basic Neuroradiology*. St. Louis: Warren H. Green, 1983:486–534.
3. Wylie EJ, Ehrenfeld WK. *Extracranial Occlusive Cerebrovascular Disease: Diagnosis and Management*. Philadelphia: W.B. Saunders, 1970.
4. Drayer BP. Diseases of the cerebral vascular system. In: Rosenberg RN, Heinz ER (eds). *The Clinical Neurosciences*, Vol. 4. New York: Churchill Livingstone, 1984:247–360.
5. Taveras JM, Wood EH. *Diagnostic Neuroradiology*. Baltimore: Williams and Wilkins, 1976:850–876.
6. Hunt JR. The role of the carotid arteries in the causation of vascular lesions of the brain with remarks on certain special features of the symptomatology. *Am J Med Sci* 1914;147:704–713.
7. Fisher M. Occlusion of the internal carotid artery. *Arch Neurol Psychiat* 1951;65:346–377.
8. Lyons C, Galbraith G. Surgical treatment of atherosclerotic occlusion of the internal carotid artery. *Ann Surg* 1957;146:487–498.
9. Stehbens WE. *Pathology of the Cerebral Blood Vessels*. St. Louis: C.V. Mosby Co., 1972:98–130.
10. Imparato AM, Riles RS, Gorstein F. The carotid bifurcation plaque: Pathologic findings associated with cerebral ischemia. *Stroke* 1979;53:665–674.
11. Kricheff II. Arteriosclerotic ischemic cerebrovascular disease. *Radiology* 1987;162:101–109.

12. Osborne AG. *Introduction to Cerebral Angiography*. Hagerstown: Harper and Row, 1980:41.
13. Wood EH, Correll JW. Atheromatous ulceration in major neck vessels as a cause of cerebral embolism. *Acta Radiol* 1969;9(special issue):520–536.
14. Fisher CM. Transient monocular blindness associated with hemiplegia. *Trans Am Neurosurg Assoc* 1951;154–158.
15. Grady PA. Pathophysiology of extracranial cerebral arterial stenosis: A critical review. *Stroke* 1984;15: 224–236.
16. Kilgore BB, Fields WS. Arterial occlusive disease in adults. In: Newton TH, Potts GD (eds). *Radiology of the Skull and Brain*. St Louis: C.V. Mosby Co., 1974: 2310–2343.
17. Kannel WB, Wolf PA, Verter J. Risk factors for stroke. In: Smith RR (ed). *Stroke and the Extracranial Vessels*. New York: Raven Press, 1984:47–85.
18. Sandok BA, Furlan AJ, Whisnant JP, Sundt TF Jr. Guidelines for the management of transient ischemic attacks. *Mayo Clin Proc* 1978;53:665–674.
19. Whisnant JP. A population study of stroke and TIA: Rochester, Minnesota. In: Gillingham FJ, Mawdsley C, Williams AE (eds). *Stroke*. New York: Churchill Livingstone, 1976:21–39.
20. Whisnant JP, Matsumoto N, Elveback LR, Transient cerebral ischemic attacks in a community: Rochester, Minnesota, 1955 through 1969. *Mayo Clin Proc* 1973; 48:194–198.
21. Matsumoto N, Whisnant JP, Kurland LT, Okazaki H. Natural history of stroke in Rochester, Minnesota, 1955 through 1969: An extension of a previous study, 1945 through 1954. *Stroke* 1973;4:20–29.
22. Marshall J. The natural history of transient ischemic cerebro-vascular attacks. *Q J Med* 1964;33:309–324.
23. Baker RN, Ramseyer JC, Schwartz WS. Prognosis in patients with transient cerebral ischemic attacks. *Neurology* (Minneapolis) 1968;18:1157–1165.
24. Easton JD, Sherman DG. Carotid endarterectomy. *Mayo Clin Proc* 1983;58:205–206.
25. Hachinski V. Carotid endarterectomy. *Arch Neurol* 1987;44:654.
26. Jonas S. Can carotid endarterectomy be justified? No *Arch Neurol* 1987;44:652–654.
27. Patterson RH. Can carotid endarterectomy be justified? Yes. *Arch Neurol* 1987;44:651–652.
28. Whisnant JP, Sandok BW, Sundt TM Jr. Carotid endarterectomy for unilateral carotid system transient cerebral ischemia. *Mayo Clin Proc* 1983;58:171–175.
29. Riles TH, Lieberman A, Kopelman I, Imparato AM. Symptoms, stenosis and bruit: interrelationships in carotid artery disease. *Arch Surg* 1981:218–220.
30. Thompson JE, Autin DJ, Patman RD. Carotid endarterectomy for cerebrovascular insufficiency: long-term results in 592 patients followed up to thirteen years. *Ann Surg* 1970;172:663.
31. Mohr JP. Asymptomatic carotid artery disease. *Stroke* 1982;13:431–433.
32. Yatsu FM, Hart RG. Asymptomatic carotid bruit and stenosis: a reappraisal. *Curr Concepts Cerebrovasc Dis* 1982;17:21–25.
33. Hart RE, Easton JD. Management of cervical bruits and carotid stenosis in preoperative patients. *Stroke* 1983;14:290–297.
34. Ropper AH, Wechsler LR, Wilson LS. Carotid bruit and the risk of stroke in elective surgery. *N Engl J Med* 1982;307:1388–1390.
35. Barnett HJM, Plum F, Walton JN. Carotid endarterectomy: an expression of concern. Editorial. *Stroke* 1984;15:941–943.
36. Heyman A, Wilkinson WE, Heyden S, Helms MJ, Bartel AG, Karp HR, Tyroler HA, Hames CG. Risk of stroke in asymptomatic persons with cervical arterial bruits: a population study in Evans County, Georgia. *N Engl J Med* 1980;302:838–841.
37. Wolf PA, Kannel WB, Sarlie P, McNamara P. Asymptomatic carotid bruit and risk of stroke: the Framingham study. *JAMA* 1981;245:1442–1445.
38. Berry SM, O'Donnell JA, Hobson RW II. Capabilities and limitations of pulsed Doppler sonography in carotid imaging. *J. Clin Ultrasound* 1980;7:405–412.
39. O'Leary DH, Persson AV, Clouse ME. Noninvasive testing for carotid artery sclerosis: 1. prospective analysis of three methods. *AJR* 1981;137:1189–1194.
40. Doorley TPG, Atkinson PI, Kingston V, Shanik DG. Carotid ultrasonic arteriography combined with real time spectral analysis: a comparison with angiography. *J Cardiovasc Surg* 1982;23:243–245.
41. Wasserman DH, Hobson RW II, Lynch TG, Berry SM, Jamil Z. Ultrasonic imaging and oculoplethysmography in diagnosis of carotid occlusive disease. *Arch Surg* 1983;117:1161–1163.
42. Manga P, Dhurandhar RW, Stockard B. Doppler frequency ratio and peak frequency in the assessment of carotid artery disease: a comparative study with angiography. *Ultrasound Med Biol* 1986;12:573–576.
43. Bandyk DF, Levine AW, Phol L, Towne JB. Classification of carotid bifurcation disease using quantitative Doppler spectrum analysis. *Arch Surg* 1985;120:306–314.
44. Ginsberg MD, Greenwood SA, Goldberg HI. Noninvasive diagnosis of extracranial cerebrovascular disease: oculoplethysmography–phonoangiography and directional Doppler ultrasonography. *Neurology* (New York) 1979;29:623–632.
45. Berguer R. Carotid artery noninvasive testing is being overused. *Stroke* 1983;14:819–821.
46. Bluth EI, Kay D. Merritt CRB, Sullivan M, Farr G, Mills NL, Foreman M, Sloan D, Schlater M, Stewart J. Sonographic characterization of carotid plaque: detection of hemorrhage. *AJR* 1986;146:1061–1065.
47. Reilly L, Lusby R, Hughes L, Ferrell L, Stoney R, Ehrenfeld W. Carotid plaque histology using real-time ultrasonography. *Am J Surg* 1983;146:188–193.
48. O'Donnell TF Jr, Erdoes L, Mackey WC, McCullough J, Shepard A, Heggerick P, Isner J, Callow AD. Correlation of B-mode ultrasound imaging and arteriography with pathologic findings at carotid endar-

terectomy. *Arch Surg* 1985;120:443–449.
49. Wolverson MK, Heiberg E, Sundaram M, Tantanasirviongse S, Shields JB. Carotid atherosclerosis: high-resolution real-time sonography correlated with angiography. *AJR* 1983;140:355–361.
50. Jacobs NM, Grant EG, Schellinger D, Byrd MC, Richardson JD, Cohan SL. Duplex carotid sonography: criteria for stenosis, accuracy and pitfalls. *Radiology* 1985;154:385–391.
51. Zwiebel WJ, Austin CW, Sackett JF, Strother CM. Correlation of high-resolution, B-mode and continuous-wave Doppler sonography with arteriography in the diagnosis of carotid stenosis. *Radiology* 1983;149:523–532.
52. Clark WH, Hatten HP. Noninvasive screening of extracranial carotid disease: Duplex sonography with angiographic correlation. *AJNR* 1981;2:443–447.
53. Cardullo PA, Culter BS, Wheeler HB. Accuracy of duplex scanning in the detection of carotid artery disease. *Bruit* 1984;8:181–186.
54. Heinz DR, Pizer SM, Fuchs J, Fram EK, Burger P, Drayer BP, Osborne DR. Examination of the extracranial carotid bifurcation by thin section dynamic CT. *AJNR* 1984;5:355–359.
55. Heinz DR, Fuchs J, Osborne D, Drayer B, Yeates H, Fuchs H, Pizer S. Examination of the extracranial carotid bifurcation by thin section dynamic CT. *AJNR* 1984;5:361–366.
56. Tress BN, Davis S, Lavain J, Kaye A, Hopper J. Incremental dynamic computed tomography. *AJNR* 1986;7:49–54.
57. Goldberg HI, Spognoli MV, Berkowitz HD, Hamilton BT, Alexander J, Grossman RI. MR imaging evaluation of carotid artery atheromatous disease Scientific exhibit, 72nd scientific assembly and annual meeting of the Radiological Society of North America, Chicago, Nov 1986.
58. Vonruden WJ, Blaisdell WF, Hall AD, Thomas AN. Multiple arterial stenoses: effect on blood flow, an experimental study. *Arch Surg* 1964;89:307–351.
59. Carmody RF, Smith JRL, Seeger JF, Ovitt TW, Capp MP. Intracranial applications of digital intravenous subtraction angiography. *Radiology* 1982;144:529–534.
60. Chilcote WA, Modic MT, Pavlicek WA, Little JR, Furlan AJ, Duchesneau PM, Weinstein MA. Digital subtraction angiography of the carotid arteries: a comparative study in 100 patients. *Radiology* 1981; 139:287–295.
61. Christenson PC, Ovitt TW, Fisher HD, Frost MM, Nudelman S, Roehrig H. Intravenous angiography using digital video subtraction: intravenous cervicocerebrovascular angiography. *AJNR* 1980;1:379–386.
62. de Lahitte MD, Marc-Vergnes JP, Rascoi A, Guirad B, Manelfe C. Intravenous angiography of the extracranial cerebral arteries. *Radiology* 1980;137:705–711.
63. Modic MT, Weinstein MA, Chilcote WA, Pavlicek ¨, Duchesneau PM, Furlan AJ, Little JR. Digital subtraction angiography of the intracranial vascular system: comparative study in 55 patients. *AJR* 1982;138:299–306.
64. Strother CM, Sackett JF, Crummy AB, Lilleas FG, Zwiebel WJ, Turnipseed WD, Javid M, Mistretta CA, Kruger RA, Ergum DL, Shaw CTG. Clinical applications of computerized fluoroscopy. The extracranial carotid arteries. *Radiology* 1980;136:781–783.
65. Earnest F IV, Houser OW, Forbes GS, Kispert DB, Folger WN, Sundt TM Jr. The accuracy and limitations of intravenous digital subtraction angiography in the evaluation of atherosclerotic cerebrovascular disease: angiographic and surgical correlation. *Mayo Clin Proc* 1983;58:735–746.
66. Turnipseed WD, Sackett JF, Strother CM, Crummy AB, Mistretta CA, Kruger RA. Computerized arteriography of the cerebrovascular system. *Arch Surg* 1981;116:470–473.
67. Meaney TF, Weinstein MA, Buonocore E, Pavlicek W, Borkowski GP, Gallagher JH, Sufka B, MacIntyre WJ. Digital subtraction angiography of the human cardiovascular system. *AJR* 1980;135:1153–1160.
68. Russell JB, Watson TM, Modi JR, Lambeth A, Sumner DS. Digital subtraction angiography for evaluation of extracranial carotid occlusive disease: comparison with conventional arteriography. *Surgery* 1983;94:604–611.
69. Cronqvist S. Total angiography in evaluation of cerebro-vascular disease. A correlative study of aorto-cervical and selective cerebral angiography. *Br J Radiol* 1966;39:805–810.
70. Grabrielsen TO, Seeger JF, Knake JE, Burke DP, Stillwill EW. The nearly occluded internal carotid artery: a diagnostic trap. *Radiology* 1981;138:611–618.
71. Yonas H, Meyer J. Extreme pseudo-occlusion of the internal carotid artery. *Neurosurgery* 1982;11:681–686.
72. Goldberg HI. Radiology of ischemic cerebrovascular disease. In: Wilkens RH, Rengachary SS (eds) *Neurosurgery*. New York: McGraw Hill Book Co, 1985; 1219–1236.
73. Earnest F, Forbes G, Sandok BA, Piedgras DG, Faust RJ, Ilstrup DM, Arndt LJ. Complications of cerebral angiography: prospective assessment of risk. *AJNR* 1983;4:1191–1197.
74. Ausman JI, Diaz FG, de las Reyes RA. Vertebral artery atherosclerosis. In: Wilkens RH, Rengachary SS (eds). *Neurosurgery*. New York: McGraw Hill Book Co., 1985:1248–1254.
75. Sheehan S, Bauer RB, Meyer JS. Vertebral artery compression in cervical spondylosis: arteriographic demonstration during life of vertebral artery insufficiency due to rotation and extension of the neck. *Neuroradiology* 1960;10:968–986.
76. Edwards JH, Kricheff II, Riles T. Angiographically undetected ulceration of the carotid bifurcation as a course of embolic stroke. Radiology 1979;132: 369–373.
77. Edwards JH, Kricheff II, Gorstein F, Riles T, Imparato A. Atherosclerotic subintimal hematoma of the

carotid artery. *Radiology* 1979;133:123–129.
78. Eikelboom B, Riles TR, Mintzer R, Baumann FG, Defillip G, Lin J, Imparato AM. Inaccuracy of angiography in the diagnosis of carotid ulceration. *Stroke* 1983;14:882–885.
79. Spencer MP, Rein JM. Quantitation of carotid stenosis with continuous-wave (C-W) Doppler ultrasound. *Stroke* 1979;10:326–330.
80. Crawford ES, Wukasch DW, Debakey ME. Hemodynamic changes associated with carotid artery occlusion: an experimental and clinical study. *Cardiovasc Res* 1962;1:3–10.
81. Kishore PRS, Kricheff II. Carotid stenosis and intracranial emboli. *Radiology* 1971;100:351–356.
82. Liebeskind AL, Cohen S, Anderson R, Schechter MM, Zingesser LH. Unusual segmental cerebrovascular changes. *Radiology* 1973;106:119–122.

5. NONATHEROSCLEROTIC LESIONS OF THE EXTRACRANIAL VASCULATURE

Solomon Batnitzky
John H. McMillan
Glendon G. Cox

Atherosclerosis is by far the most common cause of stenosis and occlusion of the extracranial arteries. However, other pathological conditions can cause stenosis and occlusion of these vessels. These include:

1. Fibromuscular dysplasia
2. Dissecting aneurysms
3. Takayasu's arteritis.

Fibromuscular Dysplasia

PATHOLOGICAL AND CLINICAL FEATURES

Fibromuscular dysplasia (FMD) is a multivessel, segmental nonatheromatous angiopathy. The term *dysplasia* does not imply a specific etiology but is derived from the Greek *dys* (bad) and *plasia* (molding). The disease was first described by Leadbetter and Burkland in 1938 [1] and usually involves primary branches of the aorta.

FMD predominantly attacks middle-aged females (the female-to-male ratio is 3:1) but has been described from four to 71 years [2]. FMD is primarily a disease of Caucasians, although it has been found in other racial groups as well. Although a clear geographic distribution has not been described, a familial disposition probably exists.

Mettinger and Ericson have suggested that FMD is a congenital mesenchymal disorder [3]. They point out an increasing number of reports of familial occurrence, family history of related problems such as stroke, hypertension, and migraine, and the association of FMD with various congenital abnormalities (discussed below).

Three histologic types of FMD have been described: intimal fibroplasia, medial fibroplasia or fibromuscular hyperplasia, and subadventitial or periarterial hyperplasia [4]. In some patients, these types may coexist with involvement of two or all three layers of the arterial walls. The most common histologic type is the medial form (greater than 88%) [5].

ANGIOGRAPHIC FEATURES

Three major angiographic patterns are recognized [2] (figure 5–1). The *string of beads* appearance (Type I) is seen in about 89% and is characterized by multiple, irregularly spaced, concentric constrictions with normal or dilated intervening segments (figures 5–2, 5–3). The dilatations are always wider than the normal lumen of the vessel. The localized constrictions usually narrow the lumen by less than 40%. Medial fibroplasia is the most common histologic type seen with this angiographic finding.

Unifocal or multifocal tubular stenosis represents a second angiographic type (Type II) and is much less common (7%) (figure 5–4). The smooth, concentric tubular narrowing is less specific than the string of beads appearance and can be associated with any type of FMD.

Type 3 FMD has been termed atypical FMD by Houser et al. [4]. Atypical FMD usually affects only one wall of the involved segments with a diverticulumlike smooth or corrugated ovoid out-pouching with noncircumferential weblike narrowings (figure 5–5). Type 3 lesions are rare (4%) and are sometimes associated with atypical Type 1 lesions, possibly as a complication.

Angiographically, FMD is found most often in the renal arteries (85%) [6]. The second most common location is in the internal carotid artery, where FMD constitutes the second most common cause for extracranial carotid narrowing [4, 7], atherosclerosis being the most common cause. About 1% of all carotid angiograms show FMD [4]. The great majority of cervical FMD is located in the high cervical internal carotid artery. Usually, the C1-2 segment of the internal carotid artery is involved, and frequently (65%) the involvement is bilateral [2, 4, 8]. Involvement of the common carotid bifurcation is exceedingly rare [2], and the proximal portion of the internal carotid artery is characteristically spared. Sparing of the petrous internal carotid artery is typical, although on very rare occasions the petrous portion may be involved (figure 5–4). Coiling and tortuosity of the internal carotid artery are also prominent features.

Vertebral artery involvement by FMD is seen in about 10–15% of cases and usually at the C2 level [4]. Most patients with vertebral involvement also had internal carotid artery lesions. FMD has been described involving numerous other vessels, including superior mesenteric, celiac, hepatic, external iliac, axillary, occipital, and external carotid arteries [9]. Involvement of intracranial vessels is very rare but has been reported [9–12]. Renal and mesenteric vein involvement have also been reported [13].

Although FMD of the cervicocephalic arteries was formerly regarded as primarily an incidental angiographic finding, many patients present with significant clinical symptomatology. These can include headaches, TIAs, acute cerebral infarction, or subarachnoid hemorrhage from a coexisting aneurysm. Other reported symptoms include carotodynia, pulsatile tinnutis, vertigo, epilepsy, and Horner's syndrome [3]. Hypertension is frequently observed, particularly with renal FMD.

Several associated abnormalities and complications of FMD have been described. FMD is known to be associated with spontaneous dissection of the carotid artery with all its ischemic complications. However, in a significant number of cases, FMD is an incidental finding. Intracranial aneurysms are associated with FMD in over one third of cases. These are most often ipsilateral to the more severely affected extracranial carotid artery. Subarachnoid hemorrhage is a well-known complication, particularly when associated with hypertension. Development of carotid-cavernous fistula has been well described (10, 11, 14, 15]. Hieshima et al. also described two cases of vertebral venous fistula [10]. Spontaneous arterial dissection occurs not uncommonly [5, 11]. Fenestration of the vertebral artery and intracranial arteriovenous malformation have also been reported [3, 16]. Skeletal deformities and dextroposition of the heart are unusual associations [3].

Differential diagnostic considerations depend on the angiographic classifications. The Type 1 string of beads appearance is usually characteristic, but differential considerations would include stationary waves or circular spasmodic contractions. In vascular spasm, the constrictions are more regular and evenly spaced and occur without the dilatation beyond the normal vessel lumen of intervening segments so

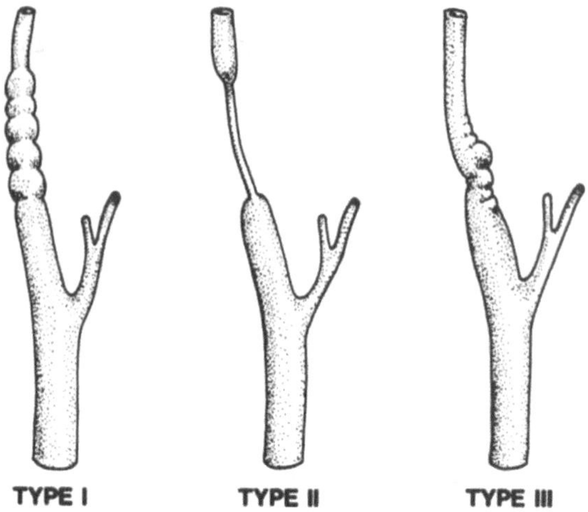

CAROTID FIBROMUSCULAR DYSPLASIA

FIGURE 5–1. Diagramatic representation of the three major characteristic angiographic patterns seen in cervical ICA FMD.

typical of FMD (figure 5–6). The constrictions are probably related to vascular irritation produced by the catheterization and subsequent injection of contrast or may occur when contrast is injected into a vessel with reduced run-off.

Differential considerations in the tubular stenotic Type 2 form would include Takayasu's arteritis, arterial hypoplasia, diminished luminal caliber secondary to decreased distal blood flow, and vascular spasm. Atherosclerosis can frequently be differentiated due to its characteristic location and the rate involvement of vessel origins by FMD. Type 3 FMD can be indistinguishable from atherosclerosis or post-traumatic aneurysm. Location is again helpful [2].

TREATMENT

Although definite progression of FMD has been observed in some cases [2, 5, 17], complete obstruction is rare [4]. Thus, treatment obviously requires careful consideration regarding the degree of symptomatology and the development of associated complicatioins.

Treatment of FMD has included graduated dilatation [18, 19], graduated dilatation with endarterectomy [19], resection with reanastomosis [18], resection with vein graft interposition [18], and percutaneous transluminal angioplasty [20, 21]. Treatment of spontaneous complications such as carotid-cavernous fistula by balloon embolization has also been described [14].

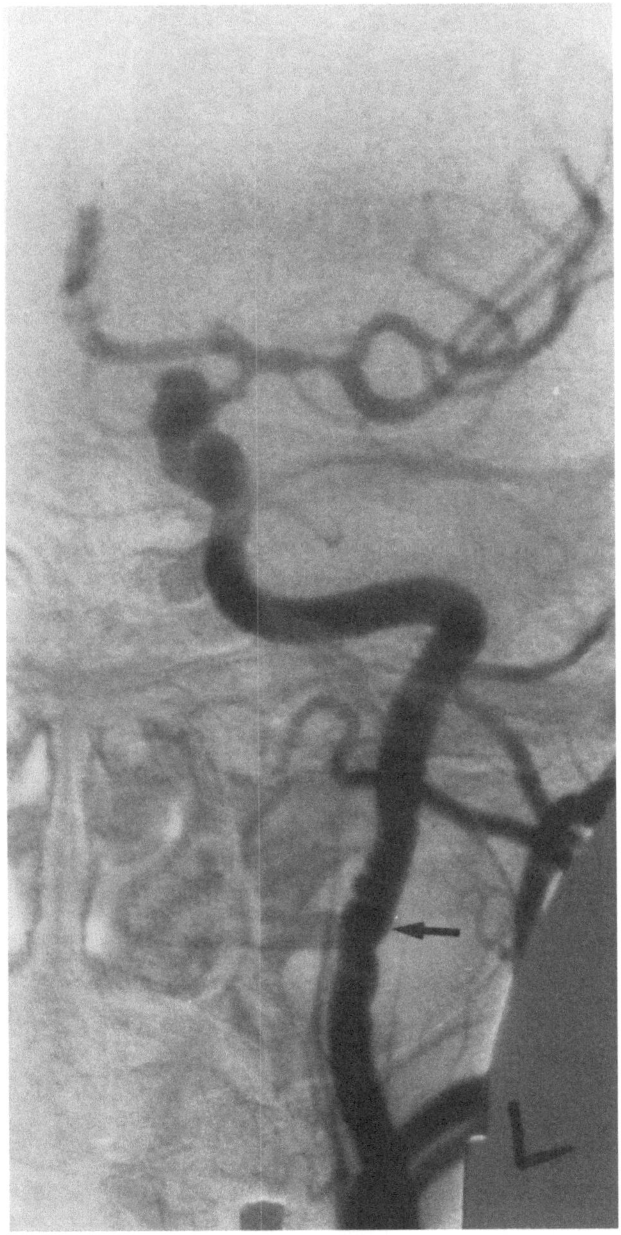

FIGURE 5–2. A 42-year-old female with FMD. The arterial phase of the left carotid angiogram demonstrates multiple, irregularly spaced, concentric stenoses of the midcervical portion of ICA (arrow). Alternating with these constrictions are normal or dilated intervening segments. This angiographic finding is the so-called *string of beads* appearance that is considered pathognomonic of FMD. The right ICA also was involved. In addition, right carotid angiogram disclosed a sausage-shaped aneurysm arising from the anterior communicating artery (not shown).

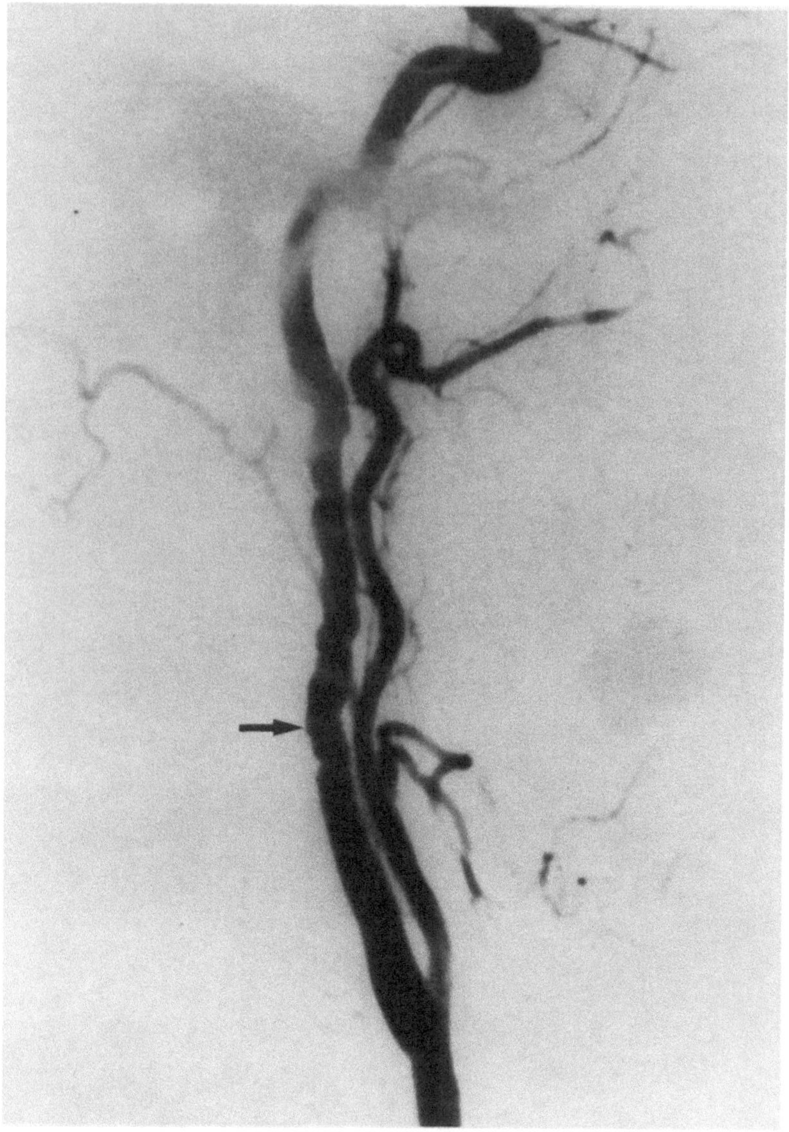

FIGURE 5–3. Cervical ICA FMD. Arterial phase of the lateral right carotid angiogram demonstrates the string of beads (arrow) appearance involving the midportion of the cervical ICA. Note sparing of the common carotid artery bifurcation and the origin of the ICA. The petrous portion is spared by the disease process. Identical changes were demonstrated in the left ICA in its cervical portion.

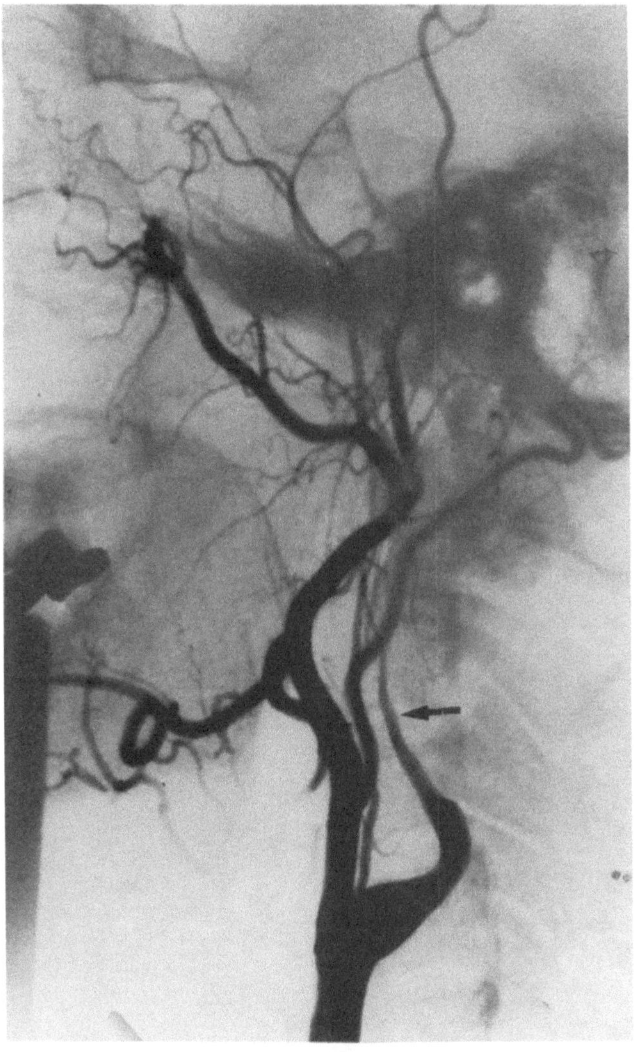

FIGURE 5–4A

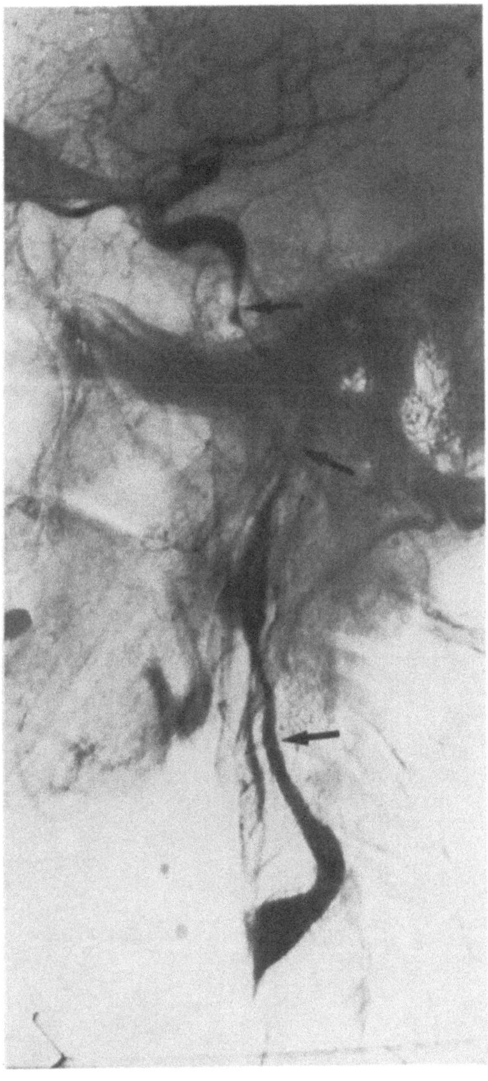

FIGURE 5–4B

FIGURE 5–4. Type 2 FMD. Early (A) and late (B) lateral arterial phases of left carotid angiogram demonstrate a long segment of smooth tubular stenosis (arrows). The petrous portion of the ICA is involved by the disease process. This latter finding is very rare and not typical of FMD.

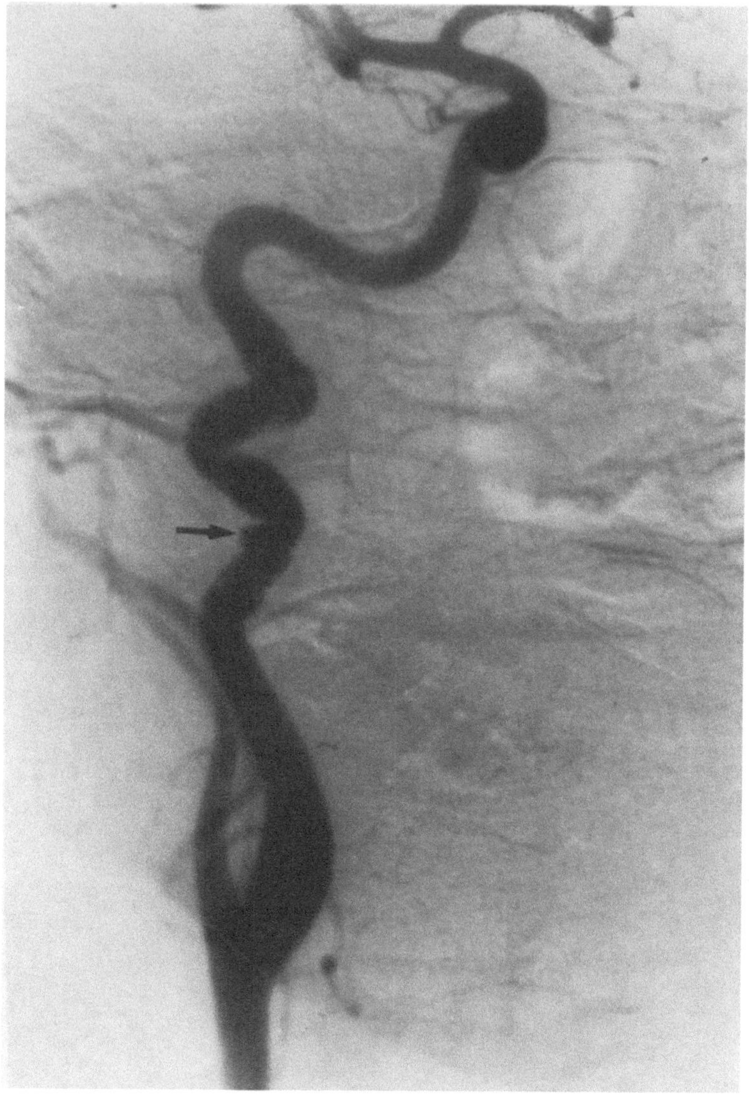

FIGURE 5–5. Type 3 or atypical FMD. Frontal view of right carotid angiogram demonstrates a lateral diverticulumlike out-pouching (arrow) that probably represents atypical FMD. The medial wall of the ICA is also involved, indicating that an atypical Type 1 lesion is also present. The string of beads pattern of FMD was present in the contralateral left ICA.

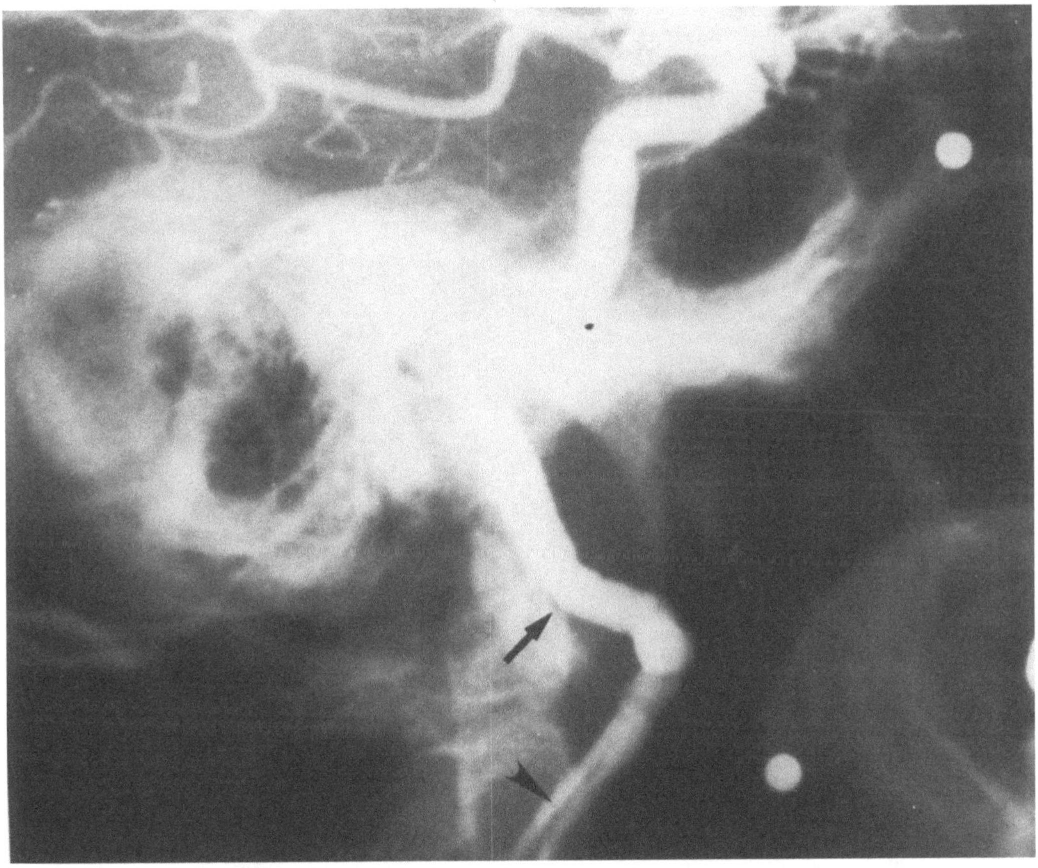

FIGURE 5–6. Vascular spasm. Note the high position of the catheter (arrowhead) in the ICA. The spastic constrictions (arrow) are fine and regularly spaced with no evidence of dilatation beyond the normal vessel lumen.

Dissecting Aneurysms

PATHOLOGICAL AND CLINICAL CONSIDERATIONS

Arterial dissection occurs when blood is forced between the tissue planes of a vessel wall. Arterial dissections (dissecting aneurysms) of the cervical internal carotid artery and/or the vertebral artery are now recognized with increasing frequency as a cause of stroke.

Dissection may result from vessel damage caused by cystic medial necrosis, fibromuscular dysplasia, Marfan's disease, or syphillitic arteritis [22]. Although some disease states may predispose to a vessel dissection, a dissecting aneurysm may also occur spontaneously in a seemingly normal artery [22, 23]. Dissecting aneurysms may also result from direct arterial trauma [24, 25]. The internal carotid artery (ICA) is relatively unprotected in its cervical course and is therefore vulnerable to traumatic injury from a variety of causes. Blunt, intraoral, or penetrating trauma to the ICA may result in thrombosis, intimal tears, dissection, stenosis, and occlusion. A sudden deceleration or hyperextension injury produces strong shearing forces at the point where the relatively flexible, unprotected ICA enters the rigid petrous carotid canal. The ICA is tensioned and stretched over the bony mass of the first and second cervical vertebrae, tearing the intima in one or more places (figure 5–7). Most traumatic occlusions of the ICA result in rapid thrombosis of the entire vessel between the carotid bulb and the ophthalmic artery, regardless of the site at which the initial damage originated (figure 5–8).

In 90% of cases, the diagnosis of carotid dissection is not made until the patient has suffered a neurological deficit in the form of a transient ischemic attack (TIA), a reversible ischemic neurological deficit, or a stroke [24]. In the majority of patients,

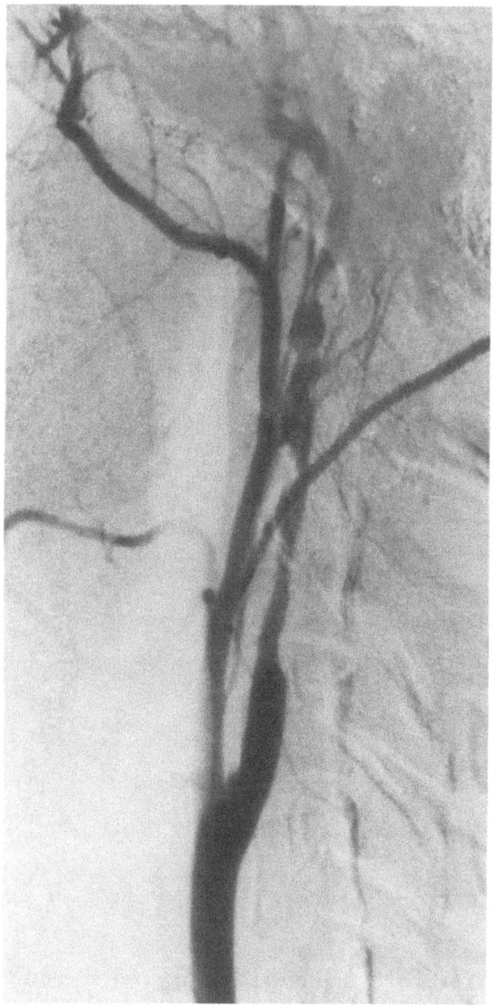

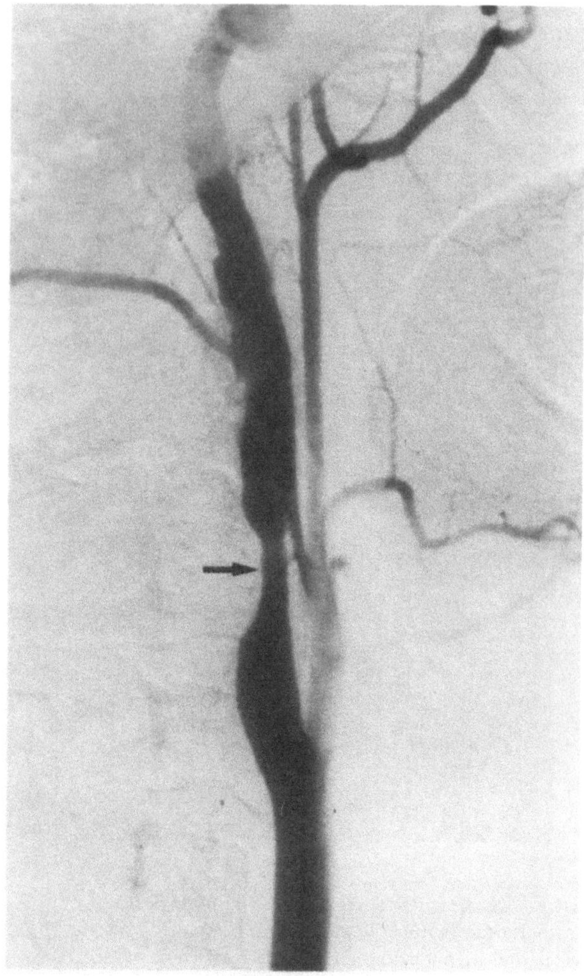

FIGURE 5–7B

FIGURE 5–7A

FIGURE 5–7. Bilateral traumatic dissecting aneurysms in a 25-year-old woman who was involved in an automobile accident. The patient was admitted with a diagnosis of intracerebral trauma. CT head scan was normal. Left (A) and right (B) lateral carotid angiograms demonstrate marked irregularity and narrowing of the lumen of both cervical ICAs. In (B), note that the dissection involves only the posterior wall (arrow) of the artery, with the anterior wall appearing normal and smooth.

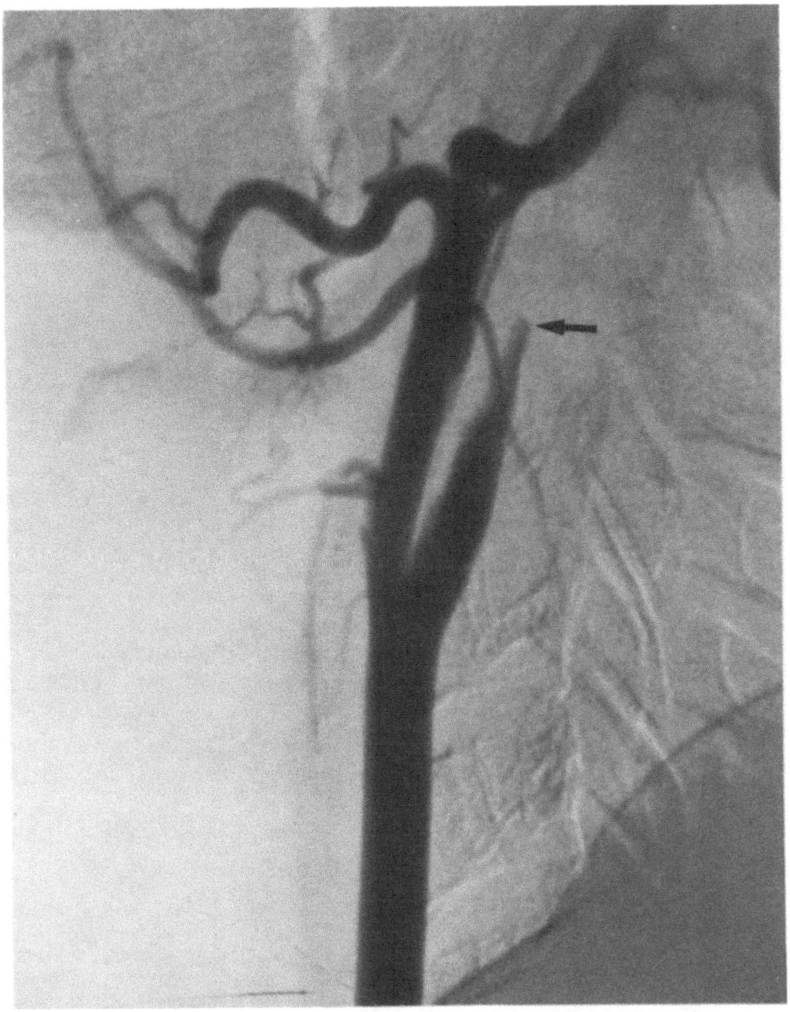

FIGURE 5–8. Traumatic occlusion of ICA (arrow). Left lateral carotid angiography in a 17-year-old male who was injured in a motorcycle accident demonstrates complete occlusion of the ICA 3 cm distal to its origin.

these neurological symptoms are delayed. Only 10% of patients develop symptomatic cerebral ischemic incidents with the onset of the dissection.

The diagnosis of arterial dissection should be considered in young patients who present with cerebral ischemia as well as in trauma patients who have both a focal neurologic deficit and a normal computerized tomography (CT) scan.

A definite gender predilection has yet to be confirmed. In some series, males predominate, while in others females predominate. O'Dwyer et al. [26] and Ehrenfeld and Wylie [27] have reported that the genders are affected equally.

The initial complaint of two thirds of patients with carotid dissection is neck pain that frequently radiates into the mastoid or suboccipital area [22]. A Horner's syndrome resulting from injury to the pericarotid sympathetic plexus is present in one half of the cases. A high cervical lesion may be associated with a bruit, and a lower cervical lesion may sometimes present as a pulsatile mass in the neck. Hypertension is not a factor in the majority of cases [28].

Although symptoms may result solely from vessel occlusion, emboli to the intracranial vessels have also been documented [29].

ANGIOGRAPHIC FEATURES

Houser et al. [30] reviewed 42 patients with spontaneous dissection of the cervical internal carotid and

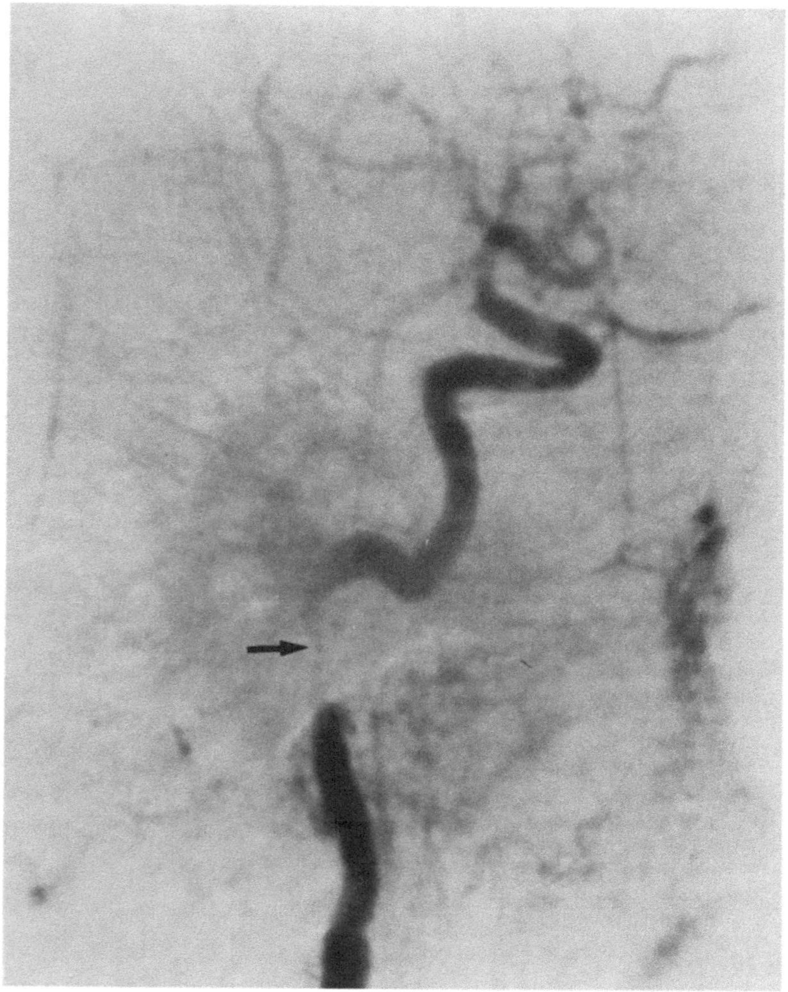

FIGURE 5–9A

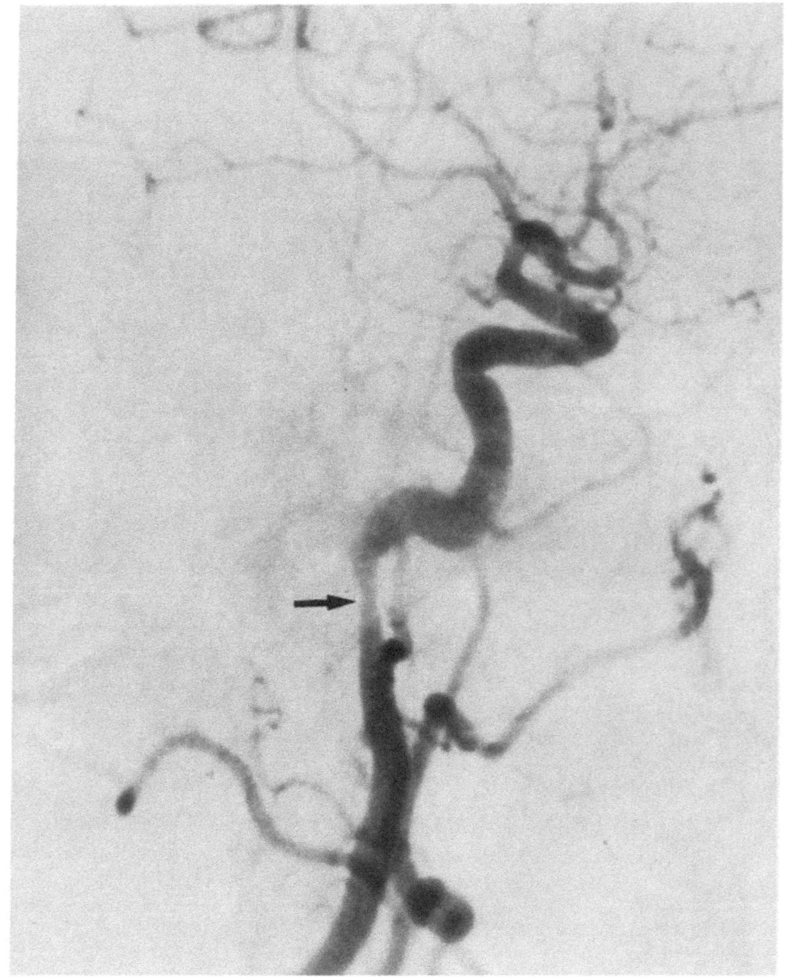

FIGURE 5–9B

FIGURE 5–9. Stenotic dissecting aneurysm with interval resolution. Initial right lateral carotid angiogram (A) shows a stenotic segment (arrow) in the high cervical area. Repeat right carotid angiogram (B) nine weeks later demonstrates interval improvement in the stenotic segment (arrow). However, some irregularity and stenosis persists.

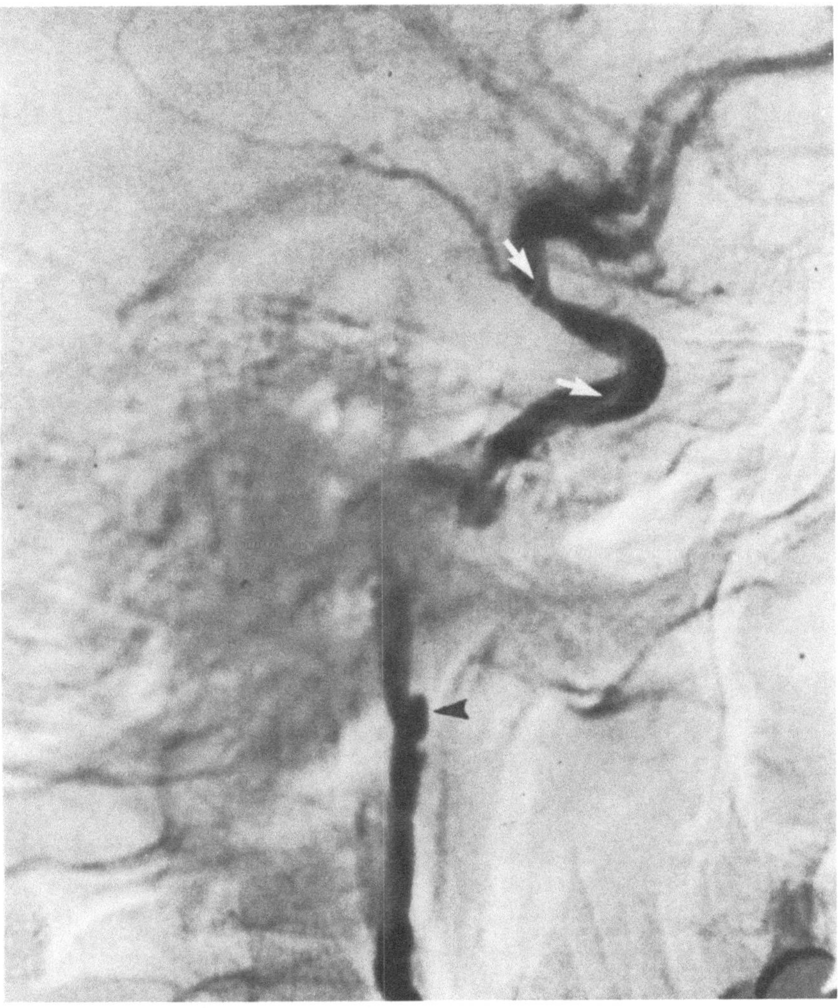

FIGURE 5–10. Dissecting aneurysm. Right lateral carotid angiogram demonstrates narrowing and irregularity of the cervical ICA. Note the small fingerlike aneurysmal dilatation (arrowhead) parallel to the vessel, with flattening of the surface that faces away from the artery. The dissection extends distally into the petrous and cavernous segments of the artery. The spiral filling defect (arrows) within the opacified vessel represents the intima between the false and true lumen of the vessel.

vertebral arteries. In their series, they found that usually a single ICA, predominantly the right, was affected. However, simultaneous multivessel dissections were present in about one third of their patients.

Angiographically, the appearance of the dissection varies depending on its severity, extent, and the interval between onset and angiography. In most cases, the true lumen of the vessel is compressed by clotted blood lying within its arterial wall and is, therefore, markedly narrowed over the involved arterial segment. This segmental luminal constriction is known as the angiographic *string sign* [28] (figure 5–9). Since the dissection normally involves only a portion of the arterial wall, the true lumen will not be concentrically narrowed, but rather will be asymmetrically compressed, off to one side. In Houser's series [30], the disruption was manifested initially by eccentric tapered stenosis in 47% of cases, tapered stenosis and aneurysmal dilatation in 28% occlusion in 18% and aneurysm dilatation alone in 7%.

The angiographic hallmark of dissection is a long, tapered, usually eccentric stenosis that begins distal to the carotid bulb and is associated with irregularity of the lumen. The tapered narrowing may extend

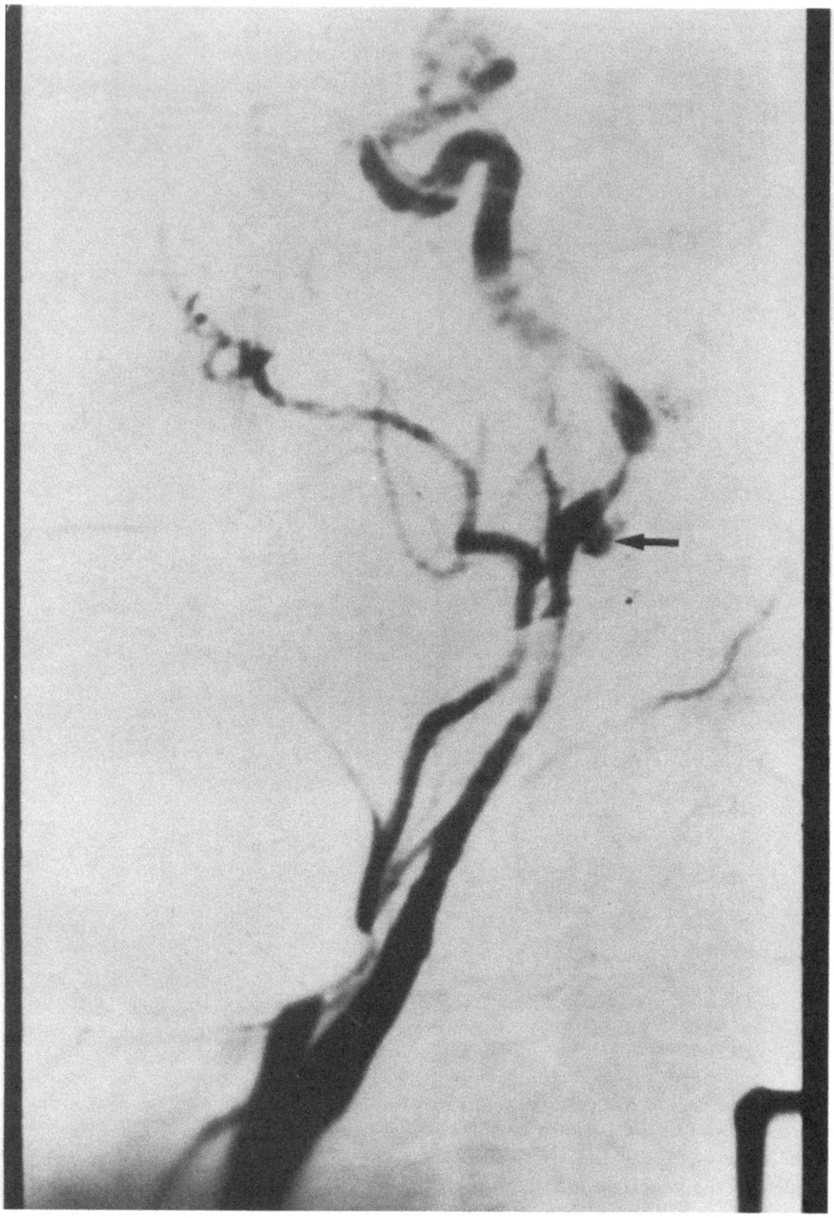

FIGURE 5–11. Spontaneous dissecting aneurysm. Left lateral carotid angiogram reveals eccentric tapered stenosis of the cervical ICA with an area of aneurysmal dilatation (arrow) just proximal to the petrous carotid canal.

into the skull base or beyond. The irregularity is more prominent along one surface of the vessel. Round, saccular aneurysms may be small with a wide neck, or less often, large. These aneurysms may extend like fingers, parallel to the vessel, and appear to be flattened along the surface that faces away from the artery (figures 5–10, 5–11).

Occasionally, the false channel caused by the dissecting blood will rupture back into the true lumen, and both channels will be patent to blood flow. On angiography, a spiral filling defect within the opacified vessel may be seen. This represents the intima between the false and true lumen of the vessel (figure 5–10).

Patients with spontaneous dissection differ from

those with atherosclerosis or fibromuscular dysplasia without concurrent dissection because these latter disorders predominate in patients who are usually more than 50 years old, while patients with spontaneous dissection are often less than 50 years old. The asymmetrical focal constriction seen in dissecting aneurysms helps differentiate the luminal narrowing of dissection from that of fibromuscular dysplasia, which concentrically narrows the vessel lumen. The angiographic appearance of alternating areas of widening and concentric narrowing—the so-called string of beads appearance—is considered pathognomonic of fibromuscular dysplasia. Atherosclerosis usually involves the carotid bifurcation and carotid siphon, whereas dissecting aneurysms generally do not.

Stenotic dissections of the vertebral artery may also occur. Vertebral dissection is most commonly associated with cervical hyperextension and rotation such as may occur with trauma or with chiropractic manipulation of the neck [31]. Vertebral dissections may occur spontaneously or in association with FMD and other ischemic diseases that affect the arterial wall [32]. They are usually 2–4 cm long and are localized to the level of the first and second cervical vertebral bodies, although they may occur anywhere along the cervical portion of the artery. Aneurysmal dilatation may also occur in the vertebral artery [22].

Carotid and vertebral arterial dissections tend to resolve (figure 5–12), sometimes progress, but seldom recur. In Houser's series [30], stenotic dissections resolved in 60%, improved in 20% and progressed in 15%, while those with aneurysm formation diminished in half and resolved in one fourth of patients. An angiographic residuum was evident in 25% of dissections.

Takayasu's Arteritis

PATHOLOGICAL AND CLINICAL FEATURES

Takayasu's arteritis is a panarteritis characterized by inflammation and stenosis of the large and intermediate-sized arteries, and usually proceeds to vascular occlusion. It is named after a Japanese ophthalmologist who described the retinal manifestations of the disease without recognizing the major vascular abnormalities of the condition. Takayasu's arteritis involves primarily the brachiocephalic vessels but may also affect the aorta and any of its primary branches [33]. It has also been called pulseless disease, aortic arch arteritis, and obliterative brachiocephalic arteritis. Takayasu's arteritis is a rare form of giant-cell arteritis, and the histological changes of the affected vessels are similar to those of temporal arteritis, except infiltration by giant cells is less common in Takayasu's arteritis.

There is a marked female predominance, with a female-to-male ratio of 9:1. The disease usually affects young women between the ages of 10 and 45 years. The disease was initially described in young Japanese women, but it is now recognized as involving all racial groups and is not geographically restricted. Nonetheless, the disease is more common in Japanese and Indians in the Western Hemisphere than it is in Caucasians [33].

The etiology of the disease is unknown. An autoimmune process directed against the aorta and its major branches has been proposed as the underlying etiology. High titers of antiaorta antibodies, demonstrated in a significant number of proven cases, support this hypothesis.

The disease has been divided into acute and chronic phases [34]. In the acute or systemic phase, low-grade fever, myalgia, cough, and an elevated erythrocyte sedimentation rate may be seen.

The signs and symptoms associated with the chronic phase reflect ischemia of the involved organ system. Not infrequently, the systemic symptoms may be absent and symptoms related to ischemia caused by occlusive elements may be the presenting complaint. The disease is usually relentlessly progressive, and surgical excision of a stenotic segment and grafting may be necessary.

Hypertension, which occurs in about half of the patients, may be secondary to narrowing of the thoracic or abdominal aorta or may be due to renal artery involvement. To detect the hypertension, the evaluation of leg blood-pressure readings is essential.

RADIOLOGIC FEATURES

Plain films may demonstrate calcification of the aortic arch and the origins of the brachiocephalic vessels.

The diagnosis is established by finding the characteristic changes on angiographic studies, and biopsy of the affected vessel or vessels confirms the diagnosis.

Angiographically, the disease may present occlusive or aneurysmal changes or a combination of the two [35].

The brachiocephalic vessels are the most frequently affected arteries. Smooth, long segment fusiform stenosis of the proximal arch vessels is the classic angiographic finding in Takayasu's arteritis. In severe cases, progressive stenosis may result in complete occlusion of one or more arch vessels. Any single arch branch or any combination of the arch branches may be affected. In 80% of cases, two or

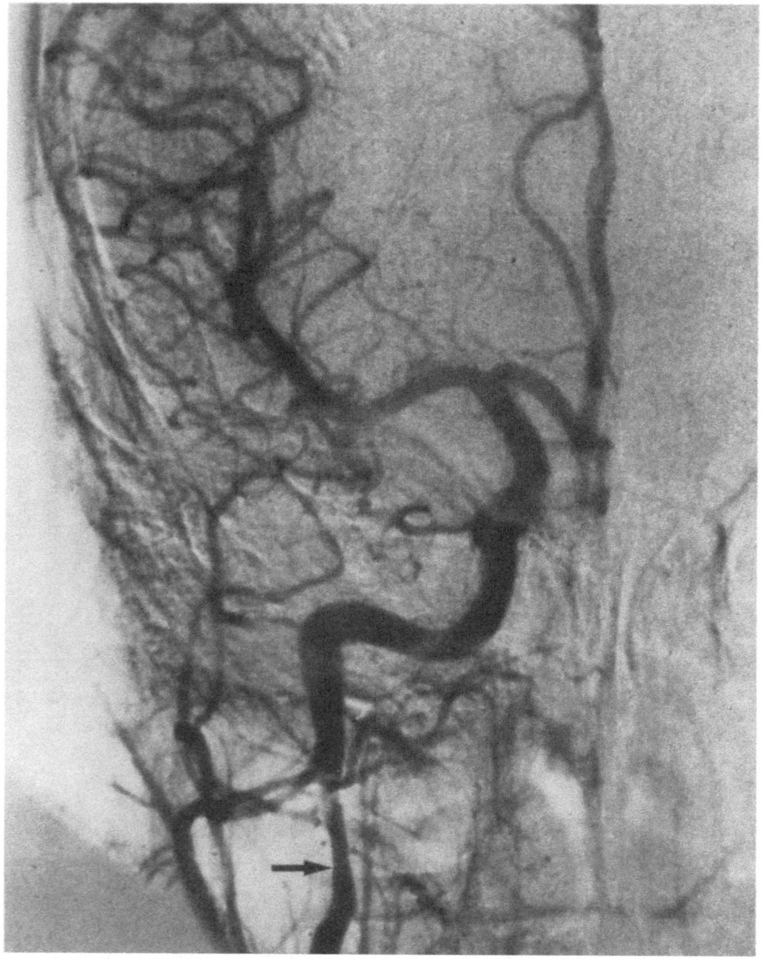

FIGURE 5–12A

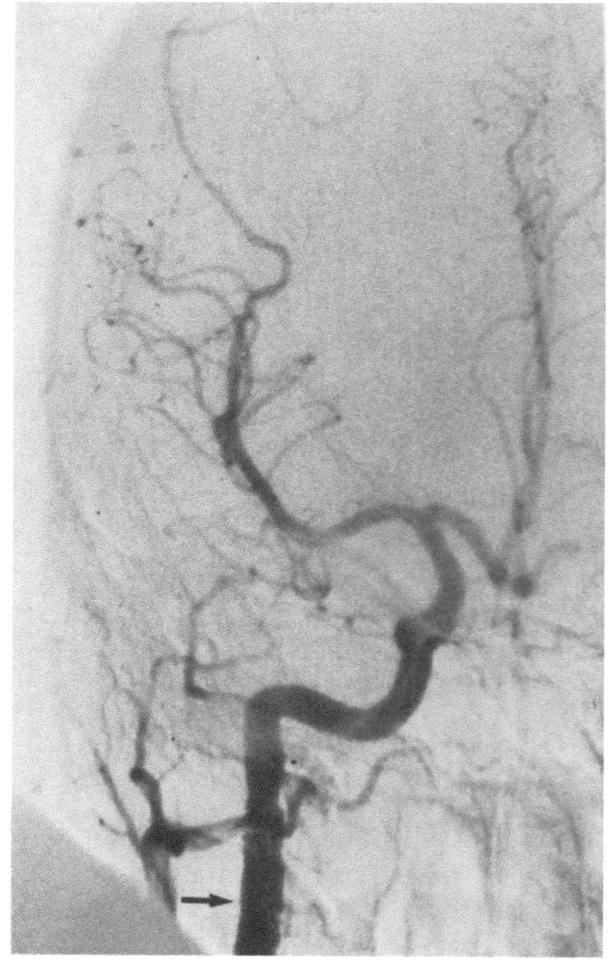

FIGURE 5–12B

FIGURE 5–12. Dissecting aneurysm undergoing significant resolution. Initial right frontal carotid angiogram (A) in 25-year-old woman demonstrates marked narrowing (arrow) of the midportion of the cervical ICA. Right frontal carotid angiogram (B) six weeks later shows marked resolution of this stenotic segment (arrow). The arterial wall at the site of the dissection demonstrates some minimal residual irregularity.

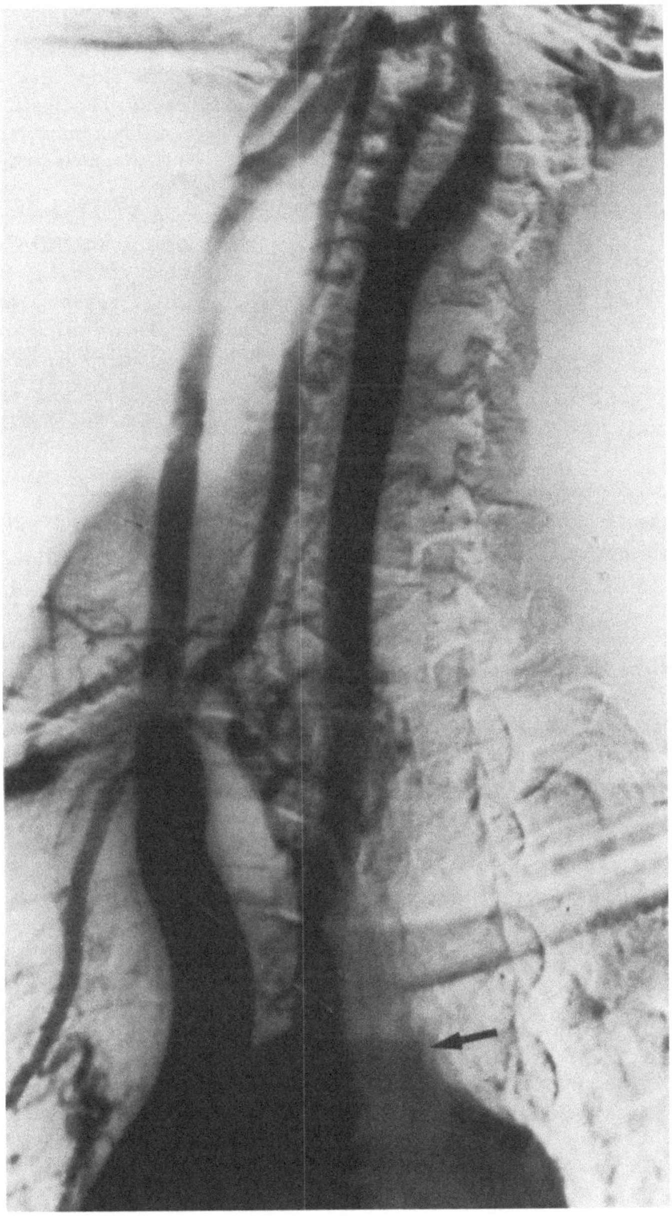

FIGURE 5–13. Takayasu's arteritis. A 42-year-old Japanese woman known to be pulseless in left upper extremity for 20 years. Arch aortogram. The left subclavian artery is totally occluded at its origin (arrow). The other brachiocephalic vessels are not affected. There was segmental narrowing in the aortic arch and descending thoracic aorta (not shown).

more vessels will be involved. The left subclavian artery is the most frequently involved vessel [35] (figure 5–13). Ninety percent of the lesions are located proximal to the vertebral artery origin. The left common carotid artery, followed by the right subclavian artery, is the next most frequent vessel involved. The majority of right subclavian arterial lesions occur distal to the origin of the vertebral artery.

The affected vessels are involved for a variable distance along their length. The degree of narrowing is not uniform for all the vessels. Patterns of collateral blood flow secondary to the stenotic and occlusive lesions may develop. In Takayusu's arteritis, saccular fusiform aneurysmal dilatation may develop in the aortic arch or at the origins of its branches [34, 36].

In one third of patients, there may be evidence of involvement of the descending thoracic and abdominal aorta and its branches.

The intracranial vessels are spared by the disease process.

Atherosclerosis may also cause narrowing and occlusion at the origins of the brachiocephalic vessels. Takayasu's disease primarily involves young females, and the lesions tend to be long, often smooth, and symmetrical in appearance, while atherosclerotic lesions are more irregular and predominantly affect older individuals of both sexes.

References

1. Leadbetter WF, Burkland CE. Hypertension in unilateral renal disease. *J Urol* 1938;39:611–626.
2. Osborne AG, Anderson RE. Angiographic spectrum of cervical and intracranial fibromuscular dysplasia. *Stroke* 1977;8:617–626.
3. Mettinger KL, Ericson K. Fibromuscular dysplasia and the brain. *Stroke* 1982;13:46–52.
4. Houser OW, Baker HL, Sandok BA, Holley KE. Cervical arterial fibromuscular dysplasia. *Radiology* 1971;101:605–611.
5. Kincaid OW, Davis GD, Hallermann FJ, Hunt JC. Fibromuscular dysplasia of the renal arteries. *AJR* 1968;104:271–282.
6. Appleberg M. Graduated internal dilatation in the treatment of fibromuscular dysplasia of the internal carotid artery. *South Afr Med J* 1977;51:244–246.
7. Momose KJ, New PFJ. Nonatheromatous stenosis and occlusion of the internal carotid artery and its main branches. *AJR* 1973;118:550–566.
8. Ehrenfeld WK, Stoney RJ, Wylie EJ. Fibromuscular hyperplasia of the internal carotid artery. *Arch Surg* 1967;95:284–287.
9. Rinaldi I, Harris WO, Kopp JE, Legier J. Intracranial fibromuscular displasia: report of two cases, one with autopsy verification. *Stroke* 1976;7:511–516.
10. Hieshima GB, Cahan LD, Mehringer CM, Bentson JR. Spontaneous arteriovenous fistula of cerebral vessels in association with fibromuscular dysplasia. *Neurosurgery* 1986;18:454–458.
11. Bellot J, Gherardi R, Poirier J, Lacour P, Debrun G, Barbizet J. Fibromuscular dysplasia of cervico-cephalic arteries with multiple dissections and carotid cavernous fistula. *Stroke* 1985;16:255–261.
12. Iosue A, Kier EL, Ostrow D. Fibromuscular dysplasia involving the intracranial vessels. *J Neurosurg* 1972; 37:749–752.
13. Rosenberger A, Adler O, Lichtig H. Angiographic appearance of the renal vein in the case of fibromuscular dysplasia of the artery. *Radiology* 1976;118:579–580.
14. Numaguchi Y, Higashida RT, Abernathy JM, Pisarello JC. Balloon embolization in a carotid cavernous fistula in fibromuscular dysplasia. *AJNR* 1987;8: 380–382.
15. Zimmerman RD, Leeds NE, Naidich TP. Carotid cavernous fistula associated with intracranial fibromuscular dysplasia. *Radiology* 1977;122:725–726.
16. Numaguchi Y, Fleming MS, Vargas EF. Cervicocephalic fibromuscular dysplasia and fenestration of the vertebral artery. *AJNR* 1987;8:573–574.
17. Pollock M, Jackson BM. Fibromuscular dysplasia of the carotid arteries. *Neurology* 1971;21:1226–1230.
18. Ehrenfeld WK, Wylie EJ. Fibromuscular dysplasia of the internal carotid artery: surgical management. *Arch Surg* 1974;109:676–681.
19. Morris GC, Lechter A, Debakey ME. Surgical treatment of fibromuscular disease of the carotid arteries. *Arch Surg* 1968;96:636–643.
20. Hasso AN, Berg CR, Zinke DE. Fibromuscular dysplasia of the internal carotid arteries: percutaneous transluminal angioplasty. *AJR* 1981;136:955–960.
21. Martin EC, Diamond NG, Casarella WJ. Percutaneous transluminal angioplasty in nonathersclerotic disease. *Radiology* 1980;135:27–33.
22. Friedman AH. Arterial dissections. In: Wilkens RH, Rengachary SS (eds) *Neurosurgery*. New York: McGraw Hill Book Co, 1985;1297–1300.
23. Fisher CM, Ojemann RG, Roberson CH. Spontaneous dissection of cervico-cerebral arteries. *Can J Neruol Sci* 1978;5:9–19.
24. Yamada S, Kindt GW, Youmans JR. Carotid artery occlusion due to nonpenetrating injury. *J Trauma* 1967;7:333–342.
25. Batnitzky S, Price HI, Holden RW, Franken EA. Cervical internal carotid artery injuries due to blunt trauma. *AJNR* 1983;4:292–295.
26. O'Dwyer JA, Moscow N, Trevor R, Ehrenfeld WK, Newton TH. Spontaneous dissection of the carotid artery. *Radiology* 1980;137:379–385.
27. Ehrenfeld WK, Wylie EJ. Spontaneous dissection of the internal carotid artery. *Arch Surg* 1976;111:1294–1300.
28. Taveras JM, Wood EH. *Diagnostic Neuroradiology*. Baltimore: The Williams and Wilkins Co., 1976: 913–917.
29. Bladin PF. Dissecting aneurysm of carotid and vertebral arteries. *Vascular Surg* 1974;8:203–223.
30. Houser OW, Mokri B, Sundt TM, Baker HL, Reese DF. Spontaneous cervical cephalic arterial dissection and its residium: angiographic spectrum. *AJNR* 1984;5:27–34.
31. Batnitzky S, Price HI. Traumatic nonpenetrating injuries of the cervical internal carotid and vertebral arteries. *J Kansas Med Soc* 1983;84:305–311.

32. Hugenholtz H, Pokripa R, Montpetit VJA, Nelson R, Richard MT. Spontaneous dissecting aneurysm of the extracranial vertebral artery. *Neurosurgery* 1982;10: 96–100.
33. Fauci AS. Vasculitis. In: Parker CW (ed) *Clinical Immunology*. Philadelphia: W.B. Saunders, 1980: 473–519.
34. Lande A, Rossi P. The value of total aortography in the diagnosis of Takayasu's disease. *Radiology* 1975; 114:287–297.
35. Osborne A. *Introduction to Cerebral Angiography*. Hagerstown: Harper and Row, 1980:45–46.
36. Deutsch V, Wexler L. Deutsch H. Takayasu's arteritis: an angiographic study with remarks on ethnic distribution in Israel. *AJR* 1974;122:13–28.

6. NEUROradiology OF CEREBRAL INFARCTION

Michael A. Mikhael

Although approximately 500,000 patients suffer from stroke each year in the United States, treatment of these patients to date has consisted primarily of prevention, supportive measures, and rehabilitation [1–3]. The modification of surgical procedures to restore adequate blood flow to areas of ischemia in the brain has initiated the need for a more aggressive approach for early diagnosis. It has not been possible to predict which patients with transient ischemic attacks (TIAs) are likely to have a stroke, although major treatment decisions may hinge on this information, since these high-risk groups are known to have substantial risk for developing stroke. The incidence of stroke in TIA patients has been estimated to vary between five to ten per year [4–6]. Patients with global ischemia (cardiac arrest) do not have enough collateral flow, and irreversible neuronal damage commences within four to eight minutes under normal thermic conditions, once the blood flow is acutely blocked [7–10]. However, in focal ischemia the trickle of flow from the collateral circulation leads to a more complex biochemical situation and needs to be corrected early before permanent damage results. Although the tolerance of neuronal tissue for these low flows is unknown, a few studies suggest that after three to four hours, neuronal death is inevitable [11–13]. The potential for collateral flow in focal ischemia (patients with TIAs) may facilitate possible reversal of neuronal damage if diagnosed early enough for possible correction. TIA attacks precede lacunar infarction in about 20% of cases, and among patients with TIAs, strokes are the fourth most common cause of death following myocardial infarction, cardiac failure, and arteriosclerotic aortic changes with their complications [14, 15]. The initial attack involves the carotid territory in 78%, with the vertebrobasilar territory only in 22% [16].

One of the primary acute reactions of brain parenchyma to vascular injury is swelling. Ischemic edema has been divided into an early cytotoxic (intracellular) and a late vasogenic (extracellular) phase. Cytotoxic edema is due to ischemia. When oxygen supply to the cells decreases, the sodium–potassium pump fails, causing cellular swelling [17–19]. The major effect of this edema is the impingement on extracellular space and on potential collateral flow. On the other hand, vasogenic edema occurs hours to days after vessel occlusion and is secondary to irreversible ischemic endothelial damage and breakdown of the blood–brain barrier with extravasation of plasma into the extracellular compartment [20, 21]. This allows iodinated contrast to accumulate in the extracellular space during postcontrast computerized tomography (CT) scan and paramagnetic agents to accumulate during contrast-enhanced magnetic responance (MR) imaging. Cases of acute carotid vascular occlusive disease may recover completely in 2–12%; 40–69% will have profound deficits and 16–55% will die of the ictus. Cases treated with emergency endarterectomy or embolectomy, however, have shown much better results, with complete recovery with no deficit in 26–36% and recovery with minimal deficit in 26–42% with 29% showing moderate hemiparesis; 12% showed dense hemiparesis and 2% died [22, 23]. Early diagnosis and early treatment are important for better recovery with minimal deficits [24, 25]. The recent advancement of radiological diagnostic procedures concerning the early detection of ischemia and/or infarction of the brain might help in bettering understanding of the underlying pathophysiology and in the early successful management of the affected patients before permanent neurological damage starts. Physicians should realize that neuronal damage can be reversible only when discovered early and that intervention (medical, surgical, or a combination) may be a better alternative than supportive measures alone.

The stroke syndrome can be produced by a variety of conditions involving the intracranial and extracranial vessels, as well as some systemic diseases.

Evaluation by imaging of the extracranial source of stroke is described in chapter 4. The present chapter primarily describes the computerized tomography (CT) and magnetic resonance imaging (MR) patterns of cerebral infarction and the applicable role of angiography in intracranial vascular disease.

Computerized Tomography

Structural changes of the brain resulting from cerebral ischemia and/or infarction can be detected by CT scan in 70–80% of cases. Because of well-developed collateral pathways reconstituting the circulation, one in three patients with symptoms of cerebral ischemia will have a negative CT scan in the presence of an occlusive disease of the middle cerebral artery shown on angiography. In the other two cases, a low-density lesion or lesions with or without enhancement following intravenous contrast may be seen in areas of luxury perfusion and/or possible change in the blood–brain barrier [26–30]. CT scan studies include plain (nonenhanced scans) and contrast-enhanced CT scans.

PLAIN (NONCONTRAST-ENHANCED) CT SCANS

Noncontrast-enhanced CT scans of the head should be carefully examined for the presence of a low-density lesion, mass effect, and/or acute intracranial bleed. The visualization of a lesion in areas of recent infarction depends on the time elapsed between the acute stroke and CT scan (figure 6–1), the size and site of lesion (area of damaged brain) (figure 6–2), the underlying condition of collateral circulation (figure 6–3), and the presence or absence of hemorrhage and/or edema in the area of infarct (figure 6–1) [26–32]. Depending on the size of infarction and the resolution of the CT scanner, the detectability of lesions on CT scans increases with time. The overall accuracy is lower in the first two days (less than 10%), higher in the following three to seven days (60%), still higher after the first week (70%), and highest in infarctions over three weeks (90%) (figure 6–4) [26, 32].

In the first 24 hours, a lesion is detected only in about 6–10% of cases of acute ischemic infarction (figure 6–1). More commonly (90% or more), CT scans are normal on the first days [28–32]. Radiologically, the acute ischemic infarct appears on CT scans as an ill-defined, poorly demarcated area of decreased density presumably from edema in the area of damaged brain. In the watershed area or in large infarctions, the gray matter tends to be involved to a greater degree than the white matter (figure 6–5). In the latter region, the area of decreased density assumes a triangular shape with the base toward the periphery. These lesions are typically shown within the territory of cerebral vessels. The demarcation between the anterior and middle cerebral arteries is an oblique line drawn from the corner of the frontal horn to the cortical surface of the brain, while the posterior demarcation is another oblique line drawn from the corner of the trigone to the surface of the brain (figure 6–5) [33]. In seven to ten days the infarcted area becomes more clearly demarcated, showing further decrease in density and probably in size as well. Later, in the three-to-six-week period, the area may become smaller, the edema disappears, the density further decreases, and the lesion may be identified as a small lacuna in cases of small infarctions. It may remain as a permanent area of damaged brain shown as a low-density lesion. Occasionally, the area of the infarct disappears completely with time, presumably when the damaged area of the brain is small and the amount of edema is minimal, or when the ischemic area of the brain is mostly viable and can return to normal through revascularization via collateral circulation. If the infarct is near the ventricular wall, it may give a false impression of direct continuity with the ventricle. If it is near the surface of the brain, the low absorption area will extend to the inner table of the skull in the form of a prominent sulcus, simulating focal atrophy [32, 34, 35].

Mass effect resulting from cerebral edema occurs in acute ischemic infarct in about 22%. It usually appears after the first day, with 70% appearing in the first week after the stroke (figures 6–5, 6–6) [26, 29]. The rate of disappearance of mass effect is fairly rapid; by the end of the third week, only 29% of cases show mass effect. By the eighth week, mass effect completely disappears and a negative effect may start to appear with dilatation of the ipsilateral ventricle and/or prominence of the sulci if the infarct was large.

Hemorrhage in the area of infarction will identify the lesion more obviously as early as a few hours after

FIGURE 6–1. Series of cranial CT scans. First row scan was done immediately after acute left body neurological deficit and showed old left occipital infarction. Second row scan was done a week later and showed right acute hemorrhagic occipital infarction, which was not shown on earlier scan. Third row scan was done four weeks after acute attack and showed gradual absorption of the blood leaving the edema and mass effect. The last row scan was done 3.5 months from acute ictus and showed low density in the site of right hemorrhagic infarction and dilatation of ipsilateral ventricle.

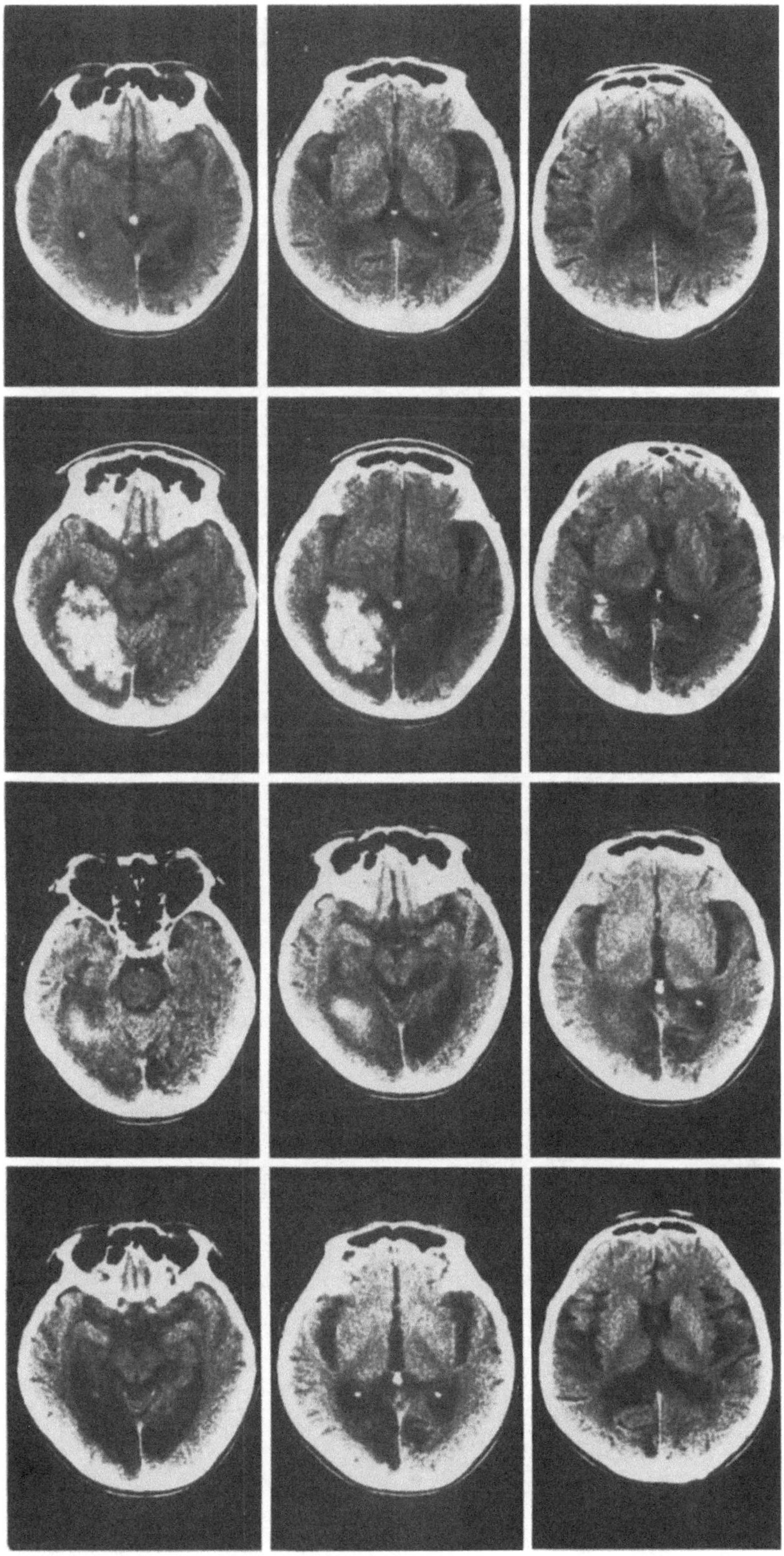

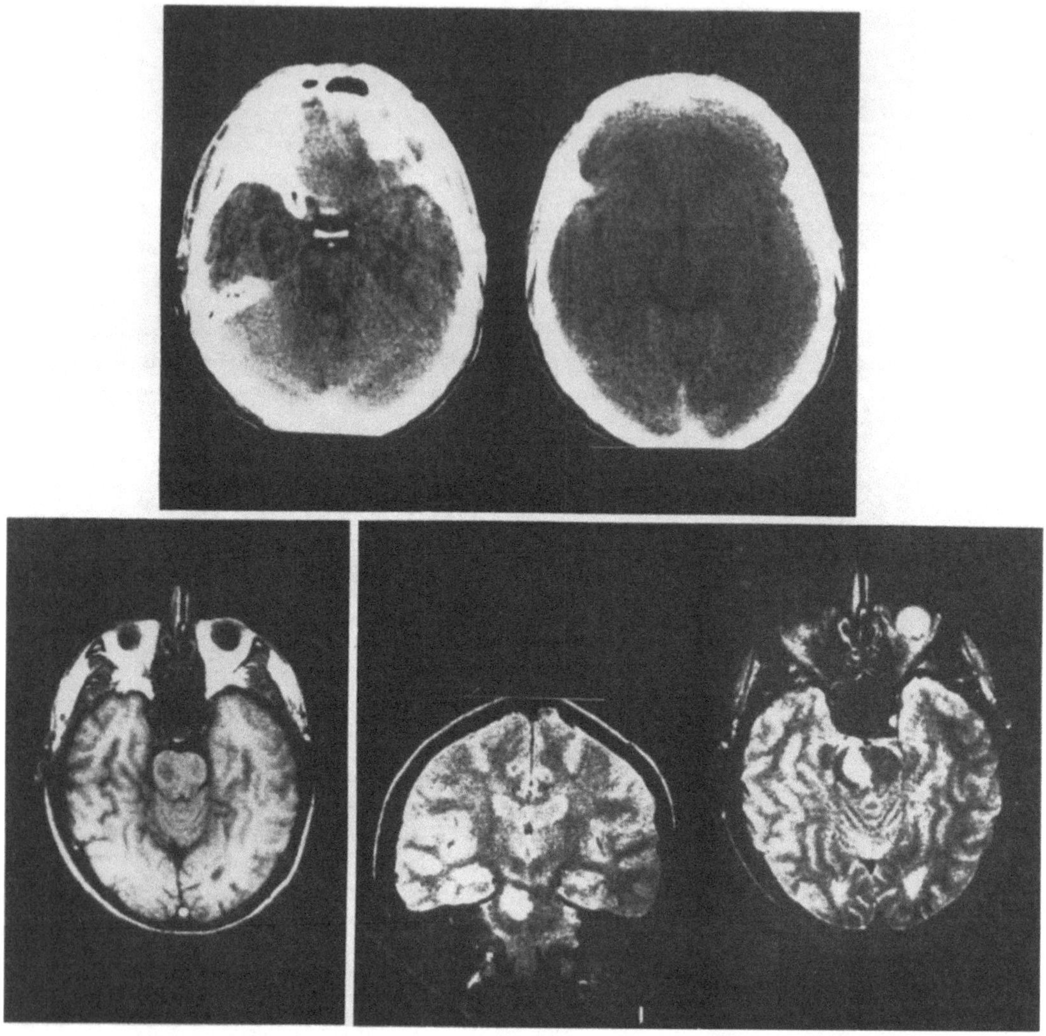

FIGURE 6–2. Contrast-enhanced cranial CT scan (upper row) showing no evidence of lesion shown on MR scan (lower row) in the right pons (arrows). Both CT and MR scans were done two days after acute left body hemiparesis. MR images obtained with spin-echo technique using TR 2,000 ms and TE 90 ms. Small cerebral infarctions and infarcts in posterior fossa are difficult to visualize on CT scans (see figures 6–11 and 6–12).

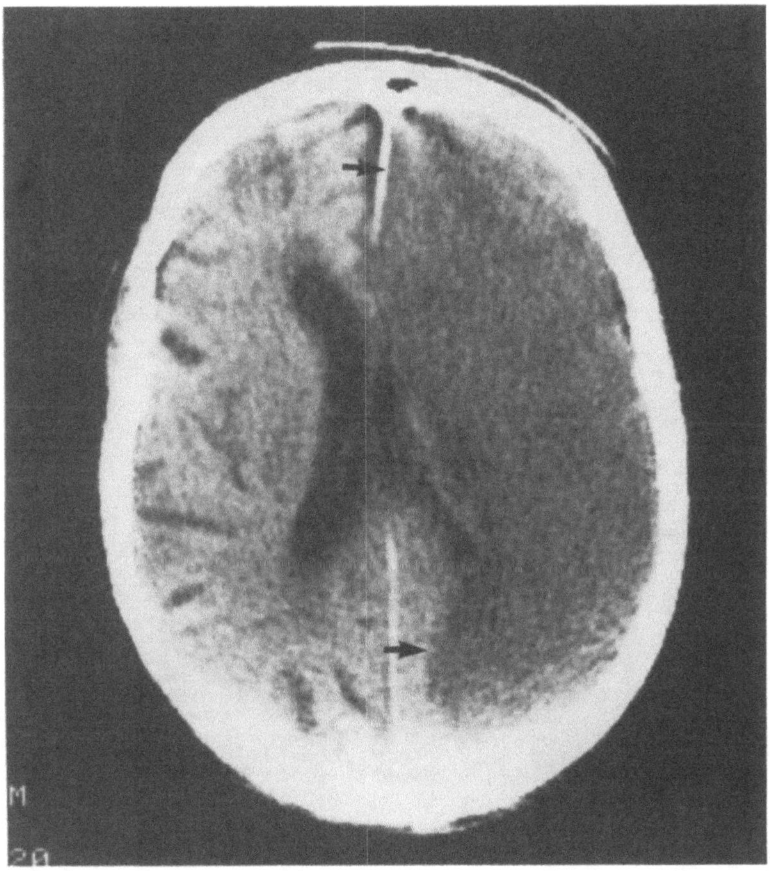

FIGURE 6–3. Contrast-enhanced cranial CT scan showing massive acute left cerebral infarction (arrows) in distribution of left anterior and middle cerebral arteries, as a result of occlusion of left internal carotid artery with no significant collateral circulation across the midline from the right side.

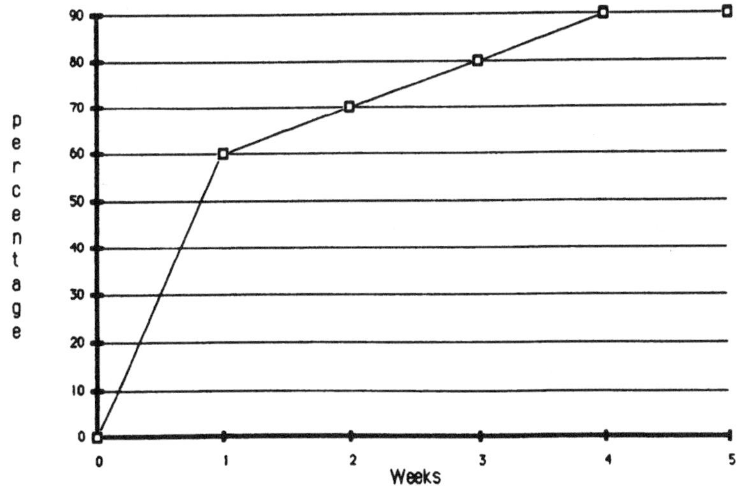

FIGURE 6–4. Incidence of positive CT scan (detecting a visual lesion in cases of acute cerebral infarcts) in relation to time of acute neurologic deficit.

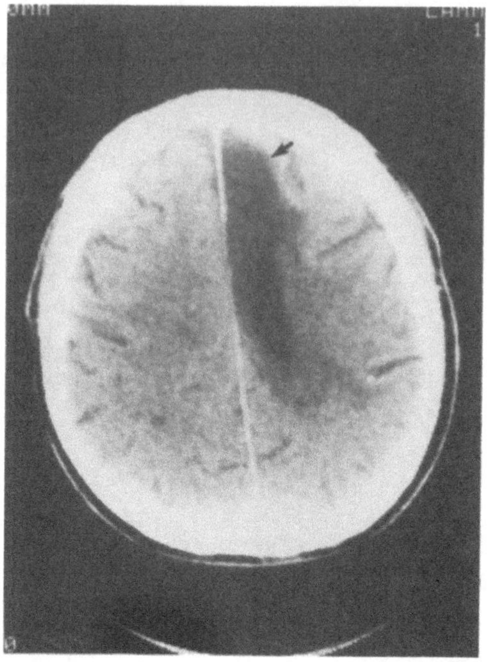

FIGURE 6–5A

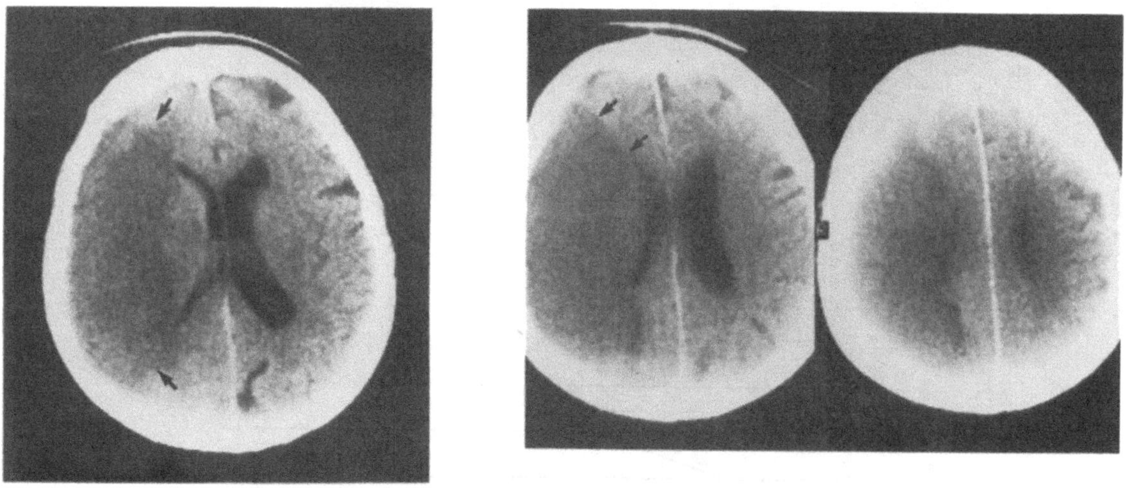

FIGURE 6–5B

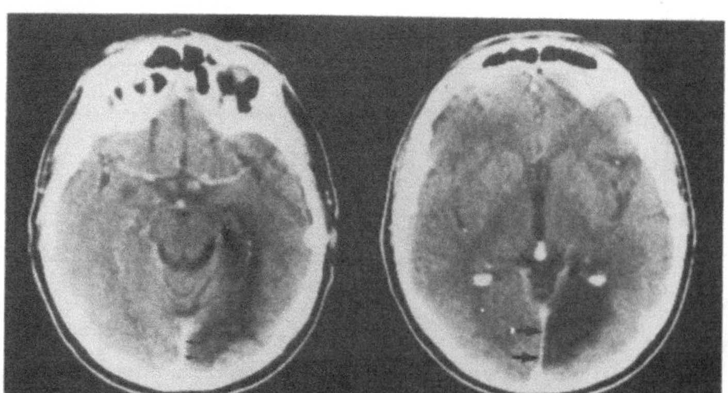

FIGURE 6–5C

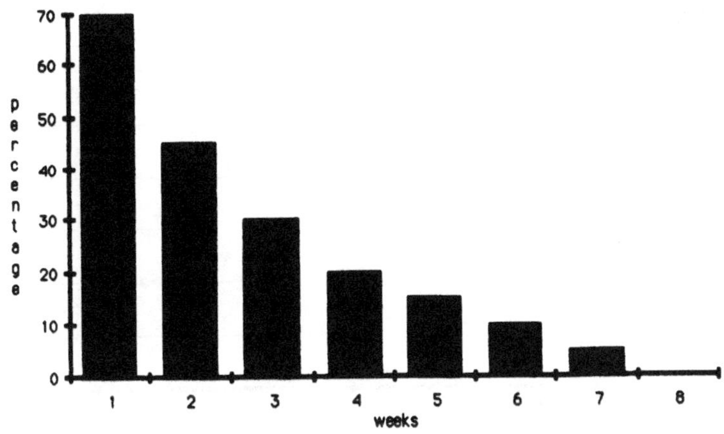

FIGURE 6–6. Incidence of mass effect in CT scans of acute cerebral infarcts in relation to time of acute neurologic deficit.

the acute ictus (figure 6–1). The amount of blood in the area of infarction varies from petechial hemorrhage to discrete hematomas. Although CT scans done with no contrast injection are commonly normal in the first day of acute infarcts, the early CT scans are important to detect hemorrhage in the brain, which can modify the medical treatment [35–37].

CONTRAST-ENHANCED CT SCANS

Contrast-enhanced CT scans increase the incidence of detecting acute infarcts. The dosage of contrast is 1 cc/lb body weight using a 60% solution of Conray 60®. Enhancement is maximum after the first week, reaching its peak around the second week after the acute stroke. Contrast-enhancement can be seen in about 50% of cases in the first week and 60–65% in the second week; then it drops back to about 45% in the third week, dropping gradually to 38% in the fourth week and diminishing gradually to less than 20% in the eighth week (figure 6–7) [30, 32, 38]. Patterns of enhancement include (1) dense, homogeneous enhancement outlining the deep-damaged brain; (2) superficial gyral enhancement (figure 6–8a); and (3) mottled or irregular enhancement within large areas of low density, probably resulting from brain liquefaction (figure 6–8b).

FIGURE 6–5. (A) Cranial CT scan showing infarction in distribution of anterior cerebral artery and watershed area (arrow). (B) Cranial CT scans (left and right) showing acute middle cerebral artery infarction. Notice the sharp demarcation between anterior, middle, and posterior cerebral arteries (arrows). (C) Cranial CT scan showing infarction in distribution of left posterior cerebral artery (arrows). Notice that infarct reaches the medial surface of occipital lobe.

Twenty percent of infarcts cannot be detected for weeks after the acute attack because of their small size (beyond the resolution of the scanner), the isodense nature of the lesion, the absence of mass effect, and/or absence of enhancement.

It has been suggested by some authors [40, 41] that contrast-enhanced scans should not be obtained in the first few days following an acute infarct because of the possibility of more damage to the brain by leakage of contrast material to the ischemic and/or infarcted area of brain.

DIFFERENTIAL DIAGNOSIS OF INFARCTS

Sometimes it may be difficult to differentiate acute infarction from brain tumors, metastases, vascular malformation, and meningitis. About 5% of stroke syndromes are caused by underlying tumors, since tumors can present as acute infarcts due to vascular occlusion, and infarcts sometimes have a gradual onset of symptoms mimicking tumors [34, 39]. The following points can be helpful in the differential diagnosis: (1) the shape and distribution of the lesion (figures 6–3, 6–5)—a wedge-shaped lesion, with its base directed against the peripheral part of brain in the distribution of a specific cerebral blood vessel, is a classic picture for infarcts; (2) follow-up scans that show gradual decrease of the edema and the mass effect (figure 6–1); and (3) cerebral angiography which may demonstrate the vascular disease (figures 6–13, 6–14, 6–16).

Serial CT scans will show that the contrast enhancement, if present, will decrease in extent and density, and the mass effect will diminish gradually

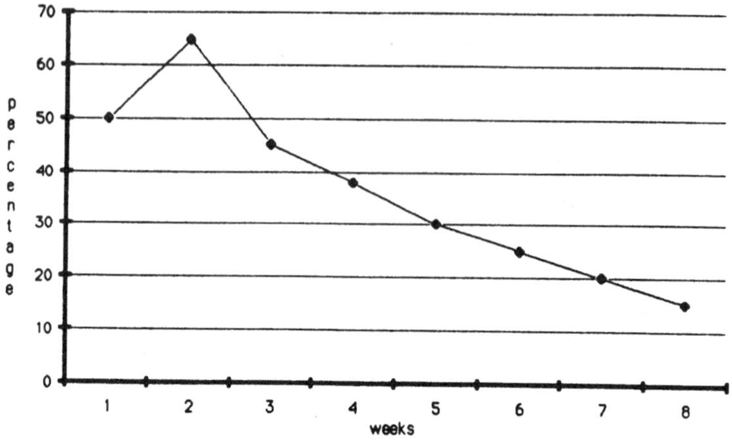

FIGURE 6–7. Incidence of enhancement of acute cerebral infarcts in CT scans in relation to time of acute neurologic deficit.

until it disappears with concomitant ipsilateral ventricular dilatation. The lesion either will shrink or disappear, depending on the original size of infarct. In cases of tumors, serial CT scans will show the size, enhancement, and the mass effect of the lesion either remaining the same or gradually increasing with time. Gyral enhancement can be shown in infarction, old subarachnoid hemorrhage, cortical vascular malformation, and also in meningitis. In acute meningitis, the CT scan will show diffuse enhancement crossing the natural boundaries between cerebral vessels, the clinical presentation of the patient will be different concerning body temperature and the state of alertness, and CSF examination may show the underlying infective organism.

Angiography is positive in about 50% of cases, showing the abnormal vascular occlusive disease in cases of infarcts (figures 6–13, 6–14). It may also show neovascularity of tumors and abnormal vessels of arteriovenous malformation [41]. The introduction of MR into the diagnostic field has helped tremendously in the diagnosis of occult vascular malformation where angiography is usually negative and CT scan may show an enhancing area similar to enhancing infarction (figures 6–9a, b).

Hemorrhagic infarcts and cerebral hematomas are detected early on CT scans after ictus because of the high density of clotted blood in the areas of brain damage (figures 6–1, 6–10) [42]. Hemorrhagic infarctions are inhomogeneous areas of increased density while hematomas appear as homogeneously dense lesions (figures 6–1, 6–10). Massive cerebral hemorrhage may result in large hematoma with massive mass effect and herniation of the brain. On the other hand, small hematomas have usually minimal surrounding edema and mass effect (figure 6–10). Follow-up CT scans show gradual absorption of the hematoma and the disappearance of the high density, passing through an isodense stage with the surrounding brain tissue, followed by a low-density lesion and a negative mass effect with dilatation of the ventricle on the same side (figure 6–1).

Magnetic Resonance Imaging

As the CT reflects a physical property (electron density) on radiographic CT scans by different degrees of absorption of the ionizing x-rays, MR image can be considered simply as a map of hydrogen intensity within the examined part of the body [43–45]. Unlike CT, MR is based on an apparently safe interaction between radio waves and hydrogen nuclei of the body in the presence of a strong magnetic field. Detection of disease in the brain by MR is generally based on alternations of normal anatomy (intensity) due to the presence of edema (hydrogen) [44]. MR can thus provide excellent gray–white matter discrimination and provide information about early changes in the normal brain pattern, detecting early ischemia and/or infarction due to changes in the dynamic state of water in the brain and its effect in altering the relaxation time on the scan (T1 and T2 relaxation times), which are the basic parameters for obtaining the MR images by the different sequences [46–48]. MR studies include plain MR and contrast-enhanced MR scans.

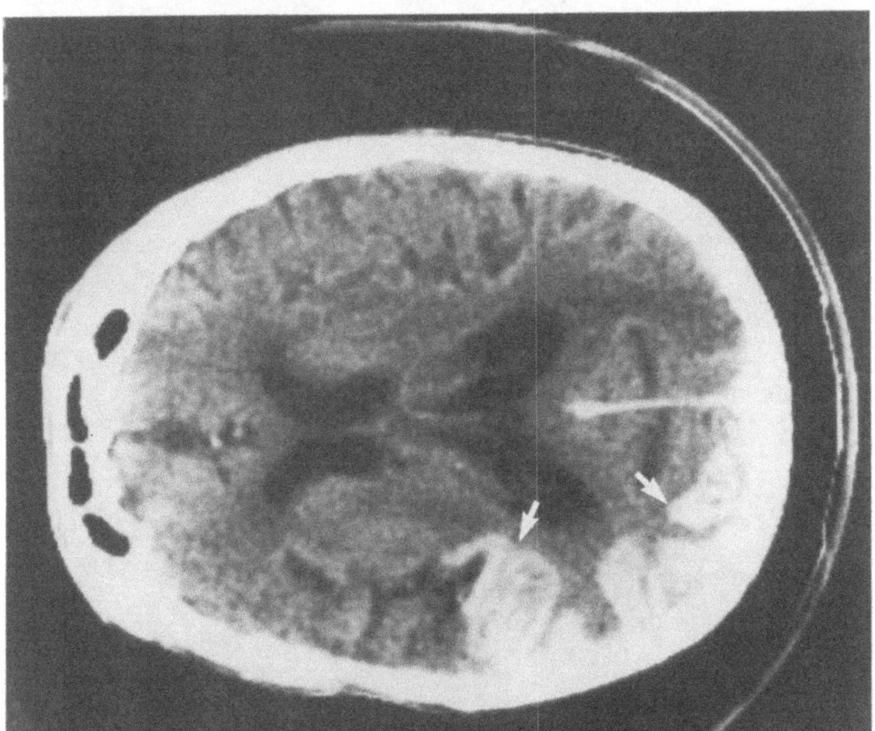

FIGURE 6–8A

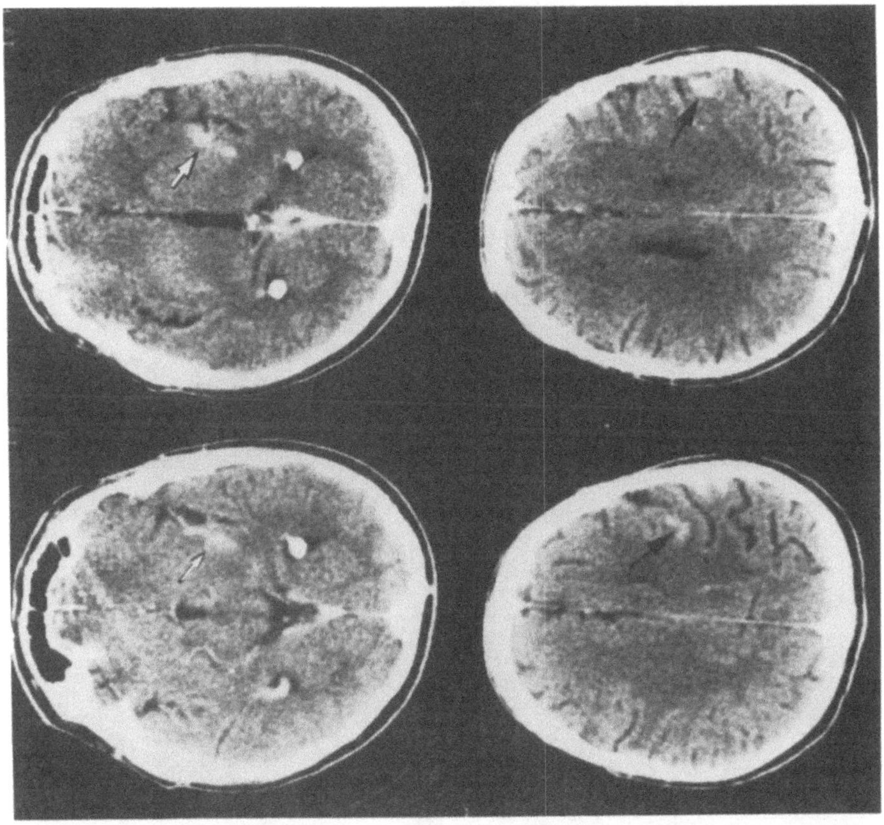

FIGURE 6–8B

FIGURE 6–8. (A) Postcontrast cranial CT scan showing the gyral pattern (arrows) of enhancement in acute cerebral infarct. (B) Postcontrast cranial CT scan showing mottled irregular pattern (arrows) of enhancement in acute cerebral infarct.

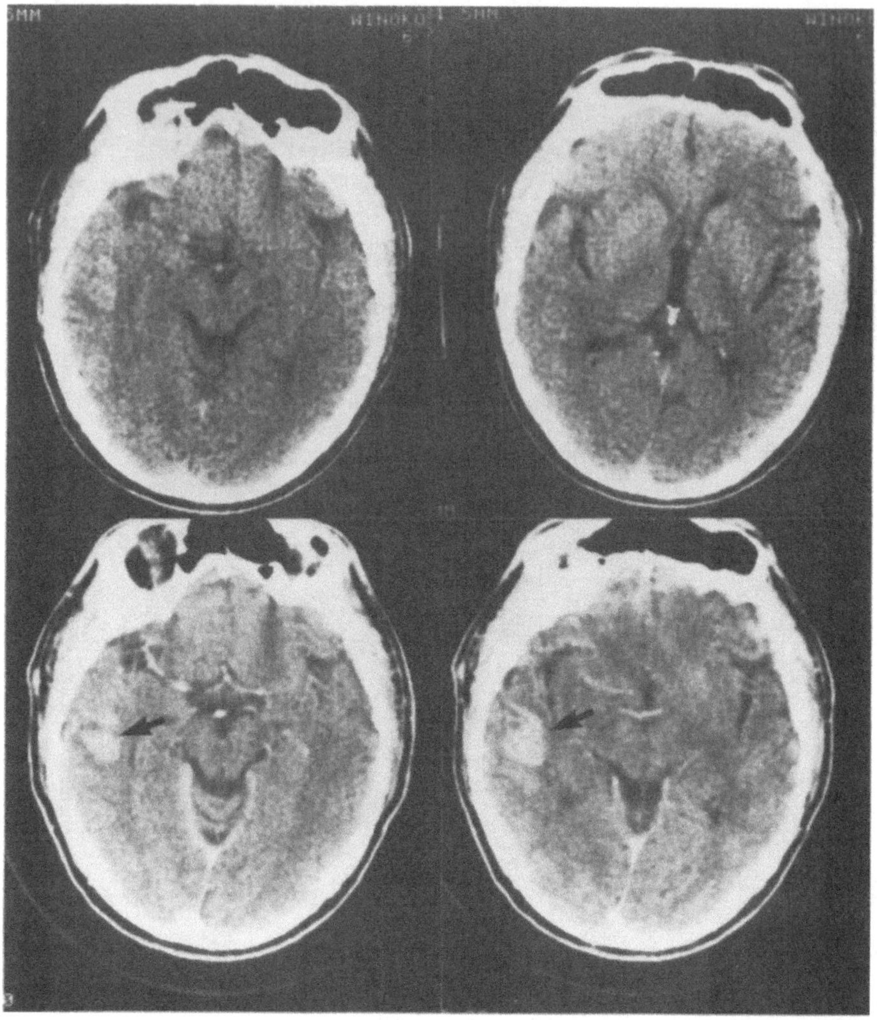

FIGURE 6–9A

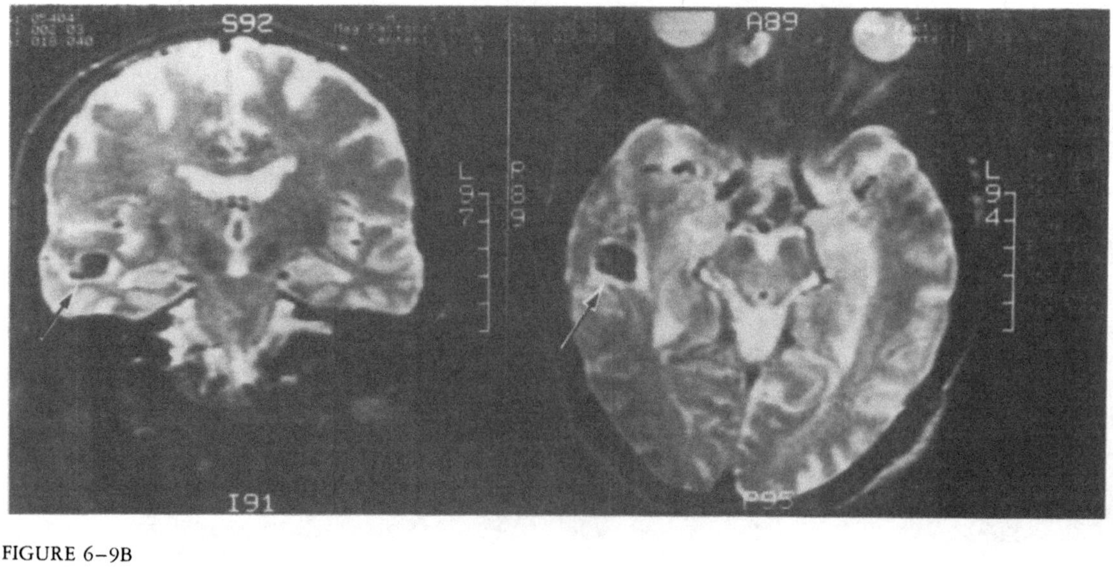

FIGURE 6–9B

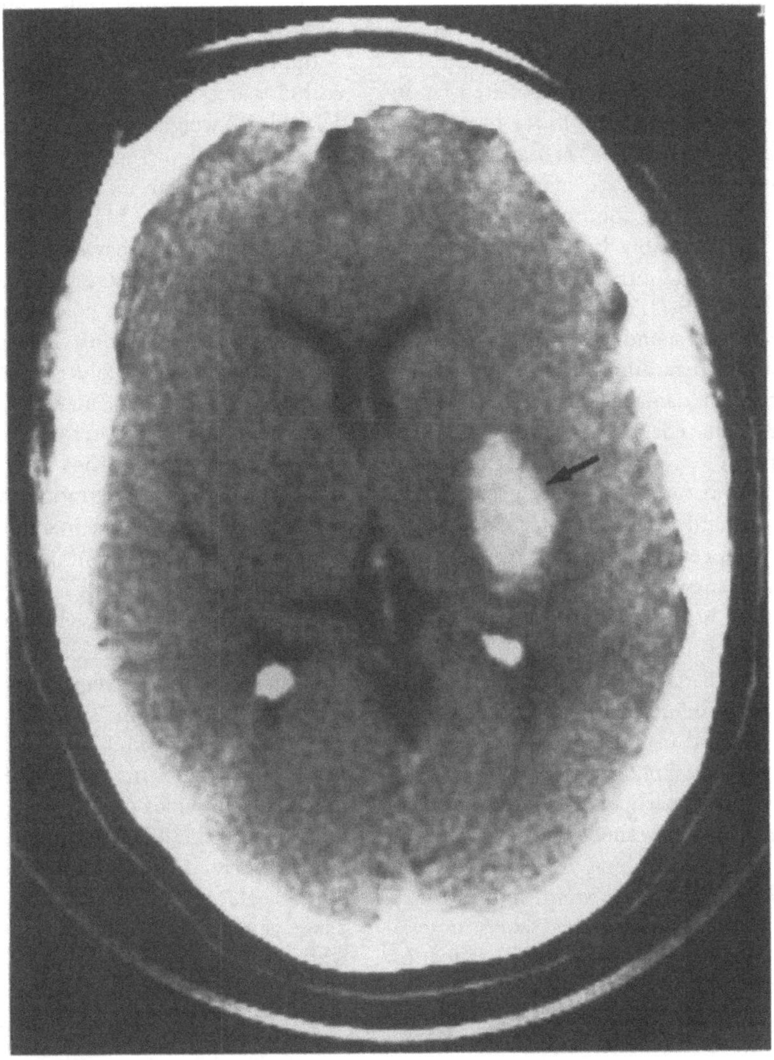

FIGURE 6–10. Cranial CT scan showing left acute deep temporal hematoma (arrow) in a patient with long-standing hypertension. Notice the minimal surrounding edema and mass effect.

FIGURE 6–9. (A) Precontrast cranial CT scan (upper row) showing a negative study and a postcontrast cranial CT (lower row) showing gyral enhancement of right temporal region (arrows) similar to enhanced acute infarct. (B) MR imaging of the same patient, coronal and axial views done with spin-echo technique using TR 2000 ms and TE 90 ms showing a right vascular malformation (arrows).

PLAIN MR SCANS

In the brain MR will show the gray and white matter standing out brightly, with marked delineation of the ventricular system and the subarachnoid spaces [49]. With the change in the amount of water that results from edema in cases of infarctions, the picture will change with different sequences. From various sequences studied, spin-echo was found to be well suited for detection of infarcts. Using echo time (TE) of 25–100 ms and repetition time (TR) of 500–2,000 ms, the study is usually both complete and diagnostic [43, 48]. T1-weighted images (spin-lattice, short time) and T2-weighted images (spin-spin, long time) are important to visualize the ischemic and hemorrhagic infarcts.

In the case of acute infarct, MR is able to show not only the mass effect of edema, if any, but also the increased intensity in affected parenchyma on T2-weighted images (white) and decreased intensity in T1-weighted images (gray to black), due to the prolongation of both T1 and T2 [50–52] (figures 6–11a, b). The T2-weighted images are more sensitive to infarction and show more abnormality than T1-weighted images, presumably because some of the T2 effect is impossible to subtract from T1-weighted images (figures 6–11, 6–12a). The abnormal areas in MR consistently correspond to retrospective histochemical observations regarding the size of the lesion and the magnitude of damage with respect to the development of brain edema [51, 53, 54]. The abnormalities shown on MR are larger than those detected on CT scan because of the high sensitivity, particularly of ischemic strokes. MR appears to effectively reflect the cerebral ischemic insult and is able to detect the changes in the ischemic brain after 30 minutes to three hours following ligation of the carotid artery in experimental animals [46, 55, 56]. It was shown in humans that an abnormality corresponding to focal neurological symptoms from stroke was clearly found in 94% of MR studies and only in 82% of corresponding CT studies [57] (figures 6–2, 6–11). In particular, CT was negative in acute ischemic infarcts scanned within two days after the onset of symptoms and in infarcts of the posterior cranial fossa and brain stem (figures 6–2, 6–4). MR performed within the same interval detected all the lesions. Moreover, CT failed to demonstrate small lesions, scanned 15–20 days after the onset of symptoms, that could be identified clearly on MR (figure 6–11). Thus, MR is an extremely valuable tool for the early detection of cerebral ischemia and is better suited than CT to detect ischemic lesions in the brain in the first 24 hours after the onset of symptoms. This is especially true for studies performed for small cortical or subcortical lesions, and for small infarcts in the white matter, posterior fossa, and brain stem [53] (figures 6–2, 6–10, 6–11, 6–12).

In MR studies, ischemic infarcts are usually clearly visible as areas of increased-signal intensity (white) in delayed T2-weighted images due to prolonged T2, independent of the age of infarcts. Thus, acute and old infarcts appear the same on delayed T2 images (figure 6–11) [48, 51]. However, old infarcts on MR usually appear having two components: those regions that behave more like CSF and those regions that behave more like edematous or gliotic brain. This reflects the changing molecular environment of water during the evolution of an infarct. The periphery of the infarct may appear more like edema or gliosis, where the water occurs in hydration layers in small cystic spaces. The center of infarct may have larger cysts and a higher ratio of water to solid tissue such that the water is the main bulk and behaves like CSF (figure 6–11). Some chronic cortical infarcts that manifested as areas of focal atrophy are well seen on T1-weighted images, which enhance the contrast between the brain and CSF. Lacunar infarcts in the basal ganglia and periventricular white matter show increased intensity in T2-weighted images (white) due to the prolonged T2 relaxation time. The white matter infarction in the periventricular region is much more obvious on MR images than on CT scan. On moderately T2-weighted (TR 2,000 ms, TE 30 ms) MR images, these infarcts appear bright (white) while the CSF is isointense with the brain. These small lesions in the periventricular region, while abnormal, are commonly seen in the asymptomatic elderly. They could be the result of silent small infarctions from arteriosclerotic vascular disease [51–59, 60].

The ability of MR imaging to separate acute hemorrhage from ischemic infarction has been suboptimal within the first few days of a clinically manifested cerebrovascular accident [61–63]. In circulation, hemoglobin exists in the oxy and deoxy forms, the relaxation times of which are similar

FIGURE 6–11. Pre- (top) and post- (bottom) contrast-enhanced cranial CT scans showing no evidence of cerebral lesions in a patient presenting with acute left hemiplegia. (B) MR imaging of the same patient showing acute right cerebral infarct as high-intensity lesion (white arrow) in T2-weighted image (TR 2000 ms, TE 90 ms) (right, top). The lesion is shown as a faint area of low intensity (black arrow) in T1-weighted image (left, top). MR scan also showed old small infarcts, just posterior to the acute infarct, shown as high-intensity lesions (small arrows) in T2-weighted images (TR 2000 ms, TE 90 ms) (right, bottom) and as areas of low intensity, behaving like CSF in small cystic spaces (small arrows) in T1-weighted images (left, bottom). (C) Top, normal CT scan of patient presenting with acute sensory deficit in left leg and foot. Bottom, MR imaging of same patient showing acute right cortical infarct as high-intensity gyral lesion in T2-weighted images (TR 2000 ms, TE 90 ms) (arrows). (D) MR imaging of another patient presenting with acute cortical blindness. Bilateral occipital infarcts are shown as high-intensity lesions in T2-weighted images (TR 2000 ms, TE 90 ms) (right). These lesions are shown as faint areas of low intensity in T1-weighted images (left).

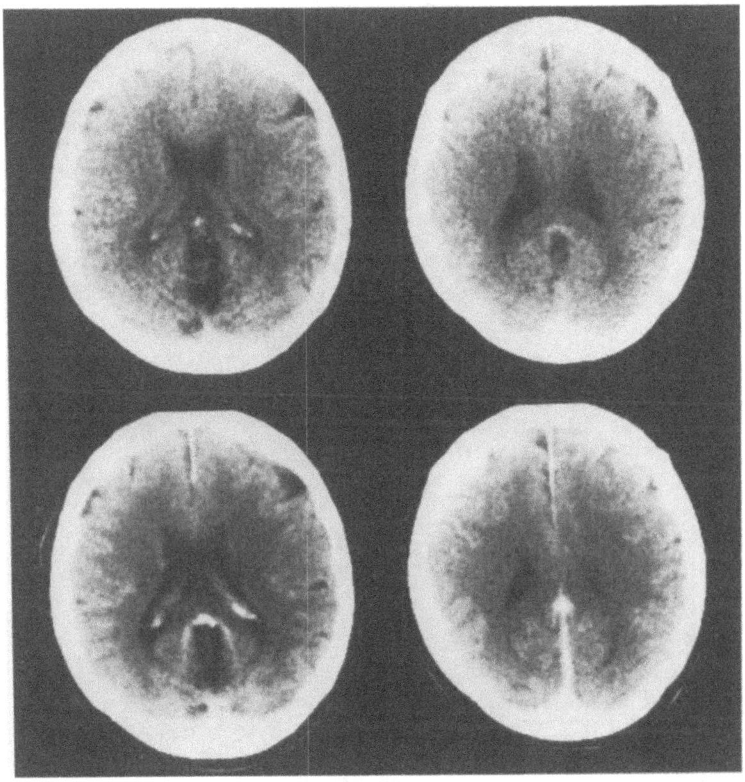

FIGURE 6–11A

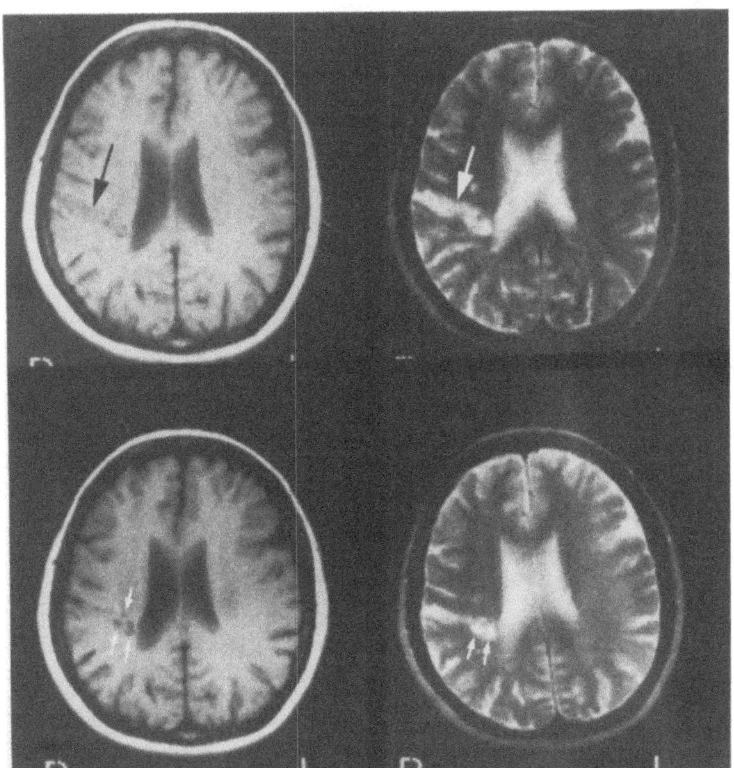

FIGURE 6–11B

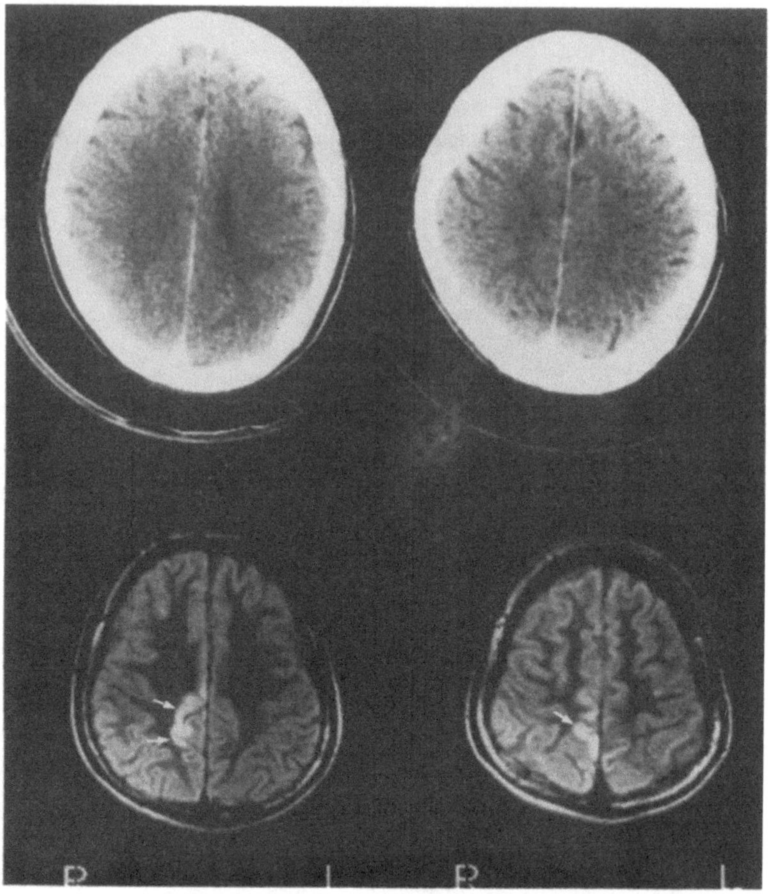

FIGURE 6–11C

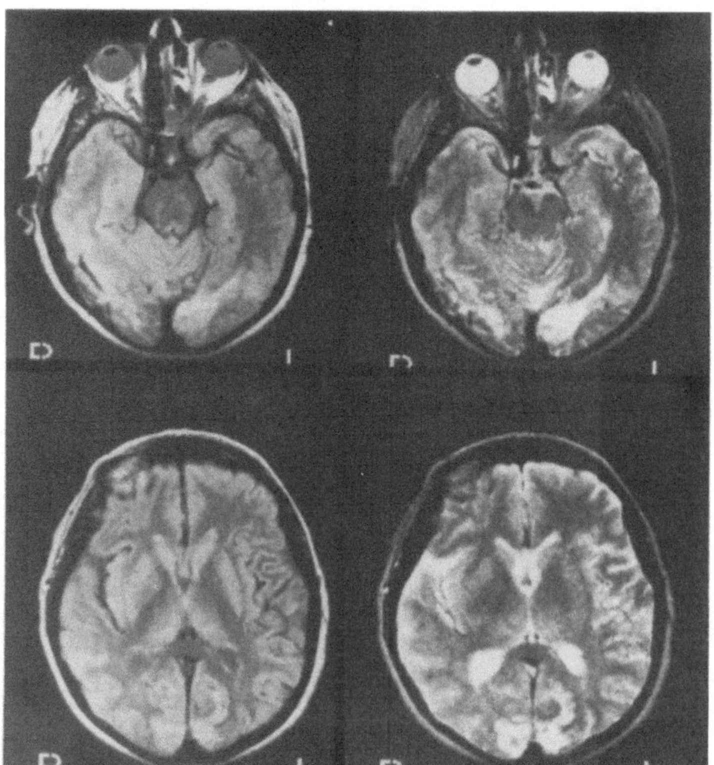

FIGURE 6–11D

to those of normal brain. Consequently, acute intracranial hemorrhage may be detected only on the basis of mass effect or the surrounding associated edema [64, 65]. CT scans are more valuable and more accurate in the detection of the acute intracranial bleed than MR. Both acute infarction and hemorrhage tend to show a high signal T2-weighted image with a variably low to isointense signal on T1-weighted images. However, when localized as hematoma, MR scans obtained in high-field scanners showed T2 shortening in acute hematomas, appearing as a low-intensity signal area amid the high-intensity signal edematous lesion, presumably because of the presence of magnetically susceptible deoxyhemoglobin. Thus, T2-weighted images are more sensitive than T1-weighted images in MR evaluation of early intracranial hemorrhage and acute ischemic infarction. After a period of several days, hemoglobin undergoes oxidation and denaturation, forming methemoglobin, which is paramagnetic. Continued oxidation over a period of months results in the formation of hemichromes, which are not paramagnetic. Paramagnetic substances have short T1 relaxation times. Shortening of the T1 relaxation time increases the intensity of the hemorrhage (white), increasing the consequent conspicuousness of subacute hemorrhage on T1-weighted MR images. Acute subarachnoid hemorrhage is also more difficult to show on MR images than on CT scans because of the minimal shortening of T1 relaxation time of the bloody CSF. A week-old subarachnoid hemorrhage will be more obvious on MR because of the formation of methemoglobin. CT, however, is more specific than MR and thus should be considered the preferred imaging technique for the evaluation of acute intracranial hemorrhage [28, 31, 42].

Many ischemic infarcts have at least some microscopic hemorrhage that is not evident on CT scans. Delayed MR scans have shown more hemorrhage in more infarctions than CT, probably because of the high sensitivity of MR to the paramagnetic effects of methemoglobin, which is formed in small foci of hemorrhage in the area of presumably ischemic infarcts on CT scan. As the controversy continues regarding the anticoagulation of patients with recent infarction, such improved sensitivity of MR imaging may add fuel to the discussion. We suggest reporting the MR scans as ischemic infarctions, infarctions with petechial hemorrhage, hemorrhagic infarctions, and cerebral hematomas.

ENHANCED MR SCANS

A paramagnetic agent (gadolinium DTPA) can be used to enhance the pathological areas of the brain before MR scanning [64–66]. Gadolinium DTPA strongly enhances proton relaxation; it has been shown to have good stability, rapid urinary excretion, and low toxicity, which makes it an ideal contrast agent for clinical application [65–67]. The design of contrast media for MR is different from that for CT scans. MR enhancement should influence the proton density or relaxation times, but the CT scan enhancement should affect the degree of absorption of the ionizing x-rays (photon density). Because of the difficulty in achieving large changes in proton density, the enhancing agents used are compounds that shorten both T1 and T2 relaxation times [67–69]. Perhaps the most important immediate application of intravenous contrast agents in MR would be for identification of the breakdown of the blood–brain barrier [67, 68). MR is more sensitive than CT for the detection of the breakdown of the permeability in acute infarction. Delayed contrast-enhanced CT has been used to visualize the same phenomenon, but it was never widely applied due to possible untoward effects of the hyperosmolar agent on already compromised neuronal tissue. It is likely that gadolinium-enhanced MR may offer a more convenient and better tolerated method of acquiring the same information regarding the blood–brain barrier integrity, which may in itself be indicative of neuronal survival and clinical reversibility. However, gadolinium MR was proven to be no more specific than enhanced CT in areas of the active permeability breakdown. As in CT, extravasation of the intravenously injected agent into the areas of the brain where the blood–brain barrier is broken by ischemia and/or infarction is the basic mechanism of the enhanced MR images. However, the process is not as simple as breakdown and extravasation, because edema when extensive can result in collapsed microcirculation and absence of perfusion to the affected area, preventing extravasation of the injected agent. This may explain why T2-weighted MR image is so sensitive to early cerebral infarction, when cytotoxic edema may dominate and effects of the blood–brain barrier breakdown are not detectable by conventional imaging. In fact, investigations thus far have shown that the fluid accumulation from early brain edema may be demonstrated by MR on T2-weighted images without contrast agents from .5 to two hours after an ischemic insult. This prolongation of T2 is expressed as an area of increased-signal intensity despite no breakdown of the permeability and no net increase in local tissue protons. Gadolinium enhancement in cases in infarction was shown in 10–35% of cases. Enhancement was found to be maximal in the median-aged and late infarcts, the mean enhancement percentages being 30% and

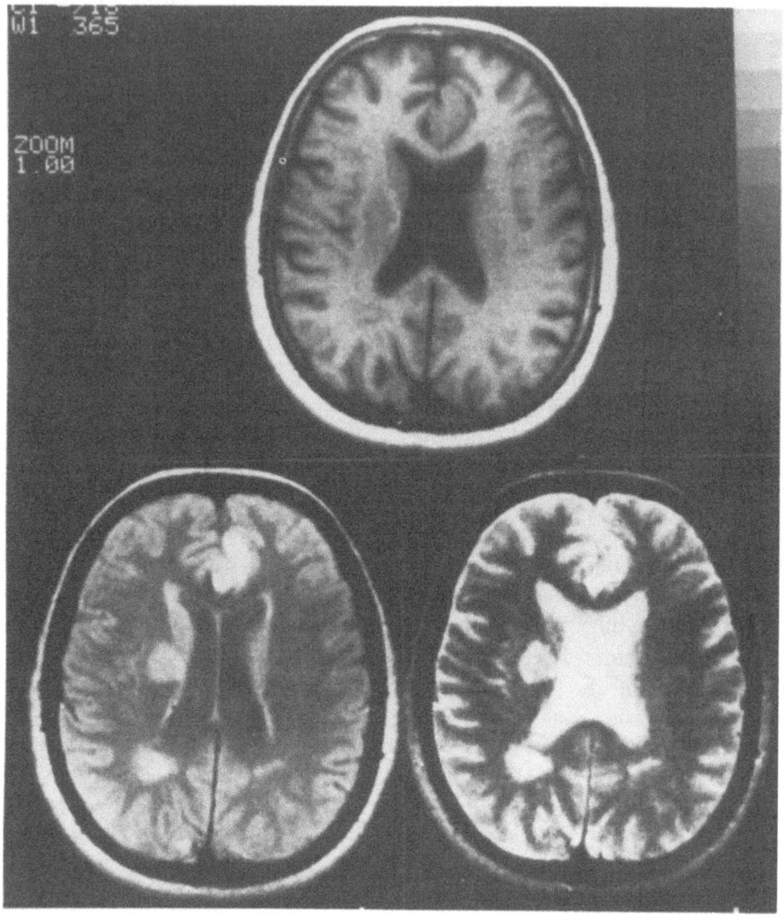

FIGURE 6–12A

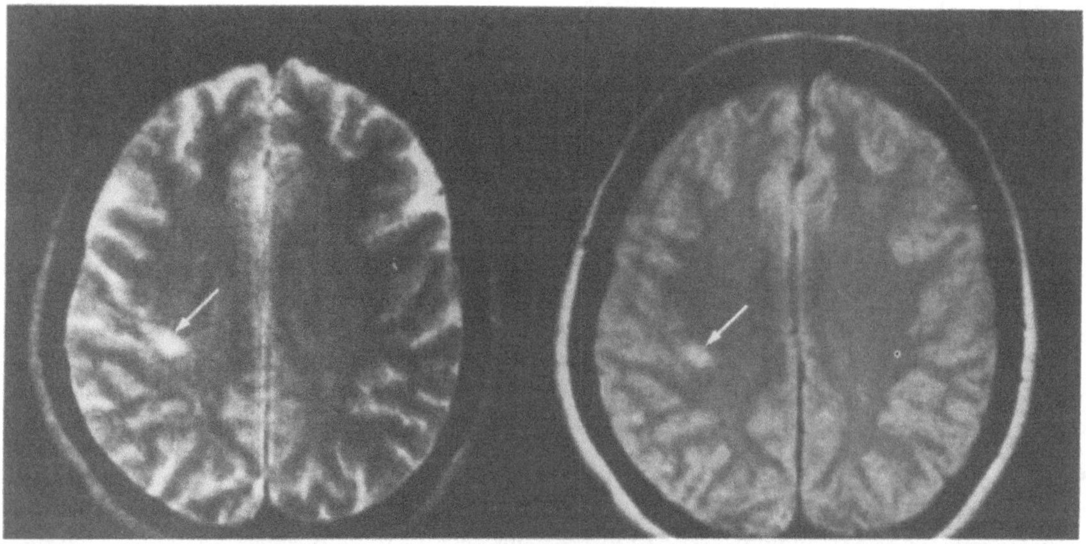

FIGURE 6–12B

29%, respectively [68, 70, 71].

The enhanced MR scans for infarctions are unlikely to exceed the sensitivity of the nonenhanced T2-weighted images in visualizing brain ischemia. At present, the clinical role of gadolinium enhancement in cerebral infarct evaluation seems as limited as that of contrast-enhanced CT scans.

Angiography

Since 1927, when Egas Moniz and colleagues laid down the basic foundation of angiography, the procedure has gone through many failures and successes, through many modifications, changes in the indicator agent (contrast), and changes in the route of injection. Recently, with the introduction of computerized radiology, the visualization of cerebral vessels can be achieved by selective arterial catheterization angiography as well as digital subtraction angiography via intraarterial or intravenous routes [34, 41, 72]. Each of these techniques has its own indications, advantages, and limitations. However, selective arterial catheterization angiography of the carotid and vertebral arteries remains the definitive investigative procedure for the accurate assessment of intracerebral vessels [41, 73, 74]. The study of the extracranial carotid and vertebral vessels will be considered in chapter 5, while in this chapter the role of angiography in studying the intracerebral vessels will be discussed.

SELECTIVE ARTERIAL ANGIOGRAPHY

Cerebral angiography is often needed for patients with suspected cerebrovascular occlusive disease presenting with transient ischemic attacks (TIAs), for patients who developed strokes, and for those who are considered for endarterectomy of the extracranial carotid arteries or for the various surgical bypass procedures (figures 6–13, 6–14). It is needed to establish the diagnosis of vascular occlusion and the site of lesion as well as the potential route of revascularization. Certain morbidity always accompanies cerebral angiography, including a local small hematoma at the site of the puncture and occasionally nausea and rarely vomiting. More serious complications are fortunately very rare and include embolic occlusions and stroke [41]. The incidence of such deficits ranges from .15–4%. Technical factors play a major role in the development of complications, including the skill of the angiographer and the time spent during the procedure, together with the condition of the patient. Complications are four times more likely to occur in patients with advanced cerebrovascular disease as compared to those having normal carotid arteries.

Both local cerebral embolism from mural thrombus formation in the extracranial carotid arteries and reduced cerebral effusion due to severe extracranial carotid stenosis or occlusion have been proposed to explain cerebral ischemia from arteriosclerotic disease. By definition a TIA is an event characterized by focal neurologic deficit for a brief time ranging from two to 30 minutes from which the patient fully recovers within a 24-hour period. The background of localized ischemia is suggested by the occurrence of symptoms in a particular vascular territory and by the reversibility, indicating the absence of cellular death [14, 75]. It has been stated that 20–40% of stroke patients have their symptoms from lesions of the extracranial vessels [14, 15]. In 149 cases of vascular disease of the head, angiography showed that the internal carotid artery was involved in 93 cases, the common carotid in 12, the innominate artery in 14, the subclavian artery in 21, and the vertebral artery in 9. The vertebral lesions are most often at the origin of the artery from the subclavian trunk. In another study done in 1967–1976, 314 patients (146 females and 168 males with a mean age of 49.2 years) were evaluated clinically and with angiography studies. The initial attack involved the carotid territory in 245 patients (78%) and the vertebral vascular territory in 69 patients (22%) [78, 79]. Fifteen patients (4.8%) had ischemic brain infarction during the follow-up and eight died. Acute myocardial infarction occurred in 40 patients (12.7%), and 24 infarctions were fatal.

There was a suggestion that recurrent cerebrovascular episodes were more common in patients with kinks in the extracranial carotid and vertebral arteries from arteriosclerosis. Metz et al. in 1961 found 16% incidence of redundancy among 1,000 angiograms reviewed. However, many others do not believe that there is any relation between redundancy

FIGURE 6–12. (A) Cerebral MR image showing multiple acute cerebral infarcts shown clearly as high-intensity lesions in T2-weighted image (TR 2000 ms, TE 90 ms) (bottom, right), and as moderately intense lesions in moderately T2-weighted image (TR 2000 ms, TE 30 ms) (bottom, left). These lesions are shown as faint areas of low intensity in T1-weighted images (top). (B) Cerebral MR image showing right acute small cerebral infarct (arrow) as a high-intensity lesion in T2-weighted image (TR 2000 ms, TE 90 ms) (left), and as a moderately intense lesion in moderately T2-weighted image (TR 2000 ms, TE 30 ms) (right). *Note*: T2-weighted images are far more sensitive in showing infarcts.

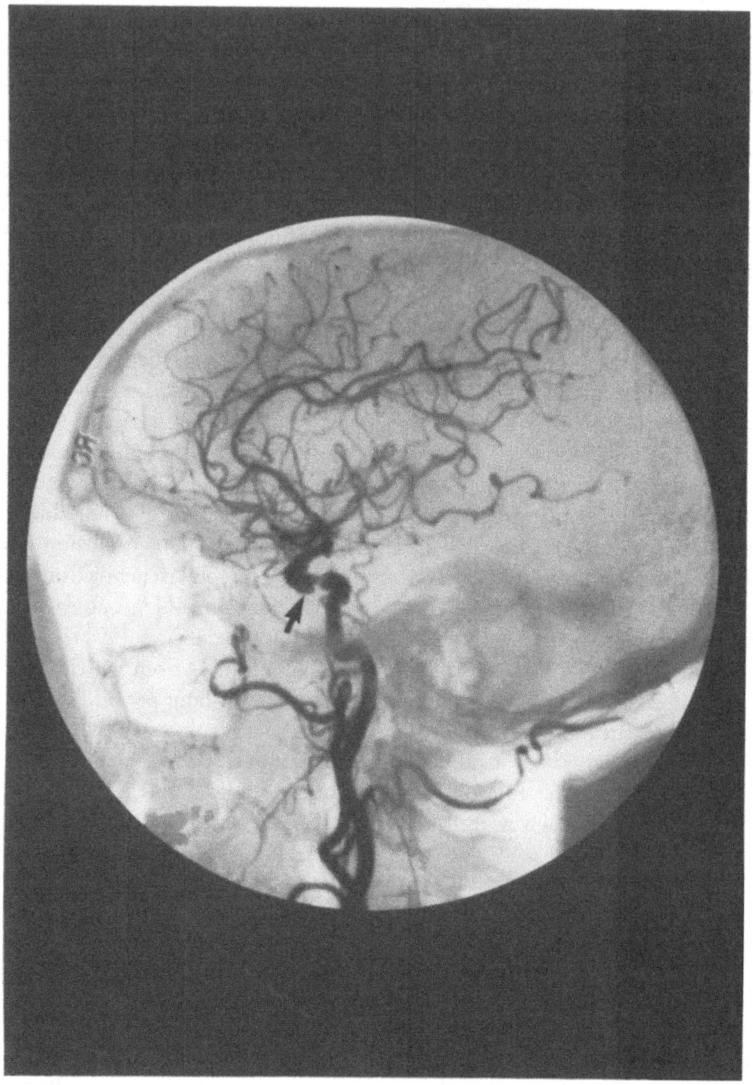

FIGURE 6–13A

FIGURE 6–13. (A) Right selective carotid angiogram showing severe arteriosclerotic changes of right carotid siphon (arrow) with irregular ulcerative plaques and stenosis. The intracranial vessels are normal. (B) Right selective carotid angiogram showing normal right carotid siphon but severe arteriosclerotic stenotic disease of intracerebral vessels (arrows). *Note*: Detailed information of carotid siphon and intracranial vessels are always helpful before considering surgical treatment and cannot be clearly shown on digital subtraction angiography. (C) Left selective carotid angiogram showing normal right carotid siphon (carotid bifurcation in neck was also normal), but showing the area of acute middle cerebral infarct (presumably embolic) as perivascular luxury perfusion or blush (asterisk).

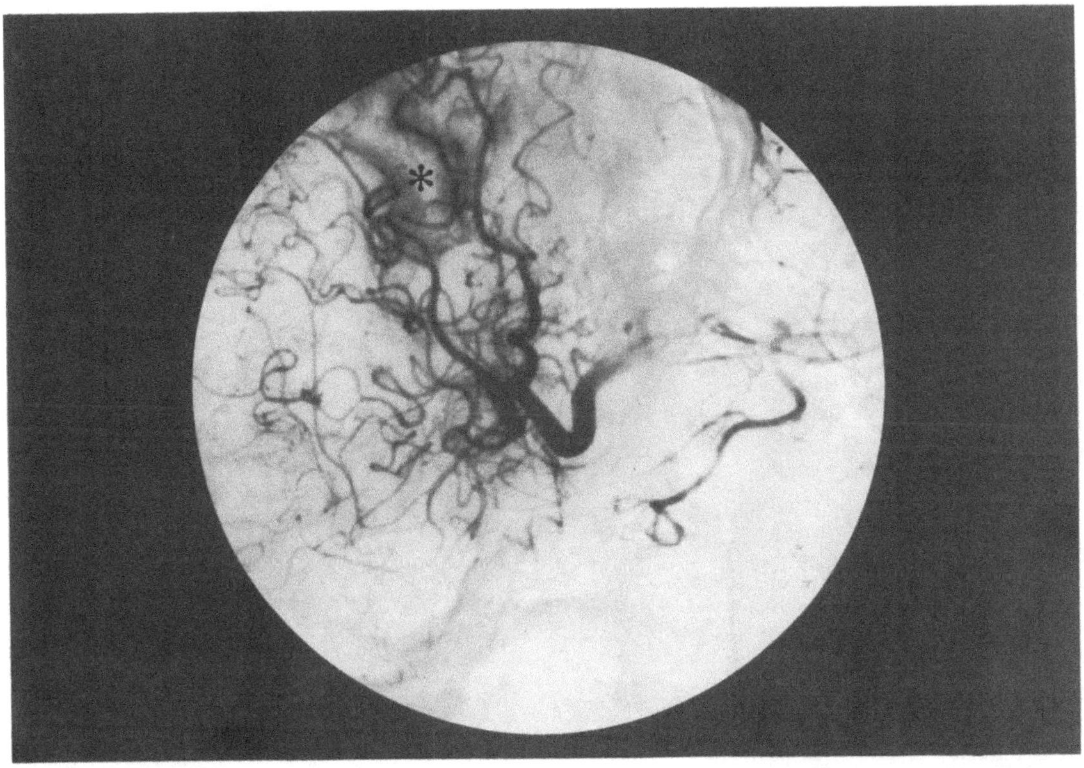

FIGURE 6–13C

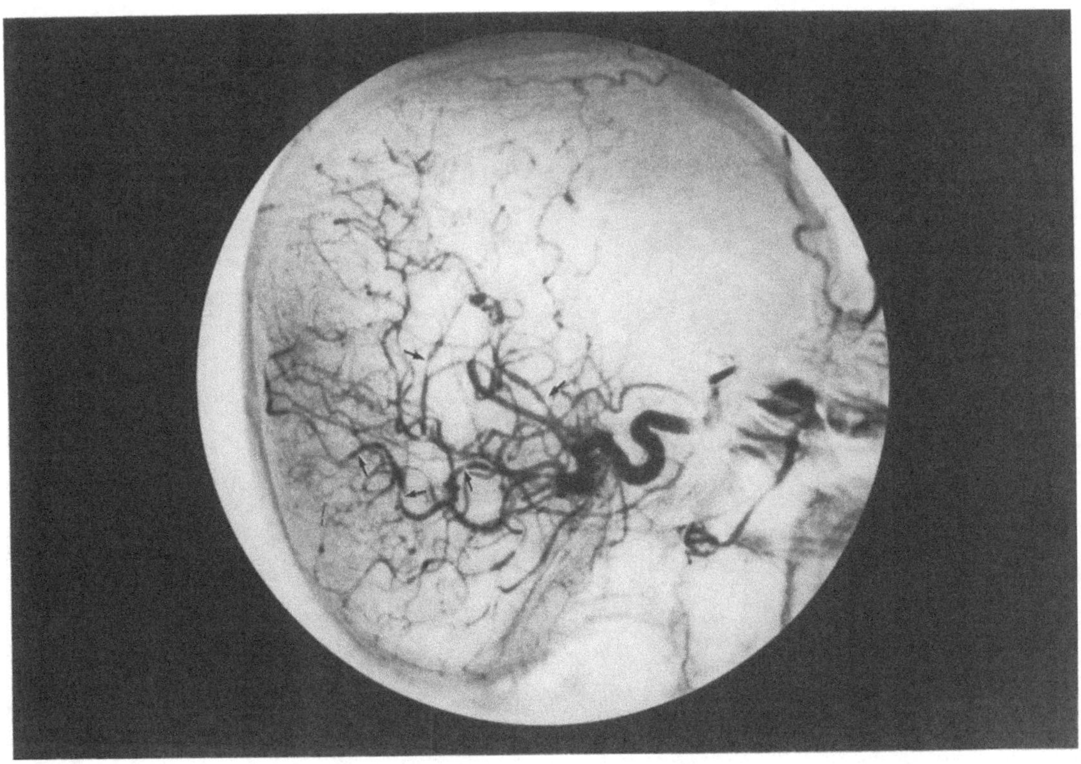

FIGURE 6–13B

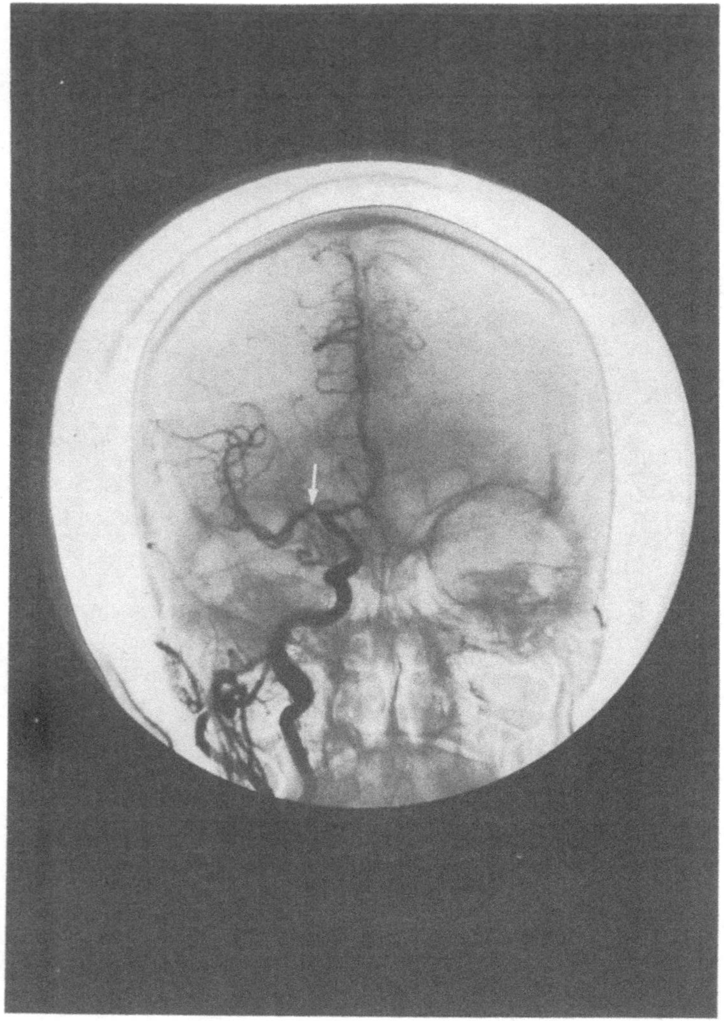

FIGURE 6–14A

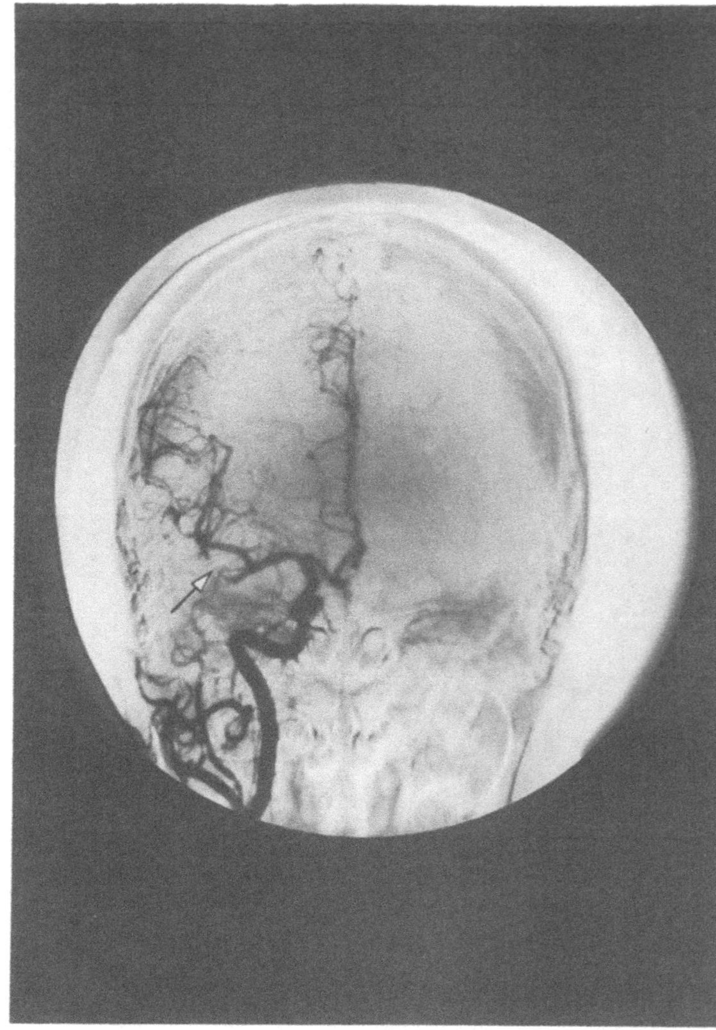

FIGURE 6–14B

FIGURE 6–14. (A) Right selective carotid angiogram showing severe stenosis of middle cerebral artery (arrow). (B) Right selective carotid angiogram showing severe stenosis of a branch of middle cerebral artery (arrow).

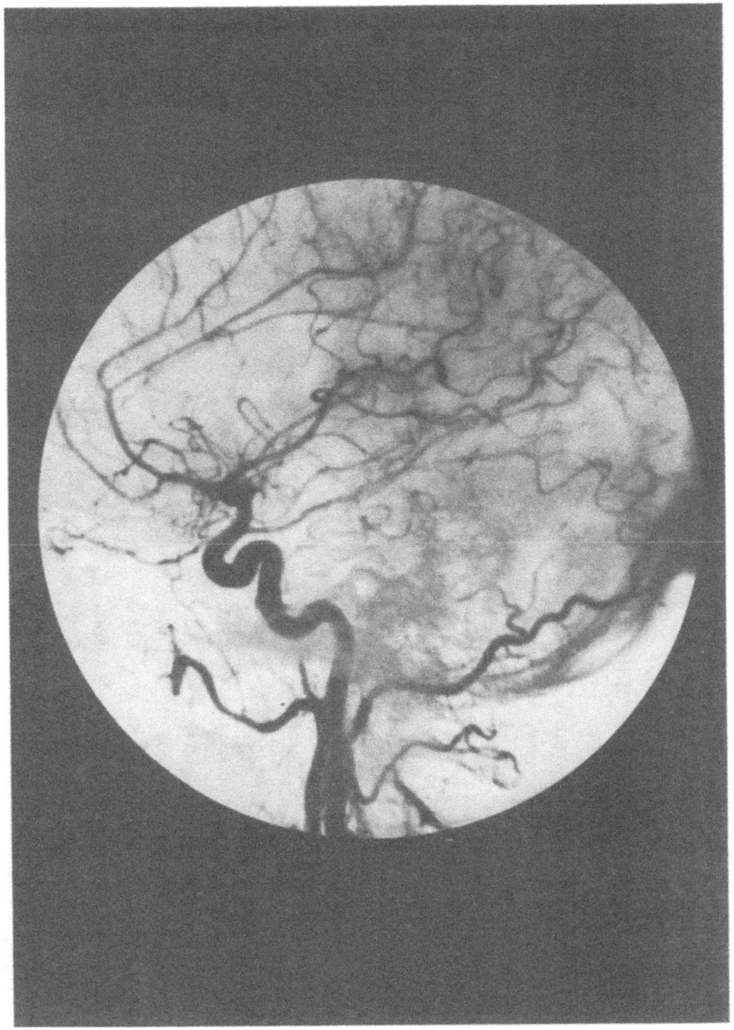

FIGURE 6–15A

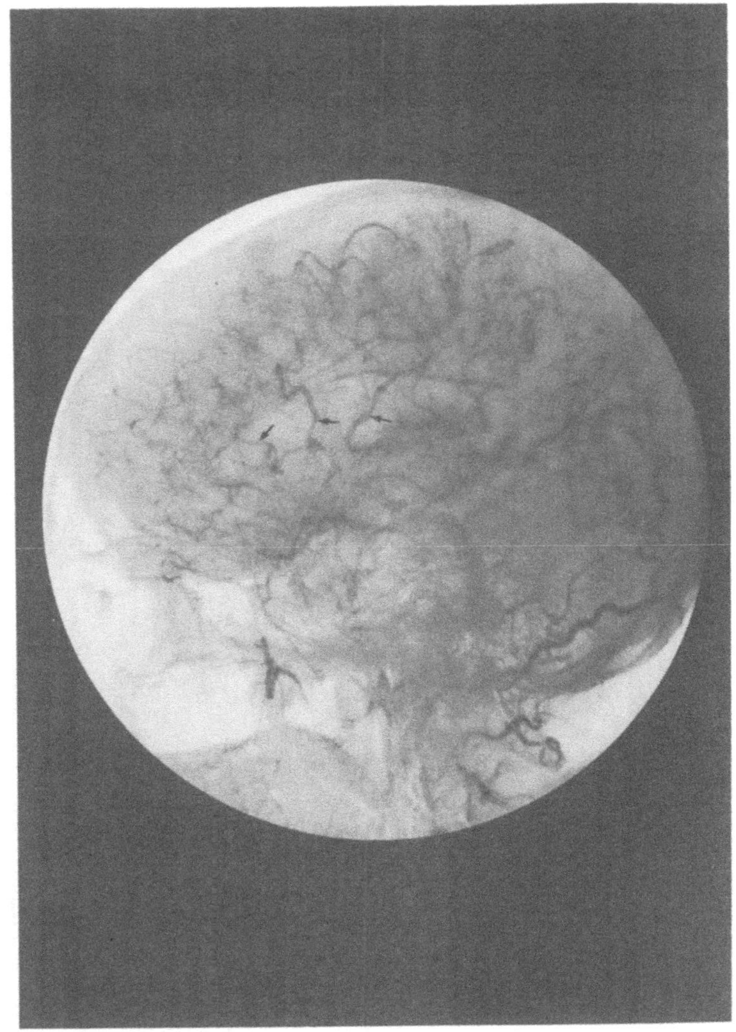

FIGURE 6–15B

FIGURE 6–15. Left selective carotid angiogram showing occlusion of frontal opercular branches of middle cerebral artery (A). These branches fill in a retrograde fashion and are visualized in delayed views (arrows) (B).

of the carotid vessels and cerebrovascular occlusive episodes [80].

For many years it has been felt that emboli are the most common direct cause of intracranial arterial occlusions of the small vessels (figures 6–13c, 6–15, 6–16). The embolic theory is logical, particularly in cases where multiple branch occlusions are demonstrated. It explains the lack of evident intracranial occlusions at autopsy in the face of severe fatal strokes. Because emboli can disappear by lysis, cerebral angiography may show vascular obstruction early in the course of the illness and patent vessels later (figure 6–17). The chance of demonstrating occluded vessels is thus greater in the early studies (figures 6–16, 6–17). The main emboli will fragment and the small fragmentations will migrate to the peripheral vessels, where they disappear by lysis.

Disappearance of arterial occlusions of the main cerebral vessels on follow-up angiogram is an uncommon but interesting phenomenon [81, 82] (figure 6–17). In our experience it occurs more commonly in younger patients (20–40 years) with segmental occlusive disease of one of the main cerebral vessels and no evidence of diffuse cerebral arteriosclerotic changes. Some of these cases show only irregularity and marked stenosis of the cerebral vessel which on follow-up angiogram opens into a normal caliber. These interval changes cannot be explained in every case on the basis of emboli with subsequent fragmentation and peripheral migration. More interesting was the case shown in figure 6–17, where the original angiogram done shortly after acute right hemiparesis of a 24-year-old man showed occlusion of the horizontal portion of the left middle cerebral artery. The follow-up angiogram showed complete opening of the previously occluded vessel with subsequent clinical recovery except for slight neurologic deficit. Because of this important group of patients showing reversible disease of the main cerebral vessels in a few weeks, it is advisable to repeat angiography within a few weeks after the acute attack before considering bypass surgery. In our experience, the bypass graft was occluded in some of these cases and the natural process of opening the previously occluded vessels occurred within a few weeks to months from the acute stroke [34, 82].

Cervicocerebral angiography studies are reviewed for the following abnormal findings: (1) stenosis or complete occlusion of the major vessels in the neck (vertebral and carotid arteries) and in the head (basilar artery, carotid siphons, and the anterior, middle, and posterior cerebral arteries) (figures 6–13, 6–14, 6–16, 6–17). The peripheral branches, including the primary, secondary, and sometimes the tertiary branches of the major cerebral vessels, should be carefully examined for evidence of disease (figure 6–13b); (2) mass effect secondary to swelling and/or edema of the brain on neighboring arteries and veins; (3) appearance of collateral arterial circulation and the extent of revascularization of the distal branches of the occluded vessel or vessels (figure 6–15). It is important to find and describe the collateral circulation through the Circle of Willis and over the watershed areas to evaluate if the collateral circulation is enough to supply the territory of the occluded vessel for normal physiologic functions; (4) local increase in the speed of circulation with shunting to local veins; (5) capillary blush or stain, which sometimes is called luxury perfusion (figure 6–13c); and (6) old infarction, which may be associated with local disruption of the normal arterial pattern.

INTRAVENOUS DIGITAL SUBTRACTION ANGIOGRAPHY

The digital subtraction imaging of cranial vasculature after intravenous (IV) injection of contrast material is an established outpatient screening procedure for vascular occlusive disease of the extracranial arteries, as well as for gross intracranial lesions like large aneurysms, arteriovenous malformation, and complete occlusion of a main cerebral vessel [83, 84]. Although selective transarterial angiography is still the most accurate radiographic visualization of the cerebral vessels and their branches, it cannot be used as a screening test, especially for high-risk patients and for serial follow-up studies. The role of intravenous subtraction angiography in the evaluation of the extracranial vasculature is discussed in chapter 4.

INTRAARTERIAL DIGITAL SUBTRACTION TECHNIQUE

Intraarterial digital subtraction angiography is becoming more and more useful in cases where there is a need for increased spatial resolution and adequate visualization for the intracranial vessels in the high-risk patients [73]. The volume of contrast agent used is much smaller than needed for IV digital angiography. The test is especially suited for high-risk patients with a fluctuating neurologic deficit or im-

FIGURE 6–16. (A) Left selective carotid angiogram showing intraluminal large embolus partially occluding left middle cerebral artery (arrows). (B) Left selective carotid angiogram showing intraluminal defects partially occluding left middle cerebral artery (arrows), proven at autopsy to be fat emboli from fractured femur. Detailed information can be shown only in selective arteriography.

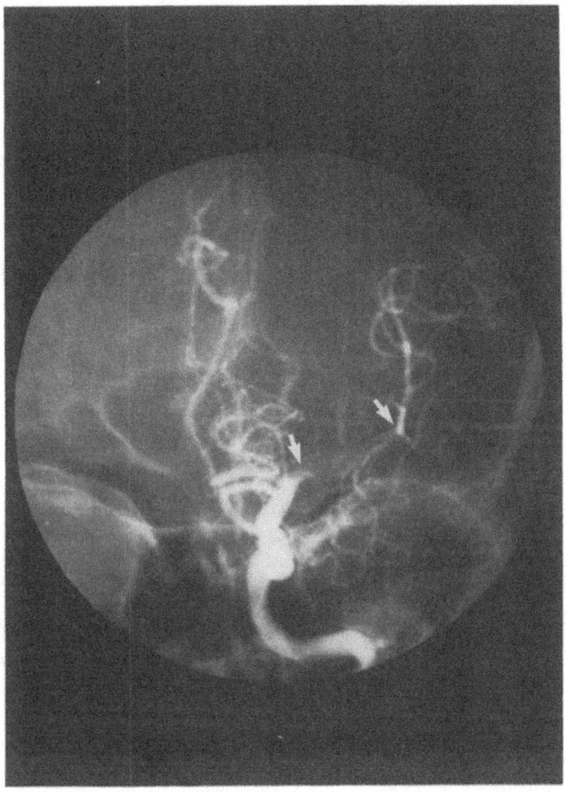

FIGURE 6–16A

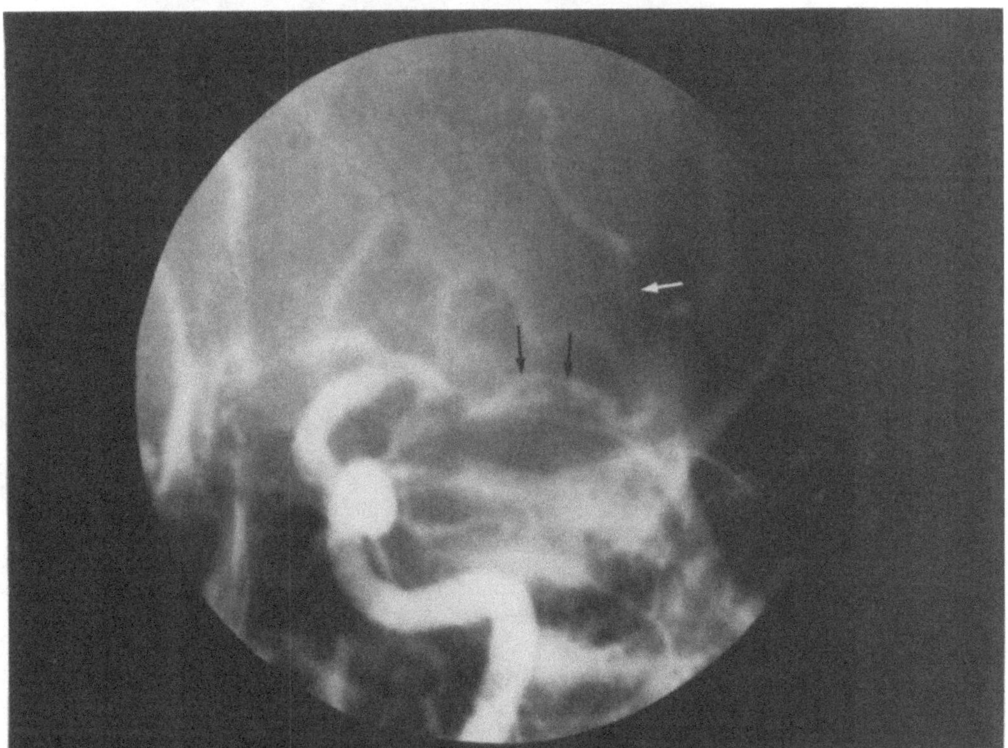

IGURE 6–16B

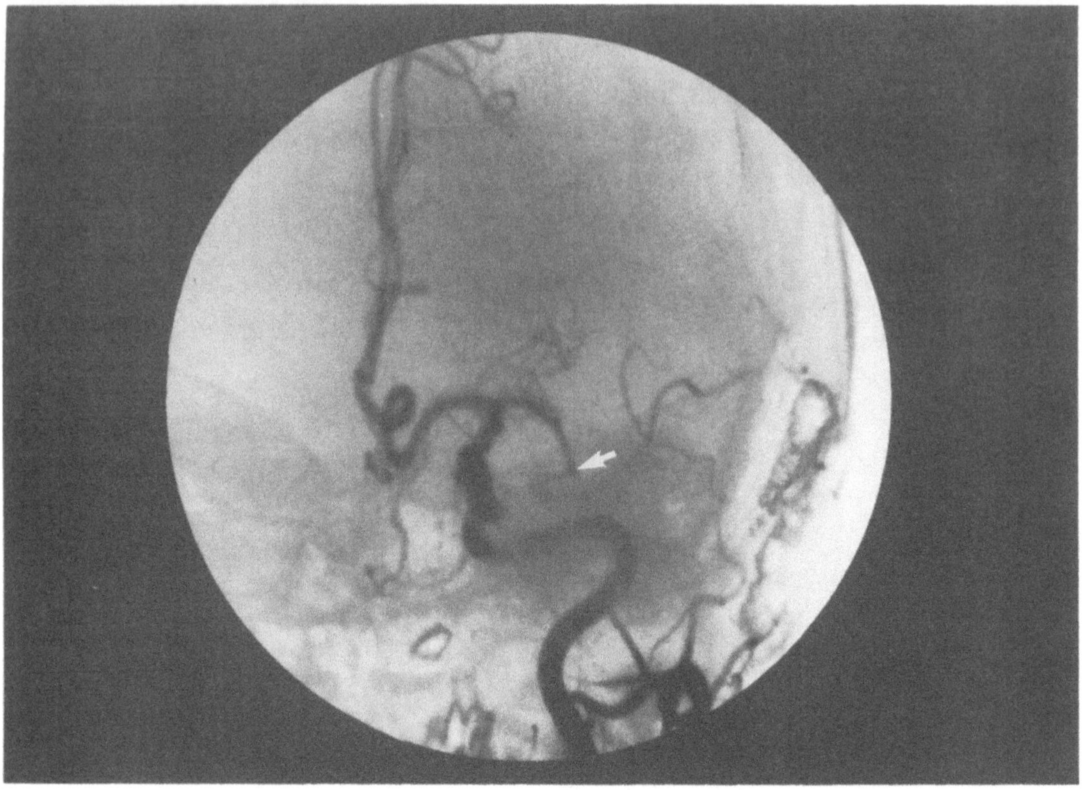

FIGURE 6–17A

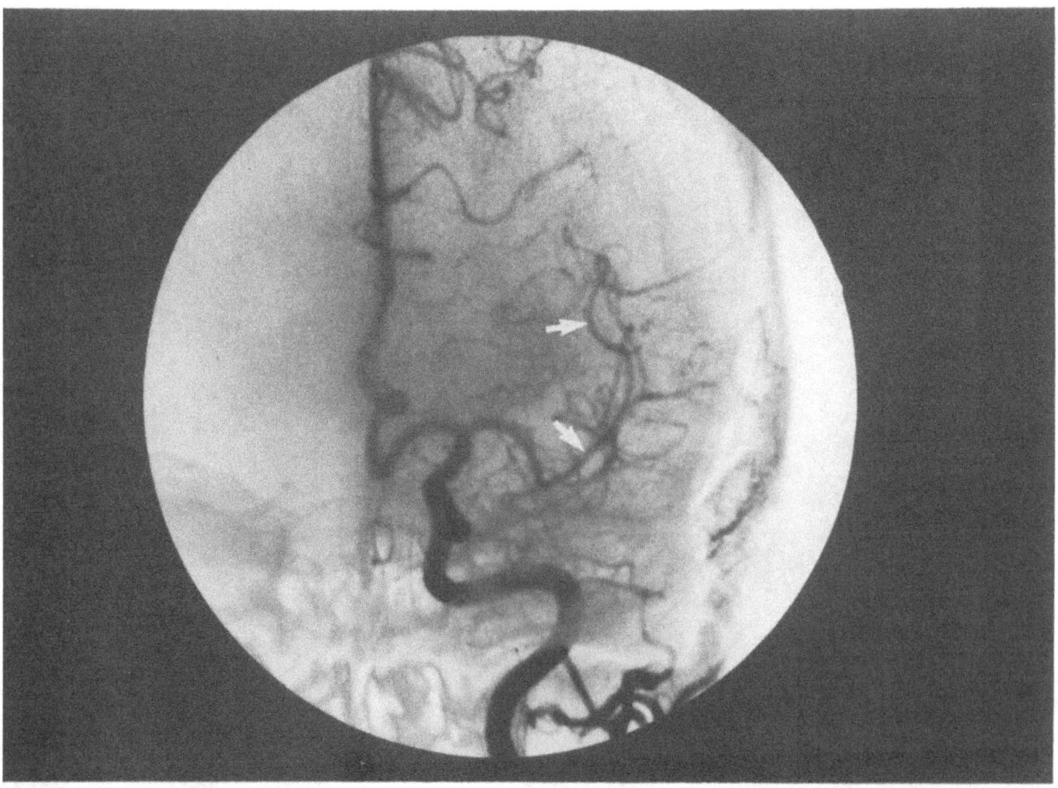

FIGURE 6–17B

paired renal function. It is especially useful in cases where repeated injections and different oblique views are necessary for better evaluation and cannot be done with IV digital studies because of the large contrast load. However, using intraarterial catheterization and the injection even of a small amount of contrast into the selected arteries obviates the main purpose of the test as being nonivasive.

With the increased diagnostic use of MR imaging with higher magnetic-field-strength scanners and the use of surface coils and thin-slices technique, the cerebral vessels can be visualized in different planes by T1-weighted images. MR will prove in the future to be of great help in the follow-up of patients for the size of aneurysms or for vascular occlusive disease of the major cerebral vessels without the use of contrast material or invasive catheterization [85–88].

Recommended Protocol for Neuroradiological Evaluation of Cerebral Infarctions

Medical and surgical treatment for patients presenting with acute cerebral stroke depends, among other factors, on the radiological visualization of 1) presence or absence of brain hemorrhage, 2) size and site of the lesion, 3) complicating factors such as mass effect, and compression and/or herniation of the brain, 4) the condition of the remaining brain, and 5) the status of the extracranial vessels.

Because cranial CT scans can detect acute intracranial hemorrhage (within a few hours), the first radiological test for patients presenting with TIAs or acute neurological deficit should be a plain CT scan of the head. CT scans can also visualize the condition of the remaining brain, including old infarcts, atrophy, and hydrocephalus, and may also detect massive acute infarct with severe edema and mass effect, as well as large tumors. Postcontrast cranial CT is not advisable in the first few days following the acute ictus [40, 41].

Once hemorrhage is ruled out, MR scan is obtained in selected cases to visualize the acute lesion within a few hours from the ictus, and to rule out other possibilities not usually detected on noncontrast-enhanced CT scan, such as multiple sclerosis, vascular malformations, and tumors with no mass effect. MR is more sensitive than CT in detecting lesions of the posterior cranial fossa, brain stem, and upper cervical spine.

FIGURE 6–17. (A) Left selective carotid angiogram showing complete occlusion of left middle cerebral artery (arrow). (B) Repeat selective carotid angiogram after four weeks showing interval opening of previously occluded left middle cerebral artery (arrows).

Follow-up CT scans are advisable particularly in cases of questionable diagnosis to rule out the possibility of tumors, and in cases of hemorrhage to follow the evolution of the hematoma and rule out fresh recurrent bleed. Postcontrast CT scan can be obtained after angiography in order to use the same injection of contrast for both studies.

Intravenous digital subtraction angiography is an outpatient study for screening the extracranial vessels in patients presenting with TIAs or acute neurological deficit. The limited resolution of this technique makes it less useful for the study of intracranial vessels. It can be repeated as a follow-up to monitor the carotid arteries in the neck after carotid endarterectomy, and in patients on long-term medical treatment.

Selective intraarterial angiography is often needed for patients with suspected cerebrovascular occlusive disease before considering surgical treatment, endarterectomy, and/or bypass procedures to study in detail the condition of the carotid siphon and intracranial vessels.

Summary

Patients with stroke syndrome should be examined first with a plain CT followed in selected cases by cranial MR, if available. Angiography is needed before considering medical and/or surgical treatment. Subtraction angiography is a screening test for extracranial vessels while selective arterial angiography is the study of choice for detailed study of the intra- as well as the extracranial vessels.

References

1. Bussuttil RW, Baker JD, Davidson RK, Machleder HI. Carotid artery stenosis: hemodynamic significance and clinical course. *JAMA* 1981;245:1438–1441.
2. Fields WS. Progress report of the joint study on extracranial arterial occlusion. In: Millikan CH, Siekert RG, Whisnant JP (eds). *Cerebral Vascular Diseases: Fifth Conference*. New York: Grune and Stratton, 1966.
3. Meyer FB, Sundt TM Jr, Yanagihara T, Anderson RE. Focal cerebral ischemia: Pathophysiologic mechanisms and rationale for future avenues of treatment. *Mayo Clin Proc* 1987;62:35–55.
4. Lyons C. Progress report of the joint study on extracranial arterial occlusions. In: Millikan CH, Siekert RG, Whisnant JP (eds). *Cerebral Vascular Diseases: Fourth Conference*. New York: Grune and Stratton, 1965.

5. Toole JF, Janeway R, Choi K, et al. Transient ischemic attacks due to arteriosclerosis. *Arch Neurol* 1975; 32:5–12.
6. Zarins CK. Carotid bruit or carotid stenosis: what is the significance? *JAMA* 1981;245:1462.
7. Ames A III, Wright RL, Kowada M, Thurston JM, Majno G. Cerebral ischemia. II. The no-reflow phenomenon. *Am J Path* 1968;52:437–453.
8. Nemoto EM, Hossmann K-A, Cooper HK. Postischemic hypermetabolism in cat brain. *Stroke* 1981; 12:666–676.
9. Kabat H, Dennis C, Baker AB. Recovery of function following arrest of the brain circulation. *Am J Physiol* 1941;132:737–747.
10. Hossmann K-A, Olsson Y. Suppression and recovery of neuronal function in transient cerebral ischemia. *Brain Res* 1970;22:313–325.
11. Morawetz RB, Crowell RH, DeGirolami U, Marcoux FW, Jones TH, Halsey JH. Regional cerebral blood flow thresholds during cerebral ischemia. *Fed Proc* 1979;38:2493–2494.
12. Meyer FB, Anderson RE, Sundt TM Jr, Yaksh TL. Intracellular brain pH, indicator tissue perfusion, electroencephalography, and histology in severe and moderate focal cortical ischemia in the rabbit. *J Cereb Blood Flow Metab* 1986;6:71–78.
13. Crowell RM, Olsson Y, Klatzo I, Ommaya A. Temporary occlusion of the middle cerebral artery in the monkey: clinical and pathological observations. *Stroke* 1970;1:439–448.
14. Sutin JA. Clinical presentation of stroke syndromes. In: Kaplan PE, Cerullo LJ (eds). *Stroke Rehabilitation.* Stoneham, MA: Butterworth, 1986:1–36.
15. Mohr JP. Lacunes. *Stroke* 1982;13:3–11.
16. Millikan CH. Transient cerebral ischemia: definition and natural history. *Prog Cardiovasc Dis* 1980;22: 303–308.
17. Iannotti F, Hoff JT, Schielke GP. Brain tissue pressure in focal cerebral ischemia. *J Neurosurg* 1985;62:83–89.
18. Little JR, Sundt TM Jr, Kerr FWL. Neuronal alterations in developing cortical infarction: an experimental study in monkeys. *J Neurosurg* 1974;40: 186–198.
19. Sundt TM Jr, Grant WC, Garcia JH. Restoration of middle cerebral artery flow in experimental infarction. *J Neurosurg* 1969;31:311–322.
20. Olsson Y, Crowell RM, Klatzo I. The blood–brain barrier to protein tracers in focal cerebral ischemia and infarction caused by occlusion of the middle cerebral artery. *Acta Neuropathol* (Berlin) 1971;18:89–102.
21. Plum F, Posner JB, Alvord EC Jr. Edema and necrosis in experimental cerebral infarction. *Arch Neurol* 1963;9:563–670.
22. McDowell FH, Potes J, Groch S. The natural history of internal carotid and vertebral-basilar artery occlusion. *Neurology* (Minneapolis) 1961;11(2):153–157.
23. Oxbury JM, Greenhall RCD, Grainger KMR. Predicting the outcome of stroke: acute stage after cerebral infarction. *Br Med J* 1975;3:125–127.
24. Meyer FB, Piepgras DG, Sundt TM Jr, Yanagihara T. Emergency embolectomy for acute occlusion of the middle cerebral artery. *J Neurosurg* 1985;62:639–647.
25. Meyer FB, Sundt TM Jr, Piepgras DG, Sandok BA, Forbes G. Emergency carotid endarterectomy for patients with acute carotid occlusion and profound neurological deficits. *Ann Surg* 1986;203:82–89.
26. Davis KR, Taveras JM, New PFJ, Schnur JA, Roberson GH. Cerebral infarction diagnosis by computerized tomography. *AJR* 1975;124:643–660.
27. Masdeu JC, Azar-Bia B, Rubino FA. Evaluation of recent cerebral infarction by computerized tomography. *Arch Neurol* 1977;34:417–421.
28. Uemura K, Fukasawa H, Takahashi S. Computed tomography and pathology of cerebrovascular disease. 3. Cerebral infarction: its CT findings and pathological background. *Progr Comput Tomogr* (Tokyo) 1981;2: 533–543.
29. Yock DH, Marshall WH. Recent ischemic brain infarcts at computed tomography: Appearance pre- and postcontrast infusion. *Radiology* 1975;117:599–608.
30. Norton GA, Kishore PRS, Lin J. CT contrast enhancement in cerebral infarction. *AJR* 1978;131: 881–885.
31. Larson EB, Omenn GS, Loop JW. Computerized tomography in patients with cerebrovascular disease; impact of a new technology on patient care. *AJR* 1978;131:35–40.
32. Davis KR, Ackerman RH, Kistler JP, Mohr JP. Computed tomography of cerebral infarction: Hemorrhagic, contrast enhancement, and time of appearance. *Comput Tomogr* 1977;1:71–86.
33. Yamamoto Y, Satoh T, Asari S, Sadamoto K. Normal anatomy of cerebral vessels by computed angiotomography in the axial transverse plane. *J Comput Assist Tomogr* 1982;6:865–873.
34. Mikhael MA. Neuroradiology of stroke. In: Kaplan PE, Cerullo LJ (eds). *Stroke Rehabilitation.* Stoneham, MA: Butterworth, 1986:37–80.
35. Caplan LR. Care of the patient with acute stroke. In: Kaplan PE, Cerullo LJ (eds). *Stroke Rehabilitation.* Stoneham, MA: Butterworth, 1986:81–93.
36. Kertesz A, McCabe P. Recovery patterns and prognosis in aphasia. *Brain* 1977;100:1–18.
37. Spetzler RF. Extracranial–intracranial arterial anastomosis for cerebrovascular disease. *Surg Neurol* 1979; 11:157–161.
38. Wing SD, Norman D, Pollock JA, Newton TH. Contrast enhancement of cerebral infarcts in computed tomography. *Radiology* 1976;121:89–92.
39. Mikhael MA. Computerized tomography and brain tumors: a retrospective–prospective review. *Semin Neurol* 1981;1:137–148.
40. Kendall BE, Pullicino P. Intravascular contrast injection in ischemic lesions: Effect on prognosis. *Neuroradiology* 1980;19:241–243.

41. Djang WT, Drayer BP. Intracranial occlusive vascular disease. In: Taveras JM, Ferrucci JT (eds). *Radiology: Diagnosis–Imaging–Intervention*. Volume 3: Neuroradiology—Head & Neck. Philadelphia J.B. Lippincott, 1987:1–17.
42. Zimmerman RA, Bilaniuk LT. Computed tomography of acute intratumoral hemorrhage. *Radiology* 1980; 135:355–359.
43. Holland GN, Hawkes RC, Moore WS. Nuclear magnetic resonance (NMR) tomography of the brain: coronal and sagittal sections. *J Comput Assist Tomogr* 1980;4:429.
44. Young IR, Burl M, Clarke GJ, et al. Magnetic resonance properties of hydrogen: imaging the posterior fossa. *AJR* 1981;137:895–901.
45. Brant-Zawadzki M, Davis PL, Crooks LE, et al. NMR demonstration of cerebral abnormalities: comparison with CT. *AJR* 1983;140:847–854.
46. Buonanno FS, Pykett IL, Kistler JP, et al. Cranial anatomy and detection of ischemic stroke in the cat by nuclear magnetic resonance imaging. *Radiology* 1982; 143:187–193.
47. DeWitt LD, Grek AJ, Buonanno FS, Levine DN, Kistler JP. MRI and the study of aphasia. *Neurology* 1985;35:861–865.
48. Levy RM, Mano I, Brito A, Hosobuchi Y. NMR imaging of acute experimental cerebral ischemia: Time course and pharmacologic manipulations. *AJNR* 1983;4:238–241.
49. Young IR, Randell CP, Kaplan PW, et al. Nuclear magnetic resonance (NMR) imaging in white matter disease of the brain using spin-echo sequences. *J Comput Assist Tomogr* 1983;7:290.
50. Bryan RN, Willcott MR, Schneiders NJ, et al. Nuclear magnetic resonance evaluation of stroke. *Radiology* 1983;149:189–192.
51. Sipponen JR, Kaste M, Ketonen L, et al. Serial nuclear magnetic resonance (NMR) imaging in patients with cerebral infarction. *J Comput Assist Tomogr* 1983; 7:585.
52. Bradley WG. NMR imaging of the central nervous system. *Neurol Res* 1984;6:91.
53. Kato H, Kogure K, Ohtomo H, Tobita M, Matsui S, Yamamoto E, Kohno H. Correlations between proton nuclear magnetic resonance imaging and retrospective histochemical images in experimental cerebral infarction. *J Cereb Blood Flow Metab* 1985;5(2):267–274.
54. Bryan RN, Willcott MR, Schneiders NJ, Rose JE. NMR evaluation of stroke in the rat. *AJNR* 1983;4: 242.
55. Levy RM, Mano I, Brito A, Hosobuchi Y. NMR imaging of acute experimental cerebral ischemia: time course and pharmacologic manipulations. *AJNR* 1983; 4:238.
56. Buonanno FS, Pykett IL, Brady TJ, et al. Proton NMR imaging in experimental ischemic infarction. *Stroke* 1983;14:177–184.
57. Heiss W-D, Herholz K, Bocher-Schwarz HG, Pawlik G, Wienhard K, Steinbrich W, Friedmann G. PET, CT, and MR imaging in cerebrovascular disease. *J Comput Assist Tomogr* 1986;10(6):903–911.
58. Kinkel WR, Jacobs L, Polachini I, Bates V, Heffner RR Jr. Subcortical arteriosclerotic encephalopathy (Biswanger's disease). Computed tomographic, nuclear magnetic resonance, and clinical correlations. *Arch Neurol* 1985;42:951–959.
59. Kistler JP, Buonanno JS, DeWitt LD, David KR, Brady TJ, Fisher CM. Vertebral-basilar posterior cerebral territory stroke—delination by proton nuclear magnetic resonance imaging. *Stroke* 1984;15:417–426.
60. Bradley WG, Waluch V, Brant-Zawadzki M, et al. Patchy, periventricular white matter lesions in the elderly: a common observation during NMR imaging. *Noninv Med Imaging* 1984;1:35.
61. Elliott D, Vacca JG, Wehrli FW. Basic principles of magnetic resonance imaging. In: Mettler FA Jr, Muiroff LR, Kulkami MV (eds). *Magnetic Resonance Imaging and Spectroscopy*. New York: Churchill Livingstone, 1986:1–41.
62. DeLaPaz RL, New PFJ, Buonanno FS, et al. NMR imaging of intracranial hemorrhage. *J Comput Assist Tomogr* 1984;8:599.
63. Wintrobe MM (ed). *Clinical Hematology*. Philadelphia: Lea & Febiger, 1981.
64. Gomori JM, Grossman RI, Goldberg HI, Zimmerman RA, Bilaniuk LT. Intracranial hematomas: imaging by high-field MR. *Radiology* 1985;157:87–93.
65. Wehrli FW, MacFall JR, Shutts D, et al. Mechanisms of contrast in NMR imaging. *J Comput Assist Tomogr* 1984;8:369.
66. Carr DH, Brown J, Bydder GM, et al. Gadolinium-DTPA as a contrast agent in MRI: initial clinical experience in 20 patients. *AJR* 1984;143:215–224.
67. Klatzo I, Piraux A, Laskowski EJ. The relationship between edema, blood–brain barrier and tissue elements in a local brain injury. *J Neuropathol Exp Neurol* 1958;17:548–564.
68. Virapongse C, Mancuso A, Quisling R. Human brain infarcts: GD-DTPA-enhanced MR imaging. *Radiology* 1986;161:785–794.
69. Carr DH, Brown J, Leung W-L, Pennock JM. Iron and gadolinium chelates as contrast agents in NMR imaging: preliminary studies. *J Comput Assist Tomogr* 1984;8:385.
70. McNamara MT, Brant-Zawadzki M, Berry I, et al. Acute experimental cerebral ischemia: MRI enhancement using Gd-DTPA. *Radiology* 1986;158:701–704.
71. Runge VM, Stewart RG, Clanton JA, Jones MM, Lukehart CM, Partain CL, James AE Jr. Work in progress: potential oral and intravenous paramagnetic NMR contrast agents. *Radiology* 1983;147:789–791.
72. Moniz E. *L'angiographie cerebrale*. Paris: Masson et Cie, 1934.
73. Brant-Zawadzki M, Gould R, Norman D, Newton TH, Lane B. Digital subtraction cerebral angiography

by intraarterial injection: comparison with conventional angiography. *AJNR* 1982;3:593–599.
74. Chilcote WA, Modic MT, Pavlicek WA, Little JR, Furlan AJ, Duchesneau PM, Weinstein MA. Digital subtraction angiography of the carotid arteries: a comparative study in 100 patients. *Radiology* 1981; 139:287–295.
75. Toole JF, Janeway R, Choi K, et al. Transient ischemic attacks due to arteriosclerosis. *Arch Neurol* 1975;32:5–12.
76. Lyons C. Progress report of the joint study on extracranial arterial occlusions. In: Millikan CH, Siekert RG, Whisnant JP (eds). *Cerebral Vascular Diseases: Fourth Conference*. New York: Grune and Stratton, 1965.
77. Martin MJ, Sayre GP, Whisnant JP. Incidence of occlusive vascular disease in the extracranial arteries contributing to the cerebral circulation. *Trans Am Neurol Assoc* 1960;85:103–105.
78. Martin MJ, Whisnant JP, Sayre GP. Occlusive vascular disease in the extracranial cerebral circulation. *Arch Neurol* 1960;3:530–538.
79. Fields WS. Progress report of the joint study on extracranial arterial occlusion. In: Millikan CH, Siekert RG, Whisnant JP (eds). *Cerebral Vascular Diseases: Fifth Conference*. New York: Grune and Stratton, 1966.
80. Metz H, Murray-Leslie RM, Bannister RG, Bull JWD, Marshall J. Kinking of the internal carotid artery. *Lancet* 1961;1:424–426.
81. Fox JL. Cerebral arterial revascularization: the value of repeated angiography in selection of patients for operation. *Neurosurgery* 1978;2:205–209.
82. Mikhael MA. Transient spasm of carotid siphon complicating ruptured cranial dermoid cyst. *Radiology* 1982;144:824.
83. O'Leary DH, Persson AV, Clouse ME. Noninvasive testing for carotid artery stenosis. 1. Prospective analysis of three methods. *AJNR* 1981;2:437–442; *AJNR* 1981;137:1189–1194.
84. Ovitt TW, Christenson PC, Fisher HD III, et al. Intravenous angiography using digital video subtraction. X-ray imaging. *AJNR* 1980;1:387–390.
85. Mistretta CA, Crummy AB, Strother CM. Digital angiography: a perspective. *Radiology* 1981;139:273–276.
86. Bradley WG, Waluch V. Blood flow: Magnetic resonance imaging. *Radiology* 1985;154:443–450.
87. Young IR, Bydder GM, Hall AS, et al. NMR imaging in the diagnosis and management of intracranial angiomas. *AJNR* 1983;4:837.
88. Worthington BS, Kean DM, Hawkes RC, et al. NMR imaging in the recognition of giant intracranial aneurysms. *AJNR* 1983;4:835.

7. RADIOLOGY OF DURAL SINUS THROMBOSIS

Chat Virapongse

Jan Malat

Mohammad Sarwar

Sagittal Sinus Thrombosis

INCIDENCE

Sagittal sinus thrombosis (SST) and SST-related cerebral sinovenous occlusive disease (SOD) are not uncommon. In one large autopsy series (8578 cases), 40 cases (.46%) of sinus thrombosis were found [1]. But in another autopsy series (189 cases), in which the dural sinuses and the venous system were examined in detail, the primary cause of death due to SOD was as high as 9% [2]. Thus far, there is no reliable documentation of the true incidence of SOD in vivo [3]; however, with the advent of computerized tomography (CT), magnetic resonance imaging (MRI), and more sensitive and specific radionuclide imaging techniques, the true incidence may eventually be known.

ANATOMY

The venous drainage of the head can be divided into three distinct groups [4]: a). superficial, which drains the scalp, b) intermediate, consisting of diploic, emissary, and meningeal veins, and c) the deep group, consisting of veins that drain the cerebral parenchyma (figure 7–1) This last group in turn can be classified into superficial and deep. Of these, the sperficial group of veins runs a course over the brain surface, usually emptying into one of the dural venous sinuses.

The deep cerebral venous system consists of small intramedullary veins, normally invisible on angiography, that drain either centrifugally or centripedally, depending on their relative depth in the brain parenchyma. Connections between the three groups of veins—superficial (scalp), intermediate, and deep veins that drain the brain—are extensive, and thus representing a large venous collateral reservoir that can be called upon in venous occlusion. The scalp veins are connected to the intermediate group of veins by the emissary veins which traverse the calvarium. Of the intermediate group, the diploic veins are located in the diploe between the two tables of the skull and drain into major dural sinuses. The meningeal veins are located between the inner table of the skull and the outer dura. They usually run parallel with the meningeal arteries, and communicate with the emissary, diploic, and cerebral veins. The meningeal veins also provide tributaries that drain the dura (including the outer and inner dura that surround the superior sagittal sinus) and falx [5].

The dural venous sinuses receive contributions from the brain, meninges, and diploe as well as emissary and scalp veins. They can be divided into a) the posterosuperior and b) the anteroinferior groups of veins.

a) The posterosuperior group comprises the superior sagittal, inferior sagittal, straight, transverse, sigmoid, and occipital sinuses (figure 7–1B). The superior sagittal sinus (SSS) begins anteriorly at the crista galli and runs posteriorly to terminate at the torcular Herophili. The superior sagittal sinus continues to grow in size as it passes posteriorly, until it reaches a maximal diameter of about 1–1.5 cm at the torcular Herophili. The most proximal portion of the superior sagittal sinus may at times be hypo- or aplastic. This finding, therefore, should not be confused with sinus thrombosis on angiography. The torcular Herophili and the transverse sinuses are subject to large individual variation. Frequently, one transverse sinus is hypo- or aplastic with one side assuming virtually the entire burden of the venous drainage. As the transverse and sigmoid sinuses run downward along the sigmoid sulcus, each sigmoid sinus lies contiguous to the mastoid, thus increasing the susceptibility of this portion of the dural sinus to thrombosis from an ear infection.

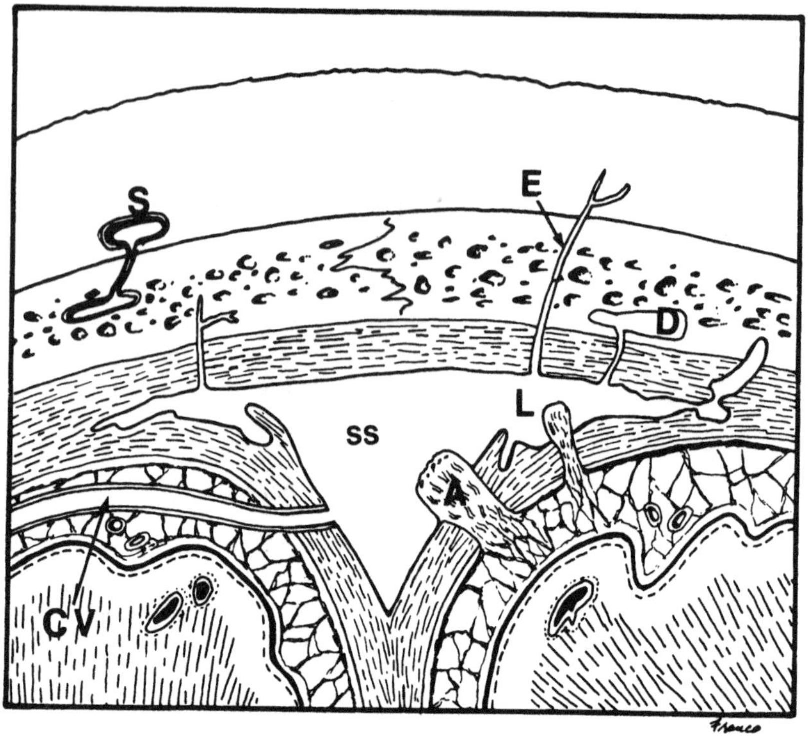

FIGURE 7–1A

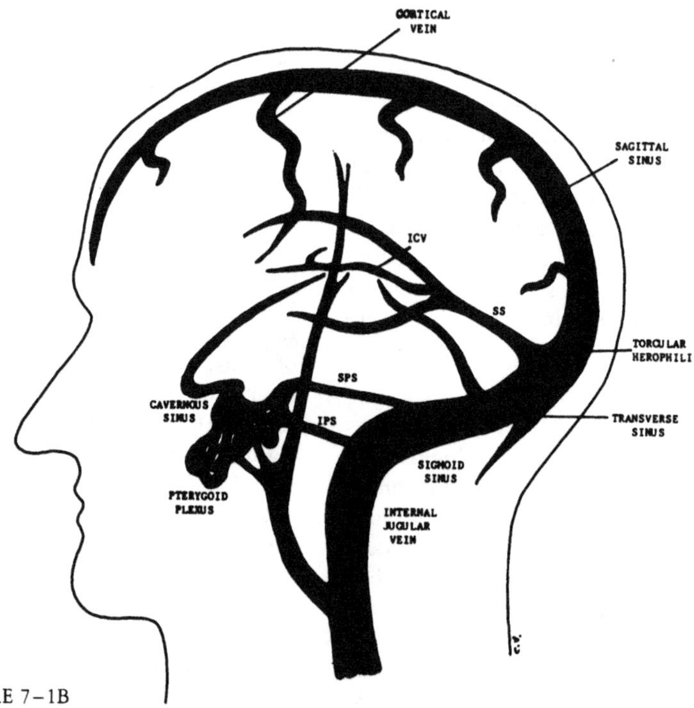

FIGURE 7–1B

FIGURE 7–1. (A) Cross-sectional diagram of the sagittal sinus (SS) showing the relationship of the sinus to all three groups of collateral venous channels in sagittal sinus thrombosis. Cortical vein = CV, lateral venous lacuna = L, scalp vein = S, diploic vein = D, emissary vein = E, arachnoid granulation = A. (B) Diagram, lateral view of the various dural sinuses. ICV = internal cerebral vein, IPS = inferior petrosal sinus, SPS = superior petrosal sinus, SS = straight sinus. (C & D) Axial CT without contrast showing the normal triangular configuration of the sagittal sinus (large arrows). (E & F) Axial CT with contrast showing enhancement of the superior sagittal sinus (large arrows), vein of Galen (arrowhead) and internal cerebral veins (white arrow). (G) Axial MRI showing lack of signal (hypointensity) emitted from the cavernous sinus (solid arrow). The signal emitted from the carotid artery (open arrow) is even less, due to faster flow in the carotid artery. (H) Sagittal MRI showing the hypointense signal arising from the sagittal sinus (large arrow), straight sinus (open arrow), vein of Galen (small arrow), and basal vein of Rosenthal (arrowhead).

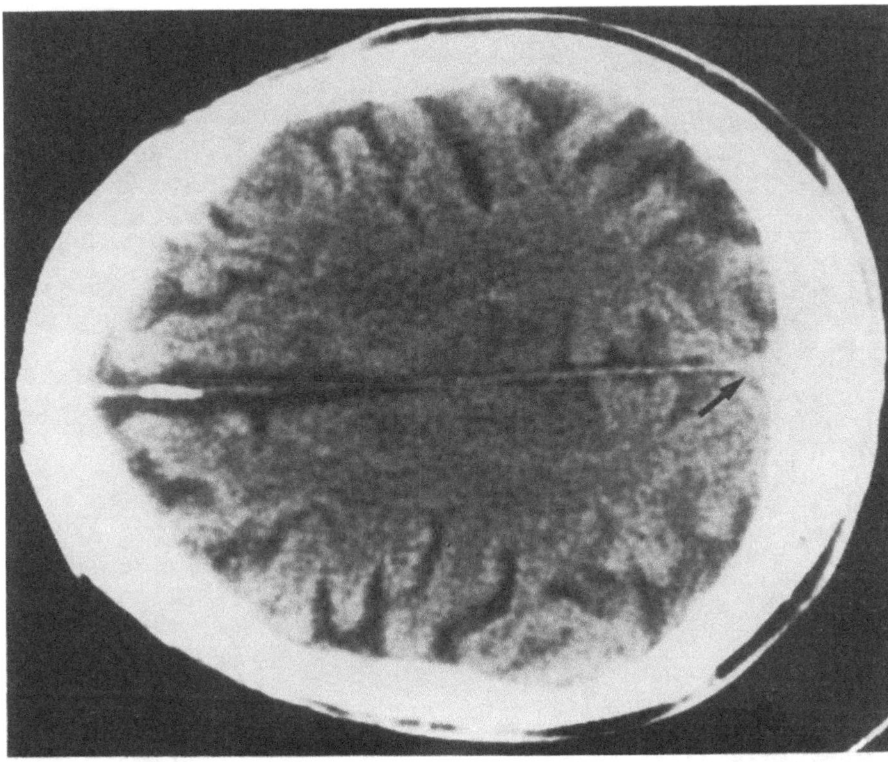

FIGURE 7–1D

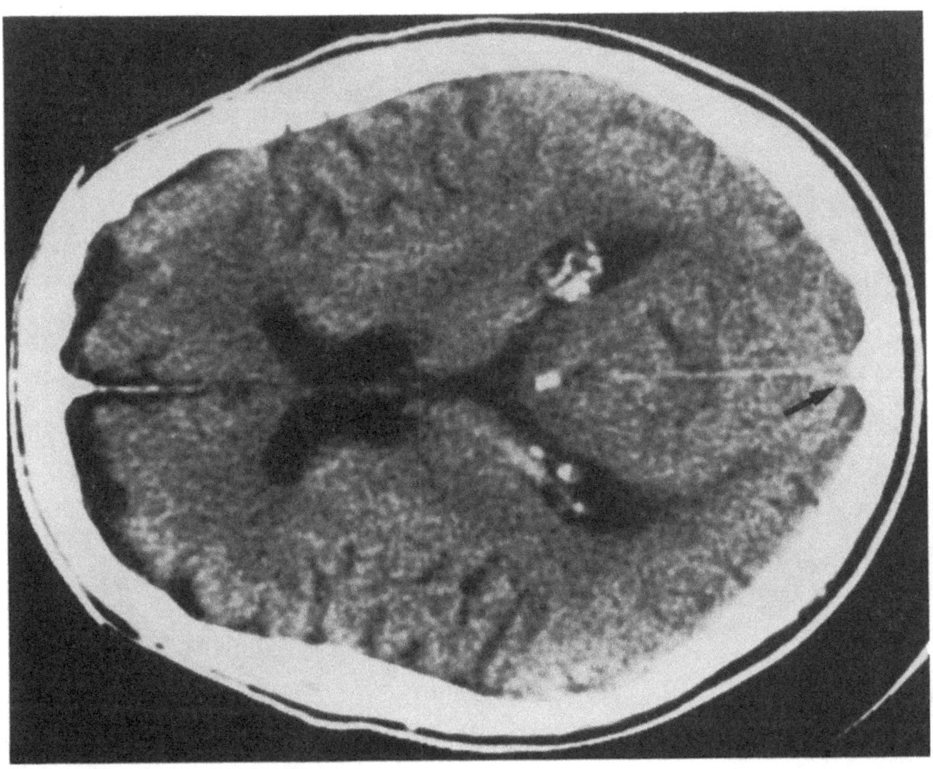

FIGURE 7–1C

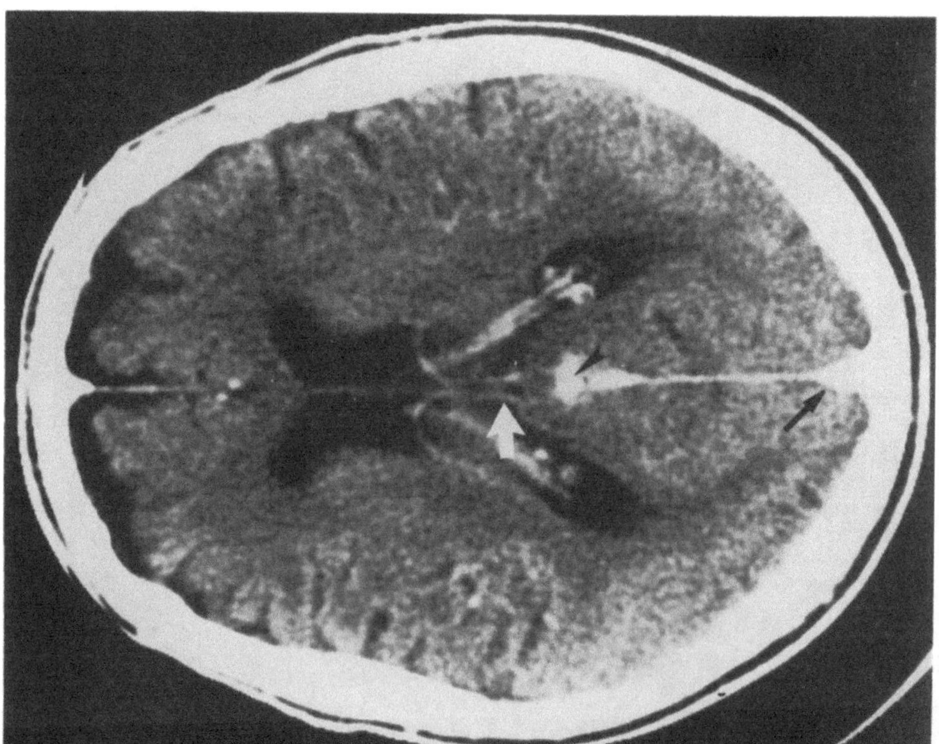

FIGURE 7–1E

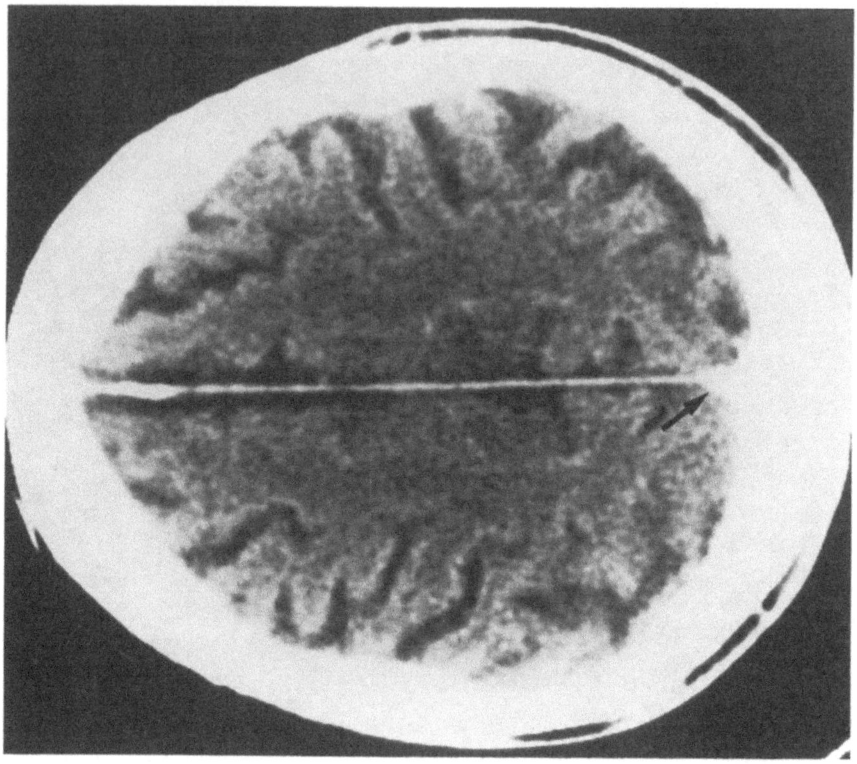

FIGURE 7–1F

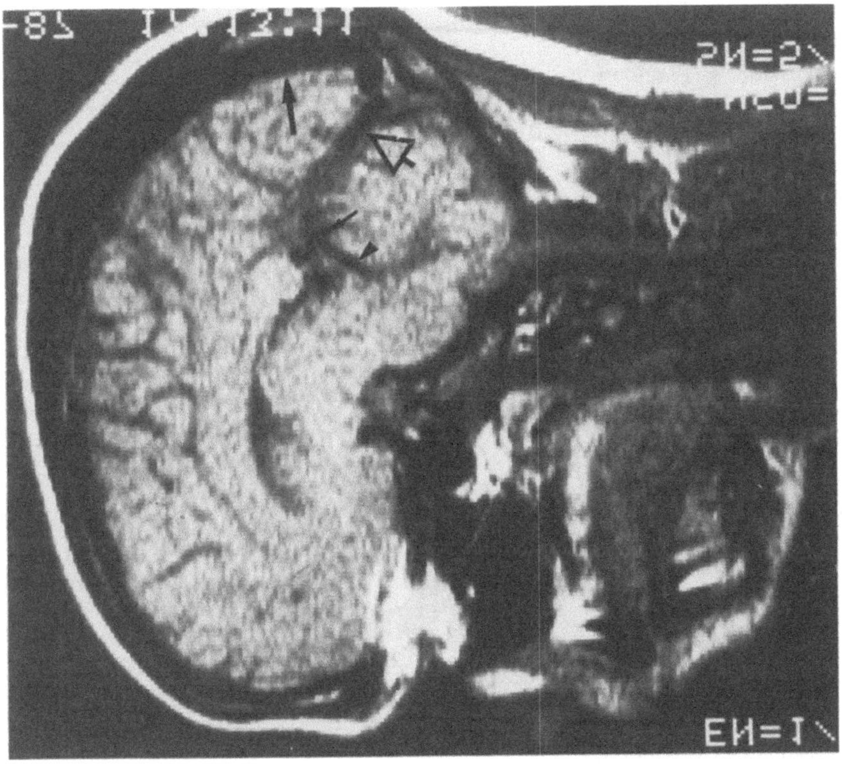

FIGURE 7–1H

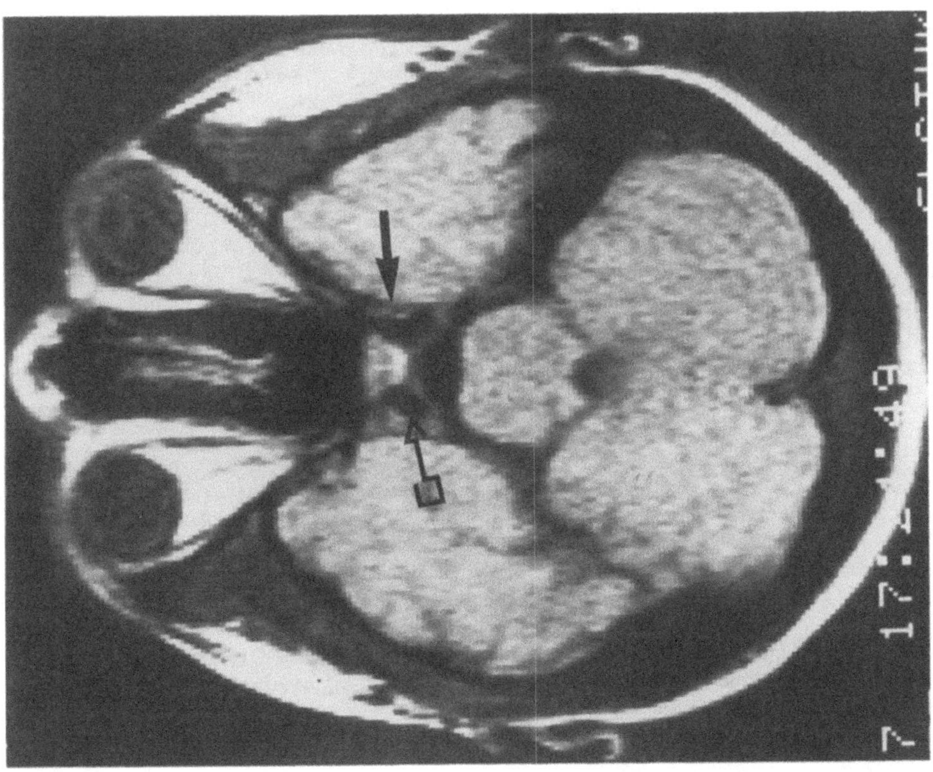

FIGURE 7–1G

b) The anteroinferior group consists of the cavernous sinuses, superior and inferior petrosal sinuses, and the basilar venous plexus (figure 7–2). (Cavernous sinus thrombosis is a rare disease. It is seldom associated with SST and is discussed in a separate section.)

The blood from the sinuses flows either posteriorly into sigmoid sinuses and then into the jugular veins, or anteriorly into the cavernous sinuses, orbital veins, and pterygoid plexus.

Superior Sagittal Sinus Morphology. The endothelial-lined lumen of the SSS is irregular in contour. Along its inferior surface there are numerous furrows, bands, and bridges referred to as the *chordae Willisii* (figure 7–3). Their function is unclear, but they are thought to help prevent reflux of blood into the cortical veins. Because the cortical veins turn acutely posteriorly (along the direction of the SSS venous blood flow) to empty their contents into the SSS, some reflux of venous blood into them may be prevented [6]. The cortical veins enter the inferior aspect of the SSS (figures 7–7, 7–3C), below but seldom connecting with lateral venous lacunae [7]. The latter are lateral extensions of the SSS (figures 7–3D and E), usually in three pairs located respectively at the frontal, parietal, and occipital portions of the SSS [8, 9]. In the elderly, the lateral lacunae can become linked, forming a chain [9]. They contain an unusual abundance of arachnoid granulations (four fifths) [7] compared to the SSS itself (figure 7–3F); consequently, the venous blood in the lacunae is probably more diluted by CSF than the SSS, which may explain why the lateral lacunae are less affected by postmortem clot than is the SSS.

Perhaps another important dural collateral pathway in SST is the *cavernous tissue*. First described by Balo [10], this spongy tissue, which resembles the corpora cavernosa of the penis, consists of non-endothelial-lined spaces about 50–100 microns in size located usually at the dorsal aspect of the SSS, and surrounding the straight sinus and torcular Herophili (figure 7–4). When filled with blood, these spaces can protrude into the lumen of the SSS to such an extent that Balo [10] thought they regulated venous blood flow in the SSS (figure 7–4C). No supportive evidence has so far been furnished to validate this speculation; however, the theory that cavernous tissue may represent another collateral pathway has been substantiated

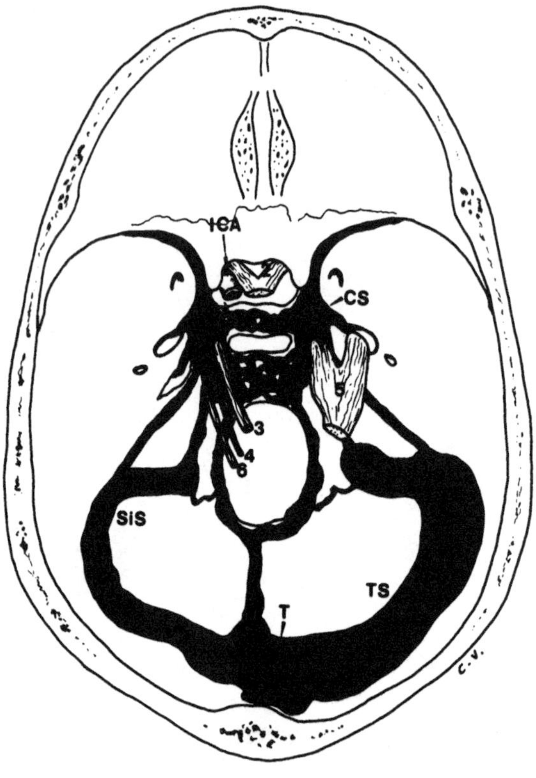

FIGURE 7–2. Diagram, superior view of the basal dural sinuses. CS = cavernous sinus, ICA = internal carotid artery, SIS = sigmoid sinus, T = torcular Herophili, TS = transverse sinus 2–6 = cranial nerves.

FIGURE 7–3. (A) Superior view of a dissected superior sagittal sinus demonstrating the multiple bands and bridges known as the *chordae Willisii* (arrows). (B) Inferior view of the superior sagittal sinus showing large accumulation of adipose tissue (arrow) surrounding the sinus. (C) Cast of the superior sagittal sinus, superior view, showing the relationship of the sagittal sinus (A & B), lateral venous lacuna (C) and the cortical veins (D). Note that the cortical veins enter the inferior aspect of the sagittal sinus below the lateral lacuna (C). (With permission. [7]) (D) Cast of the superior sagittal sinus (S) and transverse sinus (T) in a one-month-old infant, superior view, showing early development of the lateral lacuna (large arrows). Note the small meningeal veins (small arrow) communicating with the lateral lacunae. (With permission. [7]) (E) Cast of the superior sagittal sinus in a three-month-old infant showing considerable growth of the lateral lacuna (arrow). (With permission. [7]) (F) Dorsal view of dissected specimen of the superior sagittal sinus and lateral lacunae showing the great abundance of the pacchionian granulations (arrow) in the lateral lacunae compared to the sagittal sinus. (In this specimen the ratio is about 100:1.) (With permission [7]).

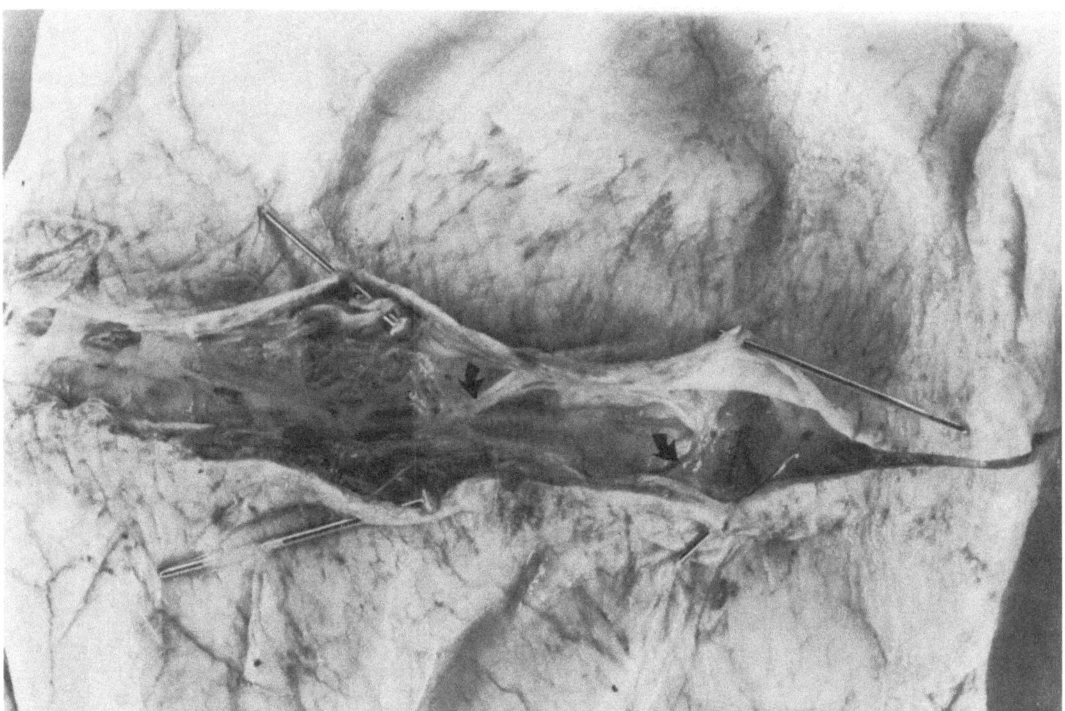

FIGURE 7–3A

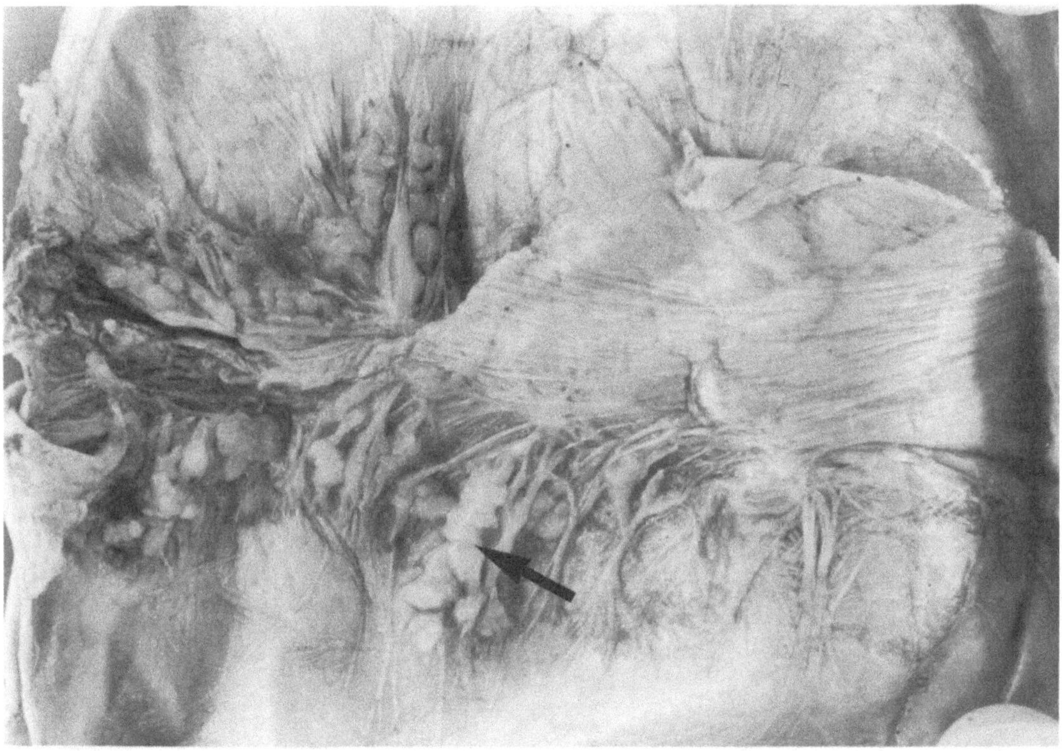

FIGURE 7–3B

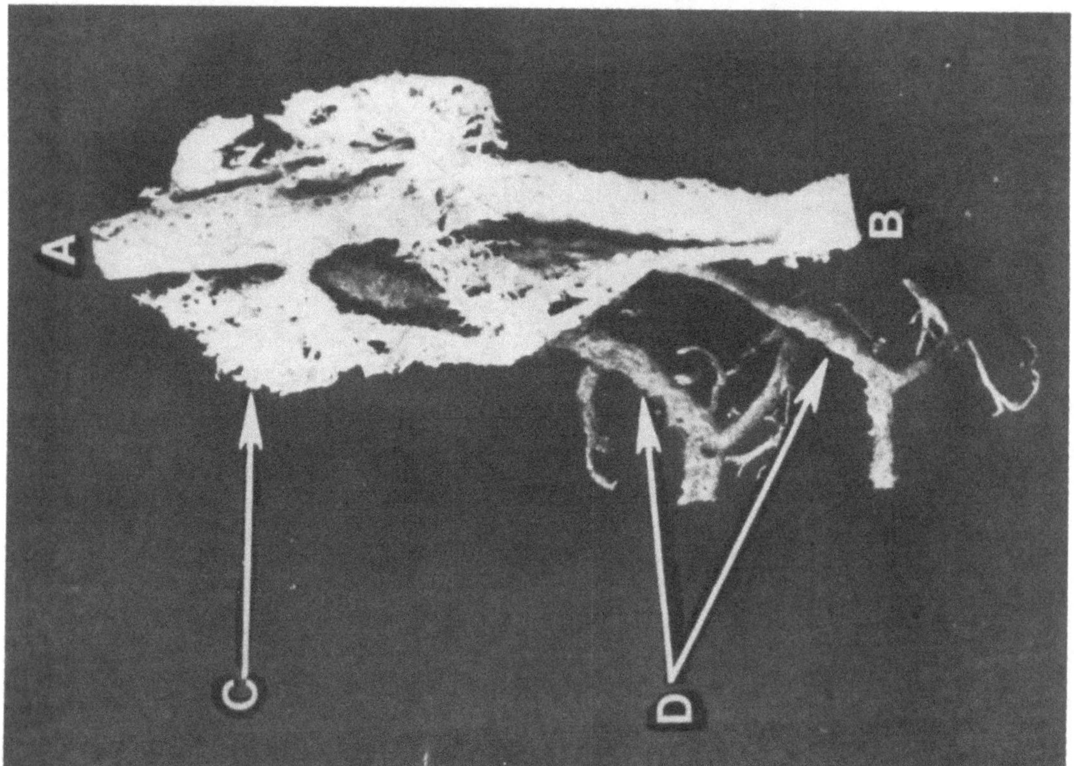

FIGURE 7–3C

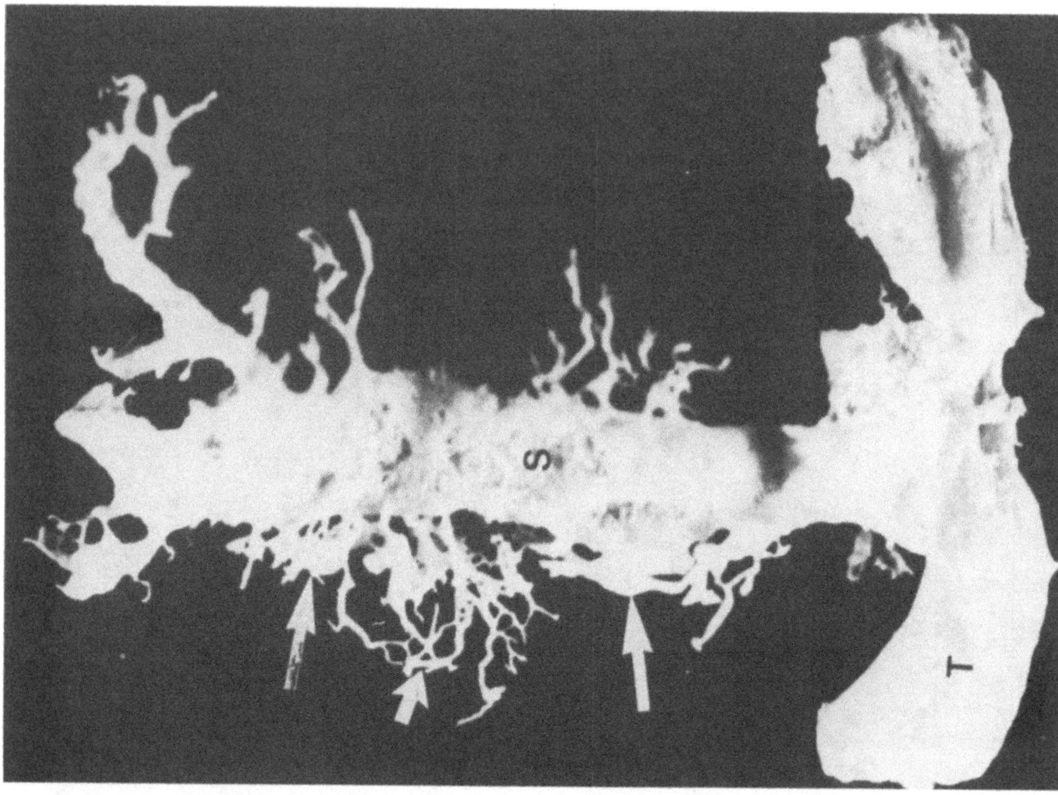

FIGURE 7–3D

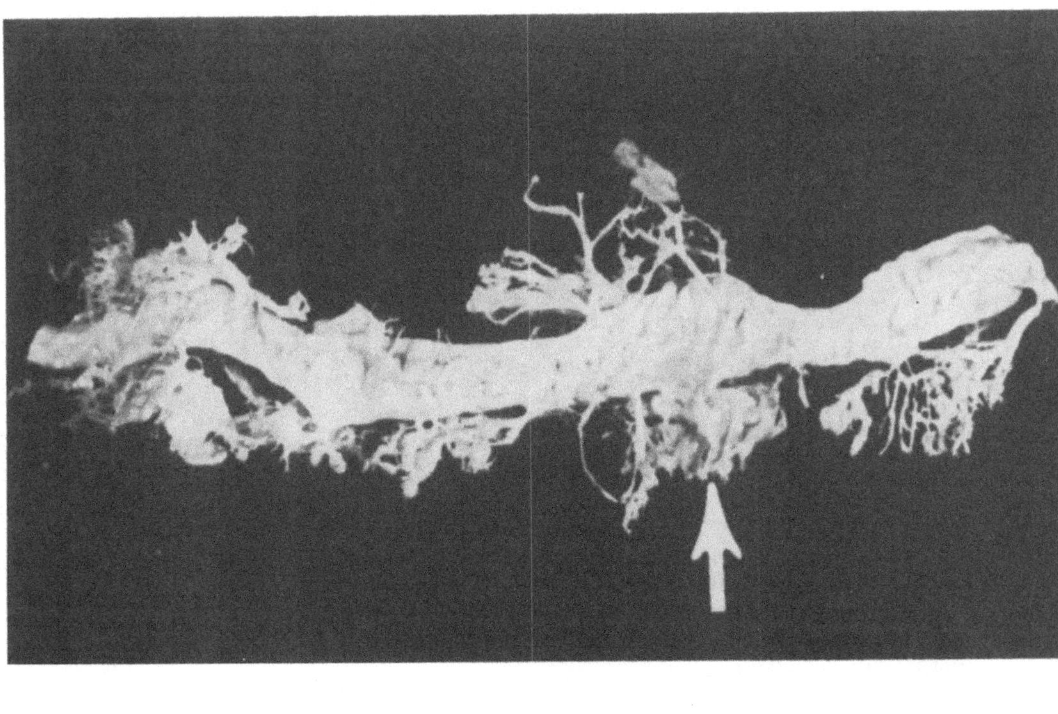

FIGURE 7–3E

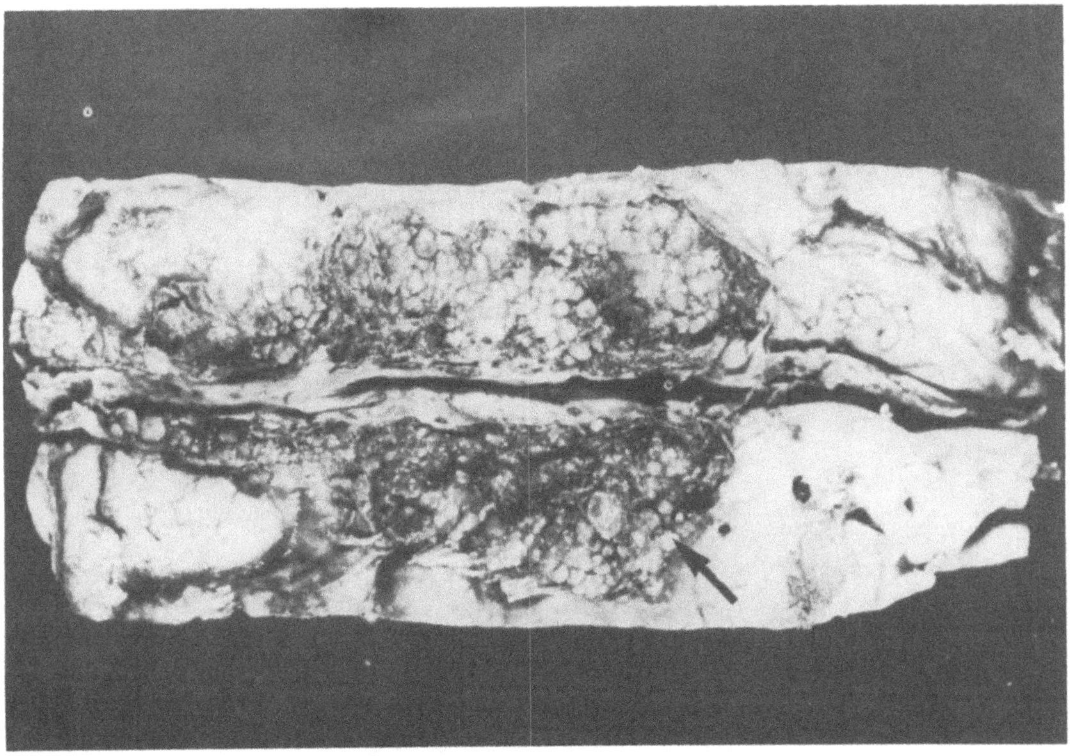

FIGURE 7–3F

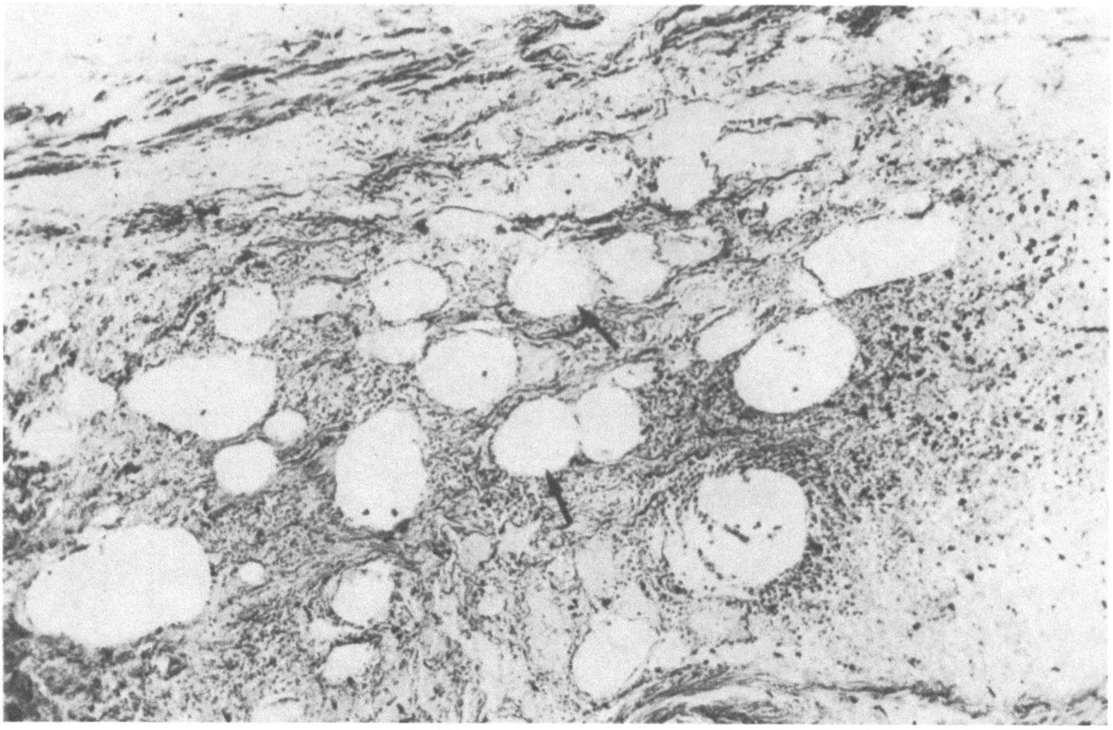

FIGURE 7–4A

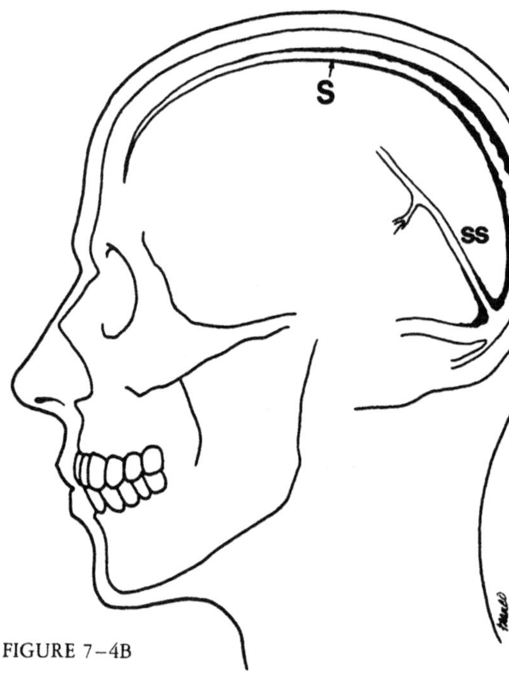

FIGURE 7–4B

FIGURE 7–4. (A) Photomicrograph of the cavernous spaces (arrows) along the dorsal wall of the superior sagittal sinus. (With permission. [7]) (B) Diagram, lateral view, showing the distribution of cavernous tissue (shaded black). Note the abundance of this tissue at the torcular Herophili and the superior sagittal sinus above the torcular. S = Superior sagittal sinus, SS = straight sinus. (C) Dorsal view of a dissected specimen of the superior sagittal (S) and transverse sinuses (T). The dorsal dural leaf has been reflected showing the degree of protrusion of the cavernous tissue into the sinus lumen (arrows). (With permission. [7]) (D) A wooden peg has been introduced into the lumen of the transverse and the superior sagittal sinus (white arrow). The dorsal dural leaf of the torcular Herophili is shown to contain numerous meningeal vascular channels (black arrow). (With permission.) [7]

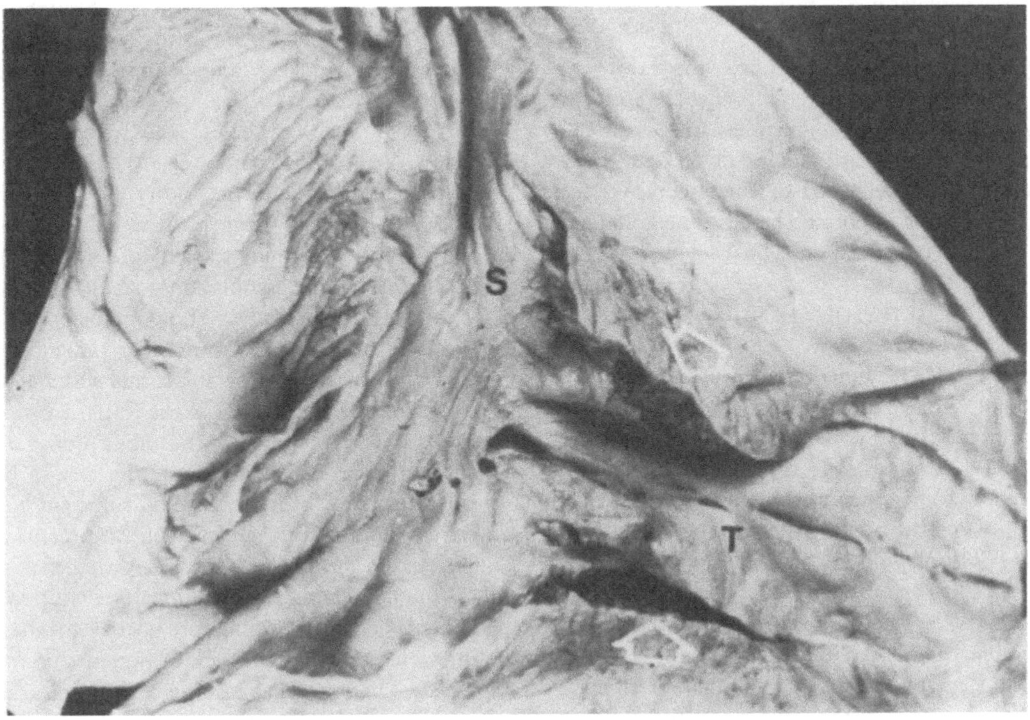

FIGURE 7–4C

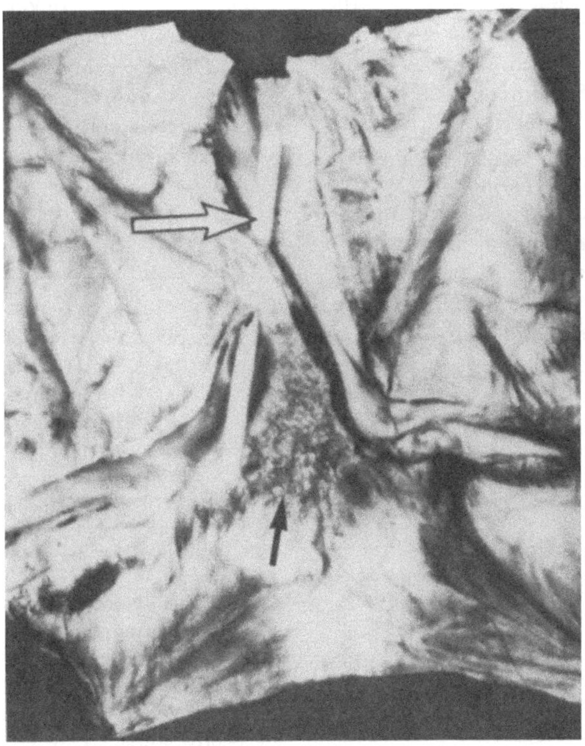

FIGURE 7–4D

by work with Vinylite casts that showed connections between the vascular mesh, meningeal veins, and the sagittal sinus [4].

In summary, the lateral lacunae, the cavernous tissue, and the meningeal veins probably represent important dural collateral channels that help to bypass a clot in the early stages of SST, thereby diminishing the effects of SST on the cerebral venous circulation. This is important because the collateral channels can be visualized on CT.

PATHOGENESIS, ETIOLOGY, AND CLINICAL FEATURES

Certain anatomic and functional features peculiar to the cerebral dural venous sinuses may predispose them to thrombosis: a) venous blood flow is relatively slow in the dural sinuses; b) the chordae Willisii of the SSS may promote flow turbulence or even stasis; and c) in infants the pliability of the cranial vault in the supine position produces an inward bony displacement, partially compressing the distal SSS [11] (figure 7–5). (In one eight-day-old infant studied with cerebral angiography, Newton and Gooding [11] found *nonfilling* of the SSS due to bony compression in the supine position. But when the SSS was examined by venous sinography it was patent, albeit markedly compressed by the parieto-occipital overlap. Other predispositions to thrombosis, common to the venous system, include a) changes in blood viscosity (such as cardiac failure), b) presence of coagulating agents (such as the release of thromboplastin in brain injury), and c) the loss of integrity of the sinus endothelium due to trauma or inflammatory disease.

In SST, some ventriculomegaly would be expected because of the venous obstruction and failure of CSF absorption. In fact, the relationship of this disease to changes in blood and CSF volume is complex [12]. The adult intracranial cavity possesses a finite volume; its contents are brain, blood, and CSF in varying proportions. When the brain volume is unchanged, then blood and CSF have an inverse relationship (the Monro–Kellie principle) (figure 7–6). The phenomenon of CSF absorption across the arachnoid granulations can be described by

Absorption (CSF bulk flow) = Pressure CSF- Pressure sagittal sinus/Resistance across the archnoid granulations [13]

Due to the presence of an unyielding adult skull, a positive CSF pressure gradient is always maintained across the archnoid granulations. This has been well documented in human and animal experiments [13]. Another variable to be considered is the fact that, with the development of raised intracranial pressure, alternative routes of CSF absorption are brought into play which also help to diminish the likelihood of hyprocephalus. Further, brain volume may increase due to brain edema as well as intraparenchymal hemorrhages. Hydrocephalus usually occurs in a child less than three years of age [13]. This fact has been attributed to the distensibility of the pediatric skull, which is able to dampen the raised intracranial pressure, thereby reducing the CSF pressure gradient across the arachnoid granulations, resulting in CSF accumulation. Also, the archnoid granulations are immature at birth and few in number. They become more widely distributed by about four years of age.

Prognosis in SST depends on the initial location of the thrombus, the rapidity of thrombus propagation, the degree of venous obstruction, the area of brain involvement, the ability to develop sufficient collateral circulation, and the rapidity of clot organization and recanalization [15]. Age is also important. The infant brain tolerates such an insult poorly; this is perhaps related to the immature, partly myelinated brain, which has the tendency to liquefy due to the high water content. The masking of the clinical signs of raised intracranial pressure due to the distensible pediatric skull is also another factor [3]. Collateral circulation occurs mainly through the large cortical veins, which drain blood in a downstream fashion (flow reversal) to opened sinuses at the base of the skull. Some of the venous blood is also rerouted to the deep venous system via medullary veins as well as the dural collateral circulation.

Tissue survival depends on whether the develop-

TABLE 7–1. Causes of Sinus Thrombosis [14]

A. Septic	
B. Nonseptic	Pregnancy/postpregnancy
	Skull trauma
	Cardiac disease
	Subarachnoid hemorrhage
	Craniotomy
	Oral contraceptives
	Miscellaneous

FIGURE 7–5. Noncontrast axial CT from inferior to superior, showing the effects of anterior displacement of the occipital bone on the superior sagittal sinus. (A & B) Note displacement of the occipital bone in the supine position (arrows). The torcular Herophili and sagittal sinus are poorly visible. (C & D) Sections above the occipital bone (same infant) showing relative prominence of the sagittal sinus (open arrow).

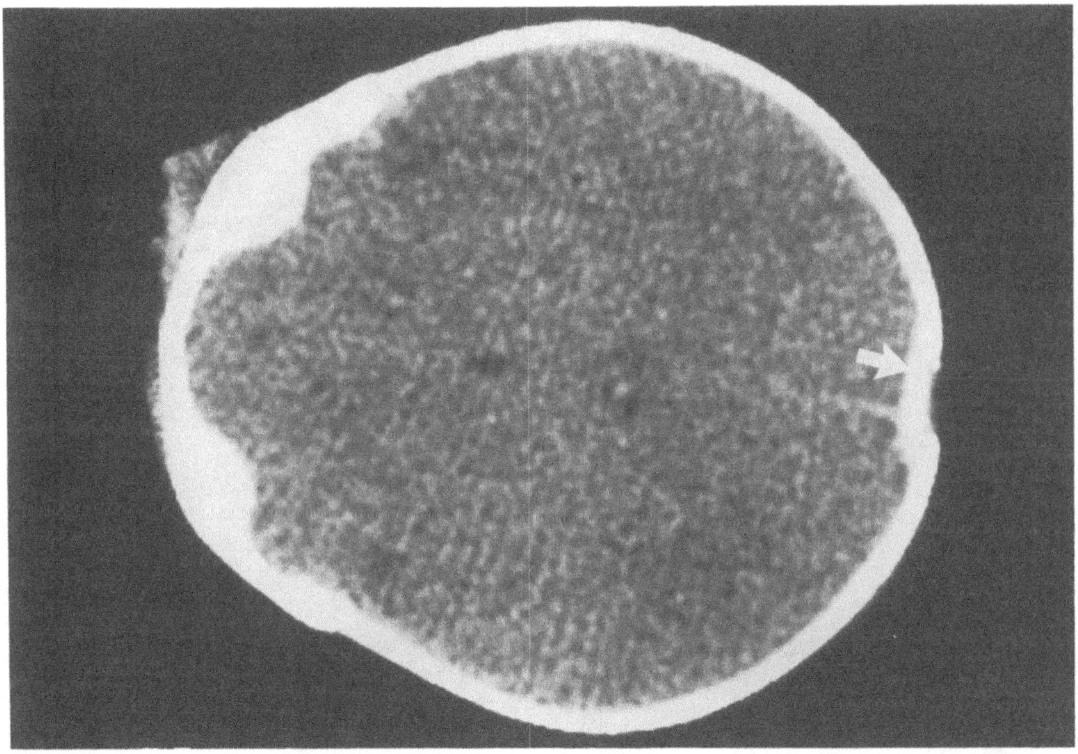

FIGURE 7–5B

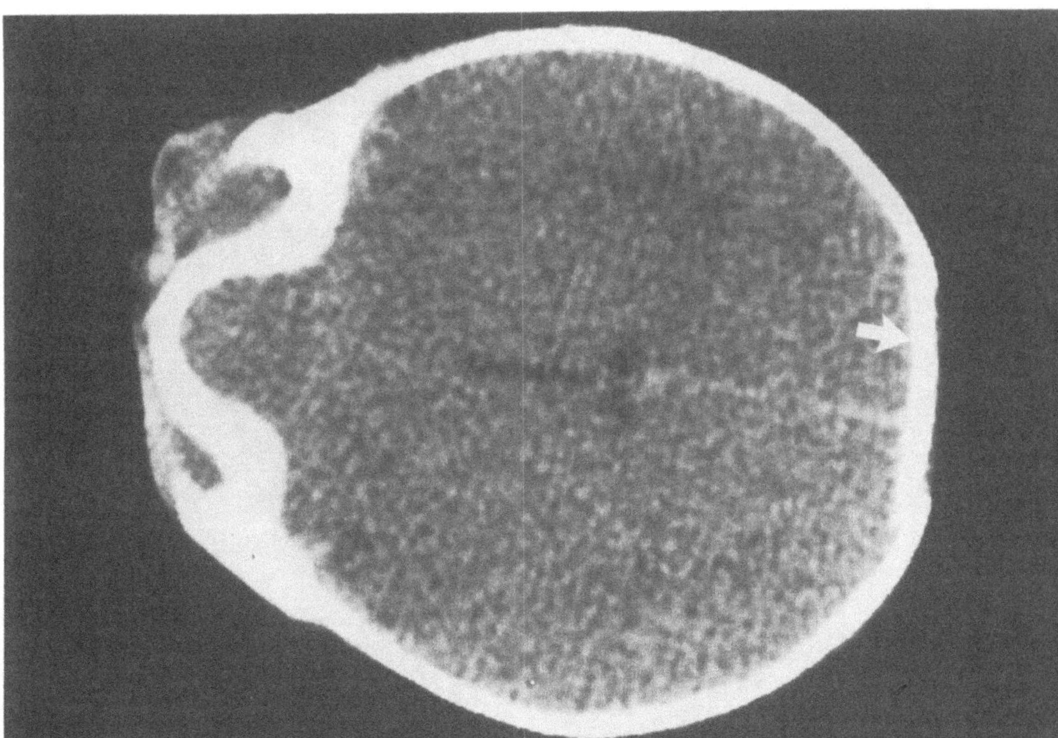

FIGURE 7–5A

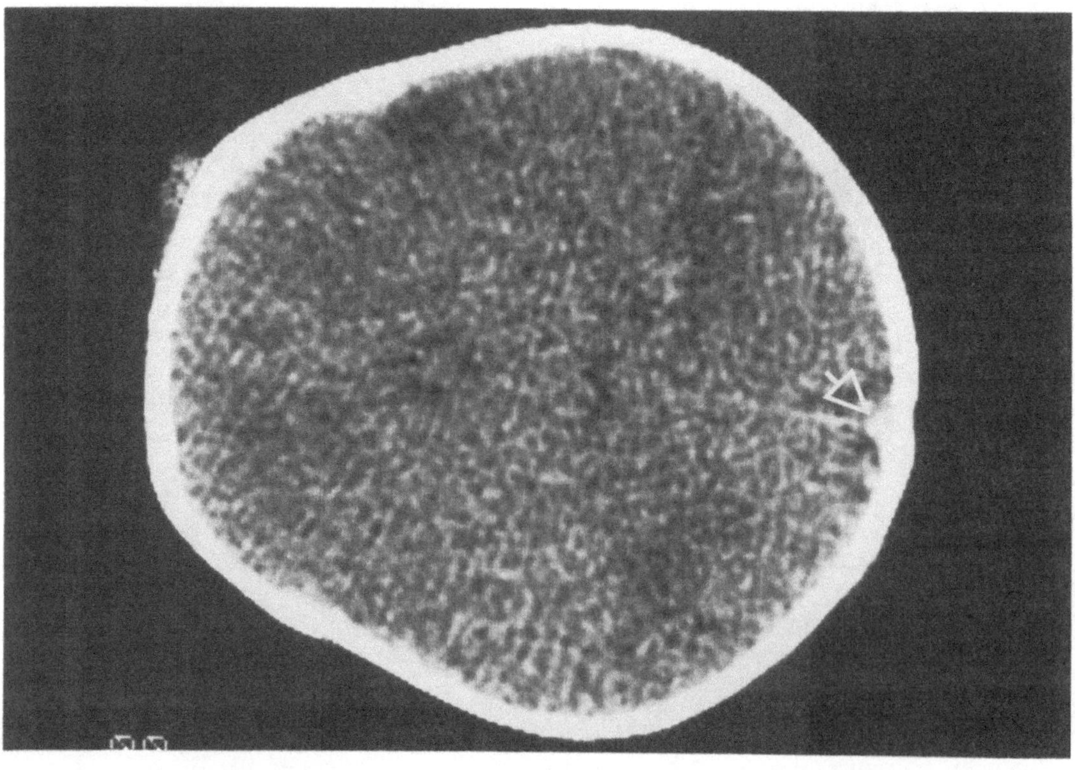

FIGURE 7–5C

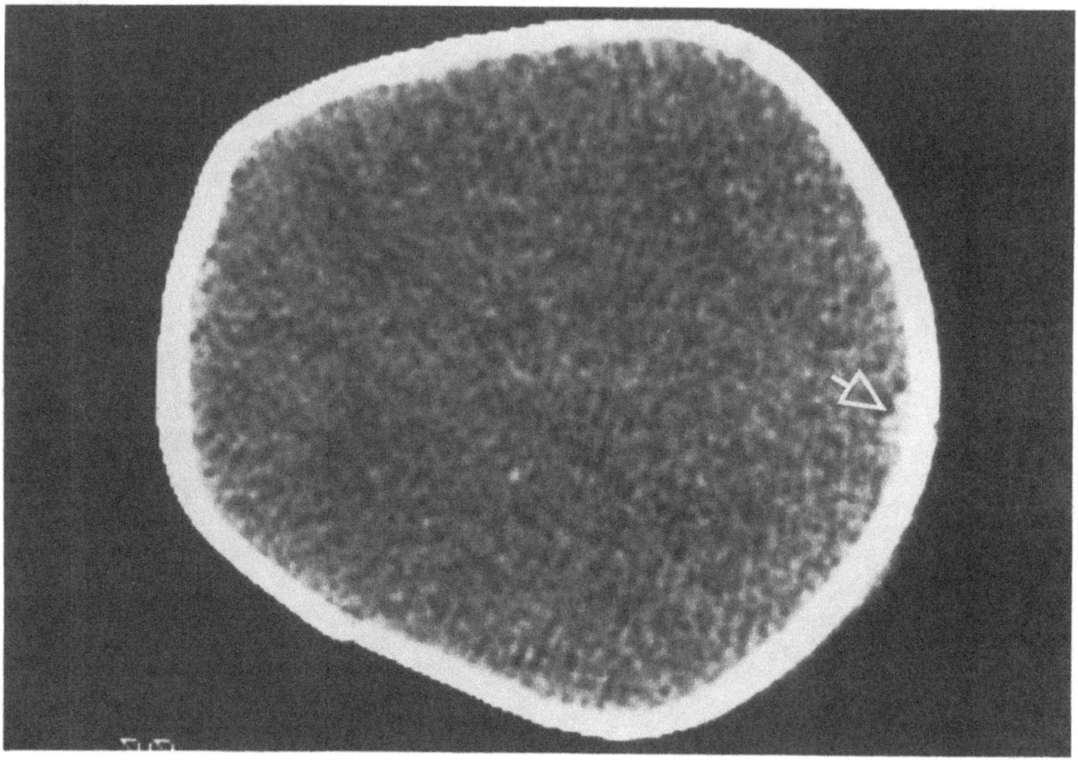

FIGURE 7–5D

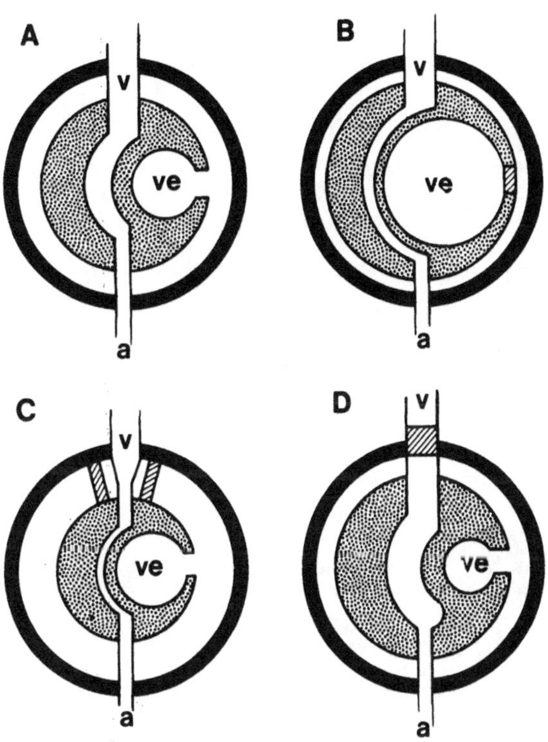

FIGURE 7-6. Diagram demonstrating the Monro–Kellie principle. (A) Normal. Dotted area = brain, darkened area = skull, A = cerebral artery, V = vein, VE = ventricle. (B) In obstructive hydrocephalus (obstructive element = oblique lines), the ventricle is expanded at the expense of the cerebral blood volume. (C) In communicating hydrocephalus, the subarachnoid space over the brain surface is expanded along with the ventricle at the expense of the cerebral blood pool. (D) In venous occlusion, cerebral blood volume is expanded at the expense of ventricular volume.

ment of collateralization is sufficiently rapid to prevent ischemia and hemorrhage due to blood stasis. Once survival is achieved, recanalization is probably the natural outcome. As shown by Beck and Russel [16] in experimental animals, regardless of any soft material placed occlusively in the SSS, recanalization will eventually take place. In humans this process may require about one to two weeks [15].

Symptomatology depends on the severity of venous stasis and its sequelae on the brain. In a series of 97 patients with sinus thrombosis, Yasargil and Damur [14] reported the most common clinical manifestation to be hemiparesis (69%), followed by seizures (45%), papilledema (41%), and coma (36%).

Bengin intracranial hypertension (BIH) refers to the condition of increased intracranial pressure in the absence of a space-occupying lesion or hydrocephalus. This condition has been linked with cerebral SOD. Patients present with features of raised intracranial pressure, i.e., headache, obscuration of vision, diplopia, tinnitus, papilledema, and occasionally sixth-nerve palsy. The prognosis is usually good, the clinical symptomatology subsiding within a few weeks or months. Patients fall into two main groups: those with *otitic hydrocephalus* and those with no known cause. The term *otitic hydrocephalus*, as coined by Symonds in 1931, is in fact a misnomer because there is absence of hydrocephalus. The condition refers to the development of BIH in a patient with an ear infection. Due to the contiguous relationship of the sigmoid sinus to the mastoid, inflammatory spread from the mastoid to produce lateral sinus thrombosis is not uncommon. In one surgical series (33 patients with ear infection), as many as 83% had lateral sinus thrombosis [12].

IMAGING TECHIQUES IN SST AND SST-RELATED SOD

Computerized Tomography (CT). The capability of CT to diagnose venous thrombosis was first recognized by Wendling (1978) [17], when he described an 18-month-old child with otitis who developed deep cerebral venous thrombosis. On CT, the vein of Galen and straight sinus were hyperdense and enhanced diffusely. Several months later, Buonanno [18] reported 11 adult patients who exhibited several CT signs of SST and SST-related SOD (including the empty delta and the cord sign). Thus far, as many as 24 separate publications have described the CT findings of this disease [15].

The CT findings of SST and SST-related SOD are best classified pathologically: a) those related to the blood clot, b) those related to the venous collateral formation, and c) those related to the effects of the sinus thrombosis on the brain.

a) On CT a blood clot is hyperdense, measuring approximately 70–80 Hounsfield units. Because the clot usually arises in the dural venous sinuses, subsequently propagating into the cerebral cortical veins, it is best seen in the dural sinus but may be found on CT in either venous system (but more commonly in the dural sinus, because of its larger size as compared to the cortical veins). This finding is referred to as the *dense vein sign*, and has most frequently been reported in the distal superior sagittal sinus superjacent to the torcular (figure 7–7). When filled with a clot, the dural sinus will usually assume a rounded configura-

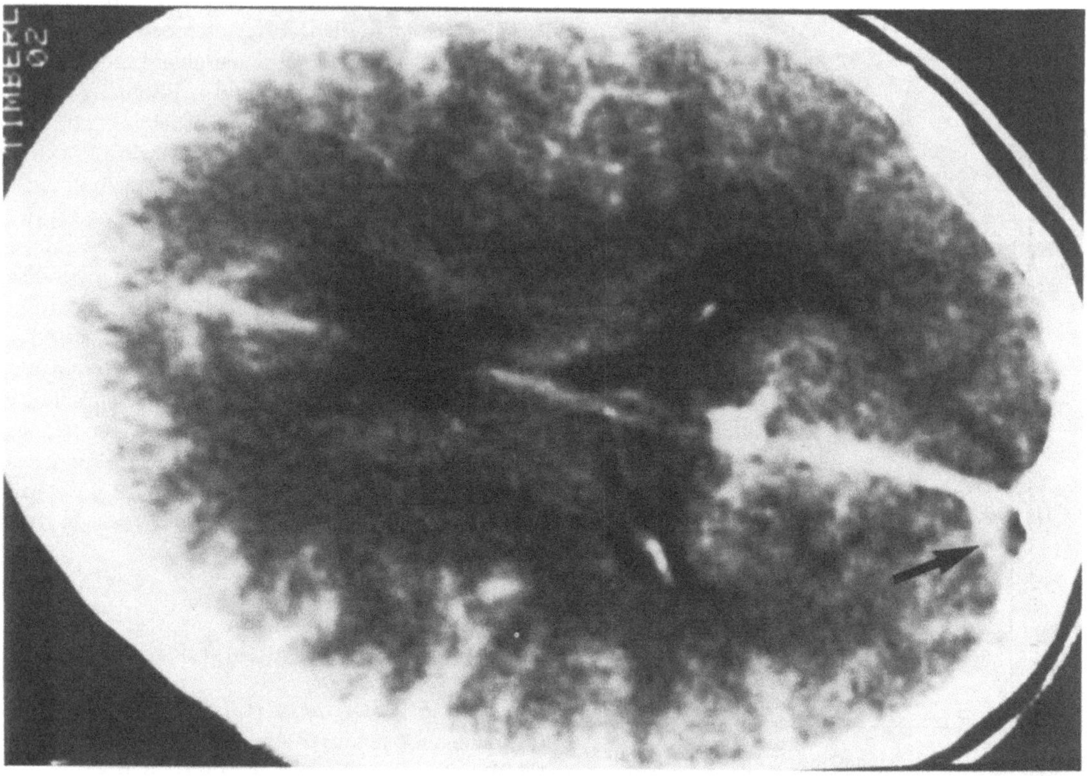

FIGURE 7–7B

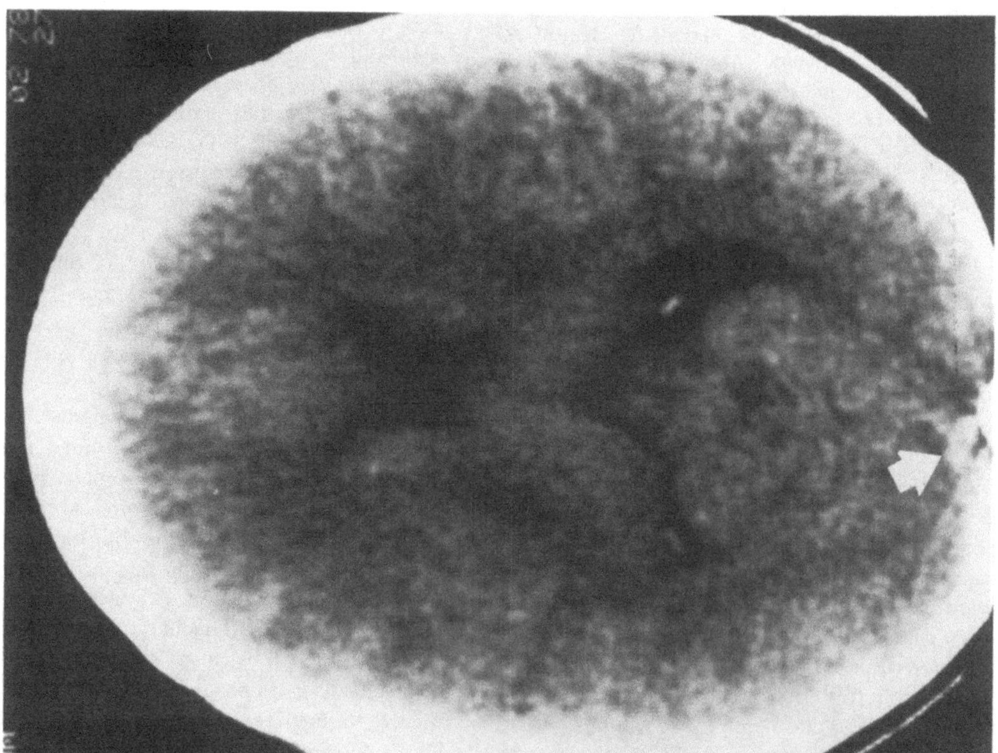

FIGURE 7–7A

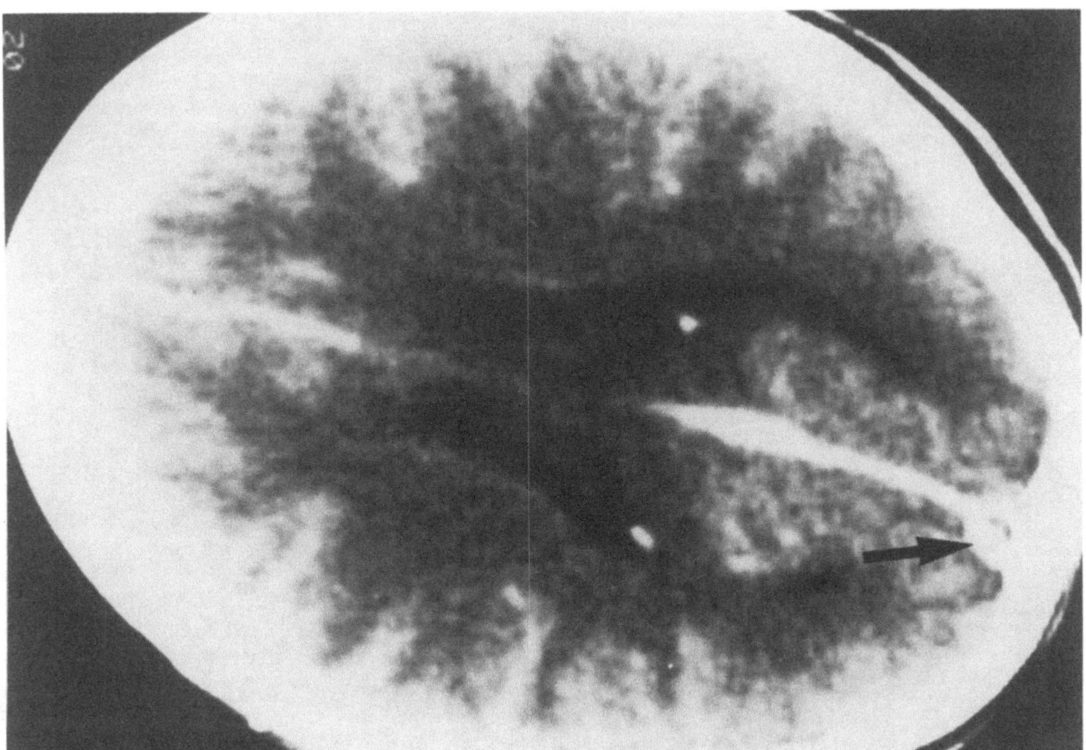

FIGURE 7–7C

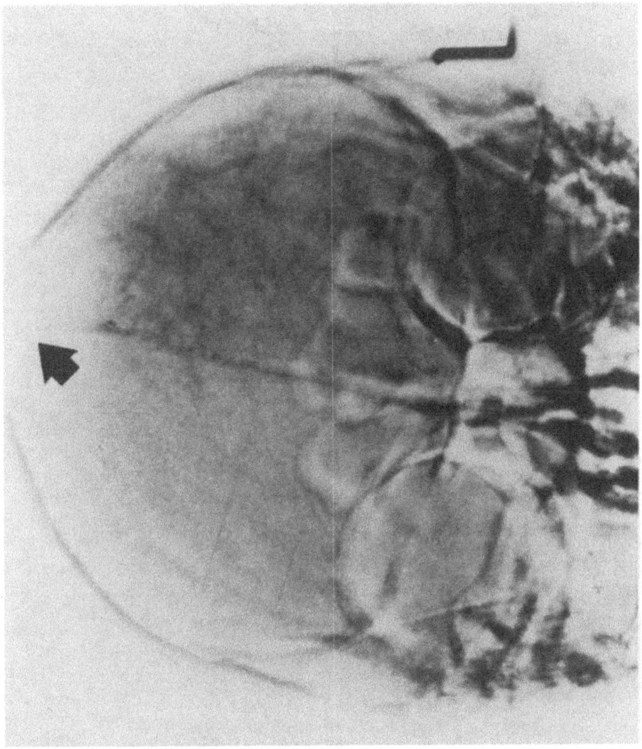

FIGURE 7–7D

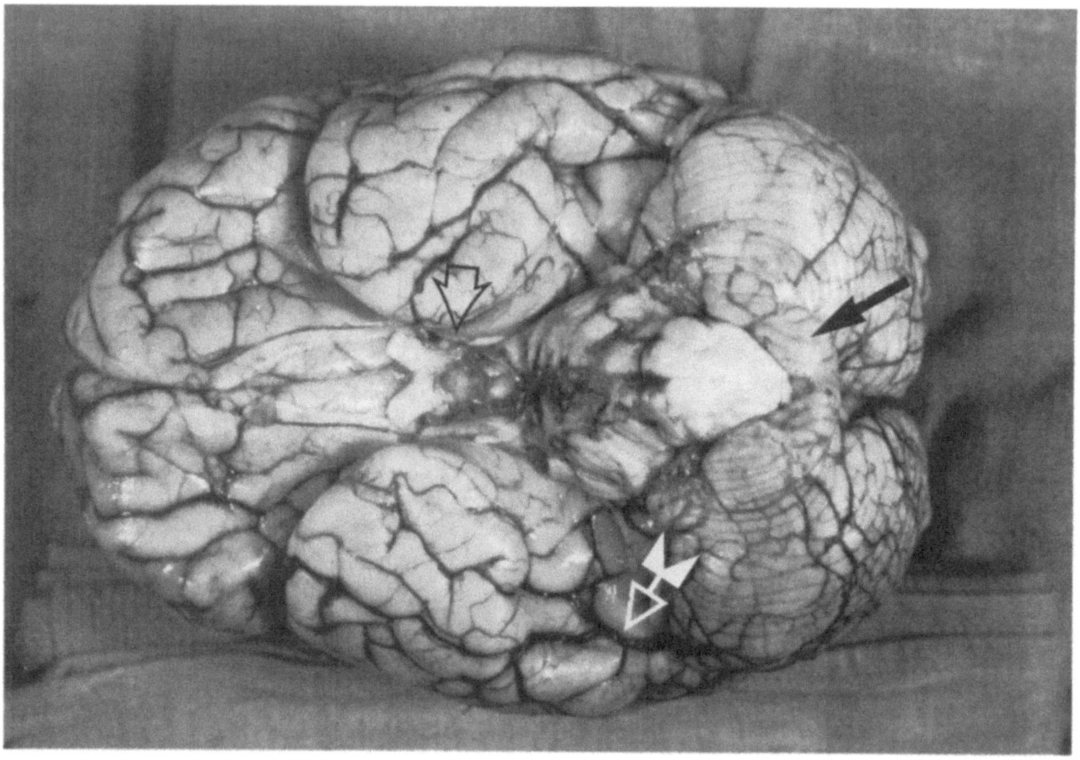

FIGURE 7–7F

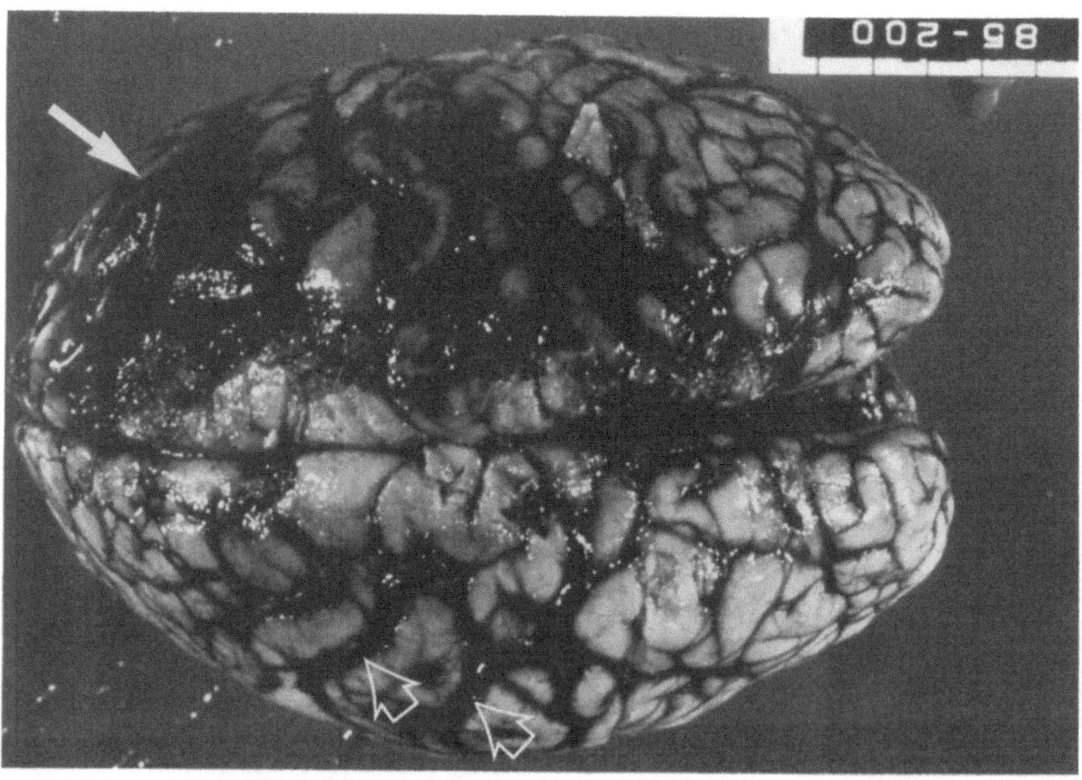

FIGURE 7–7E

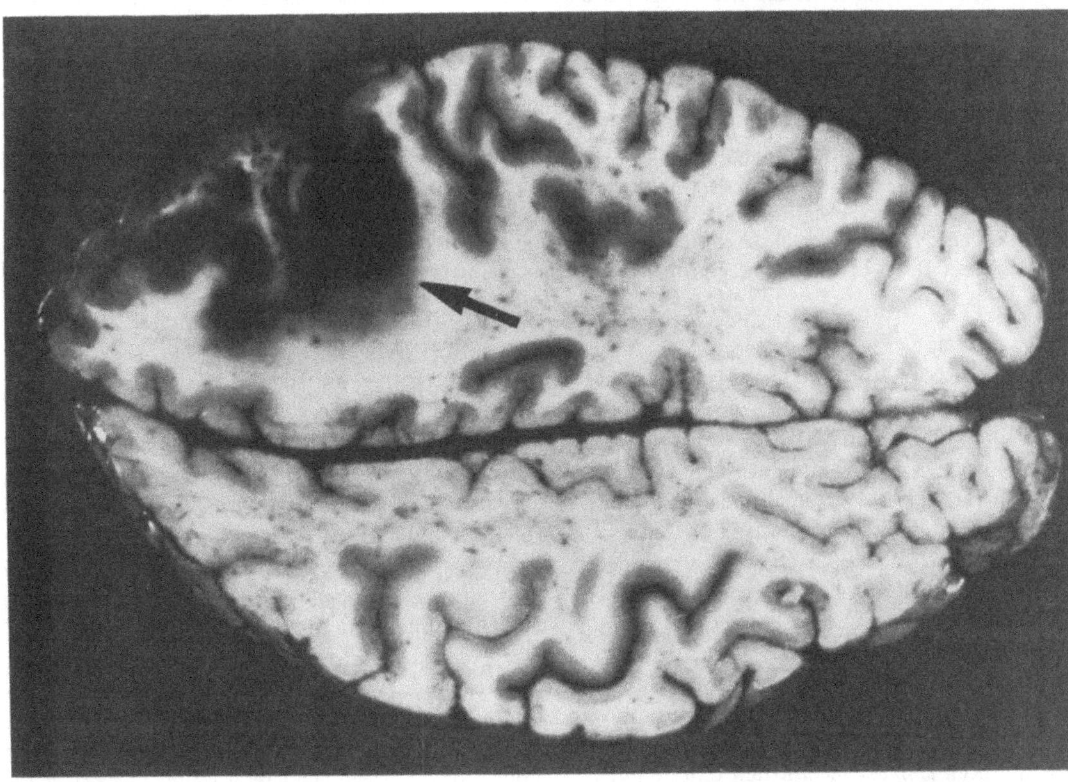

FIGURE 7–7G

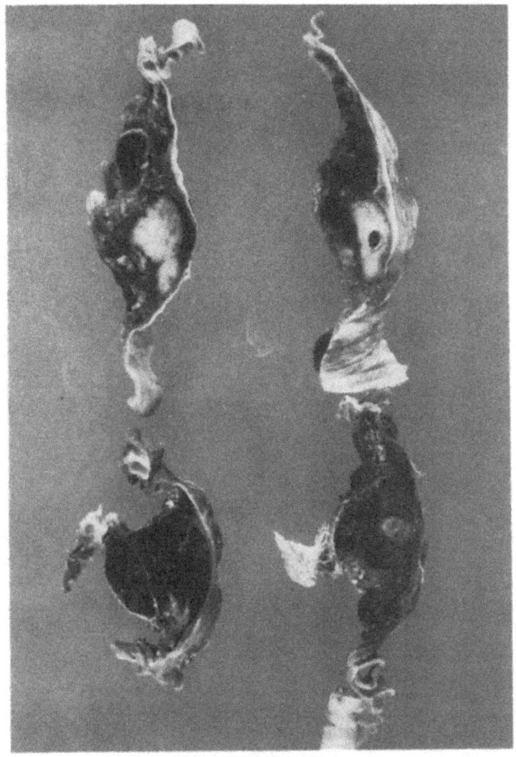

FIGURE 7–7H

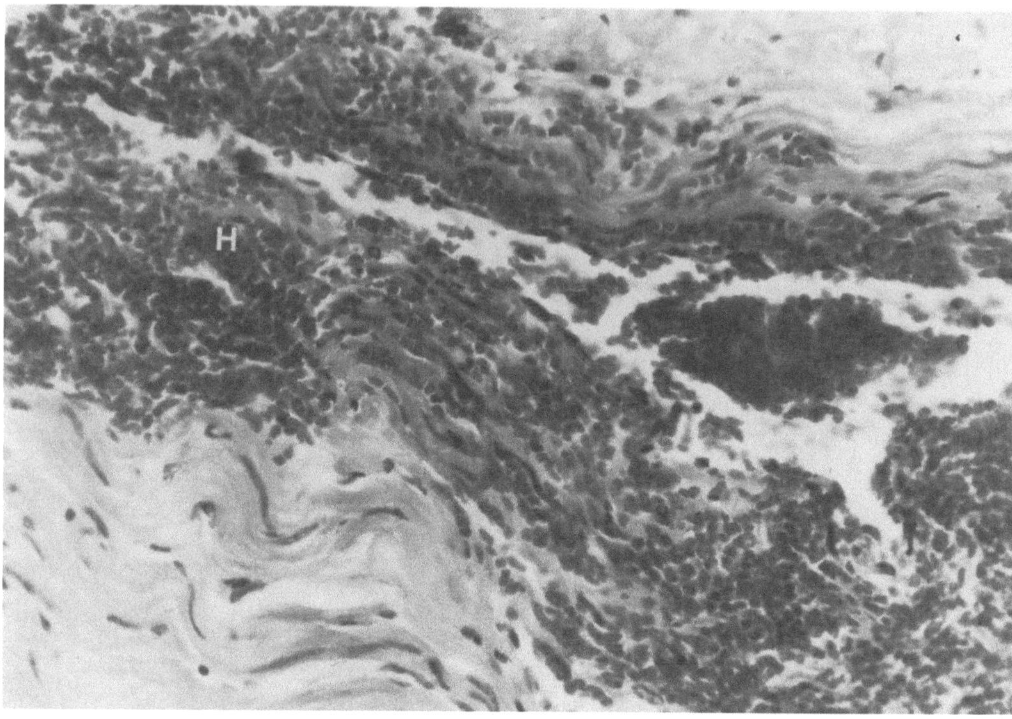

FIGURE 7–7I

FIGURE 7–7. Superior sagittal sinus thrombosis in a young female. (A) Noncontrast CT showing a rounded hyperdensity representing blood clot in the distal superior sagittal sinus (arrow). (B & C) Contrast enhanced CT performed three days after the onset of symptoms showing peripheral enhancement surrounding the clot—the empty delta sign (arrows). (D) Cerebral angiogram, anteroposterior view (agitated patient prevented proper positioning), showing absence of superior sagittal sinus opacification (arrow). (E) Gross specimen of the brain, superior view. Subarachnoid clot can be seen over the right cerebral hemisphere (closed arrow). Superficial cortical veins are distended and thrombosed (open arrows). (F) Inferior view showing extensive thromboses. Note involvement of cortical vein along the inferior aspect of the temporal lobe (open white arrow). Tonsillar (black arrow) and uncal herniation (open black arrow) can be seen. (G) Cut axial section of the brain reveals a large hemorrhagic focus (arrow) in the right frontal lobe. Note extensive subarachnoid hemorrhage as well over the fronted lobe. (H) The clot in the sagittal sinus, shown on cut section. No evidence of recanalization was seen. Note that the rounded and distended configuration of the clot-filled superior sagittal sinus corresponds closely to its appearance on CT. (I) Dural hemorrhages (H) can be seen in the superior sagittal sinus due to the rupturing of engorged dural vessels.

tion. It should be borne in mind, however, that the torcular Herophili, being partially convex, may at times show some fullness on axial CT due to venous distension as well as to the presence of cavernous spaces, which can exist in great abundance in this location [10, 15]. Also, the diagnosis of SST should only be suspected when the SSS shows this rounded configuration for a significant length (several CT slices). The dense vein sign has been reported in the deep Galenic venous system [15], particularly in children (figure 7–8). In general, the involvement of the deep venous system is the result of thrombus propagation from the SSS, a condition that is usually lethal when it occurs. The concomittant telltale signs of periventricular infarcts due to deep cerebral vein thrombosis may be seen on CT.

Propagation of the thrombus into the superficial cortical veins may result in the *cord sign* on CT, which consists of a linear density located at the paramedian high cerebral convexity extending toward the SSS (figure 7–9). However, because the cortical veins are of small size, this sign is seldom observed due to partial volume averaging.

b) Subsequent to sinus thrombosis, venous collateralization develops. This can be considered in two forms. In all likelihood, dural collateral channels are brought into play first; this is probably facilitated by connections between the lateral lacunae, dural caver-

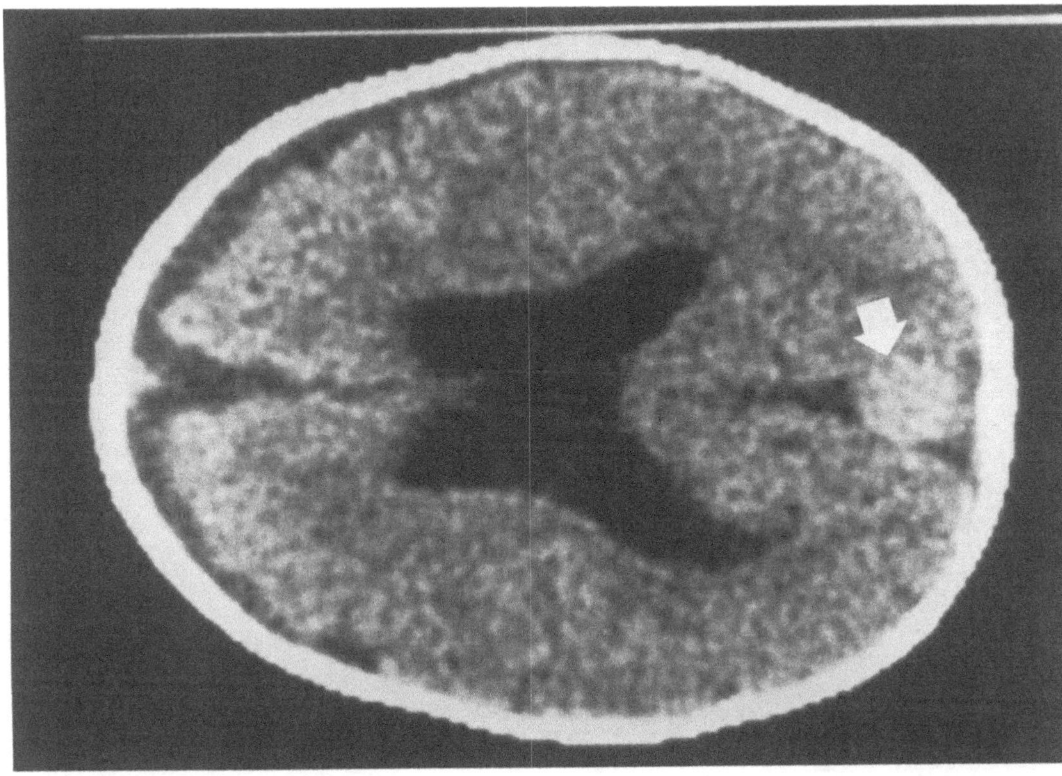

FIGURE 7–8B

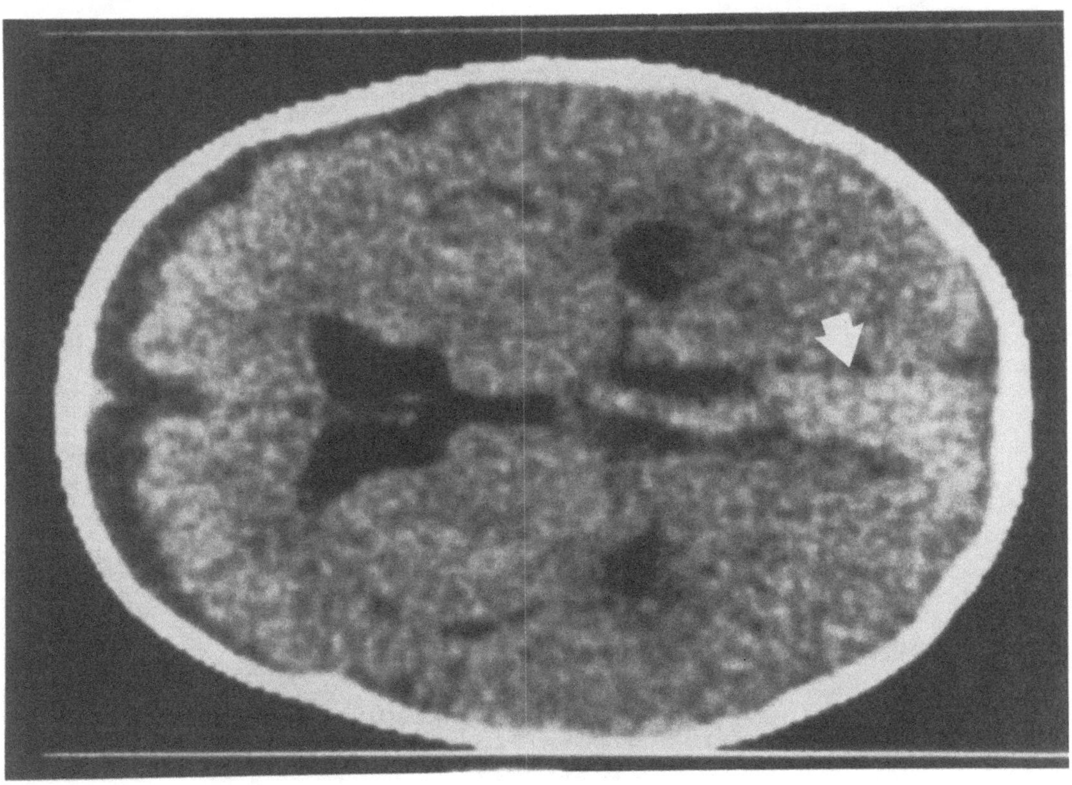

FIGURE 7–8A

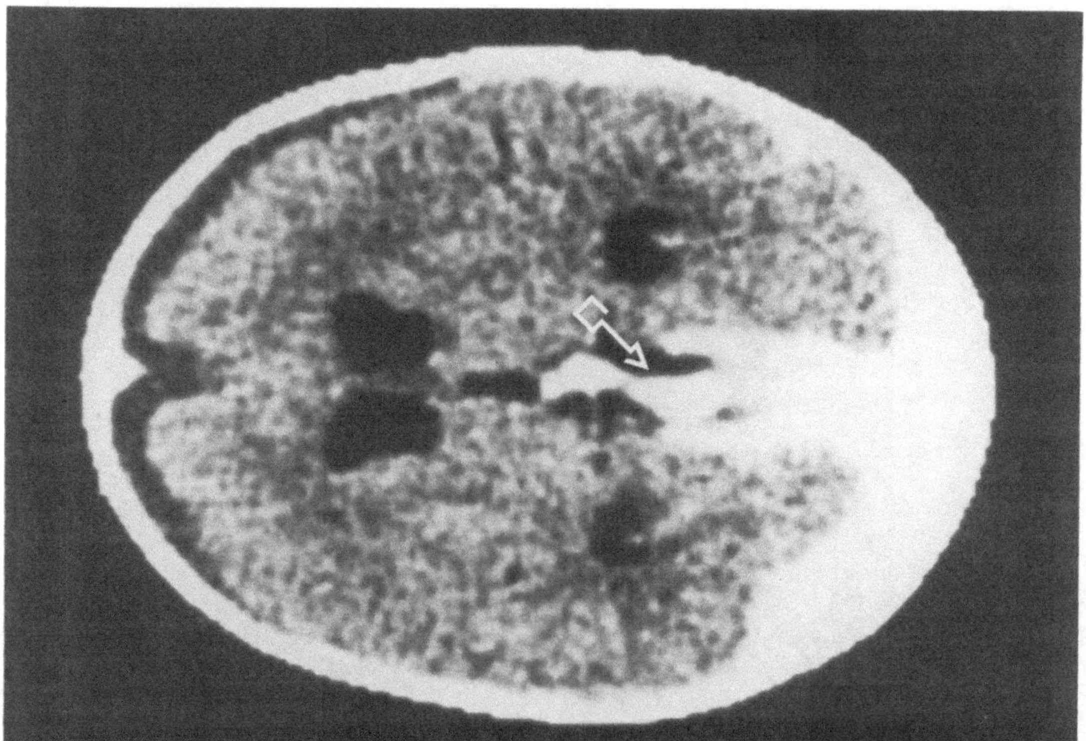

FIGURE 7–8C

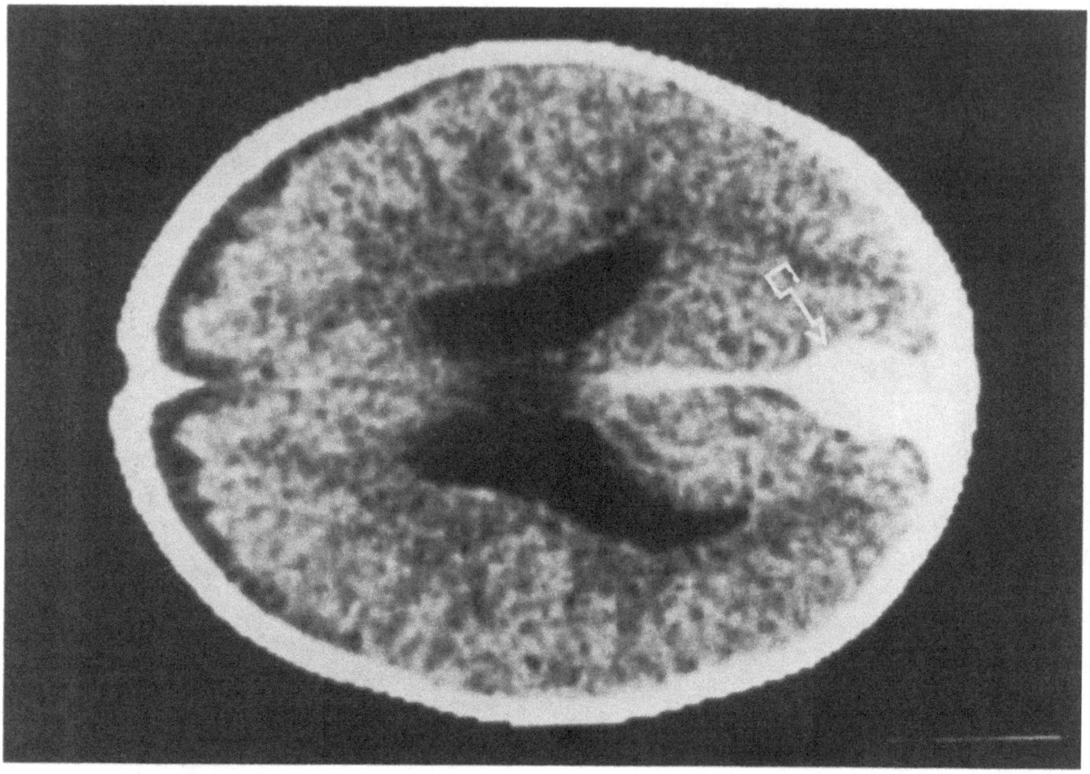

FIGURE 7–8D

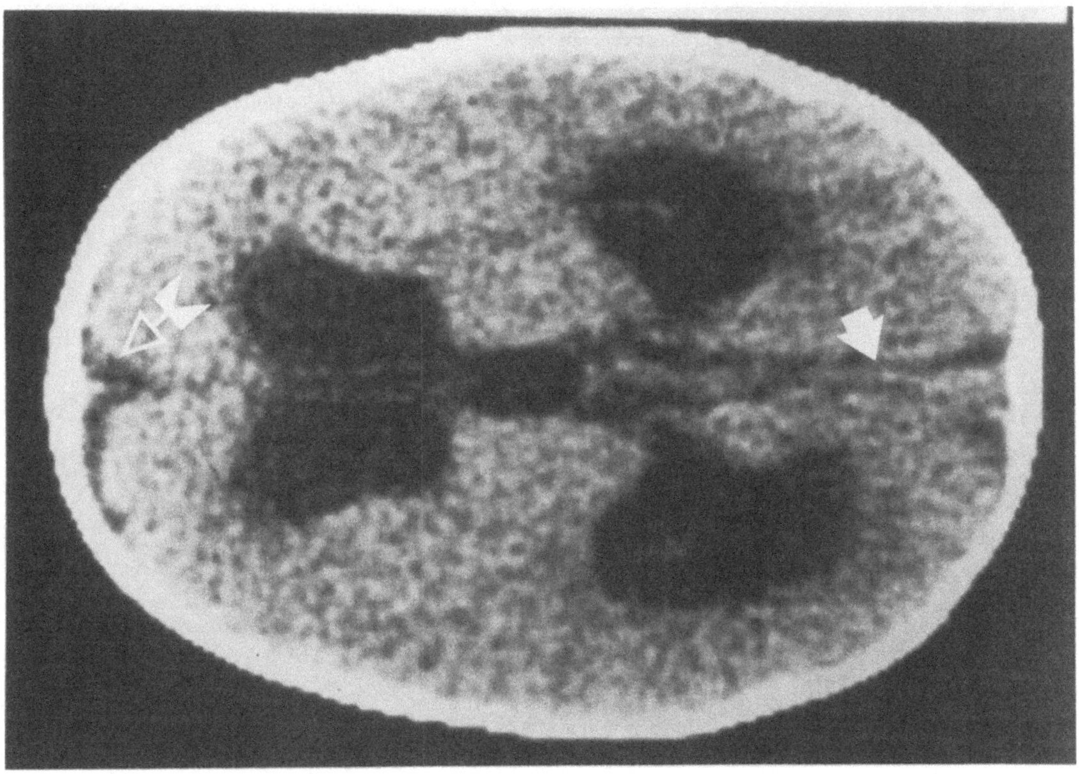

FIGURE 7–8E

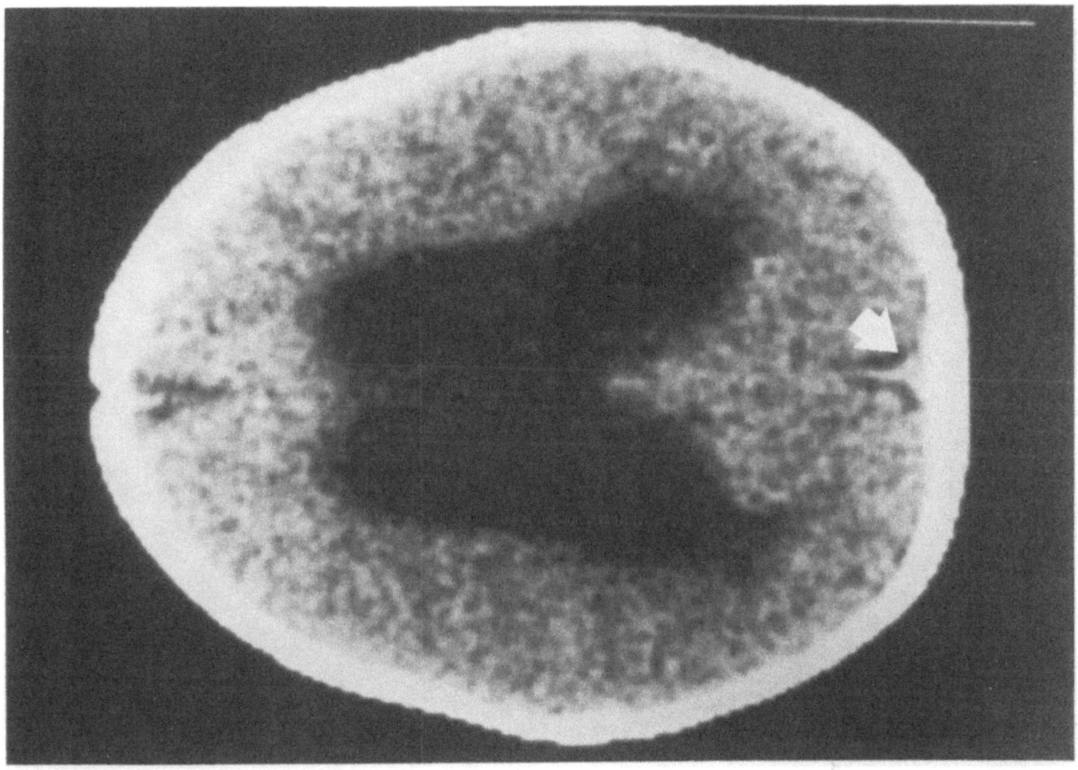

FIGURE 7–8F

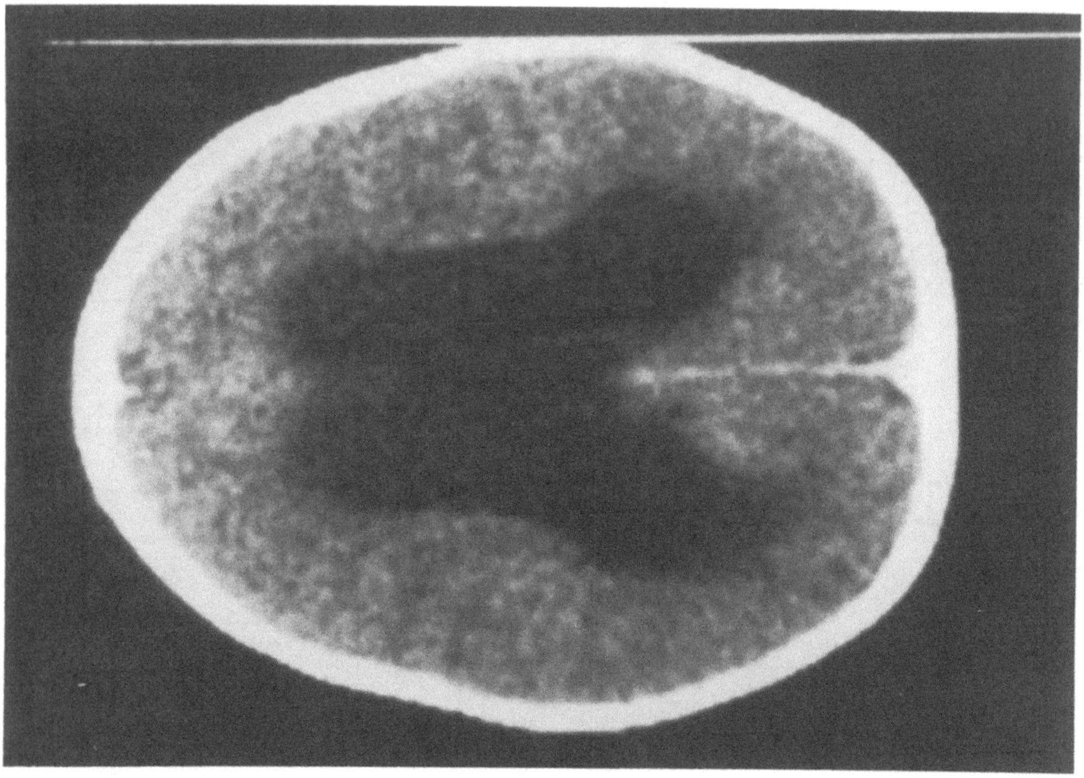

FIGURE 7–8H

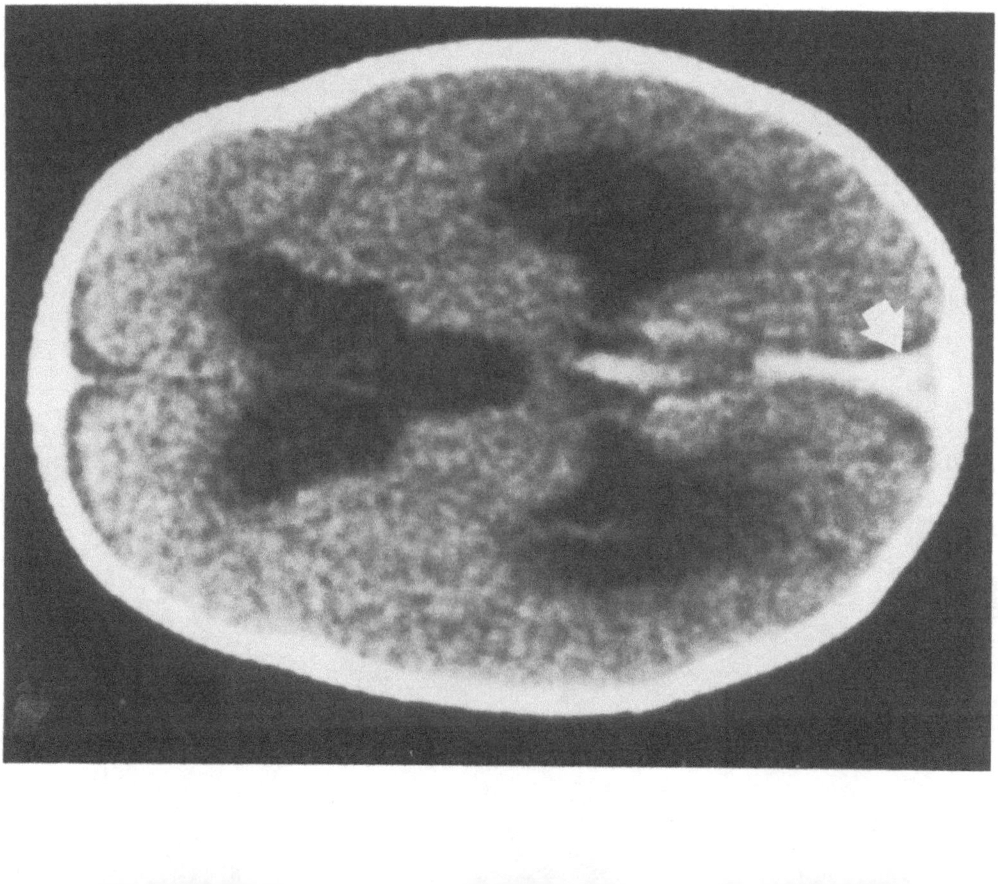

FIGURE 7–8G

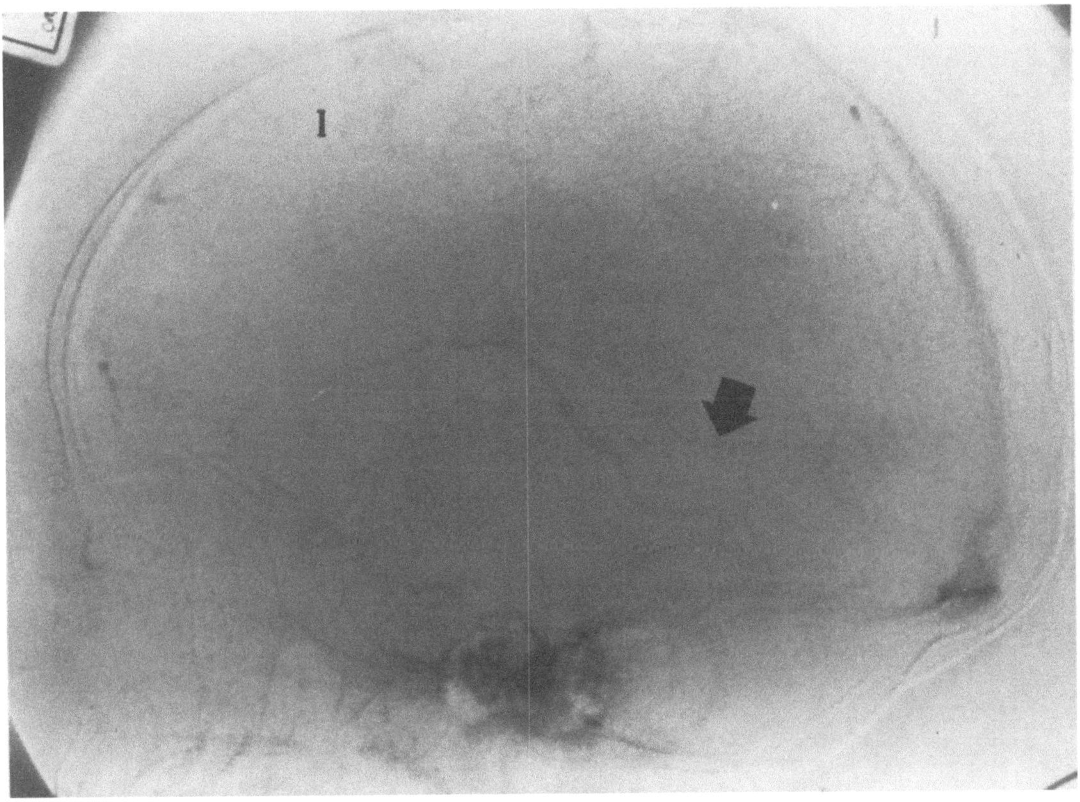

FIGURE 7–8I

FIGURE 7–8. Five-month-old black male with deep cerebral vein thrombosis due to pneumococcal meningitis. (A & B) Noncontrast CT showing a clot (the dense vein sign) in the vein of Galen, straight sinus, and torcular Herophili (arrows). Clot extended into right transverse sinus as well (not shown). (C & D) Contrast CT shows marked enhancement of these structures (arrows) as well as of the falx and tentorium. Note the widened anterior subarachnoid space as well as mild ventriculomegaly compatible with early communicating (external) hydrocephalus or subdural effusion. (E & F) Noncontrast CT one week later. Previously enlarged vessels are now almost normal in size, suggesting successful recanalization (solid arrows). But ventricles are hydrocephalic, while the widened anterior fluid collection has dimished in size (open arrow). (G & H) Corresponding contrast CT shows enhancement surrounding an intraluminal defect, which could represent a residual isodense clot (arrow). (I) Cerebral angiogram, lateral view, showing a stringy, faintly opacified vein of Galen and straight sinus suggestive of recanalized lumen (arrow).

nous tissue, and the SSS as the venous blood proximal to the obstruction attempts to bypass the block. Because a dural blood clot is lower in density then the intravascular iodine located in these dural collateral channels after CT contrast infusion the sagittal sinus takes on a triangular or rounded peripheral enhancement called the *empty delta sign*. That this sign has been seen as early as three days after the onset of symptoms of SST argues strongly against it representing recanalization and/or reorganization (which usually occur later in the clinical course of SST), and more in favor of it representing dural venous collateral channels. When the sign is unequivocally positive, it is pathognomonic of SST. Collateralization in the meningeal (and tentorial) veins may also manifest as enhancement of the tentorium on CT.

Secondly, due to back pressure the cerebral veins (superficial and deep) become engorged, and retrograde downward flow occurs toward the patent dural sinuses along the skull base. The cerebral veins may be visible on CT as serpiginous densities either along the brain surface or intraparenchymally. Because of their relatively small size and their frequent contiguity to the calvarium, these veins are rarely visible on CT, and thus are best visualized on angiography.

c) As resistance to venous blood flow develops, the overall cerebral blood volume is increased, with

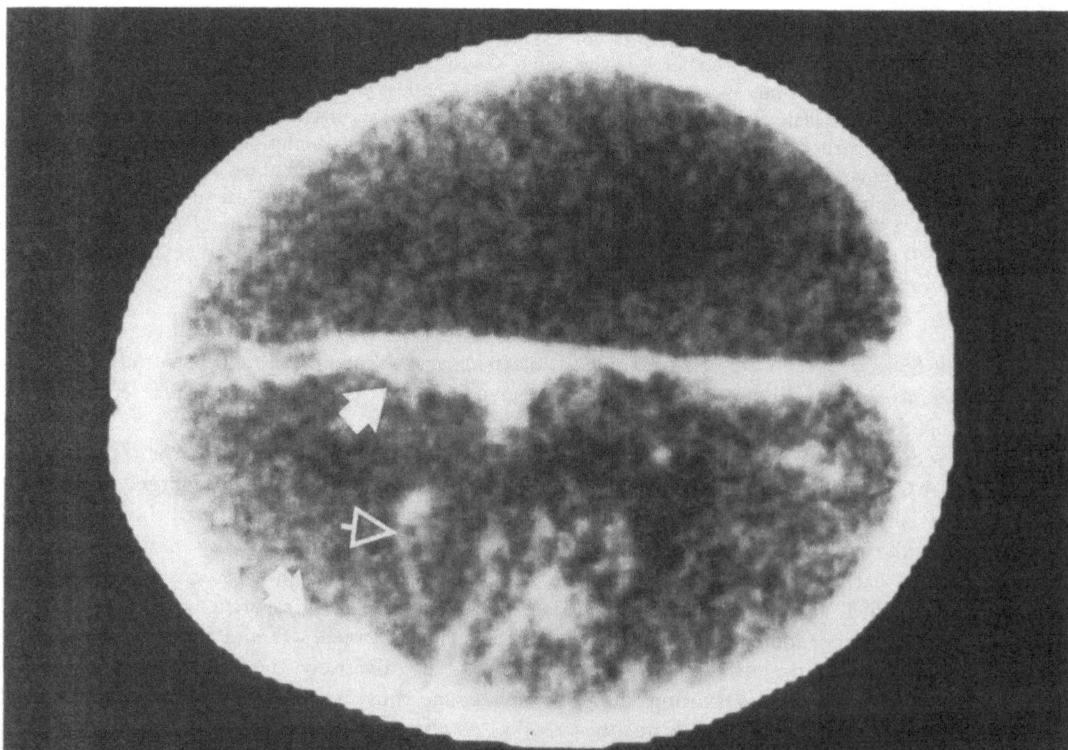

FIGURE 7–9A

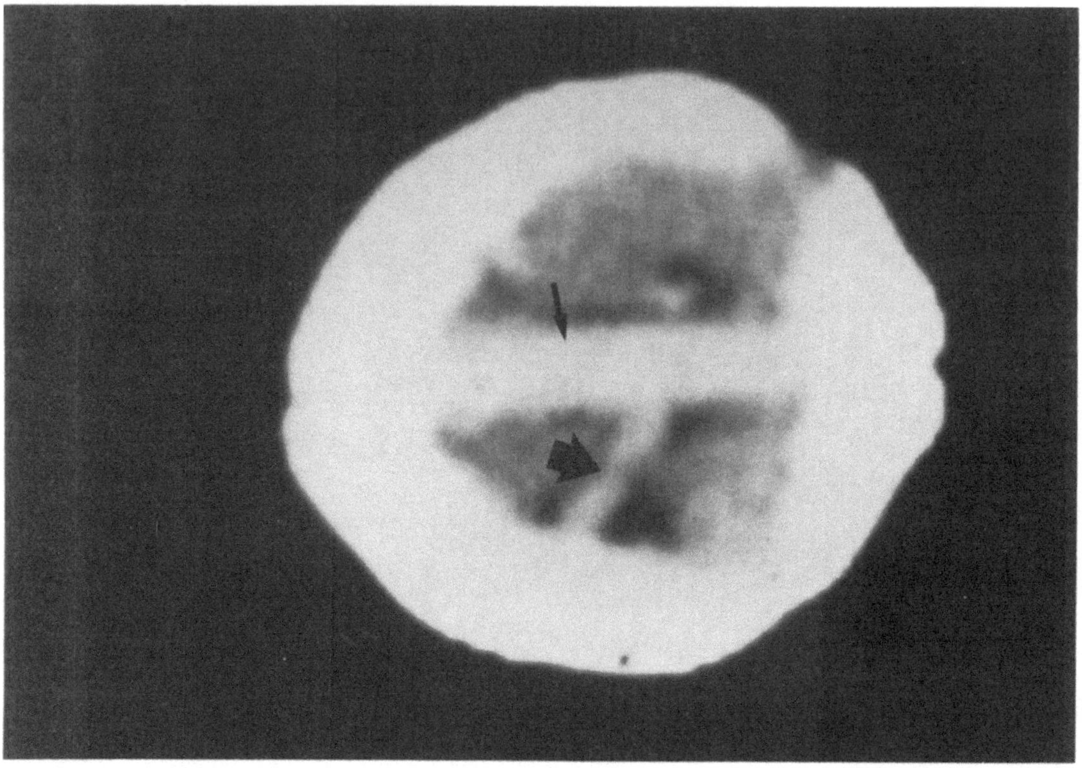

FIGURE 7–9B

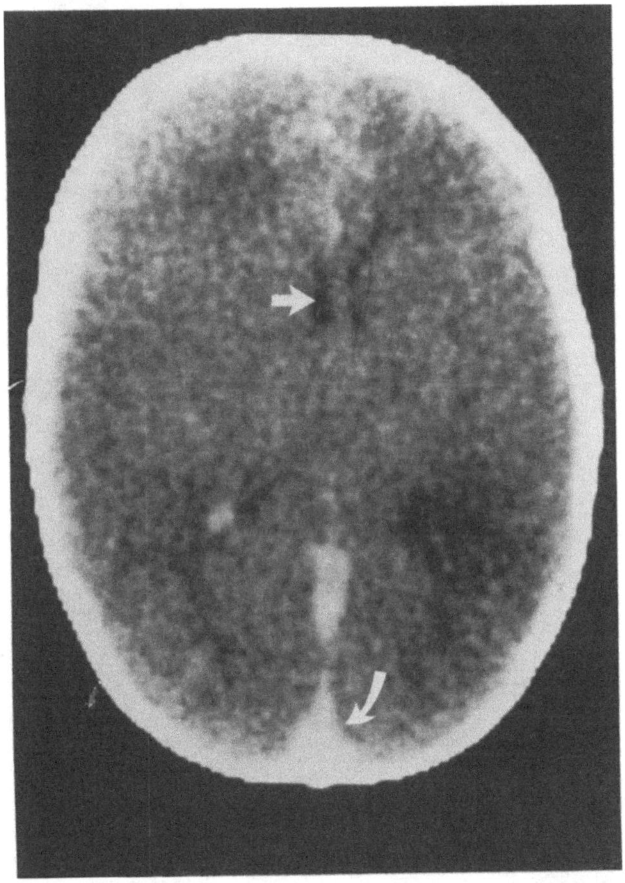

FIGURE 7–10. Noncontrast CT. Superior sagittal sinus thrombosis (curved arrow) associated with isodense right cerebral hemisphere swelling. (The relatively greater brain swelling on the right may be due to a more extensive propagation of thrombosis into veins on that side.) Note mass effect on right lateral ventricle with shift of the frontal horns (straight arrow).

distension of the cerebral vasculature. At this point, diffuse isodense swelling of the brain may be seen on CT, with compression of the lateral ventricles (figure 7–10). With further distension of these vessels and stasis of venous blood, the brain becomes ischemic, and hemorrhagic infarction may occur. The latter findings are particularly prevalent in the frontal or

FIGURE 7–9. (A) Noncontrast CT in a child with a hemorrhagic infarct (open arrow) due to superior sagittal thrombosis in the right parietal lobe. Subarachnoid blood is present on both sides of the hemisphere (solid arrows). (B) There is propagation of thrombus into a cortical vein (the cord sign, thick arrow). The superior sagittal sinus is distended and dense due to the blood clot (thin arrow).

parietal paramedian locations adjacent to the SSS. Since venous hemorrhagic infarction does not correspond to any vascular territory, it can be easily distinguished from arterial hemorrhagic infarction. Further, in a small number of patients with bilateral hemorrhages, they are frequently symmetrical, making arteriovenous malformation, amyloid angiopathy, and other primary causes of hemorrhage unlikely.

There are some pitfalls in the interpretation of the empty delta sign on CT. Subdural hematoma and empyema, subarachnoid hemorrhage, arachnoid granulations, SSS septa, and skull artifacts can simulate the empty delta sign [19, 20]. In addition, metastatic disease to the dura in this region could simulate SST (figures 7–11, 7–12). False negatives are probably common. When they are suspected, a careful search should be made by studying the SSS on the computer monitor under multiple-level and window-width settings. A high-level and window-width setting (bone window) has been suggested as a means of increasing sensitivity [21]. To minimize

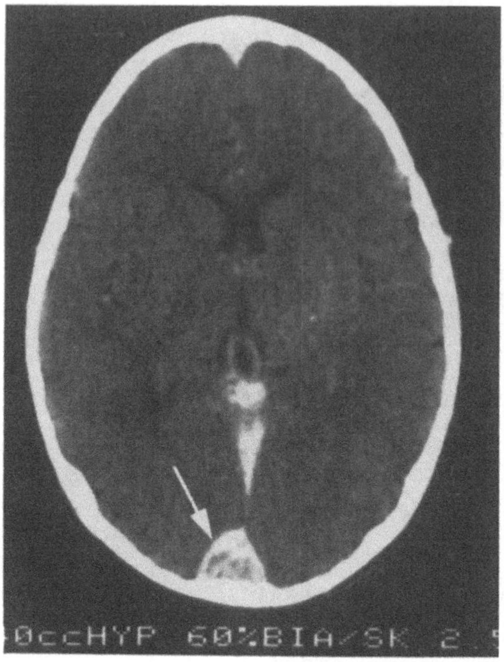

FIGURE 7–11A

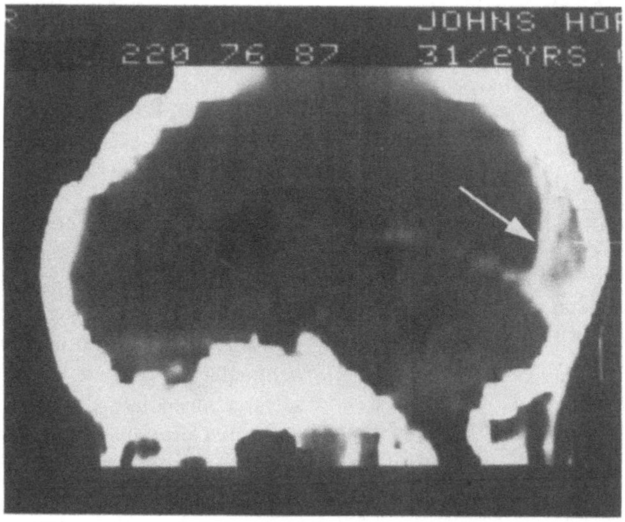

FIGURE 7–11B

FIGURE 7–11. Three-and-a-half-year-old child with metastatic neuroblastoma to the torcular Herophili. (A) Axial CT with contrast showing enhancing mass at the torcular Herophili (arrow). (B) Sagittal reformation CT demonstrating the superoinferior extent of the lesion (arrow).

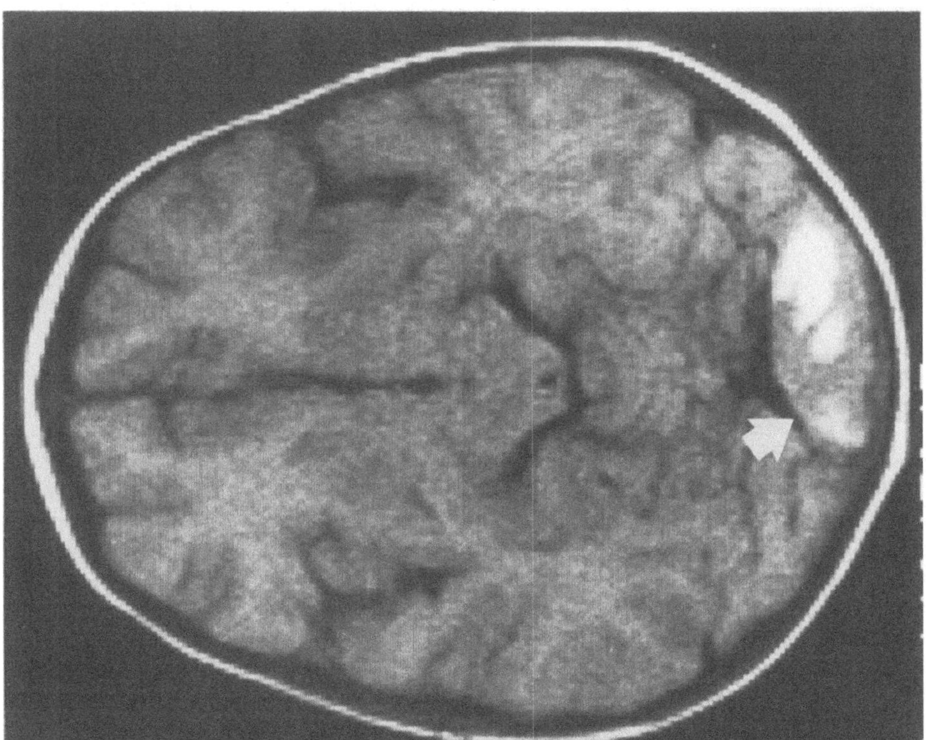

FIGURE 7–12A

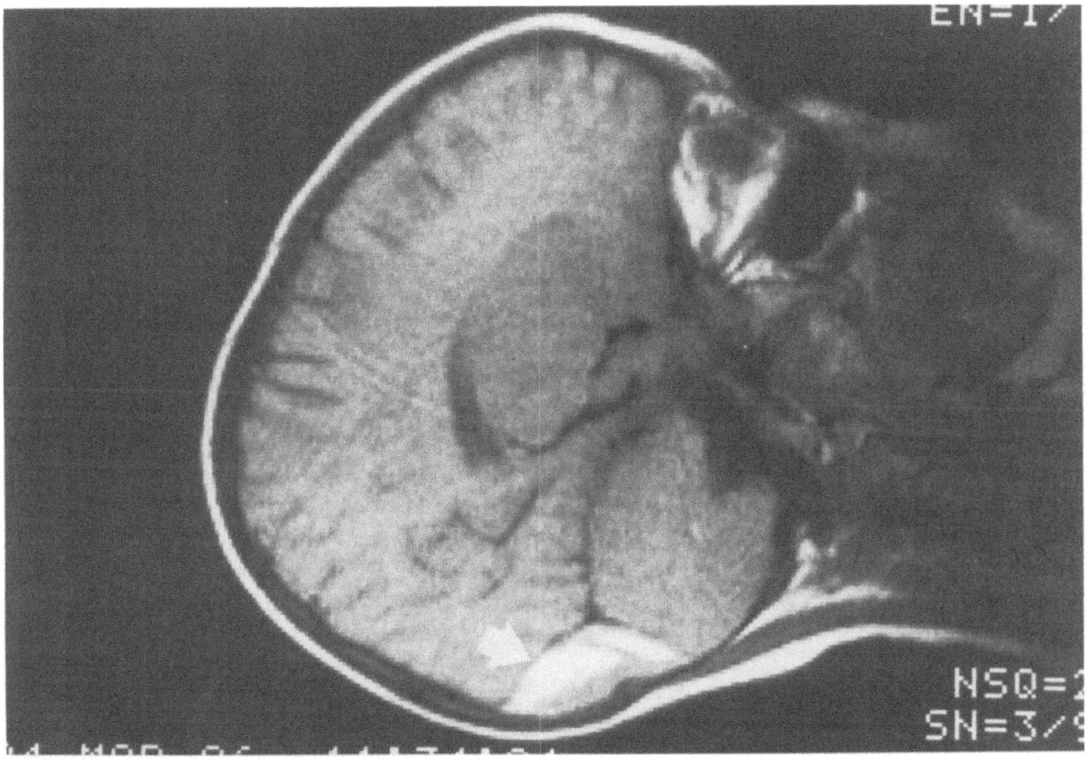

FIGURE 7–12B

FIGURE 7–12. Axial (A) and sagittal (B) T1-weighted MRI in a patient with metastatic hepatoblastoma to the torcular Herophili (arrows). The hyperintensity may be related to hemorrhagic changes in the tumor.

partial volume averaging, 5-mm thin sections should be obtained. Very thin sections, 1.5 mm or less, will result in marked reduction in contrast resolution. Goldberg et al. [22] have suggested that coronal reformations may be of value in visualizing lateral sinus thrombosis, because the lateral sinus runs a parallel course to the (axial) imaging plane (figure 7–13), and hence a central lucent defect representing the clot could be missed by partial volume effect. To be certain, the empty delta sign should also be visualized on several sections.

As discussed, the anatomic architecture of the SSS can give rise to the following effects: a) The dural collateralization in the meningeal veins, lateral lacunae, and cavernous mesh could give rise to enhancement surrounding the SSS intraluminal clot on CT (the empty delta sign); b) the cavernous mesh when filled could produce an irregular bulge or convexity in the SSS, so as to simulate a clot on noncontrast CT; and c) the chordae Willisii could give rise to spurious defects in the contrast-enhanced SSS simulating blood clot(s). These points should be borne in mind when searching for SST on CT [15].

A recent review of the CT manifestations of SST and SST-related SOD (76 cases, 83% proven by angiography, surgery, or autopsy) revealed the following [15]: the most commonly reported CT sign was the empty delta sign, (29% of all signs, table 7–2), followed by the dense vein sign (20%). Hemorrhagic infarcts were reported in 22% of cases; no bland infarcts were reported. Patients who had the empty delta sign and/or hemorrhagic infarction fared worse; and when both signs occurred together, the mortality was 100% [15, 18]. The majority of children presented with the dense vein sign, usually involving the deep venous system. The evolution of CT signs was followed in several reports [13, 21, 23, 24, 25, 26]. In general, the dense vein sign will disappear, due to venous recanalization, in about one to two weeks [23, 24] (figure 7–8), while the empty delta sign takes about one and a half to two months to diminish [19, 21, 25, 26].

TABLE 7–2. CT Signs in Superior Sagittal Sims Thrombois and Related Venosinus Occlusive Disease [15]

A. Noncontrast CT	
Cord sign	4.5%
Unilateral hemorrhage	8.5%
Bilateral hemorrhage	6.0%
Dense vein	20.0%
Edema	8.0%
Compressed ventricles	14.0%
B. Contrast CT	
Gyral enhancement	.9%
Tentorial enhancement	3.6%
Empty delta sign	28.6%
Medullary veins	1.8%
Normal	3.6%

Magnetic Resonance Imaging (MRI). The advantages of MRI over CT are numerous. MRI has better soft-tissue discrimination, has freedom to image in any plane, and lacks ionizing radiation. In contrast to CT, which is primarily dependent on the electron density of tissue, MRI is dependent on multiple parameters that influence image contrast: T1 (longitudinal), T2 (transverse) relaxation properties of tissue, proton density, and flow to create tissue contrast. Thus far, MRI of *arterial* blood flow has been the subject of several reports. Recently, however, it has become apparent that MRI is able to determine qualitatively not only CSF flow, but also venous flow. Macchi et al. [27] reported three subjects with SST diagnosed on MRI. Their early experience suggests that venous thrombosis can be easily recognized by the absence of flow void, i.e. hyperintensity instead of hypointensity of the venous sinus (figure 7–14). Normally, hypointensity represents lack of signal which can indicate the presence of flow; this signal void is created by hydrogen protons rapidly moving out of the slice prior to giving up their relaxation energy so that little or no signal can be detected.

Secondly, the chronology of the thrombus can be determined by its appearance similar to that of intracerebral hematoma. Therefore, acute hematoma would appear isointense on T1-weighted images and hypointense on T2-weighted images, whereas subacute hematoma would appear hyperintense on both T1- and T2-weighted images, probably due to the conversion of deoxyhemoglobin to the paramagnetic methemoglobin. While these speculations are yet to be proven by a study of a more extensive patient population with SOD, the elusive nature of flow and flow turbulence as well as the individual chronologic variability in thrombus formation, organization, and recanalization has created some justifiable scepticism. Erdman et al. [28] experimentally induced five venous thrombi in dogs, imaged the evolution of the thrombi; over a course of three weeks on a medium-field MRI scanner, and compared the findings to the pathologic specimen. They found that a subacute clot does indeed appear hyperintense (relative to muscle) on both T1- T2-weighted images, but its chronology cannot be readily predicted. Others have warned that, as in CT, pitfalls may exist in diagnosing SST on MRI, such as mistaking flow turbulence, paradoxical enhancement, and slow flow for SST. Attempts

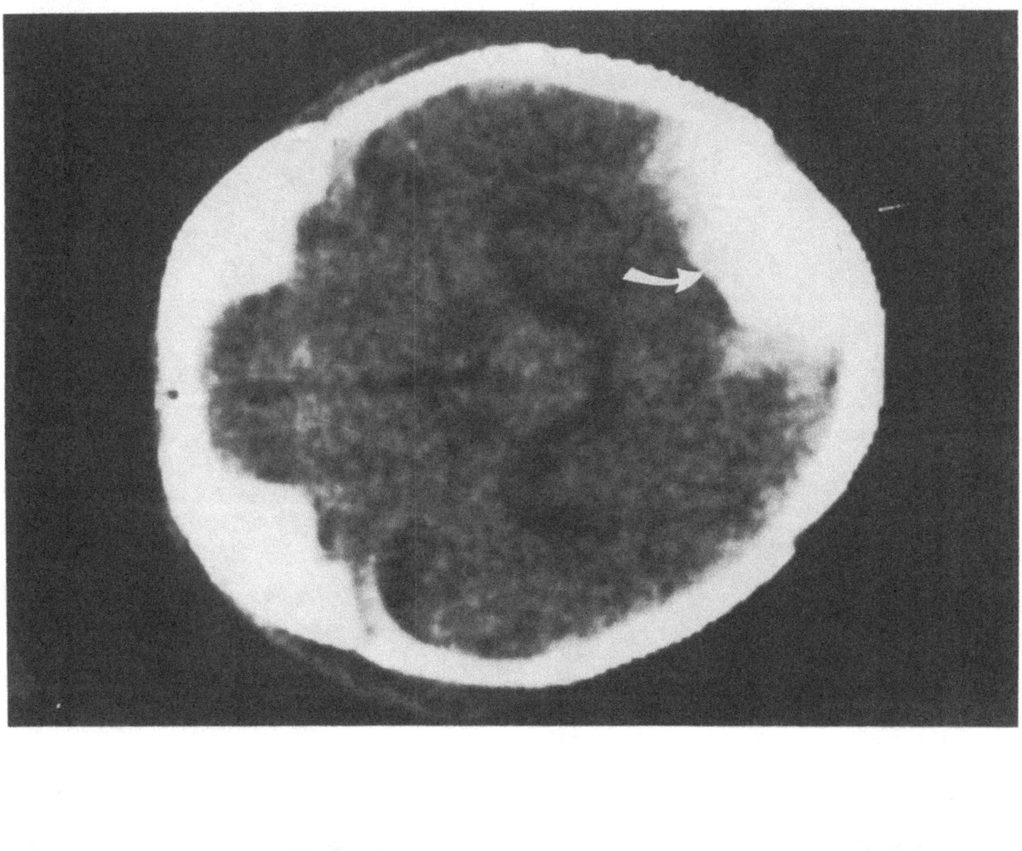

FIGURE 7–13A

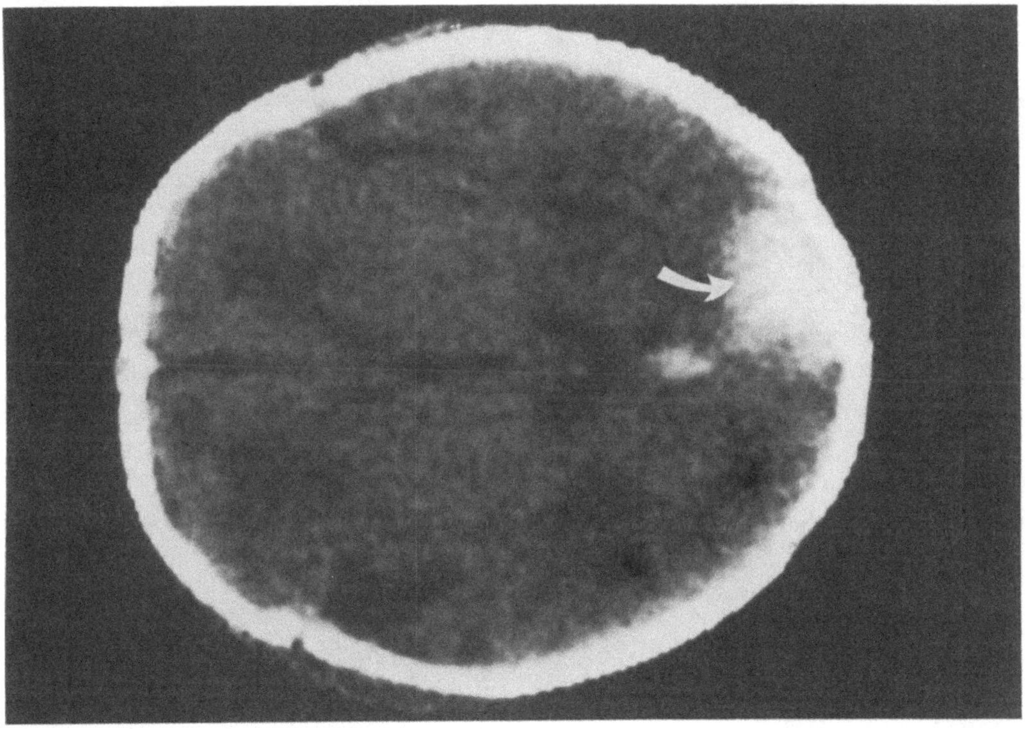

FIGURE 7–13B

FIGURE 7–13. Noncontrast axial scan in a child with lateral sinus thrombosis. Note the marked enlargement of the left lateral sinus (arrows).

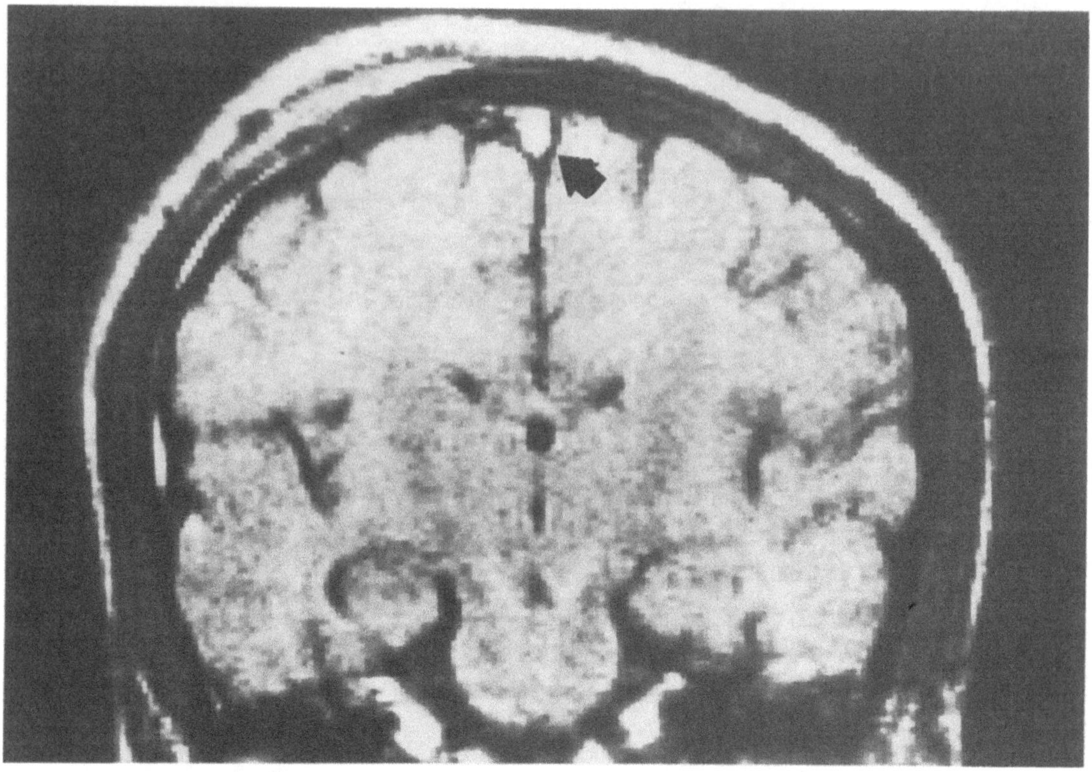

FIGURE 7–14A

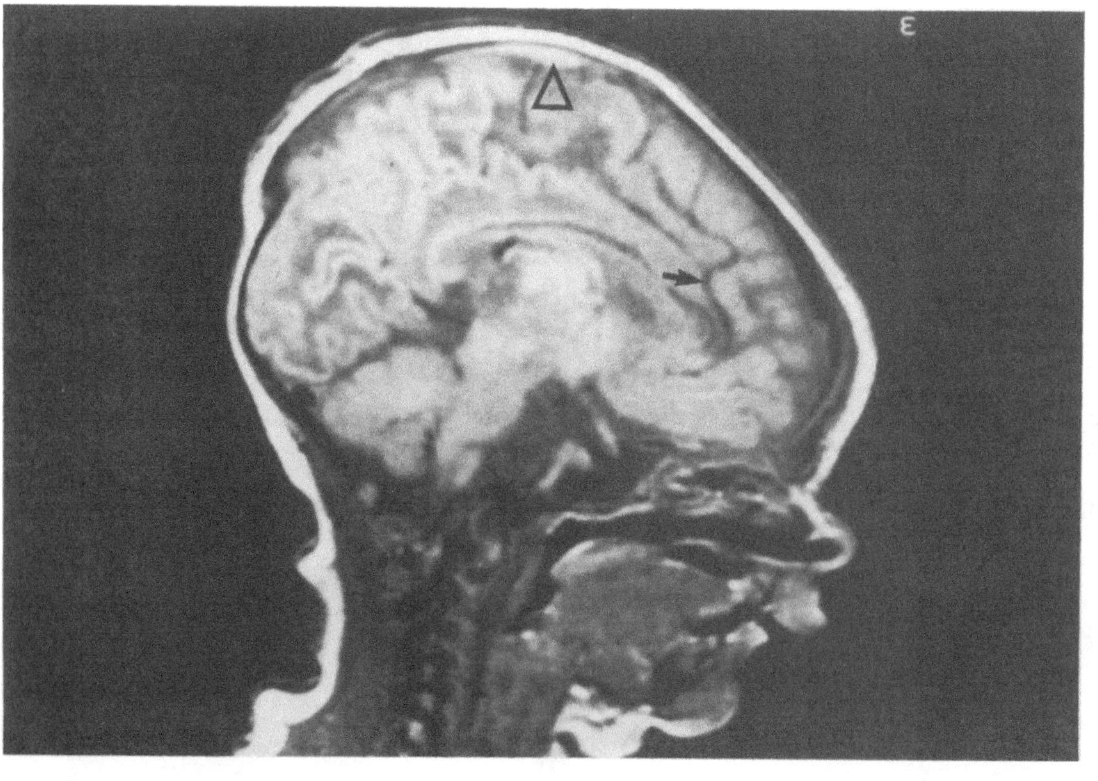

FIGURE 7–14B

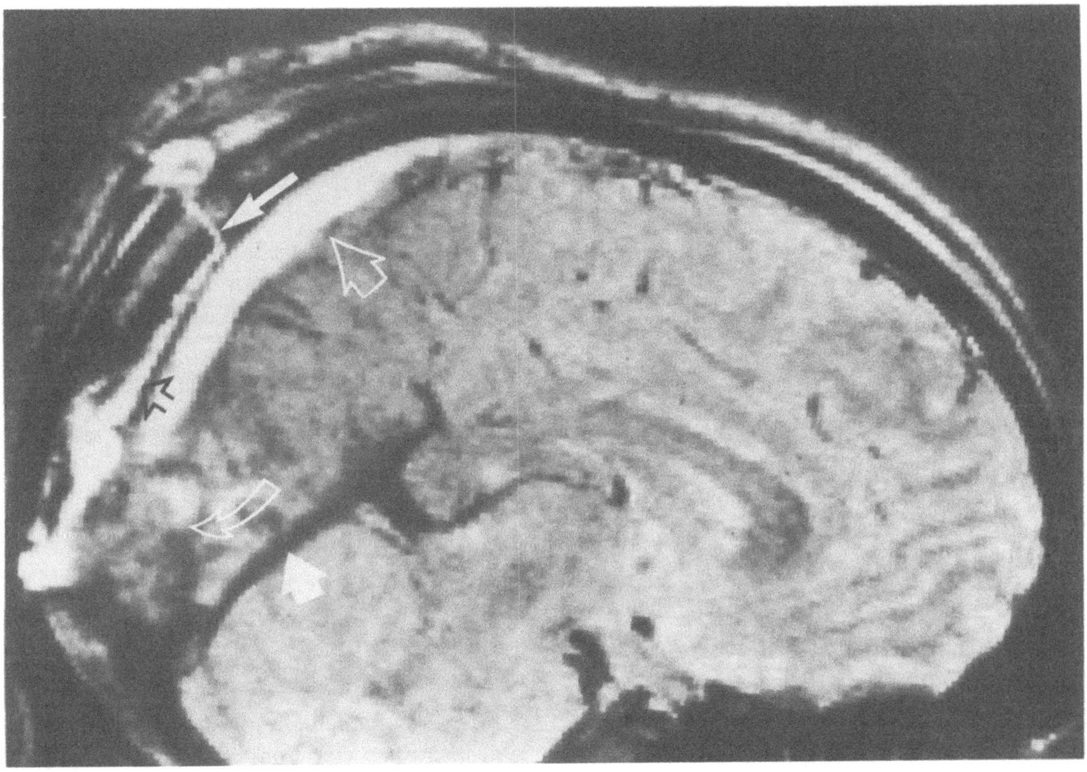

FIGURE 7–14C

FIGURE 7–14. MRI of posttraumatic SST. T1-weighted images. (A) Coronal view. An area of hyperintensity can be seen in the superior sagittal sinus representing a clot (arrow). Note the outward bulging of the superior sagittal sinus wall. (B) Sagittal view, off-midline. The clot corresponding to figure 7–11A is seen (open arrowhead). The flowing blood in the branches of the pericallosal artery is seen as black due to signal void (closed arrow). (C) Sagittal view, midline. SST extends distally (open straight white arrow), where, directly superior to the torcular Herophili, the sagittal sinus is severed (by penetrating trauma, a bullet) (open curved white arrow). Extension of blood into the extradural space is seen (open black arrow), which in turn communicates with a subgaleal hematoma, probably via a skull fracture (long white arrow). Short white arrow = hypointense echo from flowing venous blood in the straight sinus.

to predict clot chronology may be self-defeating due to recanalization [29]. Recanalization can in itself produce areas of hypointensity in addition to flow turbulence, which can all mimic the paramagnetic effects of blood breakdown products [28]. These reports seem to suggest that MRI has the greatest potential of all imaging modalities to predict SOD accurately; and in addition, MRI can accurately visualize the untoward effects of sinus thrombosis on the brain parenchyma (such as brain edema and hemorrhagic infarction), as well as collateral blood flow. But whether MRI is capable of predicting clot chronology is still subject to debate.

Radionuclide Scan. Technetium 99 m (Tc 99 m) pertechnitate planar scintigraphy dynamic brain scanning has been used in the past for the diagnosis of SOD, but with little success, due to the lack of resolution of planar scintigraphy [30]. Single photon emission computerized tomography (SPECT) has shown some promise in diagnosing cerebral SOD noninvasively. Front et al. [31] reported the utilization of technetium-labeled autologous (RBCs) in 19 patients suspected of SST. SPECT studies were performed in six patients. In 12 cases, the diagnosis was confirmed by angiography, surgery, or autopsy. Static Tc-RBC scintigraphy was found to have a sensitivity of 100%, a specificity of 86% and an overall accuracy rate of 94%; flow images were found to have a low accuracy rate (61%). SPECT scanning was useful in giving added spatial resolution so as to enable separation of other adjacent lesions such as tumors, which might otherwise create a false positive or negative reading.

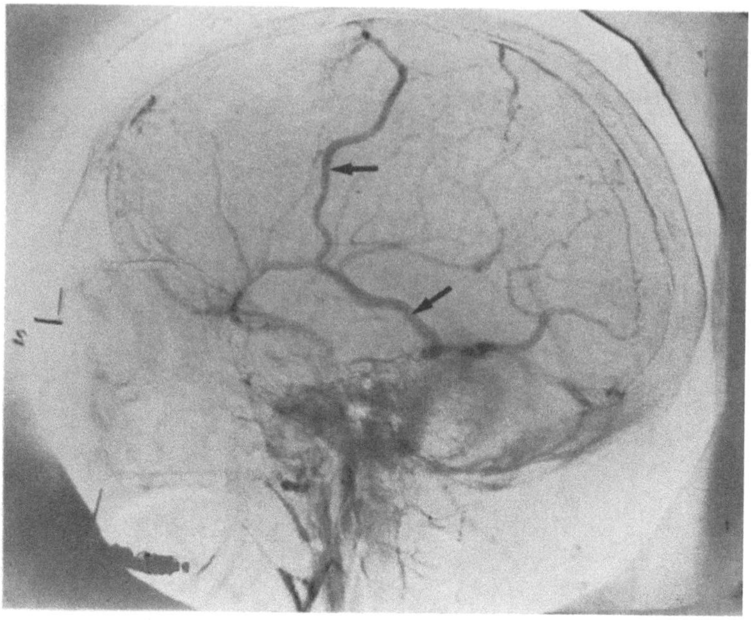

FIGURE 7–15A

Other researchers, following a similar tack by tagging blood-clot components with a radionuclide-emitting source, have been successful at imaging SST. Bridgers et al. [32] imaged a patient with proven SST with Indium-111 platelet scintigraphy, which had previously been used for detecting peripheral and intracardiac clots [33, 34]. They followed the patient for two weeks, repeating the study at the end of that period. The second study revealed decreased uptake, suggesting decreased clot activity. Thus, this early experience suggested that Indium-111 platelet scintigraphy may be used to follow platelet (and clotting) activity in SST.

Cerebral Angiography. Cerebral angiography was previously considered as the gold standard in the diagnosis of cerebral SOD. Certain points should be borne in mind in performing cerebral angiography in SST: a) Due to slow flow, the angiogram should be performed well into the venous phase (at least 15 seconds) [14]; b) To facilitate the diagnosis of SST, the head should be turned slightly on the frontal projection so that the anterior and posterior segments of the SSS will not be superimposed on one another. One view is probably adequate to diagnose complete SST, although for partial occlusion more views may be needed; c) Digital subtraction angiography is recommended to reduce the dose of the contrast agent (about half that used for cut film angiography). Digital arteriography is considered superior to digital intravenous arteriography since the latter may result in poor sinus opacification; d) Despite the purported noxious effects of hyperosmolar contrast agents on the capillary endotherlium and brain, a recent review of the complications of angiography in SST failed to establish this correlation [15]. Regardless, newly developed nonionic contrast agents should be utilized if available to maximize safety.

Findings on angiography include a) retardation of arterial and venous flow due to the overall increased resistance to flow, b) poor or nonopacification of the sinus, and c) the presence of (enlarged) collateral veins (figure 7–15).

Attempts to diagnose partial (or segmental) occlusion of the venour sinuses, in particular the transverse sinus, have been fraught with problems. One danger lies in the numerous anatomical variations that exist in this region. For example, nonopacification of one transverse sinus may be due to aplasia of the sinus on that side; clues on plain films, such as lack of occipital sinus groove and small jugular foramen, should be sought to make the diagnosis of aplasia or hypoplasia of the sinus. Another misdiagnosis of partial thrombosis may be due to filling by unopacified vein(s). Thus, Yasargil and Damur [14] have recommended that a special effort be made to identify all the veins by injecting both carotid and vertebral arteries. However, the finding of nonpacification of a venous segment plus the presence of a bypassing collateral venous formation can indicate thrombosis.

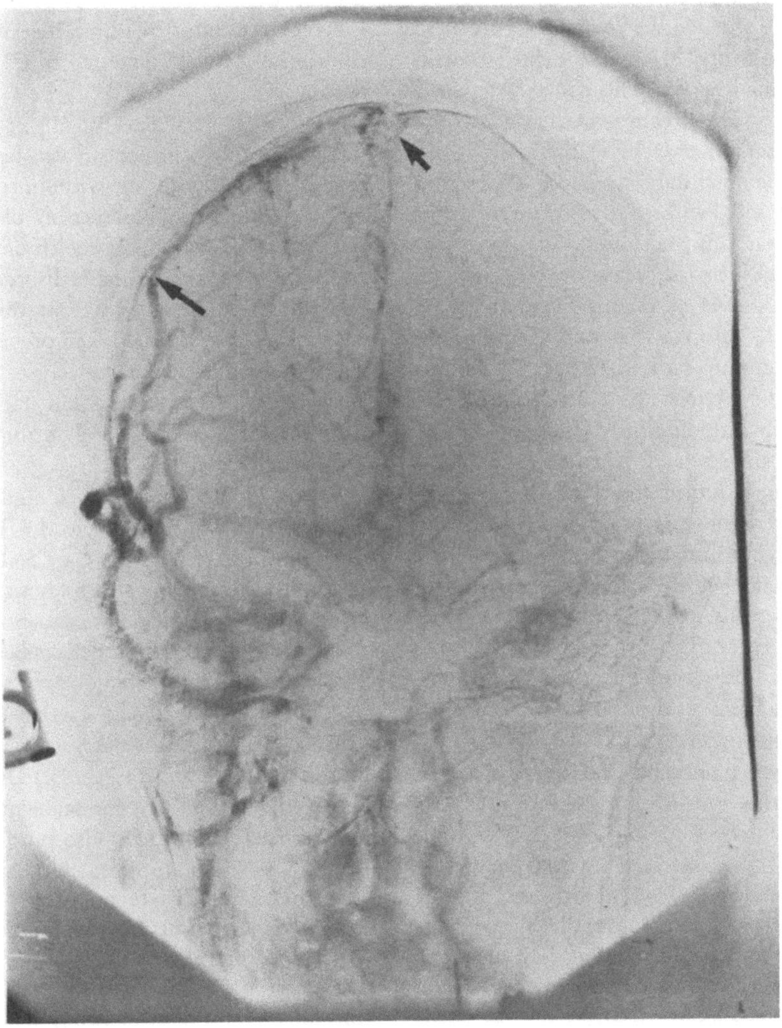

FIGURE 7–15B

FIGURE 7–15. (A) Lateral view A cerebral arteriogram in superior sagittal sinus thrombosis showing absence of the superior sagittal sinus and prominent superficial cortical veins (arrows). (B) Frontal view of the cerebral angiogram demonstrating probable dural collateral vascular channels (small arrow). Large downward draining cortical veins (large arrow) are again seen.

Likewise, a small irregular channel traversing the interrupted segment can suggest recanalization.

In an attempt to circumvent the technical problem of poor sinus opacification, jugular venography has been recommended [35]. But the venous approach is seldom utilized in practice, probably due to the interpretative difficulty of retrograde opacification.

In summary, it is our belief that despite the high accuracy of cerebral angiography in making the diagnosis of SST, the effectiveness of the noninvasive imaging modalities such as CT, MRI, and radionuclide scan represent better diagnostic alternatives. Provided that the interpreter is well acquainted with the anatomy of the venous system and the pathophysiology of SOD and its multiple radiographic manifestations, cerebral angiography can be circumvented.

Cavernous Sinus Thrombosis

NORMAL ANATOMY

The cavernous sinuses (CS) are located on either side of the pituitary fossa, extending from the superior orbital fissure to the petrosal apex from front to back. The floor is formed by the roof of the sphenoid body. Like the SSS the CS are enclosed by two (inner and outer) dura. Posterior to the CS, similar dural reflections enclose the Gasserian (fifth nerve) ganglion. Contents of the CS include the cavernous portion of the internal carotid artery (ICA) and the third, fourth, fifth and sixth cranial nerves. The ICA conveniently divides the CS into various compartments relative to the artery. The cranial nerves lie in the lateral compartment below the inner dural reflection. Numerous partitions are present within the CS. They may represent counterparts of the chordae Willisii of the sagittal sinus, or an intrinsic plexus of veins. The CS are the major recipients of venous blood from the brain, face, orbits, nasopharynx, and ears. Specifically, the CS communicate with both opthalmic veins (inferior and superior), the central retinal vein and sphenoparietal sinus anteriorly, the superior and inferior petrosal sinuses and petro-occipital sinus posteriorly, the uncal and deep sylvian veins superiorly, and the pterygoid plexus inferiorly [36]. In addition, an intercommunication exists between the two CS along the floor of the sella—a knowledge of importance to surgeons in the conduct of transsphenoidal pituitary surgery. Thus, the CS are susceptible to the direct spread of tumor, infection, and thrombosis from these aforementioned anatomical areas as well as from within themselves.

Due to the presence of blood, the CS are dense on noncontrast CT and enhance on contrast CT. The enhancement is homogenous on axial CT without distinction between the intraarterial contrast in the ICA and the intravenous iodine in the CS; but posteriorly on axial CT, the overall homogenous pattern is interrupted by the presence of the CSF-containing Meckel cave (which does not enhance). The cranial nerves passing forward from the brainstem into the CS on their way to the face lie along the horizontal (axial) plane, parallel to the CT scan plane, and are therefore difficult to visualize with axial CT due to partial volume averaging. However, on thin-section direct coronal CT, these cranial nerves are seen on cross section as small rounded defects in the lateral compartment of the CS; they are assembled in a more or less sequential vertical array, the third to sixth cranial nerves superior to inferior. On both axial and direct coronal CT, the outer contour of the CS should assume a gentle concavity outward and at times a straight line. Fullness of the CS is expressed by an outward convexity, which should immediately raise suspicion of disease. This anatomical observation is true also for MRI; however, on MRI the presence of flowing blood within the ICA and the CS produces a hypointense echo relative to the brain on both T1-and T2-weighted images (figure 7–1G), while the CSF located within the Meckel cave becomes bright on T2. The cranial nerves within the CS are usually not visualized on MRI due to the lack of inherent contrast [37] and the inferior spatial resolution of MR as compared to CT.

IMAGING IN CAVERNOUS SINUS THROMBOSIS

Cavernous sinus thrombosis can be caused by infection, trauma (with or without carotid–cavernous fistula), neoplastic obstruction of adjacent venous sinuses, and extension of the thrombus into the CS. In addition, hypercoagulable disorders such as various hematological diseases, steroid therapy, contraceptives, severe dehydration, and pregnancy can produce cavernous sinus thrombosis. Prior to the advent of antibiotics, and mortality rate for this disease was close to 100%; in the era of antibiotics this figure is considerably reduced.

The diagnosis of cavernous sinus thrombosis can be made on contrast-enhanced CT. Filling defects due to clot are seen within the CS along with indirect signs of cavernous sinus thrombosis, such as swelling of the orbital veins and extraocular muscles and scleral thickening. These filling defects bear no relationship to the course of the cranial nerves in the CS. (Not uncommonly, fat in the cavernous sinus can produce spurious filling defects. Large amounts can be found in Cushing disease [38].) Other diseases such as ICA aneurysm, trigeminal neuroma, metastasis, and fistula [36, 38], can also produce filling defects in the CS. Thus, indirect signs of cavernous sinus thrombosis on CT are important corroborative information. In addition, the CS should take on an appearance of fullness, with bulging of the lateral walls of the CS outward. Gas collection in the CS (presumably from gas-forming organisms) has also been reported in cavernous sinus thrombosis [39]. As in SST, preliminary experience with MRI has indicated that, due to a subacute clot (greater than 24 hours) in the CS, the signal emanating from the CS will be hyperintense in respect to the brain on both T1- and

FIGURE 7–16. Cavernous sinus thrombosis in a 59-year-old female who presented with blurred vision, diplopia, headache, and fever. (A) Orbital CT, axial view, showing swelling in the left eyelid (arrowhead). There is enlargement of the superior ophthalmic vein (arrow). Proptosis is present. (B) Both cavernous sinuses contain large filling defects representing clots (arrows). (C) Gallium-67 scintigraphy shows accumulation of the radionuclide in the left orbit and carvenous sinuses (arrows). (D) After 10 days of antibiotic treatment, a second CT reveals resolution of the pathology in the left orbit and cavernous sinus (not shown). There was also a corresponding decrease in Gallium-67 activity (not shown) on the radionuclide scan.

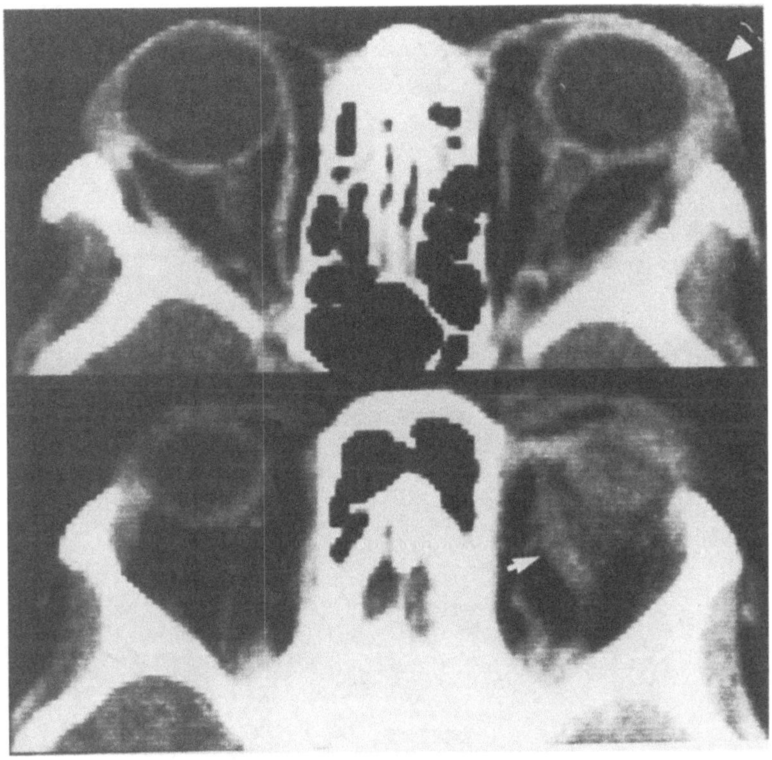

FIGURE 7–16A

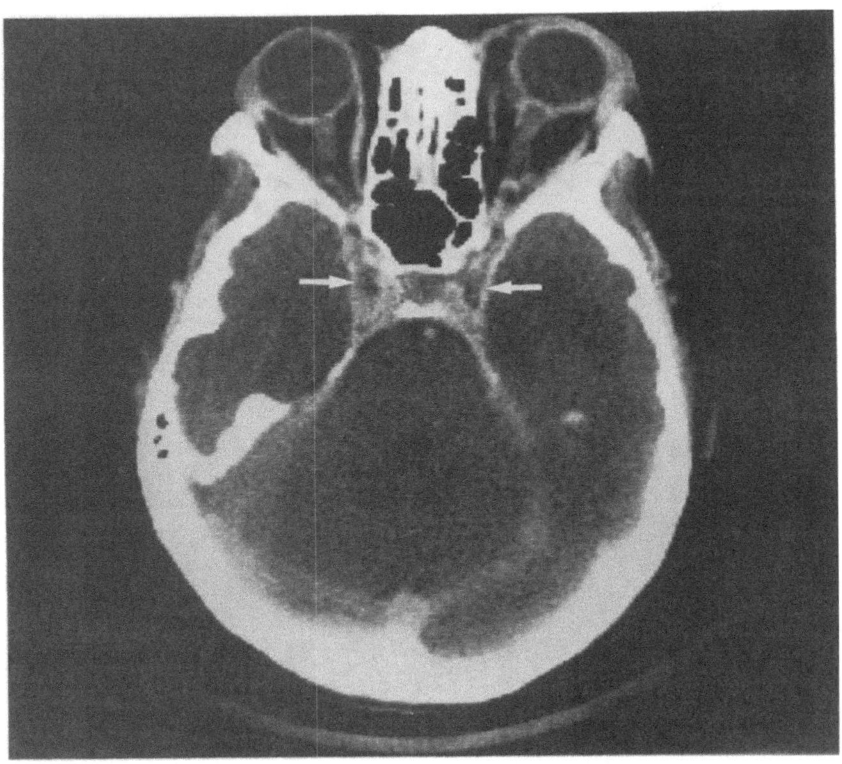

FIGURE 7–16B

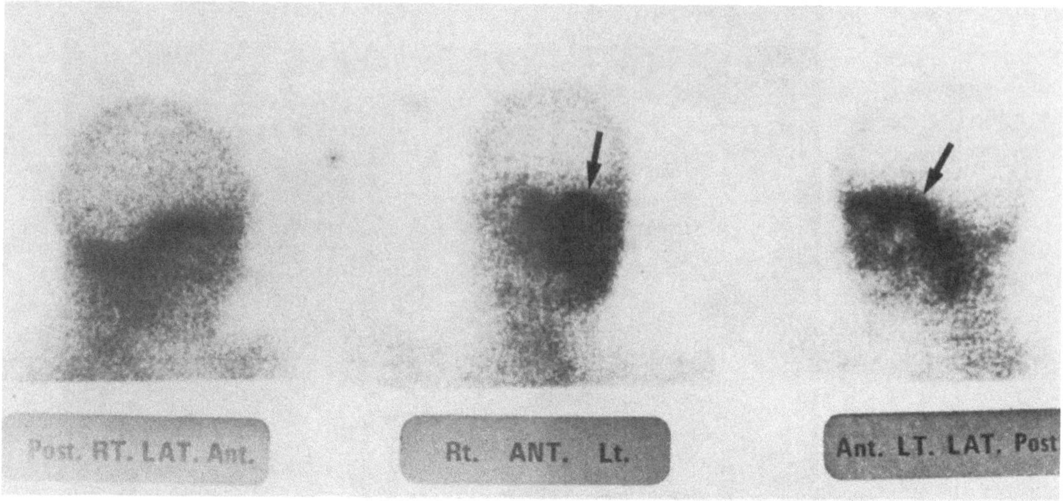

FIGURE 7–16C

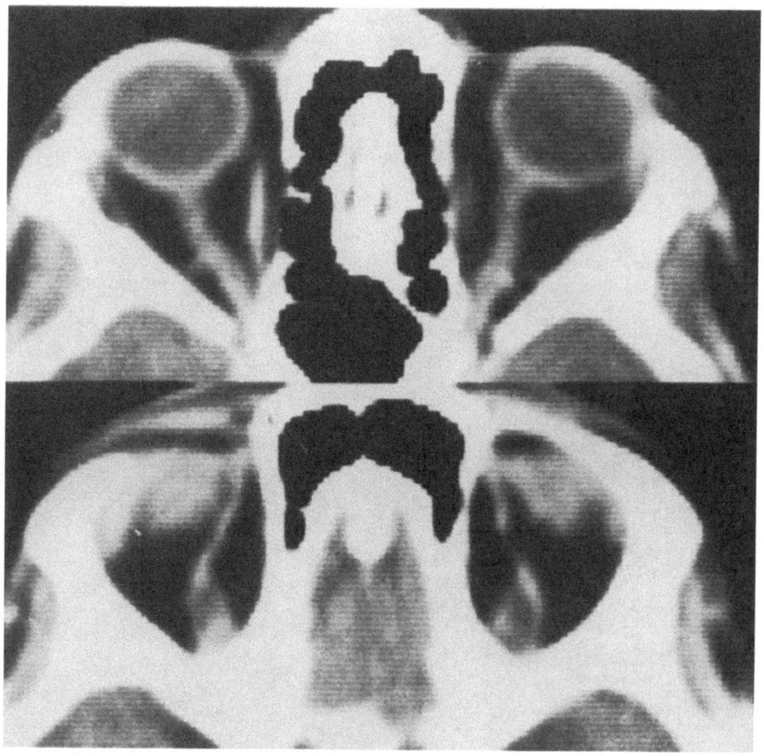

FIGURE 7–16D

T2-weighted images [40]. Because lesion conspicuity is considerably improved on MRI, this modality should be as valuable here as it is in SST diagnosis. Similarly, Indium-111 platelet scintigraphy and Gallium-67 [41] represent good adjunctive noninvasive imaging tests that can help confirm the diagnosis (figure 7–16). Given the multiplicity of imaging modalities available for the diagnosis of cavernous sinus thrombosis, angiography and superior ophthalmic venography should play a small diagnostic role (if any). Indeed, the later may in itself help to disseminate infection and extension of thrombus. On cerebral angiography, cavernous sinus thrombosis is expressed as an absence of filling of the CS; there is retrograde flow along the ophthalmic vein, when this vein is opened [42].

Summary

1. The true incidence of SST is unkown; its clinical manifestations may be more numerous and wide-ranging than realized, extending from mild to devastating types of presentations.
2. Since current imaging modalities can visualize much of pathophysiology that occurs during a major veno-occlusive event, understanding the venous anatomy and pathophysiology is crucial. After a major veno-occlusive event, several venous collateral pathways are available to reduce the physiologic impact on the brain. One of these occurs via a collateral circulation that exists in dural leaves of the sagittal sinus, the cavernous tissue, and the meningeal venules.
3. Despite an impairment in CSF absorption after SST, brain swelling is the predominant finding, due to engorgement of the vascular bed.
4. SST is a multicausal disease. Prognosis is multifactorial, depending largely on the patient's ability to develop venous collateral circulation.
5. Key imaging modalities include CT and MRI. Angiography should be performed as a last resort when the noninvasive modalities are equivocal.
6. The main finding on non-contrast CT is the dense vein sign, which represents a clot lodged in the dural sinus. The most common sign on contrast CT is the empty delta sign, which is an expression of the dural venous collaterals. Brain swelling may also be seen associated with hemorrhagic infarction.
7. On MRI, bright echoes on T1- and T2-weighted images emanating from an enlarged dural sinus are probably highly specific signs of this disease. However, as long as quantitated blood flow on MRI remains elusive, slow flow without veno-occlusion cannot be ruled out. MRI is particularly sensitive to the ancillary signs of SST, such as brain swelling and infarction, which will corroborate this diagnosis.
8. Cavernous sinus thrombosis is a rarer disease than SST clinically. Radiographic findings are similar to SST.

References

1. Kalbag RM, Woolf AL. *Cerebral Venous Thrombosis.* London: Oxford University Press, 1967.
2. Towbin A. The syndrome of latent cerebral venous thrombosis. Its frequency and relation to age and congestive heart failure. *Stroke* 1973;4:419–430.
3. Kalbag RM. Cerebral vein thrombosis. In: Kapp JP, Schmidek HH (eds). *The Cerebral Venous System and Its Disorders.* New York: Grune and Stratton, 1984.
4. Kaplan H, Browder J, Knightly JJ, Rush BF, Browder A. Variations of the cerebral dural sinuses at the torcular herophili. *Am J Surg* 1972;124:456–461.
5. Schmidek HH, Auer LM, Kapp JP. The cerebral venous system. *Neurosurg* 1985;17(14):663–667.
6. Capra NF, Anderson KV. Anatomy of the cerebral venous system. In: Kapp JP, Schmidek HH (eds). *The Cerebral Venous System and Its Disorders.* New Yourk: Grune and Stratton, 1984.
7. Browder J, Kaplan HA. *Cerebral Dural Sinuses and Their Tributaries.* Springfield, Ill: Charles C. Thomas, 1976.
8. O'Connel JFA. Some observations on the cerebral veins. *Brain* 1984;57:484–503.
9. Warwick R, Williams PL. *Gray's Anatomy.* 35th British Edition. W. B. Philadelphia: W. B. Saunders, 1973:693.
10. Balo J. The dural venous sinuses. *Anat Rec* 1950;106:319–325.
11. Newton TH, Gooding CA. Compression of superior sagittal sinus by neonatal calvarial molding. *Radiology* 1975;115:635–639.
12. Foley J. Benign forms of intracranial hypertension "toxic" and "otitic" hydrocephalus. *Brain* 1955;78: 1–41.
13. D'Avella D, Greenberg R, Mingrino S, Scanarini M, Pardatscher K. Alterations in ventricular size and intracranial pressure caused by sagittal sinus pathology in mass. *J Neurosurg* 1980;53:656–661.
14. Yasargil MG, Damur M. Thrombosis of the cerebral veins and dural sinues. In: Newton TH, Potts DG (eds). *Radiology of the Skull and Brain.* Angiography, Vol. 2, Bk 4. St. Louis: C.V. Mosby Company, 1974.
15. Virapongse C, Cazenave C, Quisling R, Sarwar M, Hunter S. The empty delta sign: frequency and significance in 76 cases of dural sinus thrombosis. *Radiology* 1987;162:779–785.
16. Beck DJK, Russel DS. Experiments on thrombosis of the superior longitudinal sinus. *Neurosurg* 1946;3: 337–347.

17. Wendling LR. Intracranial venous sinus thrombosis: diagnosis suggested by computed tomography. *AJR* 1978;130:978–980.
18. Buonanno FS, Moody DM, Ball MR, Laster DW. Computed cranial tomographic findings in cerebral sinovenous occlusion. *JCAT* 1978;2:281–290.
19. D'Avella D, Russo A, Santoro G, Pandolfo N, Cafarelli F, De Blash F, Eugenio del Vivo R. Diagnosis of superior sagittal sinus thrombosis by computerized tomography. *J Neurosurg* 1984;61:1129–1131.
20. Zilkha A, Stenzler SA, Lin JH. Computed tomography of the normal and abnormal superior sagittal sinus. *Clin Rad* 1982;33:415–425.
21. Brant-Zawadski M, Chang GY, McCarty GE. Computed tomography in dural sinus thrombosis. *Arch Neurol* 1982;39:446–447.
22. Goldberg AL, Rosenbaum AE, Wang H, Kim WS, Lewis VL, Hanley DF. Computed tomography of dural sinus thrombosis. *JCAT* 1986;10(10):16–20.
23. Patronas NJ, Duda EE, Mirfakhraee M, Wollmann RL. Superior sagittal sinus thrombosis diagnosed by computed tomography. *Surg Neurol* 1981;15(1): 11–14.
24. Kim KS, Walczak TS. Computed tomography of deep cerebral venous thrombosis. *JCAT* 1986;10(3): 386–390.
25. Imai WK, Everhart RF, Sanders JM. Cerebral venous sinus thrombosis: report of a case and review of the literature. *Pediatrics* 1982;70(6):965–970.
26. Crimmins TJ, Rockswold GL, Yock DH. Progressive postraumatic superior sagittal sinus thrombosis complicated by pulmonary embolism. *J Neurosur* 1984;60:179–182.
27. Macchi PJ, Grossman RI, Gomori JM, Goldberg HI, Zimmerman RA, Bilaniuk LT. High field MR imaging of cerebral venous thrombosis. *JCAT* 1986;10(1):10–15.
28. Erdman WA, Weinreb JC, Cohen JM, Buja CM, Chaney C, Peshock RM. Venous thrombosis: clinical and experimental MR imaging. *Radiology* 1986;161: 233–238.
29. Braun IF, Hoffman JC, Malko JA, Pettigrew RI, Dannels W, Davis PC. Jugular venous thrombosis: MR imaging. *Radiology* 1985;157:357–360.
30. Go RT, Ching LC, Neuman LA. Diagnosis of superior sagittal sinus thrombosis of dynamic and sequential brain scanning. *Neurology* 1973;23(11):1199–1204.
31. Front D, Israel O, Even-Sapir E, Feinsud M. Superior sagittal sinus thrombosis: assessment with Tc-99m labeled red blood cells. *Radiology* 1986;158:453–456.
32. Bridgers SM, Strauss E, Smith EO, Reed D, Ezekowitz MD. Demonstration of superior sagittal sinus thrombosis by Indium-111 platelet scintigraphy. *Arch Neurol* 1986;43:1079–1081.
33. Ezekowitz MD, Wilson DA, Smith EO, et al. Comparison of Indium-111 platelet scintigraphy and two-dimensional echocardiography in the diagnosis of the ventricular thrombosis. *N Eng J Med* 1982;306: 1509–1513.
34. Bowers WJ, Siegel BA. Thombosis imaging with Indium-111 platelets. *Semin Thromb Hemost* 1983; 9:115–131.
35. Vines FS, Davis DO. Clinical–radiological correlation in cerebral venous occlusive disease. *Radiology* 1971;98:9–22.
36. Hasso AN, Pop PM, Thompson JR. High resolution thin section computed tomography of the cavernous sinus. *Radiographics* 1982;2:83–99.
37. Daniels DL, Peck P, Mark L, et al. Magnetic resonance imaging of the cavernous sinus. *AJNR* 1985;6:187.
38. Ahmadi J, Kaene JR, Segall HD. CT observations pertinent to septic cavernous sinus thrombosis. *AJNR* 1985;6:755–758.
39. Curnes JT, Creasy JL, Whaley RL, Scatliff JH. Air in the cavernous sinus: a new sign of septic cavernous sinus thrombosis. *AJNR* 1986;8:176–177.
40. Savino PJ, Grossman RI, Schatz NJ, et al. High field magnetic resonance imaging in the diagnosis of cavernous sinus thrombosis. *Arch Neurol* 1986;43:1081–1082.
41. Palestro CJ, Malat J, Gladstone AG, et al. Gallium scintigraphy in a case of septic cavernous sinus thrombosis. *Clin Nuc Med* 1986;11:636–639.
42. Hasso AN, Lasjaunias P, Thomson JR, Hinshaw DB. Venous occlusions of the cavernous area—a complication of crushing fractures of the sphenoid bone. *Radiology* 1979;132:375–379.

8. NEONATAL INTRACRANIAL HEMORRHAGE

Stanton J. Rosenthal
Joy A. Johnson
Solomon Batnitzky

The institution of neonatal intensive care units has resulted in a revolutionary improvement in the survival of severely premature infants in the past 10 years. This chapter reviews the diagnostic evaluation of intracranial hemorrhage (ICH), which along with respiratory distress syndrome and necrotizing enterocolitis represents the major source of morbidity and mortality in this patient group.

ICH is present in up to 71% of premature neonates undergoing autopsy [1, 2]. However, most published studies of in vivo imaging using computerized tomography (CT) or ultrasound quote an incidence of 35–45% for some degree of ICH in this population [3–13]. The incidence of ICH is most marked in neonates with a gestational age of less than 33 weeks or birth weights of less than 1500 grams. Infants under 30 weeks of gestational age have twice the incidence of those born at 31–34 weeks [14, 15].

The germinal matrix of the subependymal lining of the lateral ventricles is the source of almost all of these hemorrhagic events [14, 16]. The germinal matrix is a highly vascularized area composed of blood vessels and pleoblastic neuronal elements. It forms during early embryologic development and serves as a source of nerve cells that migrate into superficial cortical locations during fetal brain development. This process diminishes during the third trimester, and this change is associated with marked diminution and, ultimately, the disappearance of the germinal matrix. In early development, the germinal matrix lines much of the body of the lateral ventricles, extending from the head of the caudate nucleus backward as it lies adjacent to the body and proximal tail of the caudate. The germinal matrix is largest adjacent to the caudate nucleus throughout gestation, and as it atrophies, at between 30 and 35 weeks of gestation, this remnant in the caudothalamic notch remains as a tangle of fragile vessels with a marked propensity to hemorrhage under the stresses associated with the premature's early postnatal life. The major arterial supply of the germinal matrix is through the recurrent artery of Heubner anteriorly and through the anterior and posterior choroidal arteries posteriorly [14, 16, 17].

An extensive literature has been developed evaluating precipitating causes of germinal matrix hemorrhage in prematures. These have in common either a loss of autoregulation which results in poorly controlled blood flow, an increased blood pressure secondary to hypoxia, or increasing blood flow through the unstable capillary network in the germinal matrix due to resuscitative efforts. This may alternate with episodes of hypotension, predisposing to local ischemic changes that compound the effect of subsequent hemorrhagic episodes. Acute hypoxia such as that associated with pneumothorax has a particularly high association with the sudden development of germinal matrix hemorrhage [8, 18–25]. Less commonly, coagulation abnormalities such as decreased platelet adhesiveness secondary to maternal aspirin use, thrombocytopenia, or the systemic heparinization required in extracorporeal membrane oxygenation (ECMO) may also give rise to germinal matrix hemorrhage. In ECMO therapy, an artifical lung is used as a temporary measure in severe but potentially reversible respiratory failure when other therapeutic modalities have failed. In one series, 25% of the patients developed ICH after placement on ECMO. In addition, preexisting hemorrhages may extend during this therapy [25–30].

The early literature on neonatal ICH emphasized the catastrophic clinical presentation of those cases coming to autopsy. In the most extreme cases, rapid development of stupor, respiratory distress, and seizures are associated with hypotension, bradycardia, bulging of the fontanels, and dropping hemoglobin level. Subsequently, the routine investigation of premature neonates by ultrasound has demonstrated that this overwhelming hemorrhagic catastrophe is an uncommon clinical presentation of the pathology. Over 50% of the cases are clinically silent. If clinical signs are present, the most common is an unexplained drop in hematocrit. Other common signs are alteration of the level of consciousness, hypotonia, and abnormal eye movements [3, 4, 14].

Ninety percent of prenatal intracranial hemorrhages occur within the first three days of life; 36–50% occur in the first postnatal day, 25–32% on the second day, and 15–18% on the third day. Progression of the lesions occurs in 20–40% of infants suffering ICH. The maximum extent of the disease usually occurs within three to five days after the initial hemorrhage [31–33].

Neonatal ICH is conventionally divided into four grades. In Grade I, the hemorrhage is localized to the subependymal region in the locus of the remnant of the germinal matrix in the caudothalamic groove, immediately posterior to the foramen of Monro. Rupture of the hemorrhage into the ventricular system without significant ventricular dilatation represents a Grade II hemorrhage. Intraventricular hemorrhage with significant ventriculomegaly is a Grade III hemorrhage. Intraventricular hemorrhage associated with extension into periventricular brain parenchyma advances the grading to IV [5, 6, 14].

Minor hemorrhages (Grades I and II) are reported in 41–60% of the cases documented in major series. The same reports indicate a 40–49% incidence of major hemorrhages (Grades III and IV) [9, 14, 34]. The highest incidence of major hemorrhages is found in infants with birth weights between 750 and 1250 grams [35, 36]. Ventricular dilatation occurs in just over half of those infants who survive the initial hemorrhage [17, 37]. Hydrocephalus is often transitory [14, 38]. Ventriculomegaly is rare in children demonstrating only Grade I hemorrhages. It is found in 40% of those surviving Grade II, 62% of Grade III, and 80% of those after Grade IV hemorrhage [14]. In addition, porencephaly characteristically develops in infants who survive Grade IV hemorrhages [12].

Although the presence of even a Grade I hemorrhage has been reported to lower survival rates from 94% to 55% in one series [14], most reports indicate that the survival rates of infants with Grade I hemorrhage are similar to those of prematures without neurologic pathology, i.e. a mortality rate of approximately 15% [39, 40]. Grade II hemorrhages add only slightly to the mortality rate (20%) [41]. The major hemorrhages are associated with 40% mortality rates in Grade III and 60% in Grade IV [3, 14, 17]. Major neurological sequelae occur in approximately 15% of the survivors of Grade I hemorrhage, 30% of Grade II suriviors, 40% of Grade III survivors, and up to 90% of Grade IV survivors [3, 40, 42]. Motor deficits are the most common, particularly spastic diplegia and spastic quadraparesis [43]. These major motor deficits are more common with higher grades of hemorrhage. With Grade I and Grade II hemorrhages, mild hyperreflexia or hypertonia are more common [14, 42, 44]. Most long-term studies have followed these children only through their first two years of life. The motor deficits tend to stabilize during the first year and may improve during the second, except in Grade IV hemorrhages [45]. Significant intellectual impairment may occur but is less common than motor deficits [17, 40, 44, 46]. According to Smith (1985), even infants who survived major hemorrhages have a 45% chance of a normal outcome [14]. Additional studies with longer follow-up will be necessary to determine the ultimate result of neonatal intracranial hemorrhage.

Ultrasound Findings

Although A-mode ultrasound was successfully utilized for the noninvasive diagnosis of intracranial midline shift in the 1960s, the very limited anatomic information available with this technique led to its abandonment with the advent of CT scanning in the 1970s. Articulated arm B-mode ultrasound scanning provided more anatomic details than A-mode and was used to a limited extent in the 1970s for imaging of the neonatal brain. Both transcranial and transfontanelle images could be obtained with this equipment, but these fixed installations required transport of critically ill patients to the ultrasound suite. The anatomic detail compared poorly with that available from even early CT scanners. The advent of portable high-resolution sector real-time B-mode ultrasound imaging in the late 1970s brought high-resolution cross-sectional imaging to the patient's bedside, obviating the need to remove unstable prematures from their protective environment. This approach, with subsequent improvements in resolution, has brought portable real-time ultrasound into a predominant position in the imaging of neonatal ICH [47–50].

Primary imaging planes used in ultrasound neonatal dignosis take advantage of the acoustic window of the anterior fontanelle. From this position, multiple planes of section are possible, but angled coronal and oblique parasagittal scans are of the greatest importance in the evaluation of ICH and its sequelae. A phased array or mechanical sector transducer of 5 MHz or higher frequency gives the best results. Linear arrays are less efficacious since they allow visualization of only a narrow central portion of the brain. A number of reviews of the detailed sonographic evaluation of intracranial anatomy are avail-

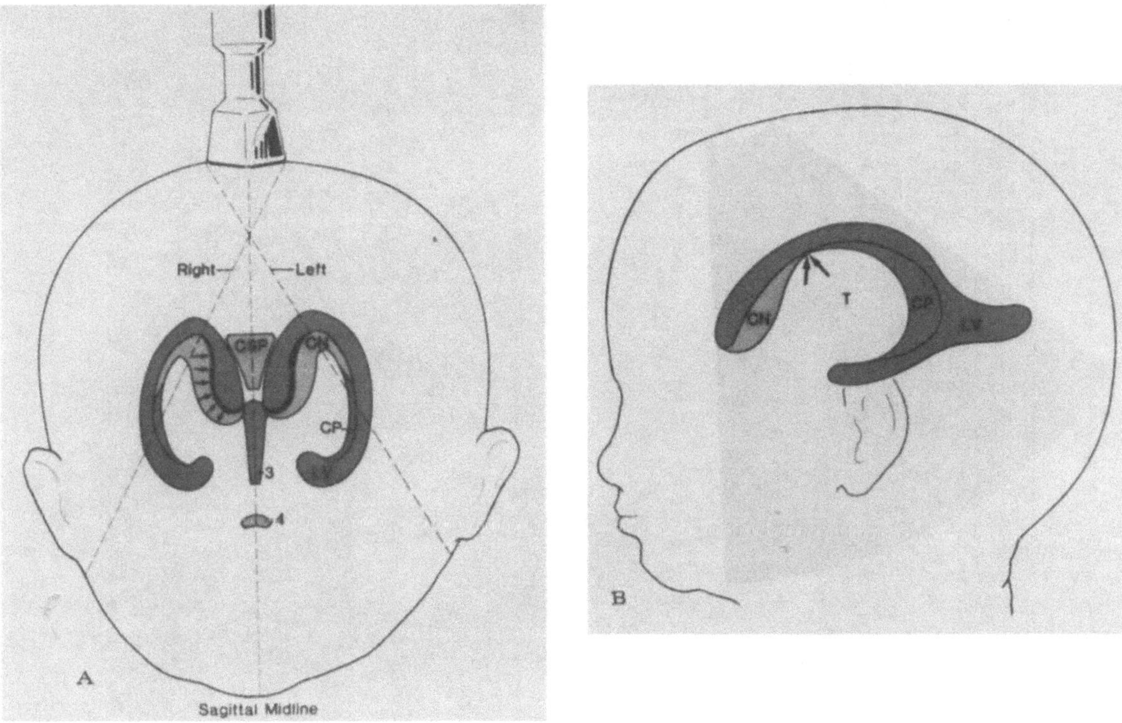

FIGURE 8–1. (A) Diagrammatic representation of coronal sonogram. Right, left, and sagittal midline indicate planes of sagittal imaging. CSP = cavum septi pellucidi; CN = caudate nucleus; CP = choroid plexus; LV = lateral ventricle; 3 = third ventricle; 4 = fourth ventricle. Black line with arrows indicates region of germinal matrix. (B) Diagrammatic representation of sagittal sonogram. CN = caudate nucleus; CP = choroid plexus; LV = lateral ventricle; T = area of thalamus. Arrows indicate typical site of germinal matrix hemorrhage.

able [4, 14, 17, 50–54], but for the purpose of diagnosis, staging, and follow-up of ICH, a few basic anatomic details are of critical importance.

From the access point at the anterior fontanelle, the echo-free ventricular system and the intensely echogenic choroid plexus dominate the central intracranial anatomy. Although there are occasionally mild variations in the size of normal lateral ventricles, the ventricular system in the vast majority of neonates is bilaterally symmetric. It is useful to begin the scan in a coronal plane to take advantage of its symmetry, which allows instantaneous comparison of the two sides for subtle discrepancies as well as the maintenance of proper orientation during the scan and its interpretation. It is helpful, therefore, to begin the scan by angling posteriorly from the anterior fontanelle in the coronal plane (figure 8–1A). With this orientation, the large concentration of choroid plexus in the trigone of the lateral ventricles is readily apparent as bright, thick, straight lines that progressively diverge posteriorly. Deep to the choroid are the thin, moderately echogenic, posteriorly divergent lines of the tentorium and the large, moderately echogenic mass of the cerbellar folia.

As the transducer is angled slowly forward in the coronal plane through the lateral ventricular bodies, the normal choroid plexus smoothly and progressively thins and ultimately plunges through the foramina of Munro into the roof of the third ventricle. Any echogenic material in or adjacent to the ventricular frontal horns is hemorrhage rather than normal choroid plexus [14, 51, 55]. In the lateral ventricular trigone, the medial and lateral walls of the ventricles are roughly parallel, but as the coronal image is brought forward, the lateral walls of the normal nondilated frontal horns become progressively medially concave (figure 8–2A).

In the parasagittal images, the transducer is angled to one side, producing a plane that transects the

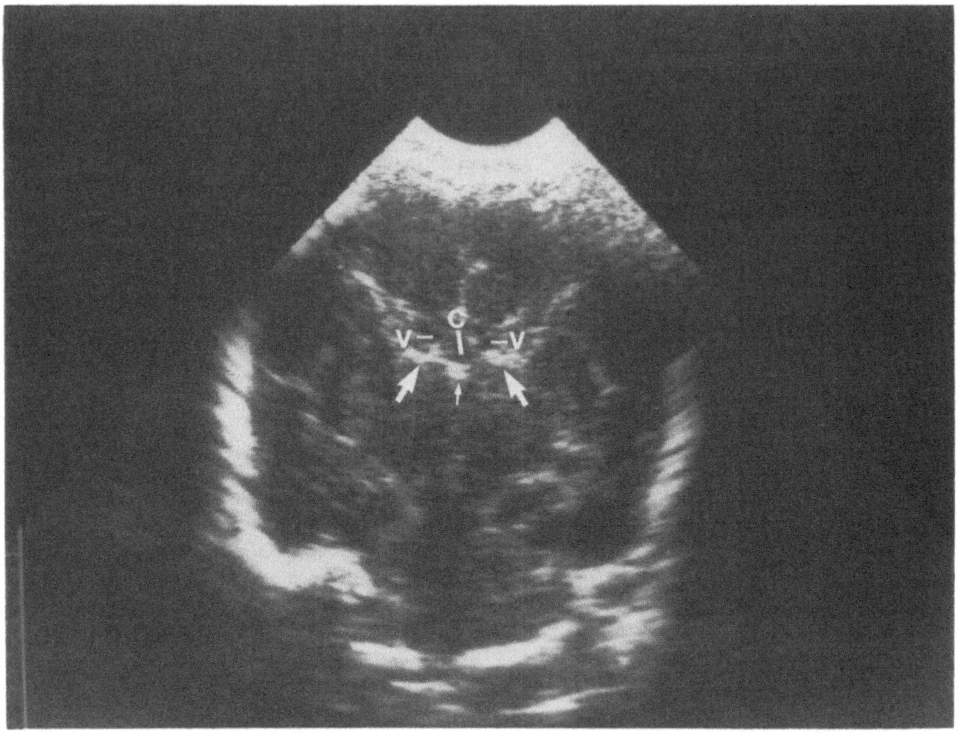

FIGURE 8–2A

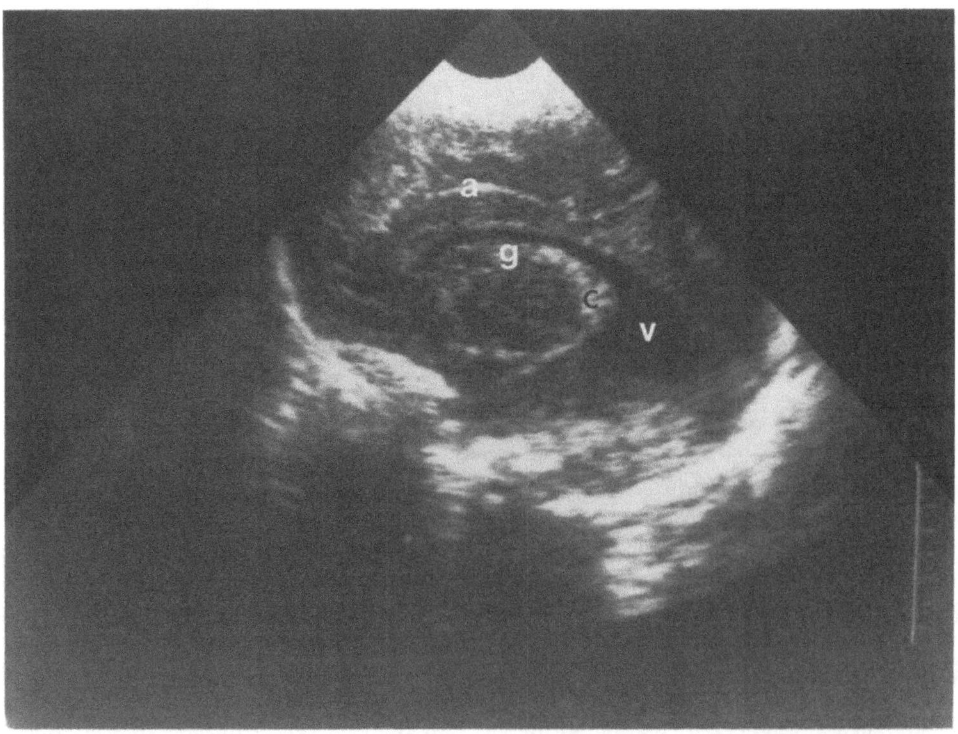

FIGURE 8–2B

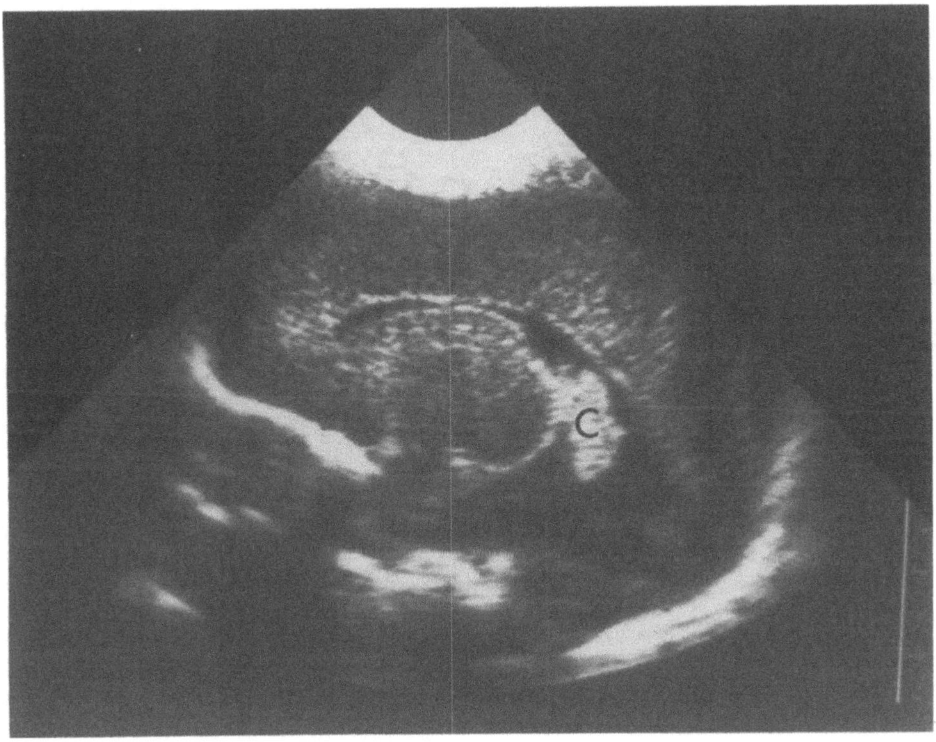

FIGURE 8–2C

FIGURE 8–2. (A) Normal coronal ultrasound scan through the midventricular bodies. Normal echogenic choroid in the lateral (large arrows) and third (small arrow) ventricles. Lucent cavum septi pellucidi (c) is flanked by normal lucent ventricles (v). (B) Normal sagittal ultrasound scan through the lateral ventricular body. Ventricle (v), normal echogenic choroid plexus (c), cingulate sulcus containing collosalmarginal artery (a), and area of germinal matrix (g) are shown. (C) Normal sagittal ultrasound scan through lateral ventricle in a 24-week gestational age patient. Note lack of sulci, larger ventricle, and prominent, bulky choroid plexus (c).

elongated curve formed by the lateral ventricle from the frontal through the temporal and occipital horns (figure 8–1B). Once this orientation has been established, additional scans should be made with as great a degree of lateral arc as possible to demonstrate the Sylvian fissure and lateral subarachnoid spaces. Arcing medially, particular care is necessary to show the frontally tapering shape of the normal choroid plexus. The inferolateral aspects of the lateral ventricular bodies just posterior to the foramina of Munro should be inspected with special care to detect limited subependymal hemorrhages. Scans in the midline show the anatomy of the third and fourth ventricles and the aqueduct. The sequence is then repeated for the contralateral side (figure 8–2B, C).

Sonographically, Grade I hemorrhages produce unilateral or bilateral echogenic foci inferolateral to the lateral ventricles at or near the caudothalamic notch. The echogenicity of the hemorrhage is similar to that of the choroid plexus. These small masses may remain echogenic for up to two months before becoming progressively isoechoic (figure 8–3) [14, 51]. A linear group of echoes near the germinal matrix seen only on parasagittal scans is most likely an artifact caused by specular reflection of sound from the wall of the lateral ventricle [56]. A genuine hemorrhage should be visible in two planes. A minority of germinal matrix hematomas regress to focal cysts at the site of the hemorrhage (figure 8–4) [54, 57].

Each of the three higher grades of ICH provides new ultrasound findings in addition to but not to the exclusion of the pathology demonstrated in the lower grades. Thus, in Grade II hemorrhage, the echogenic or lucent foci in the area of the caudothalamic notch are visualized and, in addition, echogenic clot is seen within the lumen of the cerebral ventricles. Large amounts of intraventricular thrombus are clearly visible as echogenic masses or as CSF–thrombus fluid levels, which are often fixed in position and more often found in the dependent portion of the ventri-

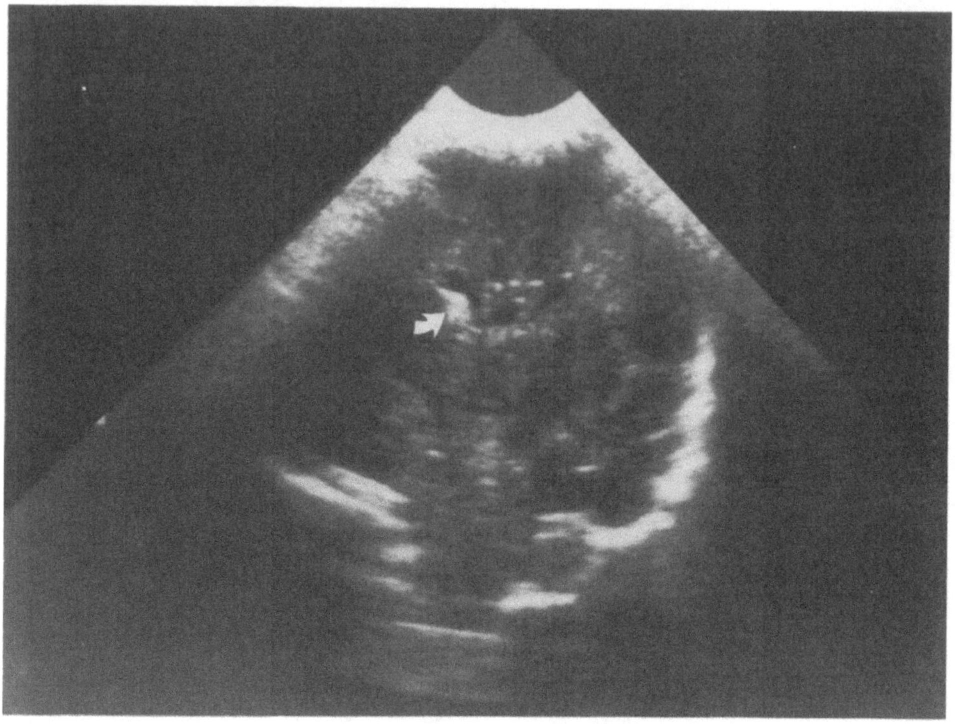

FIGURE 8–3A

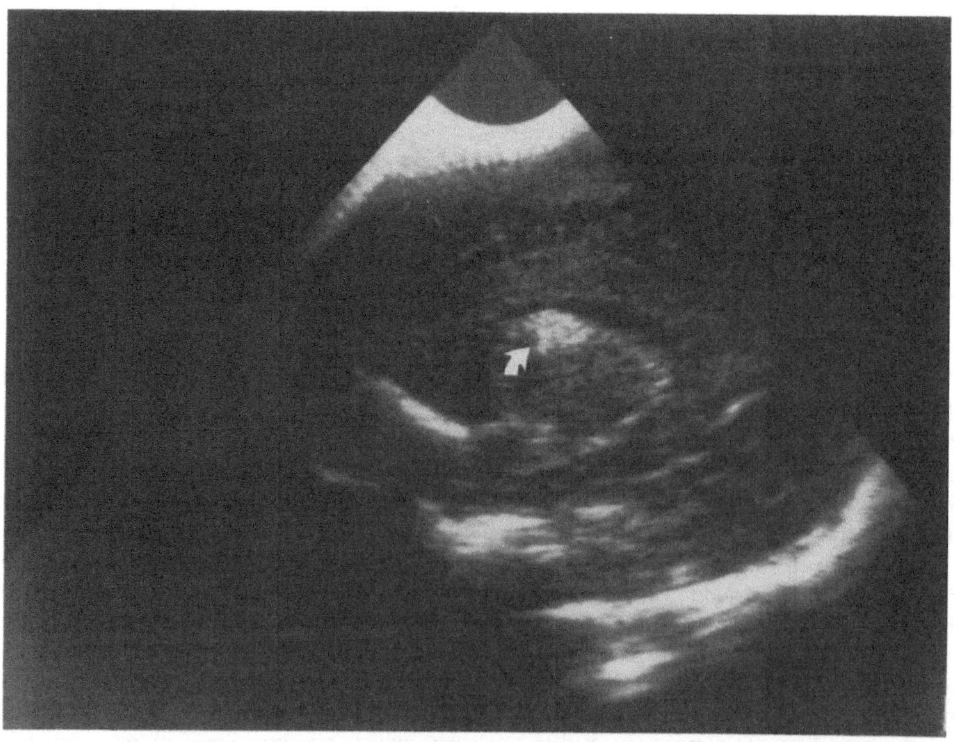

FIGURE 8–3B

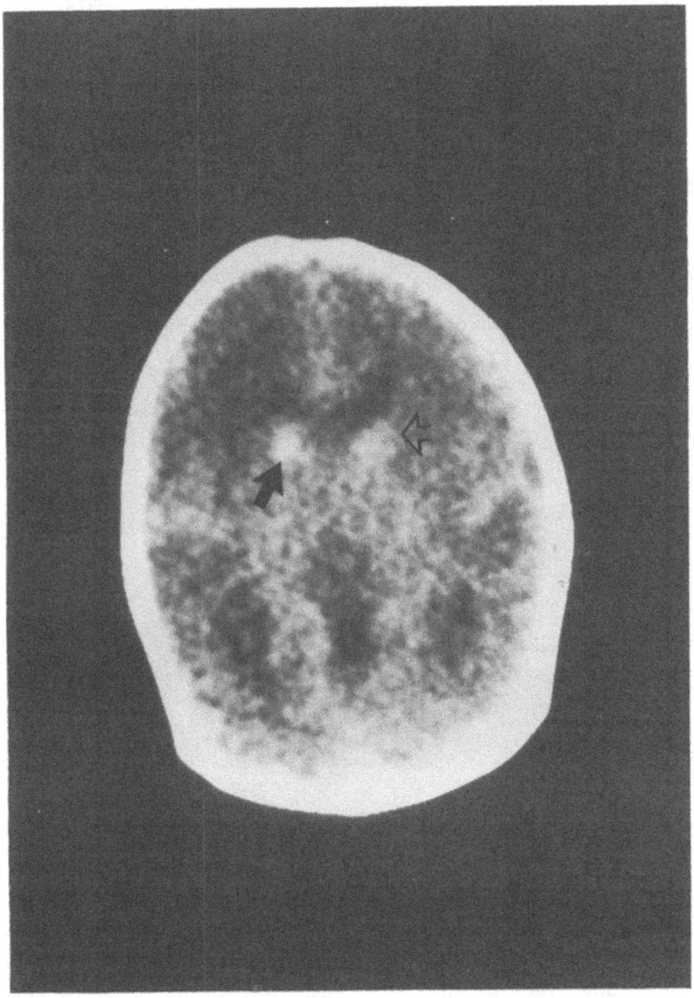

FIGURE 8–3C

FIGURE 8–3. (A) Coronal ultrasound image of right germinal matrix hemorrhage (arrow). Grade I ICH. (B) Sagittal ultrasound image of Grade I germinal matrix hemorrhage (arrow). (C) Axial CT of right germinal matrix hemorrhage (arrow). Note normal slightly dense left caudate (open arrow).

cular system, i.e., the occipital and temporal horns. Smaller amounts of clot may become adherent to the choroid plexus, or the hemorrhage may occur within the choroid plexus itself [58, 49]. In these cases, differentiation between small intraventricular thrombi and anatomic variation of the choroid plexus may be difficult. In general, the choroid is thickest in the area of the trigone and tapers smoothly as it courses anteriorly into the floor of the lateral ventricular body and the roof of the temporal horn. In general, there is little, if any, extension of normal choroid into the occipital horn [14]. A normal choroid plexus also shows little, if any, lobulation [54]. Alteration of these normal appearances by apparent major extension of the choroid plexus into the occipital horn or marked lobulation of the structure is suspicious for the presence of a small intraventricular thrombus. Normal variation can usually be excluded by follow-up scans to document change in the appearance of the clot as it resolves or extends. The normal choroid does not project anterior to the plane of the foramen of Monro. Echogenic intraventricular material in the frontal horns can be confidently identified as a hematoma (figure 8–5).

The presence of intraventricular blood is often associated with subsequent and usually transient development of ventriculomegaly, primarily secon-

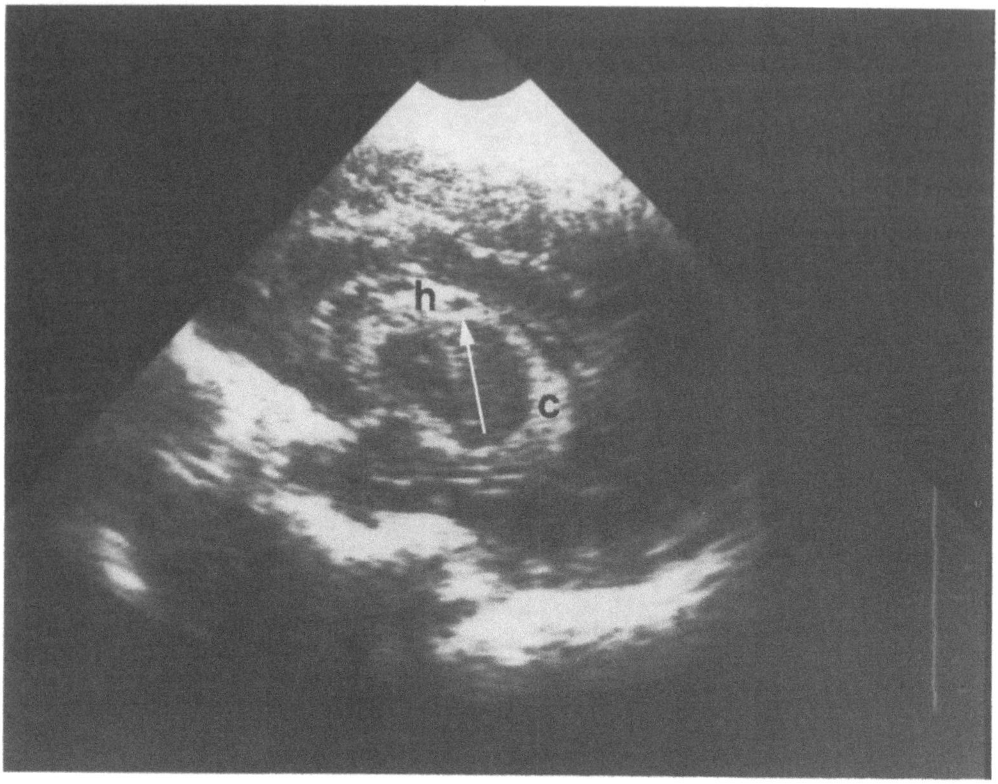

FIGURE 8–4. Sagittal ultrasound image of subependymal cyst (arrow) forming within germinal matrix hemorrhage (h). Normal choroid plexus (C).

dary to interruption of normal CSF pathways. This ventricular dilatation has been treated aggressively in the past, but more recent studies indicate that the phenomenon is transient in the overwhelming majority of cases [14, 60–65]. In general, ventricular size will either stabilize or return to normal with conservative management. In Grant's series, only 5% of patients with ventriculomegaly associated with Grade II hemorrhage required intervention because of progressive hydrocephalus [14]. Close follow-up with sonograms at weekly intervals is useful to document the course of these patients.

Marked ventriculomegaly is obvious sonographically, but early signs may be more subtle. A change in the configuration of the frontal horns from slitlike to bulbous with rounding of the lateral angles and a lateral ventricular depth of greater than 3 mm are indications of early enlargement [54].

Very rarely, a choroid plexus papilloma may mimic a Grade II hemorrhage with ventriculomegaly. This unusual benign tumor may arise congenitally. It produces an echogenic mass in the choroid plexus similar to an adherent hematoma. The absence of associated germinal matrix hemorrhage and the demonstration of chaotic bidirectional flow in both systole and diastole by Doppler sonography are characteristic of this tumor [66].

Sonographically, the large amounts of intraventricular clot in Grade III hemorrhages fill and distend the lateral ventricles and are often identified in other portions of the ventricular system as well. The echogenic clot usually obscures the choroid plexus. Either symmetric or asymmetric dilatation of the ventricles may be present. If the clot resolves, an echogenic rind is usually preserved along its periphery as progressive liquifaction occurs in the remainder of the hematoma. Eventually, the clot retracts from the

FIGURE 8–5. (A) Intraventricular hemorrhage involves the left lateral (L) and third (T) ventricles. Grade II hemorrhage. Coronal sonogram. (B) Sagittal image of Grade II (intraventricular) hemorrhage shows echogenic cast of ventricle filling frontal (F), occipital (O), and temporal (T) horns. (C) Axial CT of intraventricular hemorrhage (h) demonstrates predominately left blood with some involvement on the right.

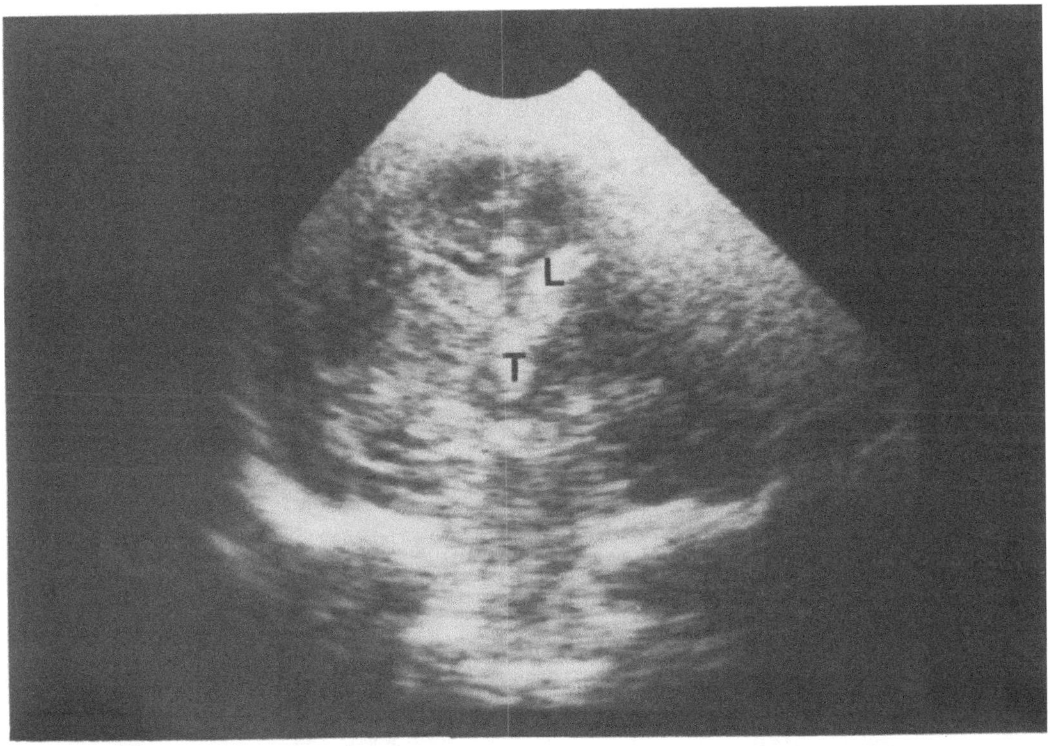

FIGURE 8–5A

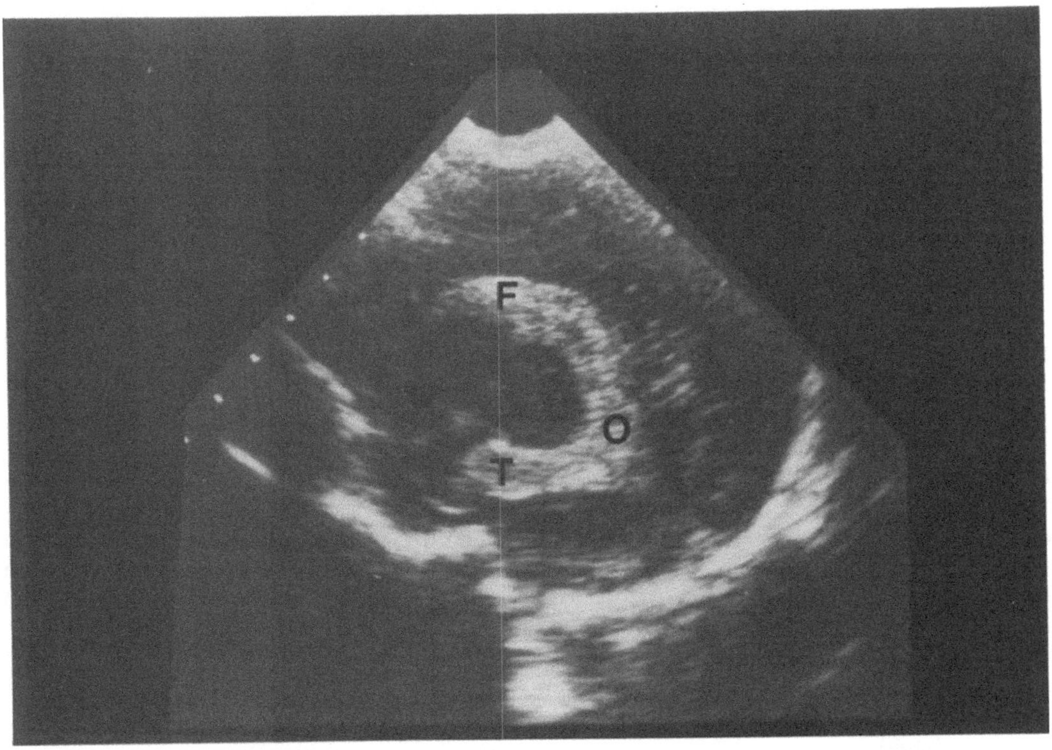

FIGURE 8–5B

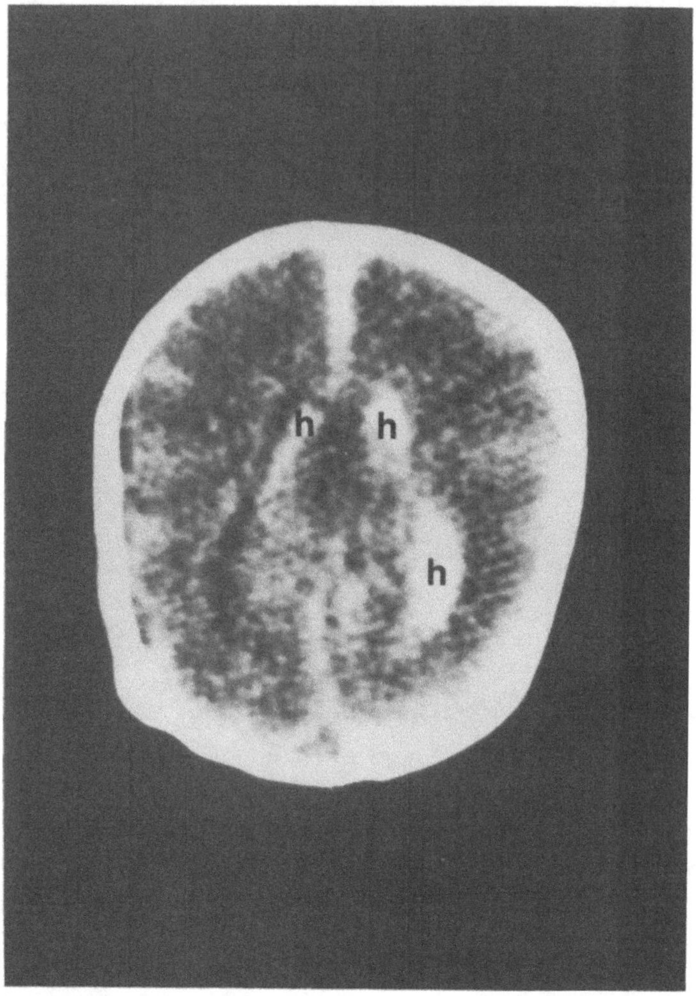

FIGURE 8–5C

ventricular wall and often produces the effect of a shrunken cast of the involved ventricles surrounded by CSF (figure 8–6) [51, 54].

The worst prognosis and highest mortality rates are associated with extension of the intraventricular thrombus through the ependyma into the adjacent brain parenchyma. These Grade IV hemorrhages may be present initially or may occur as complications of lesser degress of bleeding. The clot most often projects from the superolateral and anterior aspects of the ventricular body into the high frontal or parietal areas. The clot is intensely echogenic and usually demonstrates a similar pattern of resolution and retraction to that found in Grade III hemorrhages. The echogenic rind or cast within the damaged areas may persist for months. Because of the hemorrhagic necrosis of that portion of the brain parenchyma involved in the hematoma, the eventual production of porencephalic cysts in the necrotic area is typical (figure 8–7) [14, 67].

Recent studies have suggested that brain ischemia or infarction plays a primary role in the intraparenchymal extension of germinal matrix hemorrhage.

FIGURE 8–6. (A) Coronal ultrasound image demonstrates Grade III intracranial hemorrhage. Note right germinal matrix hemorrhage (H), intraventricular component in temporal horn (h), and enlarged ventricles (V). (B) Progression of hydrocephalus is evident on this coronal image. Same patient as (A). Enlarged lateral ventricles (V), temporal horns (T), and third ventricle (arrow) are shown. Also note persistently echogenic germinal matrix hemorrhage (H). (C) Sagittal view of Grade III ICH. Echogenic germinal matrix hemorrhage (H), enlarged ventricle (V), and blood clot in occipital horn (C) are shown.

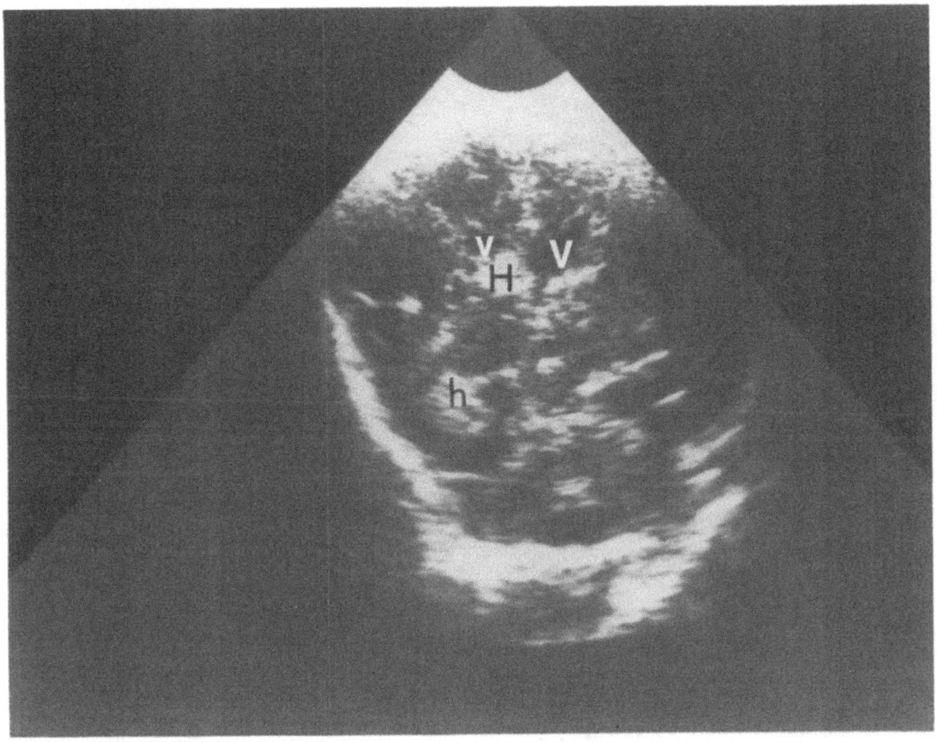

FIGURE 8–6A

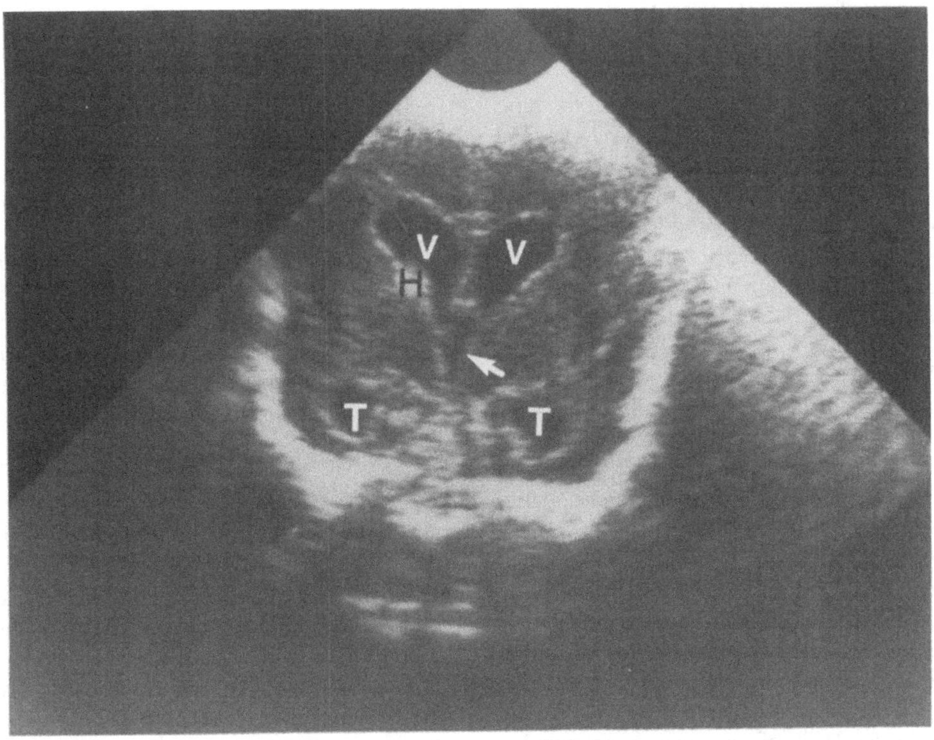

FIGURE 8–6B

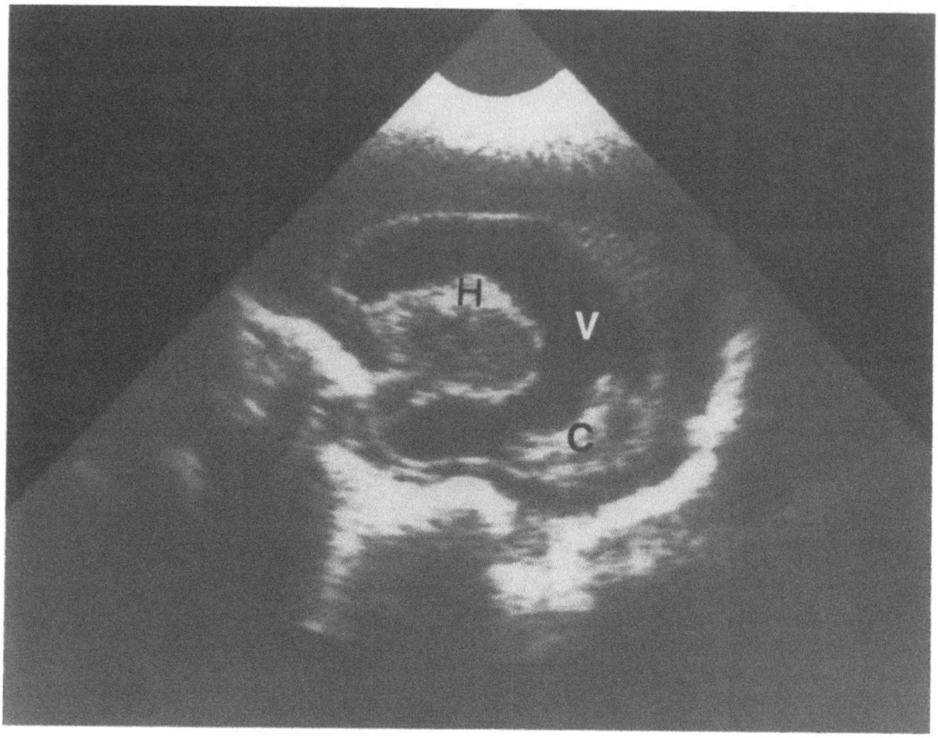

FIGURE 8–6C

The same mechanisms may lead to the development of periventricular leukomalacia (PVL). Although this lesion is much less common than germinal matrix hemorrhage, it has a much higher morbidity rate. Spastic quadriplegia or diplegia and developmental retardation are present in most survivors. Visual deficits are also common [14, 68, 69].

Sonographically, acute PVL produces a wide band of increased echogenicity adjacent to the lateral ventricles. This must be differentiated from a normal periventricular echogenic halo which is frequently visualized, especially with higher-frequency transducers. PVL is usually as echogenic as the choroid plexus. The normal halo is less bright. In some cases, definite differentiation may not be possible without follow-up songrams. Small cysts form in the infarcted periventricular parenchyma up to three weeks after the ischemic episode (figure 8–8). Identification of these characteristic cysts permits a confident diagnosis of PVL [69–75].

Although intracranial hemorrhage in prematures follows the patterns just described in the vast majority of cases, occasional variations exist. Cerebellar hemorrhage is the most frequent of these. This lesion is present in 8–21% of prematures undergoing autopsy [76, 77], but most of these hemorrhages are small. The lesion is rarely diagnosed in life unless it is large. Most often, it occurs as a complication or extension of supratentorial hemorrhage and is associated with a high mortality in prematures [3]. Sonographically, the hemorrhage is a brightly echogenic infratentorial focus often associated with obscuration of normal posterior fossa anatomic landmarks (figure 8–9) [78].

Other etiologies of neonatal parenchymal intracranial hemorrhages are extremely rare. These include coagulation defects, trauma (which usually presents as extraaxial hemorrhage), congenital aneurysms, arteriovenous malformations (usually involving the vein of Galen), and tumors such as congenital glioma [3].

CT and MRI

CT and MRI can play important roles in the evaluation of hemorrhagic disorders in infants. Ultrasound is generally obtained initially due to its advantages of cost, portability, lack of radiation, and lack of sedation required. The majority of lesions can be correctly diagnosed by ultrasound. However, some abnormalities, particularly superficial fluid collections and peripheral parenchymal lesions, are better studied by CT and MRI.

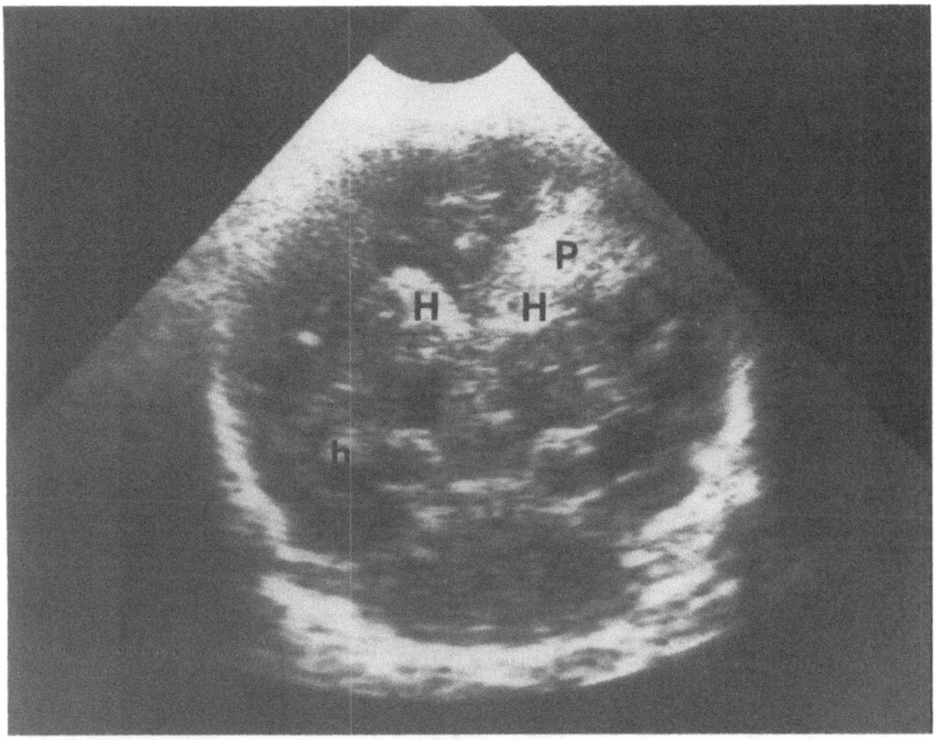

FIGURE 8–7A

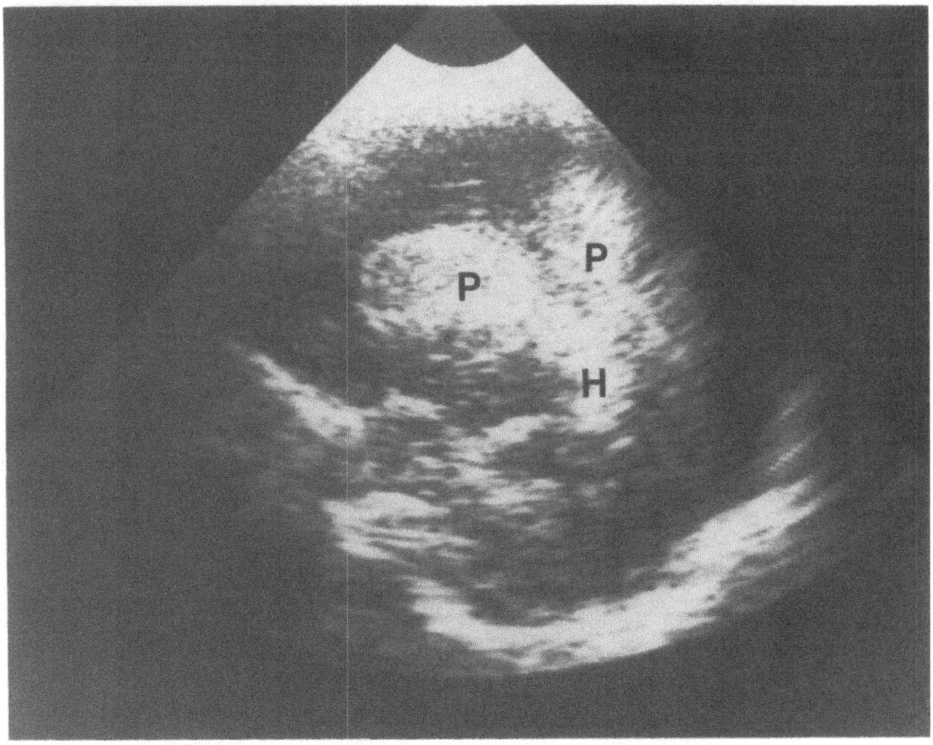

FIGURE 8–7B

FIGURE 8–7. (A) Grade IV hemorrhage on coronal ultrasound. Parenchymal hemorrhage (P) extends from left germinal matrix hemorrhage. Note bilateral germinal matrix hemorrhage (H) and clot in right temporal horn (h). (B) Sagittal image demonstrates the parenchymal (P) and intraventricular (H) blood. (C) Acute phase axial CT demonstrates intraventricular (h) and parenchymal (P) hemorrhage. (D) Follow-up scan shows progression of hydrocephalus involving frontal (F), third (arrow), and temporal (T) horns. Note retraction and change in echogenicity of the parenchymal clot as it ages (P). (E) Late sequela of Grade IV hemorrhage. Porencephaly (P) is seen on follow-up coronal ultrasound scan. Note persistent hydrocephalus and residual clot (C). Coronal image. (F) Late CT shows porencephaly (P) connecting to persistently dilated ventricles (V).

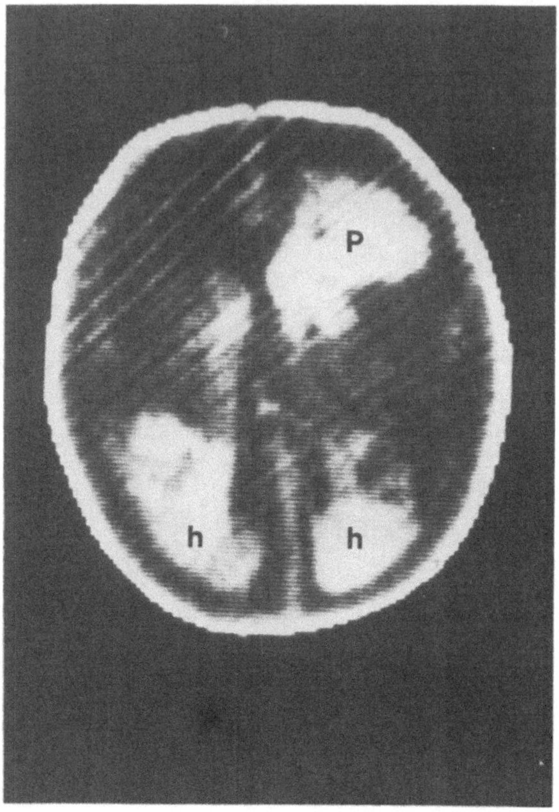

FIGURE 8–7C

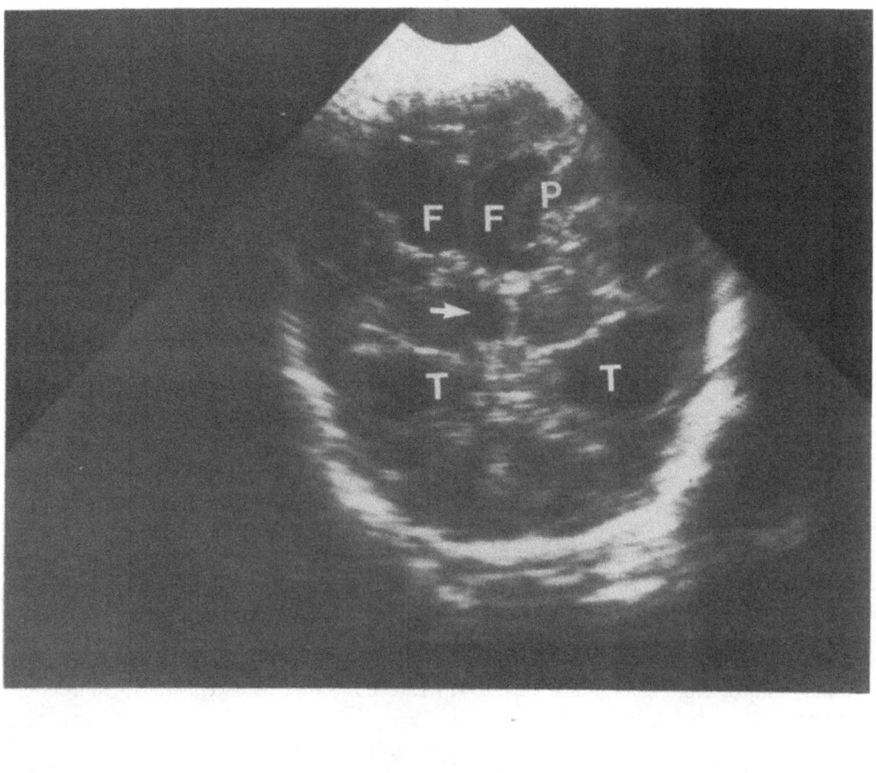

FIGURE 8–7D

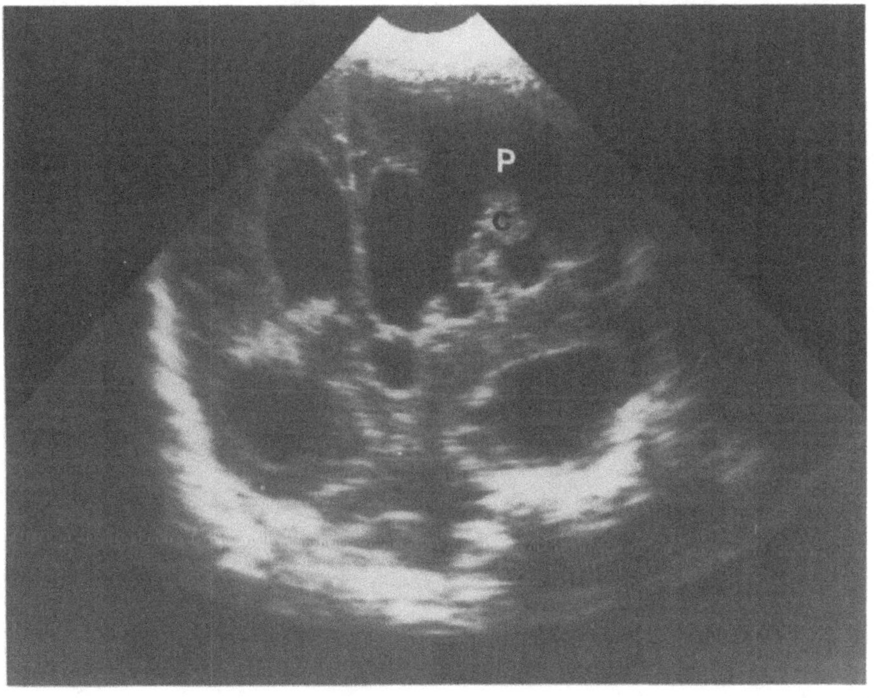

FIGURE 8–7E

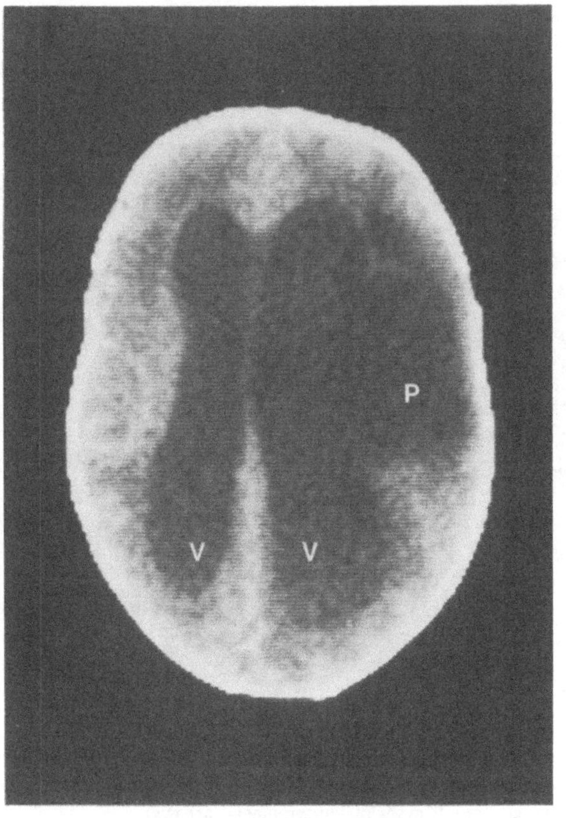

FIGURE 8–7F

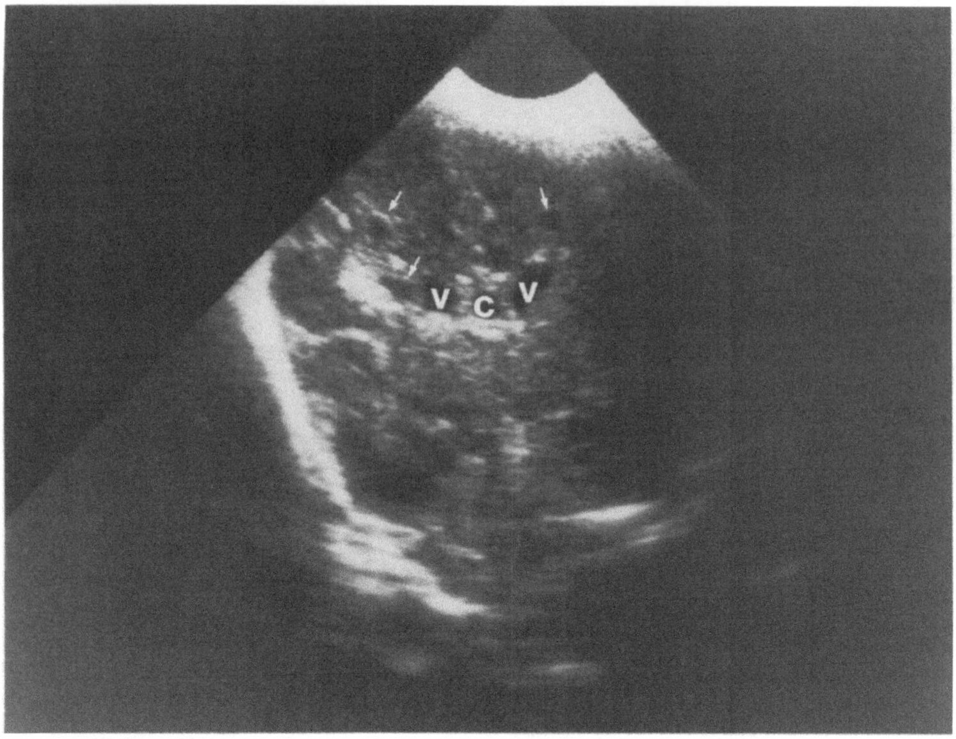

FIGURE 8–8A

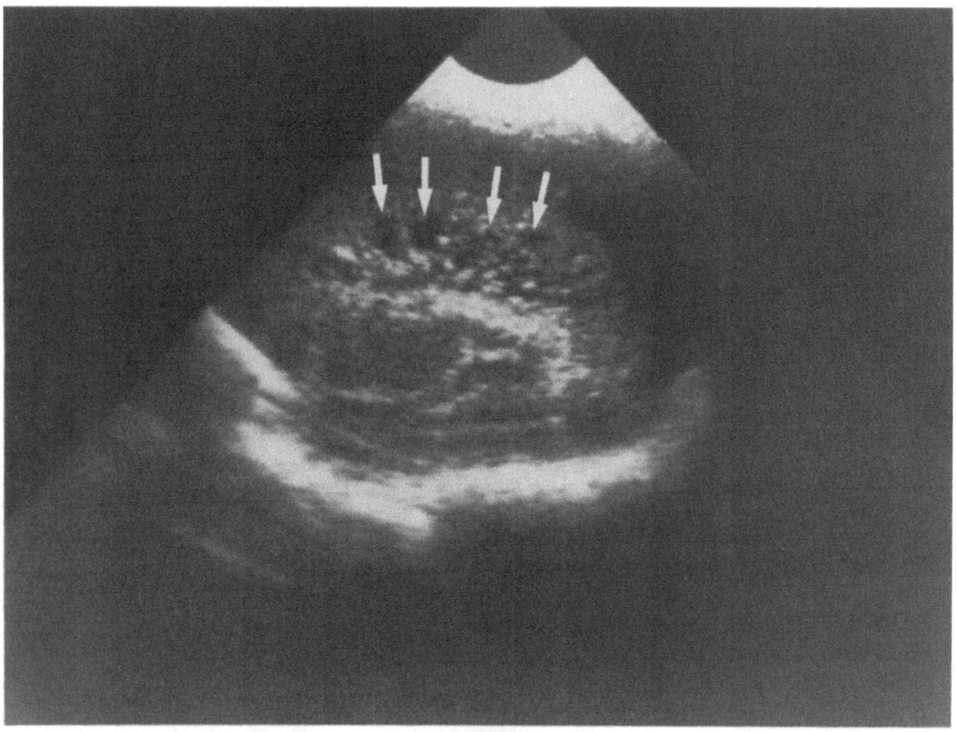

FIGURE 8–8B

FIGURE 8–8. (A) Coronal ultrasound image demonstrates multiple small periventricular cysts (arrows) typical of late periventricular leukomalacia. V-ventricles; c-cavum septi pellucidi. (B) Sagittal image just lateral to edge of ventricle, showing multiple large and small cysts of leukomalacia (arrows).

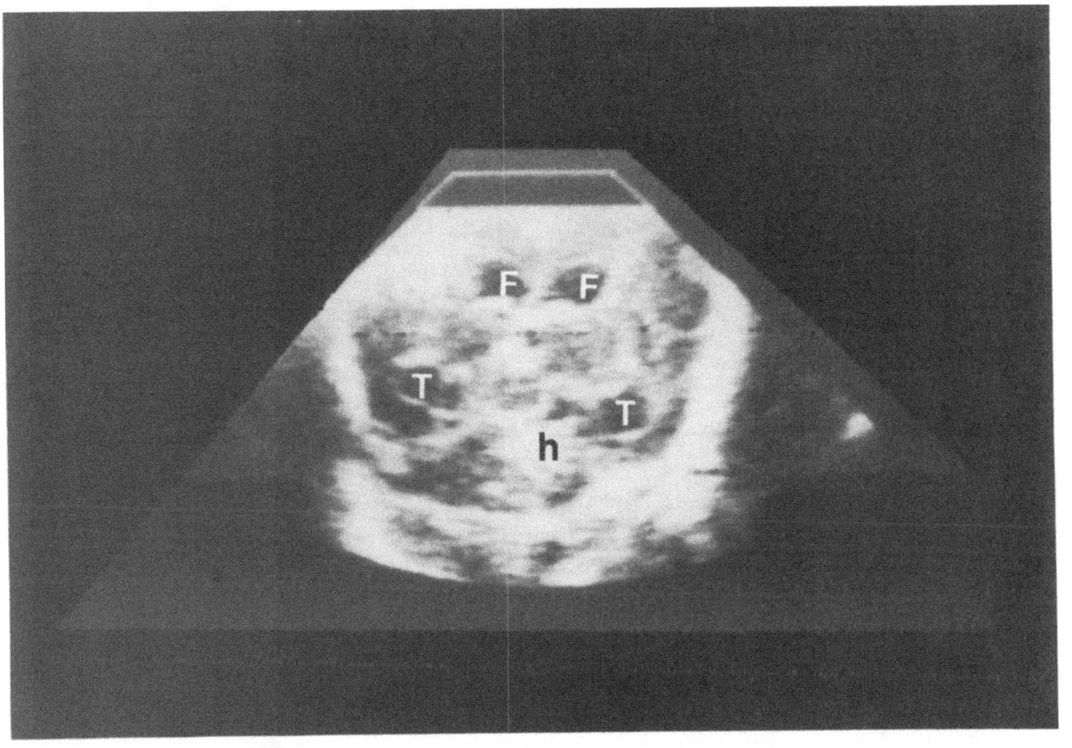

FIGURE 8–9A

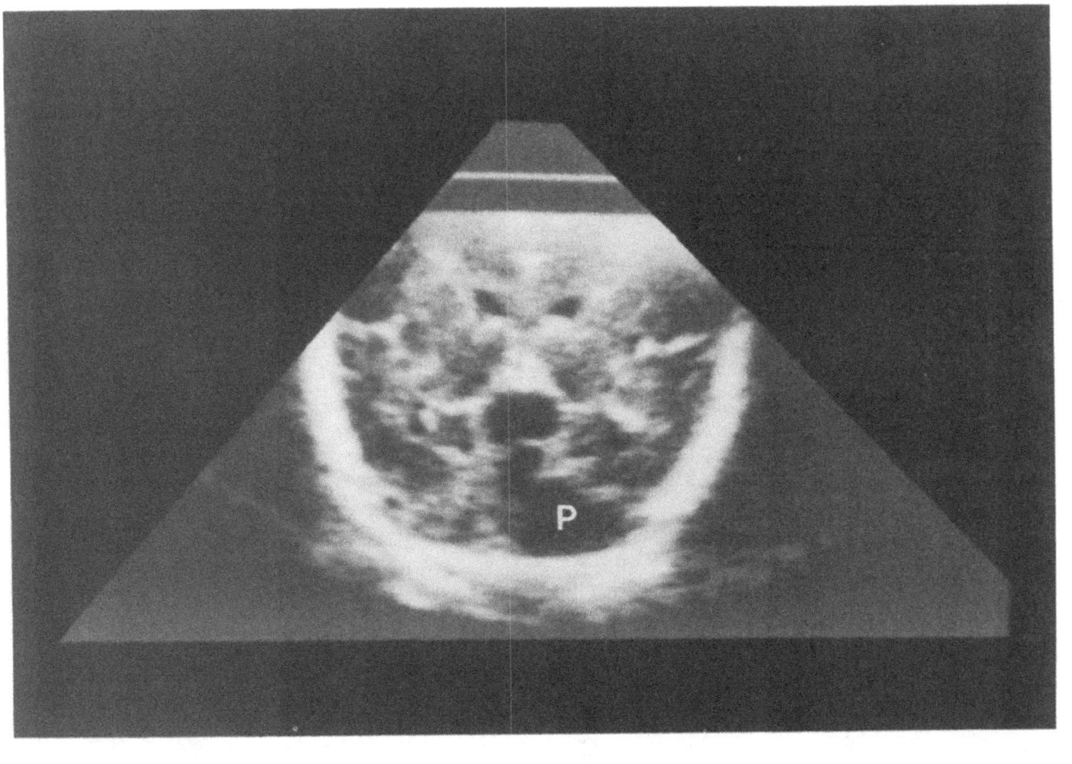

FIGURE 8–9B

FIGURE 8–9. (A) Coronal sonographic image of cerebellar hemorrhage (h). Note hydrocephalus involving frontal (F) and temporal (T) horns. (Courtesy of Carol Rumack, M.D.) (B) Follow-up scan reveals porencephaly (P) in area of cerebellar hemorrhage. Coronal scan. (Courtesy of Carol Rumack, M.D.)

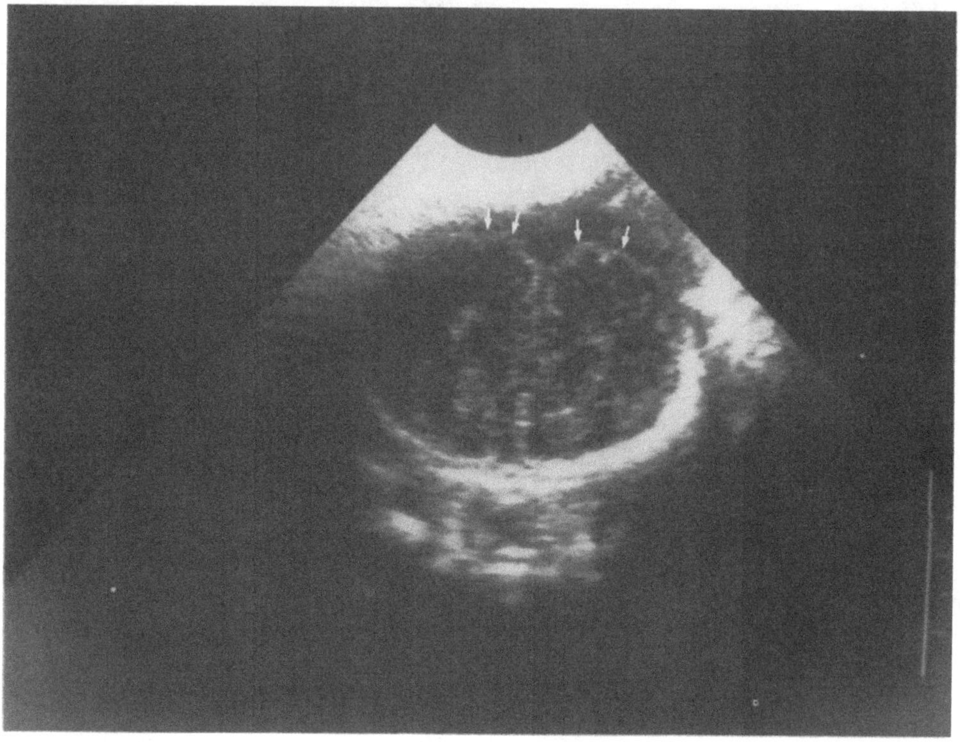

FIGURE 8–10A

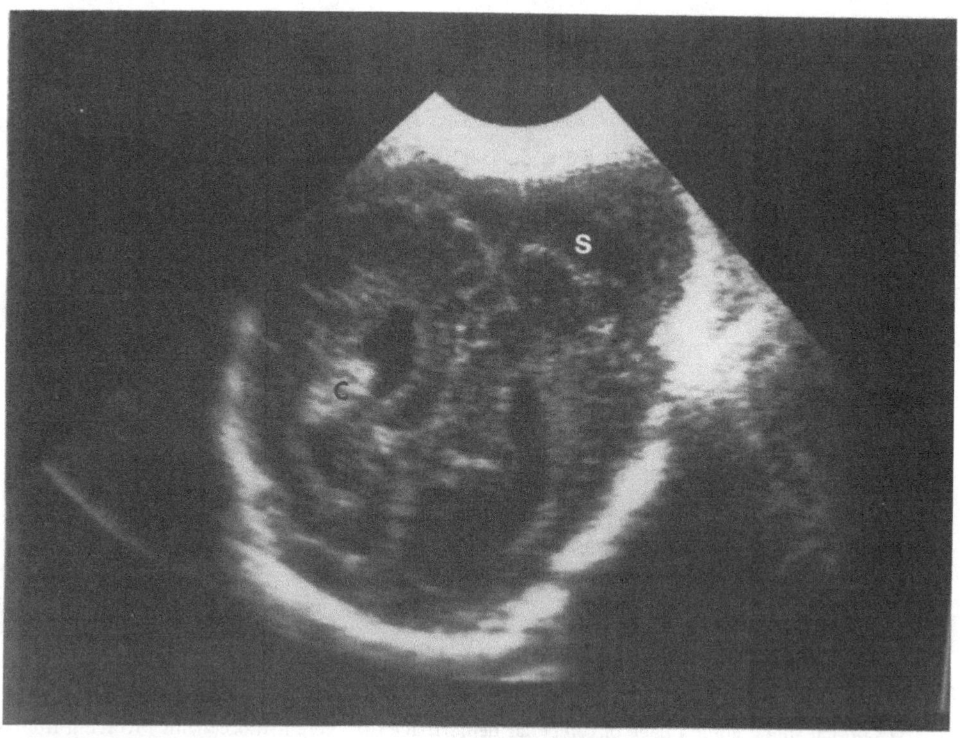

FIGURE 8–10B

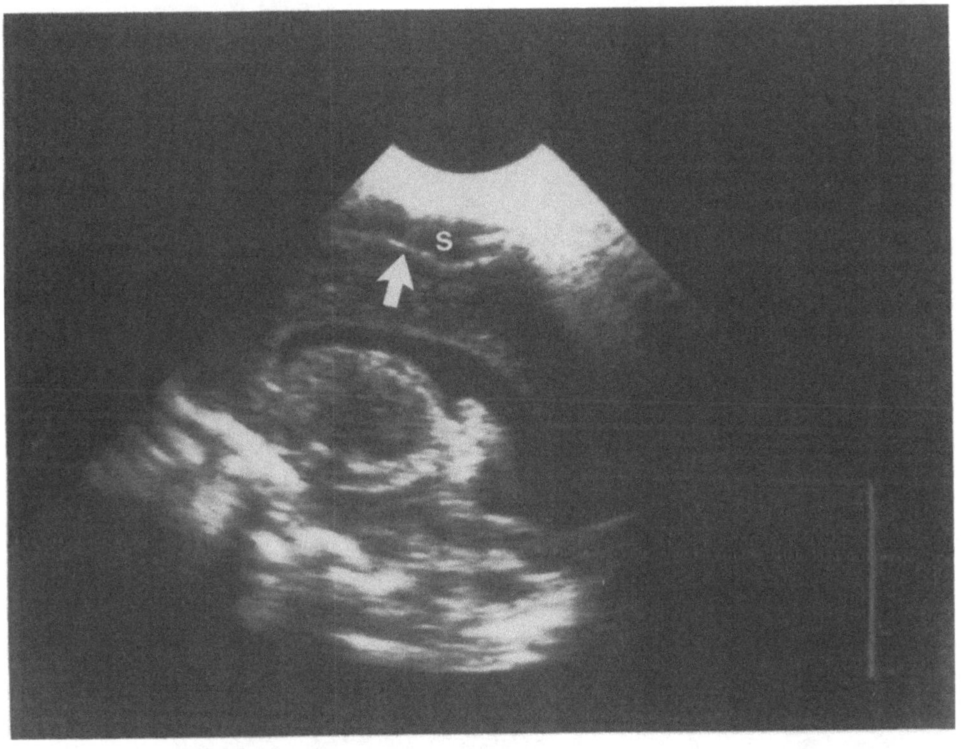

FIGURE 8–10C

FIGURE 8–10. (A) Anterior coronal image demonstrating extraaxial subarachnoid fluid. Note separation of brain from skull (arrows). (B) Posterior coronal image demonstrates extent of subarachnoid fluid (s). C-normal choroid in the right ventricular trigone. (C) Sagittal image also shows separation of brain (arrow) from skull by fluid (s).

Subdural hematomas are much more common in full-term infants than in prematures. Trauma is the most common etiology. Subarachnoid hemorrhage unrelated to the germinal matrix is also more common in full-term individuals and is usually associated with either trauma or hypoxia. These extraaxial fluid collections can be demonstrated by ultrasound if they are large (figures 8–10, 8–11), but reverberation artifacts in the near field and limited access through the anterior fontanelle make the examination incomplete [79–82].

CT is useful in evaluating peripheral lesions, whether parenchymal, subarachnoid, or subdural. Acute hemorrhage is dense, with CT numbers between 40 and 85. Some hematomas may become isodense on CT within seven to ten days, with larger ones taking up to two weeks to become uniformly isodense. When the hemorrhage begins to break down, the appearance changes from dense to isodense to low density. Ancillary findings, such as mass effect, surrounding edema, and enhancement with contrast, will help in the diagnosis. Contrast is particularly useful when dealing with chronic subdural hematomas. Membrane enhancement will differentiate between a hypodense or isodense chronic subdural hematoma and subdural hygroma (figure 8–12). CT detects calcifications very well, even when they are small. Contrast or dynamic CT may identify vascular lesions, such as AVMs, and aids in the diagnosis of hemorrhagic infarcts and abscesses. Disadvantages include radiation and the possible requirement of sedation [50].

MRI is useful in evaluating many of the same areas as CT. It can be used for small parenchymal, subdural, or subarachnoid hemorrhages. It is particularly useful for lesions near the skull, especially the skull base, since bone gives no MR signal. The ability to obtain direct sagittal and coronal images is helpful in giving clear, precise anatomical information [83]. Vascular structures with rapidly flowing blood show signal void and thus are readily identifiable. On shorter T1-weighted sequences, some signal may be seen in vascular structures, and where blood is flowing sluggishly or with turbulence, signals will be seen, particularly with varying pulse sequences [88].

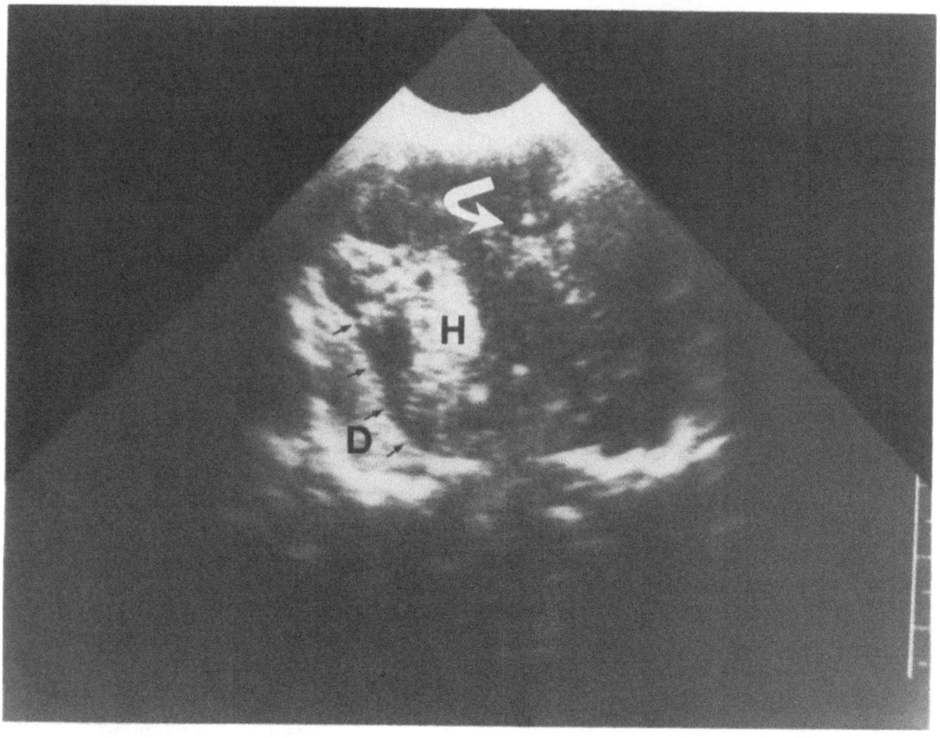

FIGURE 8–11A

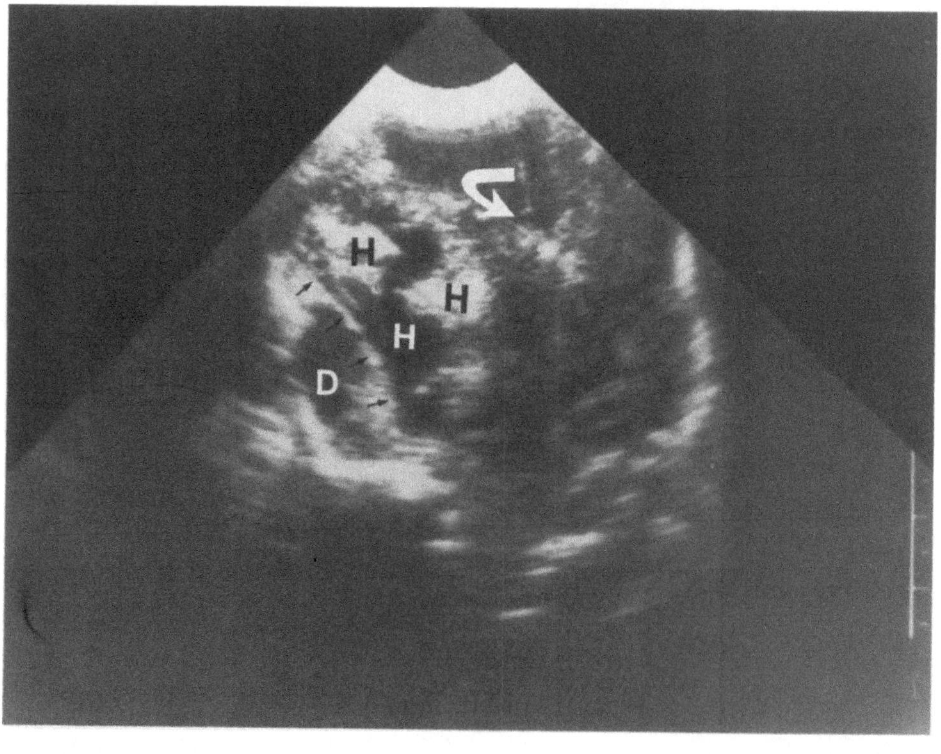

FIGURE 8–11B

FIGURE 8–11. (A) Acute subdural hematoma (D) with displacement of the brain (small arrows) from the skull. Ventricles are shifted to the left of midline (curved arrow). Coincident intraparenchymal hemorrhage is evident as well (H). Coronal image. (B) As the subdural (D) and parenchymal hemorrhages (H) mature and start to liquefy, lucent areas appear. Mass effect (arrows) and midline shift of ventricles (curved arrow) persist. Coronal image.

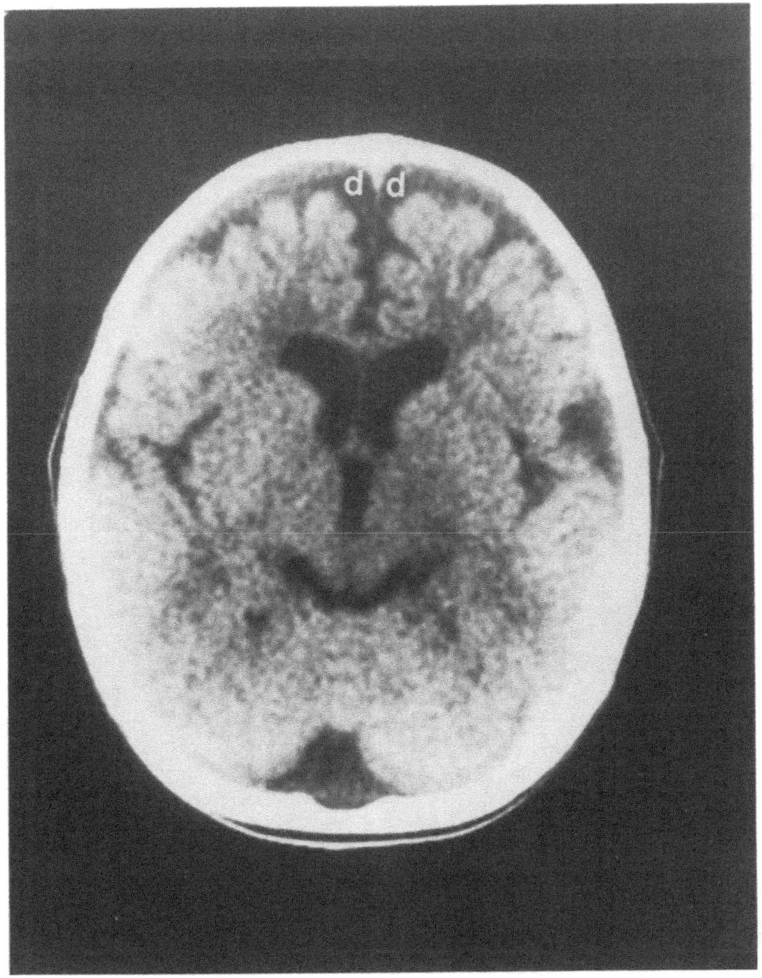

FIGURE 8–12A

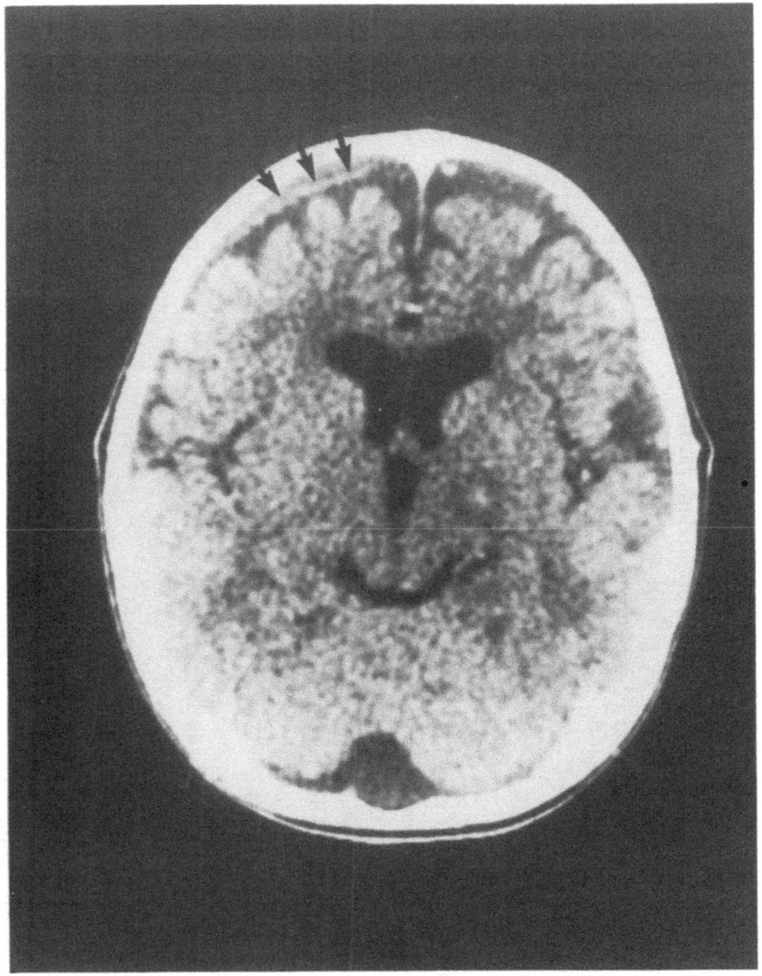

FIGURE 8–12B

FIGURE 8–12. (A) Chronic subdural hematoma is isodense on CT (d). Axial image, precontrast. (B) With contrast enhancement, the membrane (arrows) is seen, securing diagnosis of chronic, isodense subdural hematoma. Post-contrast axial image.

Clotted blood older than five days demonstrates high signal on both T1- and T2-weighted images [87].

T1 and T2 characteristics of hemorrhage vary with age. Acutely, the hemorrhage may be isointense or hypointense on T1-weighted images. The central portion may be markedly hypointense on T2-weighted images. Surrounding edema will be bright on T2. After about one week and up to several months, breakdown of the clot progresses from the periphery towards the center. This is seen as increasing intensity on T1-weighted images. On T2, the intensity also increases dramatically but with a slight temporal delay related to the T1 image. As hemosiderin from the breakdown products accumulates within macrophages, a ring of hypointensity develops around the periphery of the hematoma. This is most evident on the T2-weighted images. Chronically, the hypointensity of hemosiderin persists as the cavity fills in. A central area of cystic encephalomalacia may be present that will demonstrate MR characteristics of fluid [86].

Late sequelae of hemorrhage include porencephaly, persistent long T1 in caudate or basal ganglia, and hydrocephalus. Evaluation of hydrocephalus may be done with either CT or MR. Periventricular edema is especially well seen using MR. The exquisite anatomical detail afforded by MR is also advantageous. Appearance of small subdural hygromas can be easily detected by MR. These findings are important in determining if shunting is required and in evaluation of existing shunts [89].

Infant white matter is poorly myelinated and demonstrates much prolonged T1 and T2, possibly three to four times that of the adult brain. This changes rapidly during the first two years of life, as white matter becomes myelinated. Pulse sequences should be altered according to the patient's age to follow the changing myelination [84, 85, 89]. Edema and hemorrhage at certain stages are also bright, but by comparing T1- and T2-weighted images, they can usually be differentiated from white matter and CSF.

Disadvantages include requirement of sedation, confusion of signal void calcifications, and inability to examine with MRI extremely ill patients who require monitoring devices that contain ferromagnetic components.

Diagnostic Approach

Real-time sonography remains the primary imaging tool in neonatal ICH. High-resolution real-time transfontanelle ultrasound imaging demonstrates the central intracranial contents with extraordinary clarity. Both acute hemorrhage and sequelae such as ventriculomegaly and porencephaly can be accurately evaluated without removing the patient from the intensive care unit. Higher frequency transducers, computed image processing, and the development of color flow Doppler offer both additional anatomic detail and physiologic information that will make the portable ultrasound evaluation more complete. Infants at risk for ICH, especially those with birth weights under 1500 grams or gestational age under 34 weeks, should be screened by cranial ultrasound at four to six days of age and again one week later. Neonates with ICH should be followed at weekly intervals until both the hemorrage and secondary hydrocephalus have stabilized. Until the anterior fontanelle closes at approximately six months of age, ultrasound remains the major method of following posthemorrhagic hydrocephalus. After that time, CT or MR allows continuing evaluation of ventricular size.

Ultrasound is less sensitive for the detection of subarachnoid hemorrhage, posterior fossa lesions, and pathology of the peripheral portions of the cerebral parenchyma. CT or MR images should be obtained when there is suspicion of these lesions, a vascular abnormality not clearly defined by ultrasound, or a choroid plexus hematoma, or in any term infant with neurological signs who has a normal cranial ultrasound scan. MRI is particularly helpful in follow-up of hemorrhage, in evaluation of shunts, and in assessing the status of myelination (figure 8–13).

FIGURE 8–13. White matter underdevelopment on basis of ischemic–hypoxic encephalopathy. An eight-month-old infant. This infant was born premature at 32 weeks of gestation and weighed three pounds and four ounces at birth. He suffered severe ischemic–hypoxic insult in the neonatal period. He had presumed germinal matrix hemorrhage. (A) Proton-weighted MRI (TR/TE, 2000/20), and (B) T2-weighted MRI (TR/TE, 2000/20). There is severe ventricular enlargement on basis of neural tissue volume loss. Very thin white-matter radiations are seen, especially on T2-weighted images, indicating severe underdevelopment of the white matter. The hyperintense rim along the lateral ventricles on both pulse sequences is believed to represent extracellular methemoglobin. The slightly irregular linear hypointense area lateral to the hyperintense region along the left lateral ventricle on T2-weighted images is believed to be hemosiderin in macrophages. Evidence of old hemorrhage was present in all other images (not shown). There is also extraaxial fluid accumulation over the right frontal lobe, parietal, and occipital lobe.

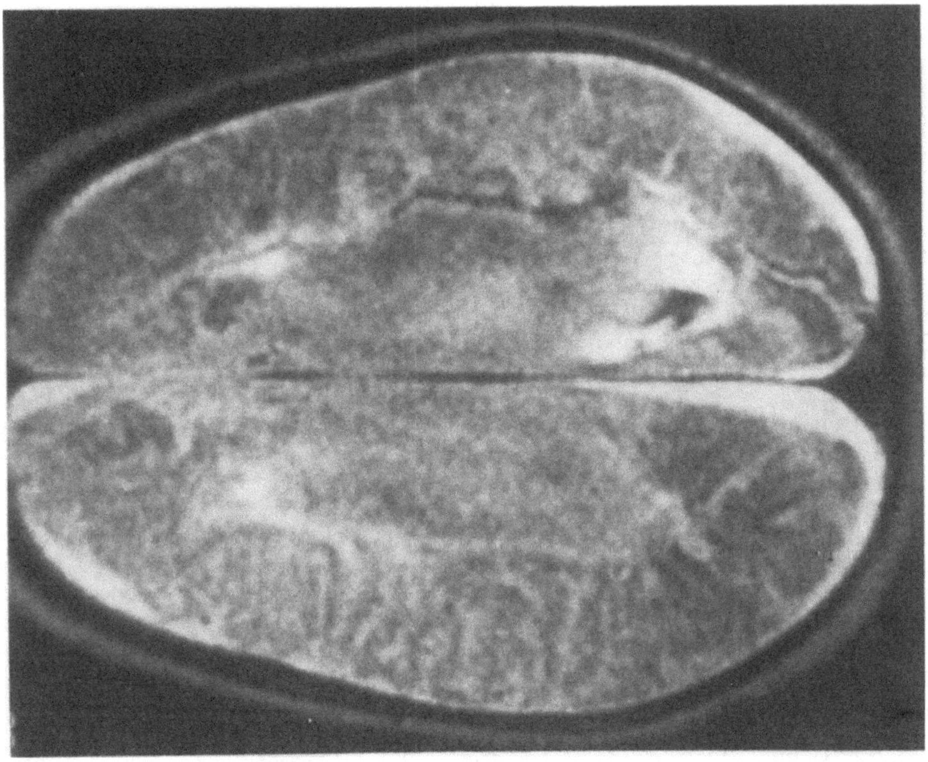

FIGURE 8–13B

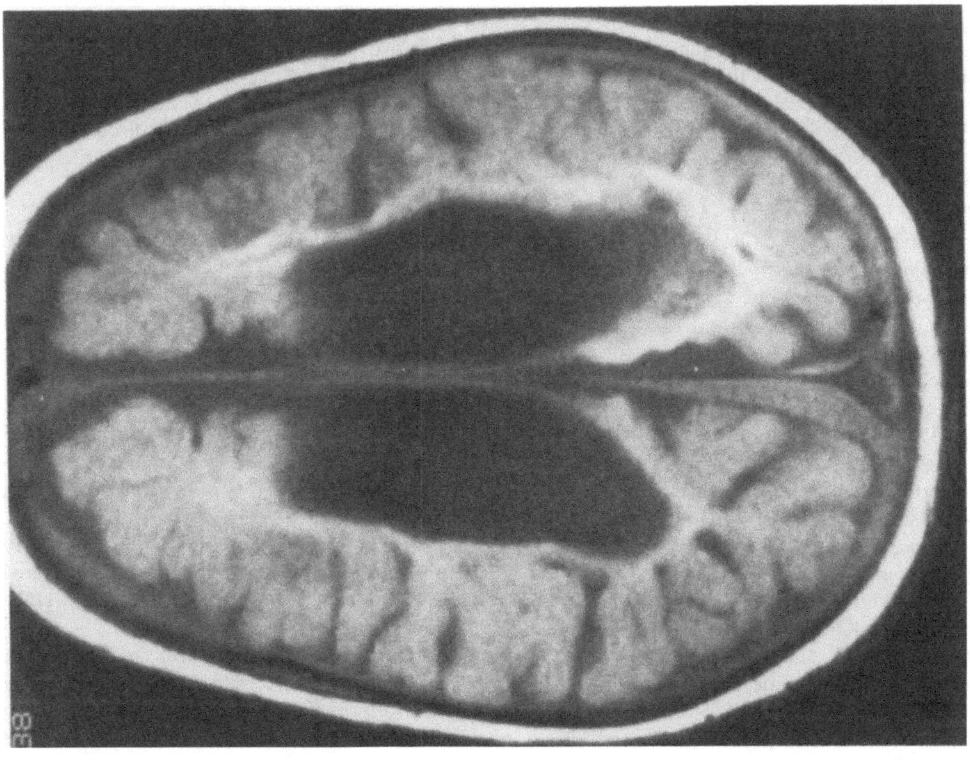

FIGURE 8–13A

References

1. Leech RW, Kohnen P. Subependymal and intraventricular hemorrhages in the newborn. *Am J Pathol* 1974;77:465–476.
2. Caen RW, Sutherland JM, Bove K, McAdams AJ. Anatomic and epidemiologic features of the stroke lesion of newborn infants. *Trans Am Neurol Assoc* 1970;95:36–40.
3. Volpe JJ. *Neurology of the Newborn*. Philadelphia: W. B. Saunders Company, 1987:282–361.
4. Grant EG. Sonography of the premature brain: intracranial hemorrhage and periventricular leukomalacia. In: Naidich TP, Quencer RM (eds). *Clinical Neurosonography, Ultrasound of the Central Nervous System*. Berlin: Springer-Verlag, 1987:110–124.
5. Papile L-A, Burstein J, Burstein R, Koffler H. Incidence and evolution of subependymal and intraventricular hemorrhage: a study of infants with birth weights less than 1,500 gm. *J Pediatr* 1978;92:529–534.
6. Burstein J, Papile L, Burstein R. Intraventricular hemorrhage and hydrocephalus in premature newborns: a prospective study with CT. *AJR* 1979; 132:631–635.
7. Lee BP, Grassi AE, Schecner S, Auld PAM. Neonatal intraventricular hemorrhage: a serial computed tomography study. *J Comput Assist Tomog* 1979;3:483–490.
8. Kosmetatos N, Dinter C, Williams ML, Lourie H, Berne AS. Intracranial hemorrhage in the premature. *AJDC* 1980;134:855–859.
9. Bejar R, Curbelo V, Coen RW, Leopold G, James H, Gluck L. Diagnosis and follow-up of intraventricular and intracerebral hemorrhages by ultrasound studies of infant's brain through the fontanelles and sutures. *Pediatrics* 1980;66:661–673.
10. Cooke RWI. Factors associated with periventricular haemorrhage in very low birthweight infants. *Arch Dis Child* 1981;56:425–431.
11. Grant EG, Borts FT, Schellinger D, McCullough D, Sirasulramanian KN, Smith Y. Real-time ultrasonography of neonatal intraventricular hemorrhage and comparison with computed tomography. *Radiology* 1981;139:687–691.
12. Smith WL, McGuinness G, Cavanaugh D, Courtney S. Ultrasound screening of premature infants: longitudinal follow-up of intracranial hemorrhage. *Radiology* 1983;147:445–448.
13. Flodmark O, Fitz CH, Harwood-Nash DC. CT diagnosis and short-term prognosis of intracranial hemorrhage and hypoxic/ischemic brain damage in neonates. *J Comput Assist Tomog* 1980;4:775–787.
14. Grant EG (ed). *Neurosonography of the Pre-term Neonate*. New York: Springer-Verlag, 1986.
15. Hayden CK Jr, Shattuck KE, Richardson CJ, Ahrendt DK, House R, Swischuk LE. Subependymal germinal matrix hemorrhage in full-term neonates. *Pediatrics* 1985;75:714–718.
16. Wigglesworth JS, Pape KE. Pathophysiology of intracranial haemorrhage in the newborn. *J Perinat Med* 1980;8:119–133.
17. Harwood-Nash DC, Flodmark O. Diagnostic imaging of the neonatal brain: review and protocol. *AJNR* 1982;3:103–115.
18. Volpe JJ. Cerebral blood flow in the newborn infant: relation to hypoxic-ischemic brain injury and periventricular hemorrhage. *J Pediatr* 1979;94:170–173.
19. Rayburn WF, Donn SM, Kolin MG, Schork MA. Obstetric care and intraventricular hemorrhage in the low birth weight infant. *Obstet Gynecol* 1983; 62:408–413.
20. Lipscomb AP, Reynolds EOR, Blackwell RJ, Thorburn RJ, Stewart AL, Cusick G, Whitehead MD. Pneumothorax and cerebral haemorrhage in preterm infants. *Lancet* 1981:414–416.
21. Goddard-Finegold J. Periventricular, intraventricular hemorrhages in the premature newborn. Update on pathologic features, pathogenesis, and possible means of prevention. *Arch Neurol* 1984;41:766–771.
22. Tejani N, Rebold B, Tuck S, Ditroia D, Sutro W, Verma U. Obstetric factors in the causation of early periventricular–intraventricular hemorrhage. *Obstet Gynecol* 1984;64:510–515
23. Hill A, Perlman JM, Volpe JJ. Relationship of pneumothorax to occurrence of intraventricular hemorrhage in the premature newborn. *Pediatrics* 1982; 69:144–149.
24. Hill A, Volpe JJ. Seizures, hypoxic–ischemic brain injury, and intraventricular hemorrhage in the newborn. *Ann Neurol* 1981;10:109–121.
25. Bartlett RH, Gazzaniga AB, Toomasian J, Corwin AG, Roloff D, Rucker R. Extracorporeal membrane oxygenation (ECMO) in neonatal respiratory failure. *Ann Surg* 1986;204:236–245.
26. Rumack CM, Guggenheim MA, Rumack BH, Peterson RG, Johnson ML, Braithwaite WR. Neonatal intracranial hemorrhage and maternal use of aspirin. *Obstet Gynecol* 1981;58:52S–56S.
27. Setzer ES, Webb IV, Wassenaar JW, Reeder JD, Mehta PS, Fitzman DV. Platelet dysfunction and coagulopathy in intraventricular hemorrhage in the premature infant. *J Pediatr* 1982;100:599–605.
28. Chaou W-T, Chou M-L, Eitzman DV. Intracranial hemorrhage and vitamin K deficiency in early infancy. *J Pediatr* 1984;105:880–884.
29. Sia CG, Amigo NC, Harper RG, Farahani G, Kochen J. Failure of caesarean section to prevent intracranial hemorrhage in siblings with isoimmune neonatal thrombocytopenia. *Am J Obstet Gynecol* 1985;153: 79–81.
30. Bowerman RA, Zwischenberger JB, Andrews AF, Bartlett RH. Cranial sonography of the infant treated with extracorporeal membrane oxygenation. *AJNR* 1985;6:377–382.
31. Rumack CM, Manco-Johnson ML, Manco-Johnson MJ, Koops BL, Hathaway WE, Appareti K. Timing and course of neonatal intracranial hemorrhage using real-time ultrasound. *Radiology* 1985;154:101–105.

32. Hecht ST, Filly RA, Callen PW, Wilson-Davis SL. Intracranial hemorrhage: late onset in the preterm neonate. *Radiology* 1983;149:697–699.
33. Ment LR, Duncan CC, Ehrenkranz RA, Lange RC, Taylor KJ, Kleinman CS, Scott DT, Sivo J, Gettner P. Intraventricular hemorrhage in the preterm neonate: timing and cerebral blood flow changes. *J Pediatr* 1984;104:419–425.
34. Hawgood S, Spong J, Yu VYH. Intraventricular hemorrhage. Incidence and outcome in a population of very-low-birth-weight infants. *AJDC* 1984;138: 136–139.
35. Lebed MR, Schifrin BS, Waffran F, Hohler CW, Afriat CI. Real-time B scanning in the diagnosis of neonatal intracranial hemorrhage. *Am J Obstet Gynecol* 1982;142:851–861.
36. Gilstrap LC III, Hauth JC, Bell RE, Ackerman NB Jr, Yoder BA, Delemos R. Survival and short-term morbidity of the premature neonate. *Obstet Gynecol* 1985;65:37–41.
37. Baerts W, Meradji M. Cranial ultrasound in preterm infants: long-term follow-up. *Arch Dis Child* 1985; 60:702–705.
38. Allan WC, Holt PJ, Sawyer LR, Tito AM, Meade SK. Ventricular dilation after neonatal periventricular–intraventricular hemorrhage. *AJDC* 1982;136:589–593.
39. Thorburn RJ, Stewart AL, Hope PL, Lipscomb AP, Reynolds EOR, Pape KE. Prediction of death and major handicap in very preterm infants by brain ultrasound. *Lancet* 1981;1119–1121.
40. Papile L-A, Munsick-Bruno G, Schaefer A. Relationship of cerebral intraventricular hemorrhage and early childhood neurologic handicaps. *J Pediatr* 1983; 103:273–277.
41. Kirks DR, Bowie JD. Cranial ultrasonography of neonatal periventricular/intraventricular hemorrhage: who, how, why and when? *Pediatr Radiol* 1986; 16:114–119.
42. Graziani LJ, Pasto M, Stanley C, Steben J, Desai H, Desai S, Foy PM, Branca P, Goldberg BB. Cranial ultrasound and clinical studies in preterm infants. *J Pediatr* 1985;106:269–276.
43. Bennett FC, Chandler LS, Robinson NM, Sells CJ. Spastic diplegia in premature infants. *Am J Dis Child* 1981;135:732–737.
44. Krishnamoorthy KS, Shannon DC, DeLong GR, Todres ID, Davis KR. Neurologic sequelae in the survivors of neonatal intraventricular hemorrhage. *Pediatrics* 1979;64:233–237.
45. TeKolste KA, Bennett FC, Mack LA. Follow-up of infants receiving cranial ultrasound for intracranial hemorrhage. *Am J Dis Child* 1984;139:299–303.
46. Williamson WD, Desmond MM, Wilson GS, Murphy MA, Rozelle J, Garcia-Prats JA. Survival of low-birth-weight infants with neonatal intraventricular hemorrhage. *Am J Dis Child* 1983;137:1181–1184.
47. Levene MJ, Wigglesworth JS, Dubowitz V. Cerebral structure and intraventricular haemorrhage in the neonate: a real-time ultrasound study. *Arch Dis Child* 1981;56:416–424.
48. Rose WS, Wolfson M. Real-time ultrasonography of the neonatal head. *JAMA* 1983;250:3212–3215.
49. Allan WC, Roveto CA, Sawyer LR, Courtney SE. Sector scan ultrasound imaging through the anterior fontanelle. *Am J Dis Child* 1980;134:1028–1031.
50. Rumack CM, Johnson ML. *Perinatal and Infant Brain Imaging: Role of ultrasound and Computed Tomography* Chicago: Yearbook Medical Publishing Co., 1984.
51. Bowerman RA, Donn SM, Silver TM, Jaffee MH. Natural history of neonatal periventricular/intraventircular hemorrhage and its complications: sonographic observations. *AJR* 1984;143:1041–1052.
52. Warren P, Garrett W, Lam A, Berry A, Kossoff G. Neonatal echoencephalography. *Ultrasound in Med & Biol* 1984;10:117–131.
53. Maklad NF. Ultrasound of the neonatal brain. In: *Ultrasound in Perinatology*. New York: Churchill Livingstone, 1986:129–156.
54. Sauerbrei EE, Digney M, Harrison PB, Cooperberg PL. Ultrasonic evaluation of neonatal intracranial hemorrhage and its complications. *Radiology* 1981; 139:677–685.
55. Fiske CE, Filly RA, Callen PN. The normal choroid plexus. Ultrasonographic appearance of the neonatal head. *Radiology* 1981;141:467–471.
56. Siedler DE, Mahony BS, Hoddick WK, Callen PW. A specular reflection arising from the ventricular wall: a potential pitfall in the diagnosis of germinal matrix hemorrhage. *J Ultrasound Med* 1985;4:109–112.
57. Veyrac C, Couture A. Normal and pathologic choroid plexus ultrasound. *Ann Radiol* 1985;28:215–223.
58. Reeder JD, Kaude JV, Setzer ES. Choroid plexus hemorrhage in premature neonates. *AJNR* 1982; 3:619–622.
59. Armstrong DW, Sauls CD, Goddard-Finegold J. Neuropathologic findings in short-term survivors of intraventricular hemorrhage. *Am J Dis Child* 1987; 141:617–621.
60. Slabaugh RD, Smith JA, Lemons J, Schreiner R, Macdonald N, Cohen MD. Neonatal intracranial hemorrhage and complicating hydrocephalus. *J Clin Ultrasound* 1984;12:261–266.
61. Kreusser KL, Tarby TJ, Taylor D, Kovnar E, Hill A, Conry JA, Volpe JJ. Rapidly progressive posthemorrhagic hydrocephalus. *Am J Dis Child* 1984; 138:633–637.
62. Levene MI, Starte DR. A longitudinal study of post-haemorrhagic ventricular dilatation in the newborn. *Arch Dis Child* 1981;56:905–910.
63. Kreusser KL, Tarby J, Kovnar E, Taylor DA, Hill A, Volpe JJ. Serial lumbar punctures for at least temporary amelioration of neonatal posthemorrhagic hydrocephalus. *Pediatrics* 1985;75:719–724.
64. Kaiser AM, Whitelaw AGL. Cerebrospinal fluid pressure during post haemorrhagic ventricular dilatation in newborn infants. *Arch Dis Child* 1985;60:920–924.
65. Marlin AE. Protection of the cortical mantle in

premature infants with posthemorrhagic hydrocephalus. *Neurosurgery* 1980;7:464–468.

66. Chow PP, Horgan JG, Burns PN, Weltin G. Taylor KJW. Choroid plexus papilloma: detection by real-time and doppler sonography *AJNR* 1986;7:168–170.
67. Levene MI, De Vries L. Extension of neonatal intraventricular hemorrhage. *Arch Dis Child* 1984;59: 631–636.
68. Banker BQ, Larroche J-C. Periventricular leukomalacia of infancy. *Arch Neurol* 1962;7:386–409.
69. Grant EG, Schellinger D. Sonography of neonatal periventricular leukomalacia: recent experience with a 7.5-MHz scanner. *AJNR* 1985;6:781–785.
70. Chow PP, Horgan JG, Taylor KJW. Neonatal periventricular leukomalacia: real-time sonographic diagnosis with CT correlation. *AJNR* 1985;6:383–388.
71. Bowerman RA, Donn SM, DiPietro MA, D'Amato CJ, Hicks SP. Periventricular leukomalacia in the pre-term newborn infant: sonographic and clinical features. *Radiology* 1984;151:383–388.
72. Grant ED, Schellinger D, Richardson JD, Coffey ML, Smirniotopoulous JG. Echogenic periventricular halo: normal sonographic finding or neonatal cerebral hemorrhage. *AJR* 1983;140:793–796.
73. Slovis TL, Shankaran S, Bedard MP, Poland RL. Intracranial hemorrhage in the hypoxic–ischemic infant: ultrasound demonstration of unusual complications. *Radiology* 1984;151:163–169.
74. Schellinger D, Grant EG, Richardson JD. Neonatal leukencephalopathy: a common form of cerebral ischemia. *RadioGraphics* 1985;5:221–242.
75. Manger MN, Feldman RC, Brown WJ, Mitchell LS, Waffarn F, Shields WD. Intracranial ultrasound diagnosis of neonatal periventricular leukomalacia. *J Ultrasound Med* 1984;3:59–63.
76. Martin R, Roessmann U, Fanaroff A. Massive intracerebellar hemorrhage in low-birth-weight infants. *J Pediatr* 1976;89:290–293.
77. Grunnet ML, Shields DW. Cerebellar hemorrhage in the premature infant. *J Pediatr* 1976;88:605–608.
78. Peterson CM, Smith WL, Franken EA. Neonatal intracerebellar hemorrhage: detection by real-time ultrasound. *Radiology* 1984;150:391–392.
79. Aoki N, Masuzawa H. Infantile acute subdural hematoma. Clinical analysis of 26 cases. *J Neurosurg* 1984; 61:273–280.
80. Fenichel GM, Webster DL, Wong KT. Intracranial hemorrhage in the term newborn. *Arch Neurol* 1984; 41:30–34.
81. Ennis MG, Kaude JV, Williams JL. Sonographic diagnosis of subarachnoid hemorrhage in premature newborn infants: a retrospective study with histopathologic and CT correlation. *J Ultrasound Med* 1985;4:183–187.
82. Dewbury KC, Bates RI. Neonatal intracranial haemorrhage: the cause of the ultrasound appearances. *Br J Radiol* 1983;56:783–789.
83. Han JS, Benson JE, Kaufman, B, et al. MR imaging of pediatric cerebral abnormalities. *JCAT* 1985; 9(1):103–114.
84. Pennock JM, Bydder GM, Dubowitz LMS, Johnson MA. Magnetic resonance imaging of the brain in children. *MRI* 1986;4(1):1–9.
85. Dubowitz LMS, Pennock JM, Johnson MA, Bydder GM. High-resolution magnetic resonance imaging of the brain in children. *Clin Radiol* 1986;37:113–117.
86. Gomori JM, Grossman RI, Goldberg HI, et al. Intracranial hematomas: Imaging by high yield MR. *Radiology* 1985;157:87–93.
87. Zimmerman RA, Bilaniuk LT. Applications of magnetic resonance imaging in diseases of the pediatric central nervous system. *MRI* 1986;4:11–24.
88. Mills CM, Brant-Zawadzki M, et al. Nuclear magnetic resonance: Principles of blood flow imaging. *AJR* 1984;142:165–170.
89. Bydder GM. Magnetic resonance imaging of the brain. *Radiol Clin NA* 1984;22:779–793.

INDEX

FSC
www.fsc.org
MIX
Papier | Fördert gute Waldnutzung
FSC® C083411

Kolibri 360 GmbH
Ferdinand-Jühlke-Straße 7
99095 Erfurt, Deutschland
produktsicherheit@kolibri360.de